Ecohydrology

Edited by Xavier Parsons

MURPHY & MOORE
www.murphy-moorepublishing.com

Published by Murphy & Moore Publishing,
1 Rockefeller Plaza,
New York City, NY 10020, USA
www.murphy-moorepublishing.com

Ecohydrology
Edited by Xavier Parsons

International Standard Book Number: 978-1-63987-174-2 (Hardback)

Cataloging-in-Publication Data

Ecohydrology / edited by Xavier Parsons.
p. cm.
Includes bibliographical references and index.
ISBN 978-1-63987-174-2
1. Ecohydrology. 2. Aquatic ecology. 3. Hydrology. I. Parsons, Xavier.
QH541.15.E19 E26 2022
577.6--dc23

TABLE OF CONTENTS

TABLE OF CONTENTS

PREFACE

The study of interactions between water and ecological systems fall under the domain of ecohydrology. The interactions between water and ecological systems occur within the water bodies like lakes and rivers. Such interactions can also occur on land, such as in forests, deserts and other terrestrial ecosystems. Ecohydrology focuses on diverse areas such as, transpiration and plant water use, effects of vegetation and benthic plants on stream flow and adaptation of organisms to their water environment. Ecohydrology studies both terrestrial and aquatic ecosystems. Its main focus in terrestrial ecosystems is on the interaction among vegetation, the vadose zone, land surface and the groundwater. In aquatic ecosystems, it emphasizes on the effects of geomorphology, water chemistry and hydrology on the structure and function of the ecosystem. This book unravels the recent studies in the field of ecohydrology. It elucidates the concepts and innovative models around prospective developments with respect to ecohydrology. It will provide comprehensive knowledge to the readers.

This book is a comprehensive compilation of works of different researchers from varied parts of the world. It includes valuable experiences of the researchers with the sole objective of providing the readers (learners) with a proper knowledge of the concerned field. This book will be beneficial in evoking inspiration and enhancing the knowledge of the interested readers.

In the end, I would like to extend my heartiest thanks to the authors who worked with great determination on their chapters. I also appreciate the publisher's support in the course of the book. I would also like to deeply acknowledge my family who stood by me as a source of inspiration during the project.

Editor

1

Effect of parameter choice in root water uptake models – the arrangement of root hydraulic properties within the root architecture affects dynamics and efficiency of root water uptake

M. Bechmann[1], **C. Schneider**[2], **A. Carminati**[3], **D. Vetterlein**[4], **S. Attinger**[2], **and A. Hildebrandt**[1,5]

[1]Friedrich Schiller University, Jena, Germany, Institute of Geosciences, Burgweg 11, 07749 Jena, Germany
[2]Helmholtz Centre for Environmental Research, Leipzig, Germany, Department Computational Hydrosystems, Permoser Straße 15, 04318 Leipzig, Germany
[3]Georg August University, Göttingen, Germany, Faculty of Agricultural Sciences, Department of Crop Sciences, Büsgenweg 2, 37077 Göttingen, Germany
[4]Helmholtz Centre for Environmental Research, Halle, Germany, Department of Soil Physics, Theodor-Lieser-Strasse 4, 06120 Halle/Saale, Germany
[5]Max Planck Institute for Biogeochemistry, Jena, Germany, Hans-Knöll-Str. 10, 07745 Jena, Germany

Correspondence to: M. Bechmann (bechmann.marcel@uni-jena.de)

Abstract. Detailed three-dimensional models of root water uptake have become increasingly popular for investigating the process of root water uptake. However, they suffer from a lack of information on important parameters, particularly on the spatial distribution of root axial and radial conductivities, which vary greatly along a root system. In this paper we explore how the arrangement of those root hydraulic properties and branching within the root system affects modelled uptake dynamics, xylem water potential and the efficiency of root water uptake. We first apply a simple model to illustrate the mechanisms at the scale of single roots. By using two efficiency indices based on (i) the collar xylem potential ("effort") and (ii) the integral amount of unstressed root water uptake ("water yield"), we show that an optimal root length emerges, depending on the ratio between roots axial and radial conductivity. Young roots with high capacity for radial uptake are only efficient when they are short. Branching, in combination with mature transport roots, enables soil exploration and substantially increases active young root length at low collar potentials. Second, we investigate how this shapes uptake dynamics at the plant scale using a comprehensive three-dimensional root water uptake model. Plant-scale dynamics, such as the average uptake depth of entire root systems, were only minimally influenced by the hydraulic parameterization. However, other factors such as hydraulic redistribution, collar potential, internal redistribution patterns and instantaneous uptake depth depended strongly on the arrangement on the arrangement of root hydraulic properties. Root systems were most efficient when assembled of different root types, allowing for separation of root function in uptake (numerous short apical young roots) and transport (longer mature roots). Modelling results became similar when this heterogeneity was accounted for to some degree (i.e. if the root systems contained between 40 and 80 % of young uptake roots). The average collar potential was cut to half and unstressed transpiration increased by up to 25 % in composed root systems, compared to homogenous ones. Also, the least efficient root system (homogenous young root system) was characterized by excessive bleeding (hydraulic lift), which seemed to be an artifact of the parameterization. We conclude that heterogeneity of root hydraulic properties is a critical component for efficient root systems that needs to be accounted for in complex three-dimensional root water uptake models.

1 Introduction

Soil–plant interactions are important factors in hydrological and ecological processes. By using soil water for transpiration, plants are the essential link in the mass and energy transfer at the soil–vegetation–atmosphere interface (Shukla and Mintz, 1982). Much of this interaction hinges upon the ability of plants to gain flexible access to soil water (Churkina and Running, 1998; Kleidon and Heimann, 2000; Feddes et al., 2001; Hildebrandt and Eltahir, 2007; Collins and Bras, 2007; Katul et al., 2012). Inversely, changes in soil water content reflect on energy partitioning and carbon fluxes at the soil surface (Kleidon and Heimann, 1998; El Maayar et al., 2009; Seneviratne et al., 2010). Furthermore, access to soil water is an important prerequisite for biomass production, including crops (Blum, 1996; Huszár et al., 1998; Cai et al., 2009).

The ubiquitous influence of root water uptake on ecological and atmospheric processes necessitates the appropriate predictionof root water uptake (Shukla and Mintz, 1982; Jackson et al., 2000). For this, together with observations, models have become vital tools that are used both in order to gain local process understanding as well as to predict macroscopic root water uptake characteristics.

Water uptake is driven by gradients in water potential, whereby water is pulled up from the soil into the root and up to the leaf (Steudle, 2001; Angeles et al., 2004). Besides soil hydraulic resistance, root tissue resistances determine the actual values of water uptake and water transport (Van Den Honert, 1948): radial resistance of soil and roots for the flow path across the soil–root interface and roots axial resistance for the flow path within the root xylem. The ratio between radial and axial resistance is of substantial importance. It shapes the distribution of xylem water potential throughout the root and thus influences root water uptake (Landsberg and Fowkes, 1978). Moreover, Zwieniecki et al. (2003) modelled a trade-off between hydraulically active root length and the corresponding water uptake in unlimited water reservoirs. The term "hydraulically active" corresponds to the portion of the root that considerably contributes to root water uptake. The proposed trade-off hinges upon the ratio of radial and axial root hydraulic resistance: when radial resistance increases, the active root length increases, whereas water uptake decreases.

For process studies of root water uptake, models that compute microscopic three-dimensional root water uptake with respect to gradients in water potential and hydraulic resistances have become more and more popular (Clausnitzer and Hopmans, 1994; Tuzet et al., 2003; Doussan et al., 2006; Javaux et al., 2008; Schneider et al., 2010). Most of these models solve water flow equations within the soil and the root system architecture at the same time. They account for the microscopic soil water flow towards individual roots, radial flow into the root xylem and the axial flow within the root xylem. The modelling scale of these process based approaches comes close to the scale at which actual root water uptake takes place. Thus, they promise an important contribution to process understanding. Indeed, they capture well-observed processes such as redistribution of root water uptake due to local limitations of soil water availability, including moving uptake fronts (Garrigues et al., 2006; Javaux et al., 2008; Schneider et al., 2010) and also hydraulic lift (Dunbabin et al., 2013). This is a major improvement compared to empirical models (Feddes et al., 1978). The inherent redistribution of root water uptake based on explicit calculations of water flow in roots is also reported to be superior to qualitative approaches (Simunek and Hopmans, 2009).

However, parameterization of small-scale models still poses a substantial challenge, since it requires detailed information that is difficult that is difficult to obtain regarding (a) root geometry and even more challenging (b) distribution of root hydraulic properties. Some progress on point (a) has already been made. Recent improvements in imaging (Oswald et al., 2008; Mooney et al., 2012) and image analysis (Leitner and Schnepf, 2012; Lobet et al., 2011; Lobet and Draye, 2013) have improved information on root system geometry such as position, orientation, branching order and root diameter. However, information on the distribution of root hydraulic properties (point b) is still extremely sparse, because the necessary measurements are tedious (Knipfer et al., 2007). Thus, an important input to three-dimensional root water uptake models, that is the exact arrangement of root hydraulic properties within the root system, remains largely unknown.

Modelling results suggest that the lack of knowledge on root hydraulic properties may be a substantial hindrance (Schneider et al., 2010; Heppel et al., 2014). As stated above, the distribution of xylem water potential and root water uptake along the root system depends strongly on the ratio between root axial and root radial resistance (Landsberg and Fowkes, 1978; Zwieniecki et al., 2003; Doussan et al., 2006; Levin et al., 2007; Javaux et al., 2008). What is more, during root maturation individual root hydraulic properties change with time (Steudle, 2000). Older suberized roots with more and mature xylem vessels have lower axial and higher radial resistance compared to younger roots. A root system contains both mature and young roots and observations show that conductivities along the radial and axial pathways vary within several orders of magnitude along root networks (Frensch and Steudle, 1989; Doussan et al., 2006). Hence a root system is a network of elements with contrasting hydraulic properties. Modellers account for this heterogeneity differently. Doussan et al. (2006) distributed hydraulic properties stepwise according to root length in taproots and root age in lateral roots. Schneider et al. (2010) translated a root developmental stage (obtained with a root generator from Pagès et al., 2004) into five hydraulic classes with distinct root hydraulic properties. However, as stated earlier, the actual arrangement of hydraulic properties within the root system is unknown most of the time and parameterization is based on scarce quantitative information, which means researchers are often left to their intuition. To our knowledge, there exists no systematic investigation on whether and how strongly the spatial arrangement of root hydraulic properties affects model results, although such an analysis would greatly help in making decisions on model parameterization.

Root hydraulic properties do not only shape root water uptake profiles (Landsberg and Fowkes, 1978) and active root length (Zwieniecki et al., 2003), but may also be important for the water relations of a plant, because they contribute to the overall resistance to water uptake of the entire soil–plant continuum and hence to evolution of xylem potential during the uptake process. Strongly negative xylem water potentials increase the danger of embolism and cavitation of xylem vessels, resulting in a progressive loss of axial hydraulic conductivity (Pockman and Sperry, 2000; McDowell et al., 2008). Research suggests that plants operate with little safety margin with regard to danger of embolism across climates (Choat et al., 2012; Choat, 2013; Manzoni et al., 2013). As a consequence, plants probably apply strategies to minimize their vulnerability to cavitation, which includes efficient distribution of resistances within their water uptake apparatus. Therefore, xylem water potential at the root collar recommends itself as a tool for distinguishing efficient from less efficient root parameterizations. On the other hand, if modelled xylem potentials are meaningful they can serve as a valuable model output for example for coupling root water uptake to stomatal control (Tuzet et al., 2003).

This modelling study aims at describing and assessing the combined role of heterogeneity of root hydraulic properties and branching topology on root water uptake dynamics. In particular, we also investigate their relation to the spatiotemporal evolution of xylem water potential, the overall efficiency of root water uptake and microscopic and macroscopic water relations, including hydraulic lift.

Background

We use a thought experiment to illustrate that root hydraulic properties inevitably shape active root length, but more importantly how they are related to the evolution of xylem potential with time.

Let us consider a single unbranched root surrounded by a soil cylinder with uniform soil and root hydraulic properties and with total soil water potential being in equilibrium at first. Let us further assume that the total amount of root water uptake is constant with time. First, water uptake occurs predominantly near the root collar, while the apical parts of the root remain inactive due to drops in xylem water potential along the root. The inactive parts of the root have also been called "hydraulically isolated" in the past (North and Peterson, 2005; Zwieniecki et al., 2003). During this stage, the active root length relates to the ratio between axial and radial resistances of the root to water flow (Zwieniecki et al., 2003), and it increases when this ratio becomes small. Next, as a consequence of the selective root water uptake, soil dries near the root collar and the soil water potential drops to more to more negative values there. In order to maintain the rate of root water uptake, the xylem water potential at the root collar has to decrease accordingly. At the same time, water uptake moves away from the collar and previously isolated regions of the root get activated, as water is easily available there. The water now has to travel a longer pathway within the xylem, which increases effective axial resistances compared to before. Over time, moving uptake fronts activate farther regions of the root, at the price that the xylem potential within the root system progressively decreases, and limits water uptake. Thus it is intuitive that roots should not be infinitely long; and that an optimum exists which balances the benefits of activating root length by moving uptake fronts and disadvantages of increased axial path length. When root length is shorter than this optimum, an increase in root length is beneficial for root water uptake, since it increases the efficiently utilizable uptake are. We will refer to this case as "radial limitation". A further increase of active root length is not efficient due to the enhanced axial resistance, and we will refer to this case as "axial limitation" in the rest of this paper.

2 Materials and methods

We conduct our investigation in two steps, using first a simple and second a complex root water uptake model. The simple model serves to describe processes of root water uptake at the single root scale that are hard to disentangle at higher levels of model complexity. Within this section we first describe those two applied models of root water uptake. Second, we explain how the root hydraulic properties were systematically varied within the different root systems. Finally, we introduce two indices that are used to quantify the efficiency of root water uptake: "water yield" and "effort". All comparisons of root hydraulic parameterizations in this paper are made using these two criteria.

2.1 Simple root water uptake model for single roots

Root water uptake along single unbranched and branched roots was calculated with a simple root water uptake model (see Fig. 1 for the considered root structures). It divides the root into n segments of equal length and treats the root as a network of porous pipes. A number of $n = 100$ segments for unbranched single roots and $n = 192$ segments for branched single roots are sufficient to prevent discretization errors. Each root segment is considered to have a cylindrical shape of radius $r^{(i)}$ (m) and length $l^{(i)}$ (m).

Each root segment is provided with a limited soil water reservoir. Water is taken up from closed soil cylinders with radius $r_{soil} = 1.2\,\text{cm}$ surrounding the root segments. The value of r_{soil} was chosen to correspond with the half

Table 1. Parameters and important features of the simple and the "aRoot" model.

Soil properties	Simple model	"aRoot" model
Limited water reservoir	Yes	
Gravitation	No	Yes
Redistribution of soil water	No	Yes (3-D Richards)
Gradients in soil hydraulic conductivity	No	Yes
Soil porosity	0.46	
Saturated hydraulic conductivity	$1.785 \times 10^{-6} \frac{m}{s}$	
n_{VG}	1.534	
α_{VG}	$1.44\,m^{-1}$	
λ_{VG}	−0.215	
Initial total soil water potential	−0.4 m	−3.7 m
Root properties	**Simple model**	**"aRoot" model**
Heterogeneous root hydraulic properties	Yes	
Critical collar potential	−150 m	
Root radius r_{root}	1 mm	
Flux boundary condition $Q(t)$	$5 \times 10^{-11}\,m^3\,s^{-1}$	$3 \times 10^{-9}\,m^3\,s^{-1}$
Total root length l_{total}	0.01–8 m	9.93 m
Branching order	≤ 1	$>> 1$
Account for root length density	No	Yes
Number of root segments	100 (unbranched)/ 192 (branched root)	1412
Root hydraulic properties	**Mature root**	**Young root**
Axial resistivity $\zeta_{Ax}\,[s\,m^{-3}]$	8×10^{10}	1×10^{12}
Radial resistivity $\rho_{Rad}\,[s]$	5×10^{8}	1×10^{8}

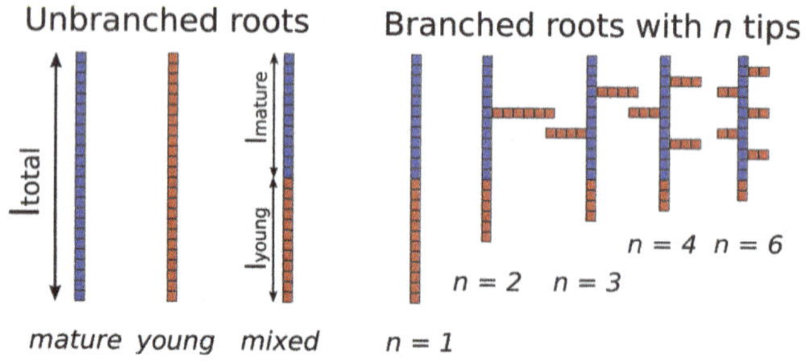

Figure 1. Schematic representation of the root topologies and parameters that were investigated with the simple root water uptake model. Young (l_{young}) and mature root length (l_{mature}) are varied independently both in unbranched and branched root structures, resulting in varying total length (l_{total}) and mature root proportion (p_{mature}). In all heterogeneous cases mature roots constitute the basal part of the root. Within branched roots, total young root length is evenly divided into n parts, which are attached to the central mature root at equal distances. A mixed root strand can equivalently be regarded as a branched root with $n = 1$. Gravity and soil water flow are neglected in the simple model.

average root distance within the complex model. The water content within each of the soil cylinders is assumed to be spatially constant, but may be different between different soil segments. Soil water flow between the soil cylinders was neglected. All soil cylinders share the same hydraulic properties. The soil water potential $\psi_{Soil}^{(i)}$ (m) within each soil cylinder i is derived from volumetric soil water content $\theta_{Soil}^{(i)}$ ($m^3\,m^{-3}$) with a van Genuchten parameterization of the soil $\theta_{Soil}^{(i)} = f(\psi_{Soil}^{(i)})$. Parameters are taken from Schneider et al. (2010) and were originally obtained for a sandy soil (see Table 1 for details). Furthermore, gravitational potential was neglected within the simple model. Thus, the change in soil water status within the soil cylinders is related entirely to root water uptake or release. Simulations are started with initially uniform total soil water potential throughout the entire soil domain (hydrostatic equilibrium).

Water transport within the roots follows an axial pathway, while water uptake (flow from the surrounding soil into the root) occurs along the radial pathway only. Water flow along each pathway is governed by gradients in hydraulic potential and resistances, similar to Ohm's law. In either direction, the water flow for a given root segment i is given as

$$Q_{Rad}^{(i)} = \frac{\psi_x^{(i)} - \psi_{Soil}^{(i)}}{R_{Rad}^{(i)}} \tag{1}$$

$$Q_{Ax,in}^{(i)} = \sum_j \frac{\psi_x^{(j)} - \psi_x^{(i)}}{R_{Ax}^{(j)}} \tag{2}$$

$$Q^{(i)}_{\mathrm{Ax,out}} = \frac{\psi_x^{(i)} - \psi_x^{(k)}}{R^{(i)}_{\mathrm{Ax}}}, \tag{3}$$

where $Q^{(i)}_{\mathrm{Ax,in}}$, $Q^{(i)}_{\mathrm{Ax,out}}$ and $Q^{(i)}_{\mathrm{Rad}}$ ($\mathrm{m^3\,s^{-1}}$) are the volumetric rates of water flow along the axial pathway into root segment i, out of root segment i and along the radial pathway from the soil into root segment i. $\psi_x^{(i)}$, $\psi_x^{(j)}$, $\psi_x^{(k)}$ and $\psi^{(i)}_{\mathrm{Soil}}$ (m) are the xylem water potentials within the root segment i, all subsequently connected root segments j and the preceding root segment k, as well as the bulk soil water potential within the soil surrounding the root segment i. Finally, $R^{(i)}_{\mathrm{Ax}}$ and $R^{(i)}_{\mathrm{Rad}}$ ($\mathrm{s\,m^{-2}}$) are the axial and radial root resistance within segment i. These resistances are derived from material properties and scale with geometric dimensions as follows:

$$R^{(i)}_{\mathrm{Ax}} = \zeta^{(i)}_{\mathrm{Ax}} \cdot l^{(i)} \tag{4}$$

$$R^{(i)}_{\mathrm{Rad}} = \frac{\rho^{(i)}_{\mathrm{Rad}}}{A^{(i)}_{\mathrm{Surf}}} = \frac{\rho^{(i)}_{\mathrm{Rad}}}{2 \cdot \pi \cdot r^{(i)} \cdot l^{(i)}}. \tag{5}$$

The factors $\zeta^{(i)}_{\mathrm{Ax}}$ ($\mathrm{s\,m^{-3}}$) and $\rho^{(i)}_{\mathrm{Rad}}$ (s) are the axial and radial root hydraulic resistivity of root segment i. Although the resistances $R^{(i)}_{\mathrm{Ax}}$ and $R^{(i)}_{\mathrm{Rad}}$ determine water flow along potential gradients in the model, the underlying axial and radial root resistivities $\zeta^{(i)}_{\mathrm{Ax}}$ and $\rho^{(i)}_{\mathrm{Rad}}$ define root hydraulic properties and can be obtained via measurements. Each root segment obtains root hydraulic resistivities corresponding to two discrete hydraulic classes taken from Schneider et al. (2010) (see Table 1). Heterogeneity of root hydraulic properties is introduced in roots by associating these different hydraulic classes with different regions of the root system (see Sect. 2.3 below).

As a consequence of mass conservation and the absence of storage capacities within the root, the water mass balance holds for each segment i:

$$Q^{(i)}_{\mathrm{Ax,in}} + Q^{(i)}_{\mathrm{Rad}} = Q^{(i)}_{\mathrm{Ax,out}}. \tag{6}$$

By substituting the axial and radial flow rates by Eqs. (1), (2) and (3) for all n root segments, by denoting $Q^{(0)}_{\mathrm{Ax}}$ ($\mathrm{m^3\,s^{-1}}$) and $\psi_x^{(0)}$ (m) as the unknown total outflow and water potential at the root collar, and by setting $Q^{(i)}_{\mathrm{Ax,in}} = 0$ at the root tips, we obtain n equations for the $n+1$ unknown xylem water potentials including $\psi_x^{(0)}$. Closure of this system of equations is achieved by fixing a boundary condition at the root collar. In our model, this can either be a prescribed (time-dependent) flux rate $Q^{(0)}_{\mathrm{Ax}}(t)$ or a constant xylem water potential $\psi_x^{(0)}$. The former represents a given transpirational demand of a plant at a given time; the latter is used to simulate a plant under water stress. At the onset of water stress transpiration reduces, as collar potential does not further decrease. All simulations are started with a flux boundary condition until collar potential drops to a critical threshold (here taken as a typical value of the permanent wilting point $\psi_{\mathrm{Crit}} = -150\,\mathrm{m}/-1.5\,\mathrm{MPa}$) upon which the boundary condition switches to the potential boundary condition $\psi_x^{(0)} = \psi_{\mathrm{Crit}} = -150\,\mathrm{m}$, thus mimicking "isohydric plants".

After all soil and xylem water potentials have been calculated, root water uptake rates can be deduced using Eq. (1) and soil water status is updated using a steady-state approach for a sufficiently short interval of time Δt (s)

$$\theta^{(i)}_{\mathrm{Soil;\,new}} = \theta^{(i)}_{\mathrm{Soil;\,old}} - \frac{Q^{(i)}_{\mathrm{Rad}} \cdot \Delta t}{V^{(i)}_{\mathrm{Soil}}}, \tag{7}$$

where $V^{(i)}_{\mathrm{Soil}}$ ($\mathrm{m^3}$) is the total volume of soil surrounding the root segment i. The soil water potential decreases correspondingly.

The strongly simplified assumptions within this model allow for investigation of feedbacks between the distribution of soil water potential and root water uptake, depending on different root hydraulic architectures. In particular, they allow for understanding the combined role of heterogeneous root hydraulic properties and branching for root water uptake dynamics, which would be hard to detect at a higher level of complexity. In order to test whether the results are reproduced in more realistic conditions, we compare them against the complex root water uptake model, which explicitly accounts for soil water flow and gravitational potential as described in the next section.

2.2 Root water uptake model for complete root systems

We modelled root water uptake in complete root systems of a single plant individual with the three-dimensional root water uptake model "aRoot", developed by Schneider et al. (2010). We simulate a pot experiment where a complete root system is embedded in one block of soil with a volume of $V_{\mathrm{Soil}} = 0.45\,\mathrm{m} \cdot 0.45\,\mathrm{m} \cdot 0.3\,\mathrm{m}$. Within this block, soil water flow is gradient driven and numerically calculated with a finite element method solving the Richards equation in fully explicit 3-D (Kolditz et al., 2012). "aRoot" accounts both for gravitational potential within the soil as well as for gradients in soil water potential in the immediate vicinity of individual roots. The model of water flow within the root system is equivalent to the simple model described above. All simulations were initialized with total soil water potential being homogeneous (hydrostatic equilibrium). For detailed information about the features of "aRoot", please refer to Schneider et al. (2010). Both the van Genuchten parameters of the soil and the root hydraulic properties are the same as in the simple model (Table 1).

2.3 Systematic variation of root hydraulic properties in roots

Both at the single root and at the single-plant scale, the complex process of root maturation is simplified by introducing two discrete root hydraulic classes. These two classes possess both different axial and radial resistivities $\zeta_{\mathrm{Ax}}^{(i)}$ and $\rho_{\mathrm{Rad}}^{(i)}$, as well as different ratios of radial and axial resistivity $\rho_{\mathrm{Rad}}^{(i)}/\zeta_{\mathrm{Ax}}^{(i)}$. Values are taken from Schneider et al. (2010) and refer to "young" and "mature" roots of a 28-day-old sorghum plant. For reasons of simplicity the root radius is set equal to 1 mm for both young and mature roots. This simplification has little influence on values for root resistances, since dependence on root radius is small compared to dependence on root length (see Eqs. 4 and 5).

In order to assess the influence of heterogeneity of root hydraulic properties, the distribution of the two hydraulic classes along the roots is varied systematically. For this, we neglect information about root age or geometry, as we do not focus on reproducing a specific plant. However, we assume that mature roots always constitute the basal parts and young roots the apical parts in all roots. This is achieved differently at the single-root and at the single-plant scale.

Single unbranched and branched roots are created using three parameters: (a) total root length (l_{total}), (b) the proportion of young or mature roots (p_{young} or p_{mature}) which have to sum up to one, and (c) the number of root tips (n). Figure 1 illustrates the construction of single roots used within the simple model. In unbranched single roots the mature root is located in the basal, the young root in the apical part of the root. We modelled unbranched single roots with a total length between 1 and 800 cm, containing between 0 and 100 % of mature roots. Branched single roots are assumed to have two, three, four or six young root branches. All of those branches are distributed evenly along a central mature root strand and have equal lengths, resulting in fishbone-like structures. For branched single roots, l_{total} is varied between 5 and 400 cm and p_{mature} varies between 10 and 90 %. We are aware that unbranched roots of great length are unrealistic. However, this artificial set-up allows for assessing the efficiency of root water uptake depending on the branching structure.

At the single-plant scale, the assignment of root hydraulic properties is somewhat different, as root geometry and topology are given a priori. The root system geometry is obtained with the root generator "RootTyp" by Pagès et al. (2004) and the location of the roots within the soil was kept the same for all simulations (see Fig. 7). The parameters used for "RootTyp" are taken from Schneider et al. (2010) and correspond to a 28-day-old sorghum plant. The resulting total root length was $l_{\mathrm{total}} = 9.93$ m. In order to investigate the influence of heterogeneous hydraulic properties on spatiotemporal root water uptake and its efficiency, we varied the proportions of young and mature roots in steps of 20 % between 0 and 100 % on this geometry as follows: first, starting at the outer ends of the root system, all tip segments were classified as young roots. Afterwards, this assignment was iterated with the immediately preceding segments. The assignment was suspended at branching points until all branches associated with this point were classified entirely (as young roots). If the desired amount of young roots is achieved, the remaining segments are classified as mature roots. This ensures that mature roots are never preceded by young roots and they therefore constitute the basal and apical root part, respectively. Please note that this manipulation of the root properties was not performed with the intention of to re-produce a natural plant, but to discover shortcomings in root parameterization.

2.4 Measuring the efficiency of root water uptake

In order to compare the efficiency of the root water uptake process between different root topologies and degrees of heterogeneity of root hydraulic properties, we define two indices: "water yield" and "effort".

Water yield $v(t)$ ($\mathrm{m^3\,m^{-1}}$) assesses how much water $V_{\mathrm{H_2O}}^{\mathrm{unstressed}}$ ($\mathrm{m^3}$) could be taken up per unit root length under unstressed conditions within a given time:

$$v(t) = \frac{V_{\mathrm{H_2O}}^{\mathrm{unstressed}}(t)}{l_{\mathrm{Total}}(t)} = \frac{\int\limits_{\tau=0}^{t} \chi(\tau)\cdot Q(\tau)d\tau}{l_{\mathrm{Total}}(t)}, \tag{8}$$

where $Q(\tau)$ ($\mathrm{m^3\,s^{-1}}$) is the transpirational demand at time τ (s) and$\chi(\tau)$ is used to indicate water stress at time τ by zero and one otherwise. Thus, root water uptake under stressed conditions does not contribute to water yield. As stated above, we assume that water stress occurs when xylem water potential at the collar $\psi_x^{(0)}$ (m) drops below $\psi_{\mathrm{Crit}} = -150$ m (-1.5 MPa). We normalize by total root length to obtain unstressed transpiration per invested metre root length, in order to reflect on the increased soil water reservoir available to longer roots.

Expression (8) simplifies for certain conditions. For all simulations presented in this paper, we will be assuming a time-constant transpiration rate $Q(t) = Q$ and a drying scenario. This ensures the existence of a unique point $\tilde{t}$ (s) in time at which water stress occurs. In that case and assuming the absence of storage capacities within the root system, water yield is directly proportional both to the transpirational demand Q and the time at which water stress occurs. If root growth is furthermore neglected ($l_{\mathrm{total}} = \mathrm{const.}$), water yield $v(t)$ can be calculated as

$$v(t) = \begin{cases} \frac{Q\cdot t}{l_{\mathrm{Total}}} & t < \tilde{t} \\ \tilde{v} = \frac{Q\cdot \tilde{t}}{l_{\mathrm{Total}}} & t \geq \tilde{t}. \end{cases} \tag{9}$$

Thus, after water stress occurs water yield remains unaltered and becomes independent of time. Within this paper, we will refer to the above stated conditions and denote "water yield" simply as $\tilde{v}$. The lowercase "v" indicates that water yield is a normalized volume of water uptake. Assuming a time-constant transpiration rate $Q =$ const. is a strong simplification made here for matters of simplicity. However, it does not limit the application of the index to transient conditions.

Effort $w(t)$ (J m^{-3}) is a time-dependent quantity that measures the average work $W(t)$ (J) necessary to take up a unit of water $V_{H_2O}(t)$, and is evaluated over a given interval of time. Following thermodynamic principles (see Appendix A), $w(t)$ can be derived from the transpirational demand $Q(\tau)$and the collar potential $\psi_x^{(0)}(\tau)$. It takes the following form:

$$w(t) = \frac{W(t)}{V_{H_2O}(t)} = \frac{\int\limits_{\tau=0}^{t} Q(\tau) \cdot \psi_x^{(0)}(\tau) d\tau}{\int\limits_{\tau=0}^{t} Q(\tau) d\tau}. \qquad (10)$$

Effort uses the temporal evolution of xylem water potential at the root collar $\psi_x^{(0)}$ to estimate the efficiency of root water uptake. According to Eq. (10), it can be interpreted as a flow-weighted average collar potential. In accordance with $\psi_x^{(0)}$ effort has units of a negative hydraulic head (m water column). Please note that the pressure of 1 MPa can alternatively be stated as a hydraulic head of 101,97 m water column, but also has the physical meaning (and units) of an energy density of 10^6 J m^{-3}. The effort $w(t)$ therefore also has units of a specific energy and we refer to the absolute values of w when saying "effort is minimized". Under the conditions stated above (time constant transpiration rate Q, a drying scenario with unique occurrence time of water stress $\tilde{t}$), Eq. (10) simplifies for $t \leq \tilde{t}$ and effort can be described with another interesting meaning:

$$w(t) = \frac{\int_{\tau=0}^{t} Q(\tau) \cdot \psi_x^{(0)}(\tau) d\tau}{\int_{\tau=0}^{t} Q(\tau) d\tau} = \frac{Q \cdot \int_{\tau=0}^{t} \psi_x^{(0)}(\tau) d\tau}{Q \cdot t}$$
$$= \bar{\psi}_x^{(0)}(t) \qquad (11)$$

in which $\bar{\psi}_x^0(t)$ (m) is the time-average collar potential between times $\tau = 0$ and $\tau = t$. In contrast to water yield, effort still changes after the onset of water stress. But as this contribution is very small (see App. A) we will approximate the effort under our specific model conditions with $\tilde{w} = w(\tilde{t}) = \bar{\psi}_x^0(\tilde{t})$. As for water yield, the lowercase "w" indicates that effort corresponds to a specific (normalized) energy. Assuming a time-constant transpirational demand $Q =$ const. is a strong assumption which is made here for reasons of simplicity, but does not limit the application of the index to transient conditions.

Figure 2 illustrates how water yield and effort can be used to compare the efficiency of root water uptake for one branched (green) and one unbranched (red) single root, both sharing the same total length. Under the above-mentioned conditions, they can be deduced from the temporal evolution of xylem water potential at the root collar. As the total root length is the same, water yield $\tilde{v}$ is directly proportional to the time at which the plant enters water stress $\tilde{t}$ (see Eq. 10). In this case, differences in the respective values of $\tilde{t}$ and $\tilde{v}$ are very small. Effort $\tilde{w}$ corresponds to the area below the two curves, divided by the respective values of $\tilde{t}$. The green area is much smaller than the red area, which indicates that on average a less negative collar potential and consequently less energy was needed for maintaining root water uptake in the branched root. As all other parameters were equal, this indicates an overall lower resistance to root water uptake experienced by the branched compared to the unbranched root.

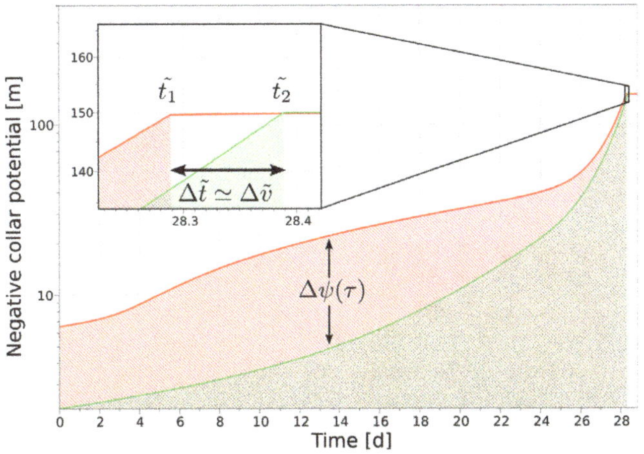

Figure 2. Evolution of collar xylem water potential over the course of time for two exemplary chosen single roots of equal total length (0.8 m): an unbranched homogeneous young root (red) and a branched root with six tips (green). Water yield measures the total amount of water that could be extracted before reaching critical xylem water potential. Effort is given by the area below the graph, divided by the respective occurrence times of water stress. Although water yield is very similar between the two root structures in this case, effort is substantially different.

In this particular case, the differences are induced by branching (see Sect. 3). Water yield is related to the total amount of water that could be extracted under unstressed conditions (unstressed transpiration), but is additionally referenced to total root length. Unstressed transpiration was used before by other researchers to evaluate root parameterizations (Schneider et al., 2010; Javaux et al., 2008). On the other hand, effort relates to the temporal evolution of xylem water potential at the root collar and the average work necessary for root water uptake. It includes information on the total resistance to root water uptake a root system has to overcome and also depends on the soil water retention. As far as we are aware, both indices are novel ways of measuring plant performance, and carry physiological as well as hydrological meaning.

Please note that the indices are related, as they both depend on the root hydraulic resistance. However, effort carries more information on plant function. Since research suggests that plants operate with little safety margin with regard to danger

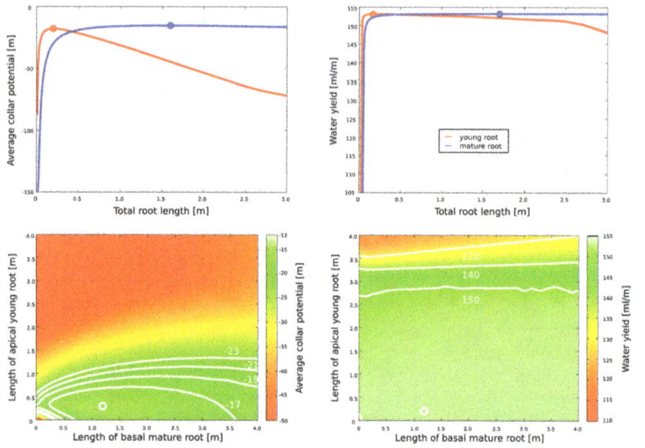

Figure 3. Effort $\tilde{w}$ (left) and water yield $\tilde{v}$ (right) in unbranched single roots, depending on the proportion of young and mature roots. Data was obtained with the simple model. Shown are effort and water yield for (top) unbranched homogeneous young (red) and mature (blue) roots over total root length and (bottom) for heterogeneous roots. Optimal values are indicated with circles.

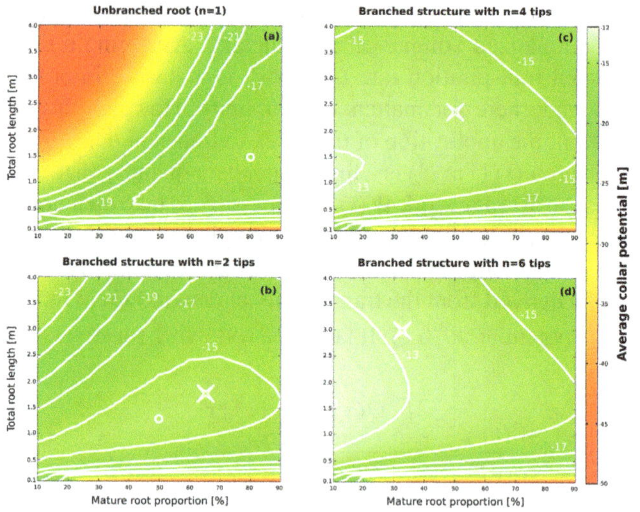

Figure 4. Effort $\tilde{w}$ depending on topology and composition of single roots, obtained with the simple model. Results are shown for **(a)** unbranched roots and branched roots (fishbone structures) with **(b)** two, **(c)** four and **(d)** six tips. Root composition is given by total root length (y axis) and the proportion of mature roots (x axis). Colours are the same as in Fig. 3 (bottom left). Optimal values of effort are denoted by white circles. The crosses in **(b–d)** indicate effort for a root that is the same as the optimal unbranched heterogeneous root from **(a)** except for containing one, three and five more equal young root tips, respectively.

for embolism across climates, plants should apply strategies to avoid very negative xylem water potentials. As lower effort is tantamount for lower average xylem water potentials, it recommends itself as a tool for distinguishing efficient from less efficient parameterizations.

3 Results

We first present results obtained from the simple model separately for single unbranched and branched roots and next the results obtained with the more encompassing aRoot model for entire root systems.

3.1 Optimal effort and water yield in unbranched single roots

Figure 3 shows effort (top left) and water yield (top right) in unbranched single roots with homogenous root hydraulic properties and increasing length. For both mature and young roots, optimal root lengths emerge. This implies that the average xylem potential (effort) assumes a minimum and the average uptake per root length a maximum at a given root length. Both indices propose similar optimal root lengths (Table 2), but different ones for young and mature roots: young roots have to be short in order to achieve optimal effort and water yield, whereas mature roots have to be long. Interestingly, the actual values at the respective optima are not much different – it is (almost) as efficient to be a short young root as it is to be a long mature root. Water yield is by far the lesser sensitive of the both measures with regard to changes in root length. Also, mature roots exhibit less pronounced differential changes in effort and water yield than young roots when changing root length.

Results for heterogeneous unbranched roots are shown at the bottom of Fig. 3. All heterogeneous single roots consist of basal mature and apical young roots. Heterogeneity affects the efficiency at the respective optimal lengths differently: optimal heterogeneous roots have decreased their effort by 15 %, but increased their water yield only slightly by 1 %. Moreover, for the optimal mixed root strand, the optimal total root lengths are shorter than expected, in that the optimal mixed root strand is not a composition of an optimal mature root strand and an optimal young root strand, but altogether shorter (Table 2). In composed roots some of the water is taken up by the basal mature root part and less water has to be transported through the apical young roots. Therefore drops in xylem potential are smaller, axial limitation is less severe and the hydraulically active young root region is extended in composed roots. For this reason, in optimal composed roots, young roots are longer and mature roots are shorter compared to their homogenous peers. This leads to overall shorter composite unbranched single roots.

3.2 Optimal effort and water yield in branched single roots

Figure 4 shows the effort of single roots with one, two, four and six tips, respectively (Fig. 4a–d, the properties of the

Table 2. Optimal compositions of single roots referring to effort (top) and water yield (bottom). Results are obtained with the simple model for different root topologies.

Structure	l_{total}	l_{mature}	l_{young}	l_{young} per branch	$\tilde{w}$
Young root strand	0.20 m	–	0.20 m/100 %	0.20 m	−18.0 m
Mature root strand	1.60 m	1.60 m/100 %	–	–	−15.3 m
Mixed root strand	1.50 m	1.20 m/80 %	0.30 m/20 %	0.30 m	−15.1 m
Branched root, 2 tips	1.30 m	0.65 m/50 %	0.65 m/50 %	0.33 m	−14.4 m
Branched root, 3 tips	0.90 m	0.09 m/10 %	0.81 m/90 %	0.27 m	−13.5 m
Branched root, 4 tips	1.20 m	0.12 m/10 %	1.08 m/90 %	0.27 m	−12.8 m
Branched root, 6 tips	1.60 m	0.16 m/10 %	1.44 m/90 %	0.24 m	−12.3 m

Structure	l_{total}	l_{mature}	l_{young}	l_{young} per branch	$\tilde{v}$
Young root strand	0.15 m	–	0.15 m/100 %	0.15 m	153.07 $cm^{-3}\,m^{-1}$
Mature root strand	1.80 m	1.80 m/100 %	–	–	153.21 $cm^{-3}\,m^{-1}$
Mixed root strand	1.60 m	1.28 m/80 %	0.32 m/20 %	0.32 m	153.21 $cm^{-3}\,m^{-1}$
Branched root, 2 tips	0.90 m	0.27 m/30 %	0.63 m/70 %	0.32 m	153.24 $cm^{-3}\,m^{-1}$
Branched root, 3 tips	0.90 m	0.18 m/20 %	0.72 m/80 %	0.24 m	153.28 $cm^{-3}\,m^{-1}$
Branched root, 4 tips	1.20 m	0.12 m/10 %	1.08 m/90 %	0.27 m	153.30 $cm^{-3}\,m^{-1}$
Branched root, 6 tips	2.00 m	0.20 m/10 %	1.80 m/90 %	0.30 m	153.32 $cm^{-3}\,m^{-1}$

optimal combinations are given in Table 2). The root composition is now given by the total root length of the respective root (y axis) and the proportion of mature roots (x axis). Colours are the same as in Fig. 3 (bottom left). While the proportion of mature roots in optimally branched roots decreases disproportionately, the total length of all young roots increases almost proportionally to the number of tips n (Table 2). When adding new tips, individual young root branches shorten only a little, allowing for the total root length to expand while also decreasing effort. In this way, branching favours soil exploration, without compromising efficiency. Notably, the effort surface becomes flatter, and hence the domain of nearly efficient hydraulic parameterizations expands with the number of tips.

Similar results are obtained for water yield but results are far less sensitive (Fig. 5). For all branched roots, water yield is nearly constant (little sensitive) within the domain of modelled root compositions and increases only very little compared to the optimal unbranched single root (see Table 2).

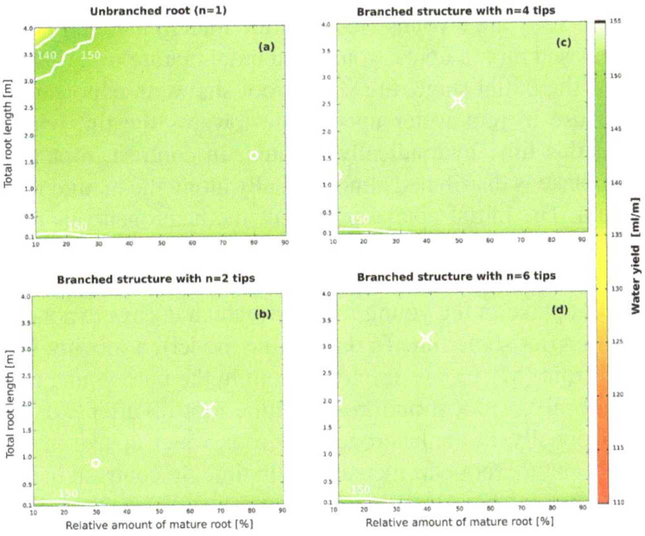

Figure 5. Water yield $\tilde{v}$ depending on topology and composition of single roots, obtained with the single model. Results are shown for **(a)** unbranched roots and branched roots (fishbone structures) with **(b)** two, **(c)** four and **(d)** six tips. Root composition is given by total root length (y axis) and the proportion of mature roots (x axis). Colours are the same as in Fig. 3 (bottom right). Optimal values of water yield are denoted by white circles. The crosses in figures **(b–d)** indicate water yield for a root that is the same as the optimal unbranched heterogeneous root from **(a)** except for containing one, three and five more equal young root tips, respectively.

3.3 Water uptake dynamics and redistribution patterns in single roots

The proportions of root hydraulic properties within a branched or unbranched single root do not only affect the efficiency of root water uptake, but also its location and dynamics. This may even be the case, if the efficiency is similar between parameterizations. Figure 6 shows root water uptake rates along three exemplarily chosen unbranched roots of equal length ($l_{total} = 0.42$ cm) and similar water yield and

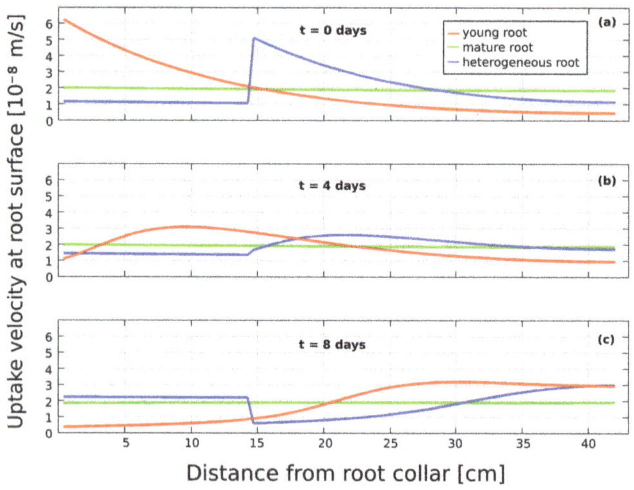

Figure 6. Velocity of radial inflow (uptake velocity) at the root surface along three unbranched single roots with equal length ($l_{\text{total}} = 0.42$ m) but different composition. Values are obtained with the simple model for roots containing young roots only (red), mature roots only (blue) or an optimal mixture with respect to water yield (green; $l_{\text{mature}} = 0.14$ m, $l_{\text{young}} = 0.28$ m). Results are depicted for **(a)** initial stage (hydrostatic equilibrium), **(b)** 4 days and **(c)** 8 days of simulation time.

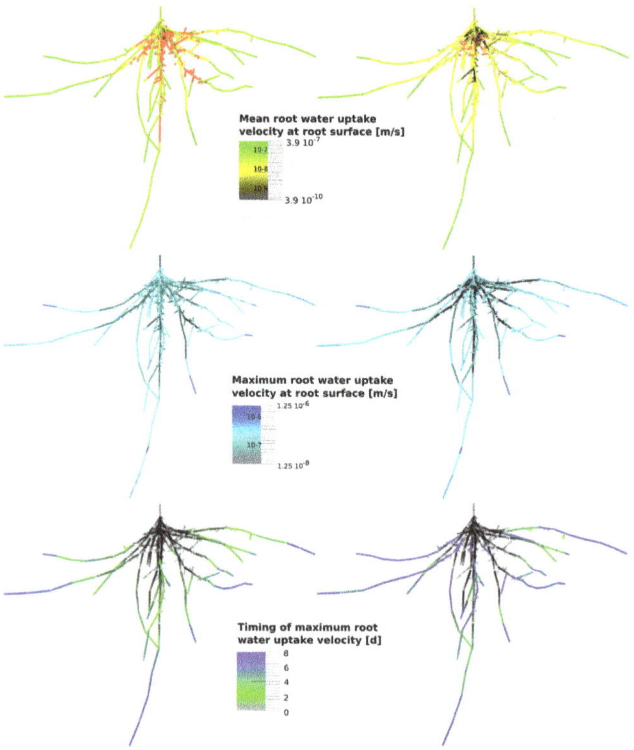

Figure 7. Root water uptake dynamics in a fixed root geometry with two different hydraulic parameterizations. Results were obtained with the "aRoot" model for one root system containing young roots only (left, least efficient) and a mixture of 40 % of basal mature and 60 % of apical young roots (right, most efficient). (Top) Time-averaged root water uptake rate along the root system. Regions with negative net uptake (hydraulic lift or bleeding) are depicted in red, independent of the actual amount of water released. (Centre) Magnitude and (bottom) timing of maximum uptake velocity along the root system. Please note the log-scale of the colour bar in the top and centre panel.

effort. They are a young (red), mature (green) and optimally composed mix of apical young and basal mature root (blue).

At the initial stage, the young root shows an exponential decrease in root water uptake rate towards the tip, which is at this time hydraulically isolated. In contrast, root water uptake is distributed almost equally along the mature root strand. The initial uptake pattern of the heterogeneous root is a combination: an almost homogeneous uptake rate in the basal mature root part is followed by an increased rate of root water uptake in the young root part, which decays exponentially. After some time (4 days in the model), a moving uptake front (MUF) has developed both in the pure young and in the mixed root strand, reaching the root tip after 8 days. Additionally, in the heterogeneous root, water uptake in the basal mature root part increases with time. In contrast, in the pure mature root, the water uptake profile is static and does not change much over the course of the simulation. Although the occurrence of moving uptake fronts is accentuated by the neglect of soil water flow and gravity within the simple root water uptake model, qualitatively the same results are obtained within the complex "aRoot" model, in which soil water flow and gravity are explicitly considered (see Sect. 3.5 and Fig. 7).

3.4 Effort and water yield in entire root systems

In order to quantify what influence the above-mentioned small-scale processes have at the scale of an individual plant, and taking soil water flow and gravitation into account, we used the detailed three-dimensional root water uptake model "aRoot". We calculated effort and water yield along with spatiotemporal root water uptake for one exemplary root system geometry, which was kept the same for all simulations (see Fig. 7 for geometry). Only the proportions of young and mature roots were systematically varied in steps of 20 % between 0 and 100 % (see Sect. 2.3).

Table 3 shows water yield and effort for these six different hydraulic parameterizations. Both criteria showed lowest efficiency in the homogeneous root systems, with the young one being the least efficient. This is in agreement with the simple models above, where long young roots were inefficient, while mature roots suffer less from radial limitation when they are sufficiently long. The most efficient root systems were heterogeneous ones (containing between 20 and 60 % of mature roots). Compared to homogenous systems, they increased water yield by about 25 % and cut the effort by one-half. Root systems with more mature roots (80 %) were less efficient, because the potential of young roots was not fully explored (Sect. 3.2).

In order to preclude that our results are subject to an artifact of the evaluation time (i.e. the different time of first occurrence of water stress at which effort is calculated), we also evaluated effort 5 days after the start of the simulation, and confirmed that the ranking of the root systems did not change (Table 3). Additionally, we repeated our analysis with a transient (sinusoidal) transpirational demand and qualitatively obtained the same results (see Supplement).

3.5 Water uptake dynamics and redistribution patterns in entire root systems

Figure 7 compares the spatial distribution of root water uptake characteristics in a homogenous (least efficient) and heterogenous (most efficient) root system. Mean root water uptake rates uptake rates (top) vary much less in the homogeneous compared to the heterogeneous root system (spanning 1 order of magnitude compared to 3 orders of magnitude). Also, within all heterogeneous root systems, water uptake of mature roots is always smaller than the mature root proportion (Fig. 8, left). This indicates the separation of root function in the heterogeneous root system between uptake roots and transport roots, and is in agreement with the earlier observations in the simple model. Apical young roots have a higher mean uptake rate than inner young roots in both hydraulic parameterizations, which is due to higher root density in the central parts of the root system.

The lower part of Fig. 7 shows the magnitude (centre) and timing (bottom) of the maximum uptake at each location of the root system. This allows for the tracking of moving uptake fronts. The timing of the maximum shows how uptake moves evenly away from the collar in the young root system as expected from the simple model (see Fig. 6). In heterogeneous root systems the uptake pattern is more complex. Maximum uptake rates occur in the young roots, irrespective of their actual position within the root system (see Sect. 2.4 for the distribution of root hydraulic properties). The timing of the maximum uptake shows that uptake fronts move not only outwards but also inwards (see the blue roots in the centre of the root system, Fig. 7, bottom right). Inner mature roots are activated late and only if the surrounding soil was not previously dried out by young roots. Together with distant young roots, mature roots contribute the majority to total water uptake after 8 days (see Figs. 7 and 8). This redistribution pattern corresponds to the one observed with the simple model in heterogeneous single roots (Sect. 3.3 and Fig. 6). In the simple model, root water uptake was redistributed in two ways: "forward" along young roots towards the root tips by moving uptake fronts; and "backward" away from distal young roots to inner mature roots. In the complex "aRoot" model, which considers root length density and soil water redistribution, a third redistribution pattern is added: redistribution between different root branches. Root water uptake is

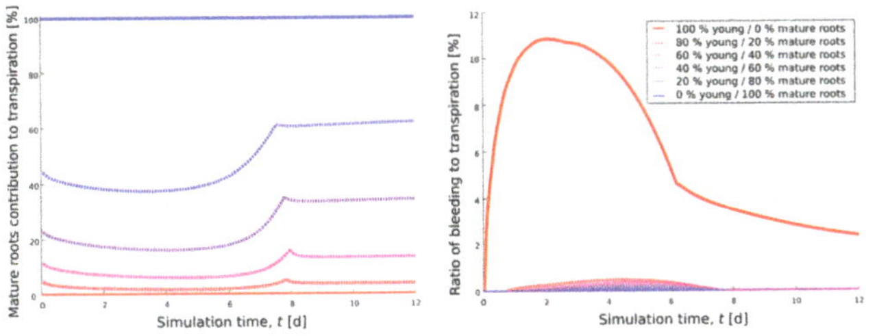

Figure 8. Evolution of mature root contribution to overall transpiration (left) and the ratio of bleeding (right) over time in the fixed root geometry for the six different hydraulic parameterizations. Results are obtained with the "aRoot" model for fractions of apical young roots between 0 and 100 %. Homogeneous root systems are depicted in solid lines; heterogeneous root systems are depicted with dashed lines.

distributed away from (inner) branches of young and mature roots, as they fall dry in the course of soil drying, and is redistributed towards roots in wetter soils. Altogether, this leads to higher efficiency in heterogeneous root systems compared to homogeneous root systems (see Table 3), which is likely due to a more efficient compensation for local water stress and enhanced soil exploration.

Uptake depth in root systems with mature roots was deeper compared to homogenous root systems for much of the simulation time. Figure 9 shows temporal evolution of the depth z_{50} (m) above which half of the root water uptake occurred. Over the course of time, z_{50} moves downwards in all hydraulic parameterizations and equilibrates at the onset of water stress, with the homogeneous young root system being most dynamical, and most shallow at the same time.

Hydraulic lift occurred in all root parameterizations. However, the domain of hydraulic lift is noticeably larger in the homogenous young root system compared to all other hydraulic parameterizations. Both the total length of bleeding roots and the absolute amount of water released decrease along with young root proportion, being smallest in the homogeneous mature root system (see also Fig. 8, right). The by far highest values of hydraulic lift are modelled for the the homogeneous young root system (up to 10 % of total root water uptake). It must be stated that bleeding usually occurs at night and may hence not be well captured with the time-constant flux boundary condition used here. However, simulations with a sinusoidal day/night cycle of transpiration showed qualitatively the same results.

4 Discussion

We used two models to examine to what extent heterogeneity of root hydraulic properties influences root water uptake at two spatial scales. In order to disentangle different processes of root water uptake redistribution acting at the same time, we simplified the model scenarios. First we presuppose soil

contrast, our results show the highest amount of bleeding in the most inefficient root hydraulic parameterization, namely in the homogeneous young root system. This result remained unaltered when a sinusoidal transpirational demand was used instead of a fixed flux boundary condition. This indicates that bleeding in this case did not act to improve the overall water status of the plant. Thus although hydraulic redistribution is frequently observed in the real world (Neumann and Cardon, 2012), its occurrence in models does not necessarily imply efficient parameterization.

5 Conclusions

In this modelling study we show that root hydraulic properties, in particular the ratio of root radial and axial resistivity, determine optimal root length for single roots in a drying scenario. We investigate this with two different indices introduced to compare the efficiency of root water uptake: water yield and effort. Water yield measures the amount of root water uptake extracted from the soil before a plant enters water stress; effort indicates the xylem water potential (average energy) necessary for this root water uptake under unstressed conditions. Both are suitable to detect efficient lengths of young and mature roots, with effort being more sensitive than water yield. Optimal lengths of unbranched young roots are some centimetres, compared to several metres for mature roots. However, the efficiency of simulated root water uptake increases, when more young root length can be activated. This is achieved in branched roots with heterogeneous root hydraulic properties, which allow for a division of function between water uptake and transport. This finding is supported by simulations in a complex three-dimensional root system, where mature roots contribute disproportionately less to overall root water uptake compared to young roots, suggesting that they act as transport roots.

As heterogeneity in root hydraulic properties leads to lower effort, increased water yield and altered root water uptake dynamics, it should be addressed in root water uptake models. Overall, parameterization of the root system has a great effect on modelled processes that are of interest for the hydrological and ecological community, such as root water uptake profiles, moving uptake fronts, temporal evolution of xylem water potential, and, and hydraulic re-distribution. As the exploration of these processes is one of the main purposes for using complex three-dimensional models, we believe that parameterization of root properties warrants more attention. Some root water uptake features are similar within a broad range of efficient heterogeneous parameterizations. Therefore the actual degree of heterogeneity may play a subordinate role for root water uptake simulations, as long as hydraulic heterogeneity is accounted for in a principal way.

Edited by: N. Ursino

References

Angeles, G., Bond, B., Boyer, J. S., Brodribb, T., Brooks, J. R., Burns, M. J., Cavender-Bares, J., Clearwater, M., Cochard, H., Comstock, J., Davis, S. D., Domec, J.-C., Donovan, L., Ewers, F., Gartner, B., Hacke, U., Hinckley, T., Holbrook, N. M., Jones, H. G., Kavanagh, K., Law, B., López-Portillo, J., Lovisolo, C., Martin, T., Martínez-Vilalta, J., Mayr, S., Meinzer, F. C., Melcher, P., Mencuccini, M., Mulkey, S., Nardini, A., Neufeld, H. S., Passioura, J., Pockman, W. T., Pratt, R. B., Rambal, S., Richter, H., Sack, L., Salleo, S., Schubert, A., Schulte, P., Sparks, J. P., Sperry, J., Teskey, R., and Tyree, M. T.: The Cohesion-Tension theory, New Phytologist, 163, 451–452, 2004.

Bauerle, T. L., Richards, J. H., Smart, D. R., and Eissenstat, D. M.: Importance of internal hydraulic redistribution for prolonging the lifespan of roots in dry soil, Plant. Cell Environ., 31, 177–186, 2008.

Blum, A.: Crop responses to drought and the interpretation of adaption, Plant Growth Regul., 20, 135–148, 1996.

Brooksbank, K., White, D. A., Veneklaas, E. J., and Carter, J. L.: Hydraulic redistribution in Eucalyptus kochii subsp borealis with variable access to fresh groundwater, Trees-Struct. Funct., 25, 735–744, 2011.

Cai, X., Wang, D., and Laurent, L.: Impact of Climate Change on Crop Yield: A Case Study of Rainfed Corn in Central Illinois, J. Appl. Meteorol. Climatol. 48, 1868–1881, 2009.

Caldwell, M. M., Dawson, T. E., and Richards, J. H.: Hydraulic lift: Consequences of water efflux from the roots of plants, Oecologia, 113, 151–161, 1998.

Choat, B., Jansen, S., Brodribb, T. J., Cochard, H., Delzon, S., Bhaskar, R., Bucci, S. J., Feild, T. S., Gleason, S. M., Hacke, U. G., Jacobsen, A. L., Lens, F., Maherali, H., Martinez-Vilalta, J., Mayr, S., Mencuccini, M., Mitchell, P. J., Nardini, A., Pittermann, J., Pratt, R. B., Sperry, J. S., Westoby, M., Wright, I. J., and Zanne, A. E.: Global convergence in the vulnerability of forests to drought, Nature, 491, 752–755, 2012

Choat, B.: Predicting thresholds of drought-induced mortality in woody plant species, Tree Physiol., 33, 669–671, 2013.

Churkina, G. and Running, S. W.: Contrasting Climatic Controls on the Estimated Productivity of Global Terrestrial Biomes, Ecosystems, 1, 206–215, 1998.

Clausnitzer, V. and Hopmans, J. W.: Simultaneous modelling of transient 3-dimensional root growth and soil water flow, Plant. Soil, 164, 299–314, 1994.

Collins, D. B. G. and Bras, R. L.: Plant rooting strategies in water-limited ecosystems, Water Resour. Res., 43, W06407, doi:10.1029/2006WR005541, 2007

Couvreur, V., Vanderborght, J., and Javaux, M.: A simple three-dimensional macroscopic root water uptake model based on the hydraulic architecture approach, Hydrol. Earth Syst. Sci., 16, 2957–2971, doi:10.5194/hess-16-2957-2012, 2012.

Domec, J.-C., King, J. S., Noormets, A., Treasure, E., Gavazzi, M. J., Sun, G., and McNulty, S. G.: Hydraulic redistribution of soil water by roots affects whole-stand evapotranspiration and net ecosystem carbon exchange, New Phytol., 187, 171–183, 2010.

Domec, J. C., Scholz, F. G., Bucci, S. J., Meinzer, F. C., Goldstein, G., and Villalobos-Vega, R.: Diurnal and seasonal variation in root xylem embolism in neotropical savanna woody species: impact on stomatal control of plant water status, Plant. Cell Environ., 29, 26–35, 2006.

Doussan, C., Pierret, A., Garrigues, E., and Pagès, L.: Water uptake by plant roots: II – Modelling of water transfer in the soil root-system with explicit account of flow within the root system – Comparison with experiments, Plant Soil, 283, 99–117, 2006.

Dunbabin, V. M., Postmam, J. A., Schnepf, A., Pagès, L., Javaux, M., Wu, L., Leitner, D., Chen, Y. L., Rengel, Z., and Digglc, A. J.: Modelling root–soil interactions using three–dimensional models of root growth, architecture and function, Plant. Soil, 372, 1–2, 93–124, 2013.

Edlefsen, N. and Anderson, A.: Thermodynamics of soil moisture, Hilgardia, 15, 31–298, 1943.

El Maayar, M., Price, D. T., and Chen, J. M.: Simulating daily, monthly and annual water balances in a land surface model using alternative root water uptake schemes, Adv. Water Resour., 32, 1444–1459, 2009.

Feddes, R. A., Kowalik, P. J., and Zaradny, H.: Simulation of Field Water Use and Crop Yield, John Wiley & Sons, 188 pp., 1978.

Feddes, R. A., Hoff, H., Bruen, M., Dawson, T. E., Rosnay, P. d., Dirmeyer, P., Jackson, R. B., Kabat, P., Kleidon, A., Lilly, A., and Pitman, A. J.: Modeling Root Water Uptake in Hydrological and Climate Models, Bull. Am. Meteorol. Soc., 82, 2797–2809, 2001.

Frensch, J and Steudle, E.: Axial and Radial Hydraulic Resistance to Roots of Maize (Zea mays L.), Plant Physiol., 91, 719–726, 1989.

Garrigues, E., Doussan, C., and Pierret, A.: Water uptake by plant roots: I – Formation and propagation of a water extraction front in mature root systems as evidenced by 2D light transmission imaging, Plant. Soil, 283, 83–98, 2006.

Heppel, C., Payvandi, S., Zygalakis, K. C., Smethurst, J., Fliege, J., and Roose, T.: Validation of a spatial-temporal soil water movement and plant water uptake model, Geotechnique, 64, 526–539, 2014.

Hildebrandt, A. and Eltahir, E. A. B.: Ecohydrology of a seasonal cloud forest in Dhofar: 2. Role of clouds, soil type, and rooting depth in tree-grass competition, Water Resour. Res., 43, W11411, doi:10.1029/2006WR005262, 2007.

Huszár, T., Mika, J., Lóczy, D., Molnár, K., and Kertész, A.: Climate Change and Soil Moisture: A Case Study, Phys. Chem. Earth A, 24, 905–912, 1998.

Jackson, R. B., Sperry, J. S., and Dawson, T. E.: Root water uptake and transport: using physiological processes in global predictions, Trends Plant Sci., 5, 482–488, 2000.

Javaux, M., Schröder, T., Vanderborght, J., and Vereecken, H.: Use of a Three-Dimensional Detailed Modeling Approach for Predicting Root Water Uptake, Vadose Zone J., 7, 1079–1088, 2008.

Kalbacher, T., Schneider, C., Wang, W., Hildebrandt, A., Attinger, S., and Kolditz, O.: Modeling Soil-Coupled Water – Uptake of Multiple Root Systems with Automatic Time Stepping, Vadose Zone J., 10, 727–735, 2011.

Katul, G. G., Oren, R., Manzoni, S., Higgins, C., and Parlange, M. B.: Evapotranspiration: A process driving mass transport and energy exchange in the soil-plant-atmosphere-climate system, Rev. Geophys., 50, RG3002, doi:10.1029/2011RG000366, 2012.

Kleidon, A. and Heimann, M.: A method of determining rooting depth from a terrestrial biosphere model and its impacts on the global water and carbon cycle, Global Change Biol., 4, 275–286, 1998.

Kleidon, A. and Heimann, M.: Assessing the role of deep rooted vegetation in the climate system with model simulations: mechanism, comparison to observations and implications for Amazonian deforestation, Clim. Dynam., 16, 183–199, 2000.

Knipfer, T., Das, D., and Steudle, E.: During measurements of root hydraulics with pressure probes, the contribution of unstirred layers is minimized in the pressure relaxation mode: comparison with pressure clamp and high-pressure flowmeter, Plant. Cell Environ., 30, 845–860, 2007.

Landsberg, J. J. and Fowkes, N. D.: Water Movement Through Plant Roots, Ann. Botany, 42, 493–508, 1978.

Leitner, D., Klepsch, S., Bodner, G., and Schnepf, A.: A dynamic root system growth model based on L-Systems, Plant. Soil, 332, 177–192, 2010.

Leitner, D. and Schnepf, A.: Image analysis of 2-dimensional root system architecture, 19th Conference on Scientific Computing, Vysoké Tatry – Podbanské, Slovakia, 9–14 September 2012, 113–119, 2012.

Levin, A., Shaviv, A., and Indelman, P.: Influence of root resistivity on plant water uptake mechanism, part I: numerical solution, Transport. Porous Media, 70, 63–79, 2007.

Lobet, G., Pagès, L., and Draye, X.: A novel image-analysis toolbox enabling quantitative analysis of root system architecture, Plant Physiol., 157, 29–39, 2011.

Lobet, G. and Draye, X.: Novel scanning procedure enabling the vectorization of entire rhizotron-grown root systems, Plant Methods, 9, doi:10.1186/1746-4811-9-1, 2013.

Manzoni, S., Vico, G., Katul, G., Palmroth, S., Jackson, R. B., and Porporato, A.: Hydraulic limits on maximum plant transpiration and the emergence of the safety-efficiency trade-off, New Phytol., 198, 169–179, 2013.

McDowell, N., Pockman, W. T., Allen, C. D., Breshears, D. D., Cobb, N., Kolb, T., Plaut, J., Sperry , J. S., West, A., Williams, D. G., and Zepey, E. A.: Mechanisms of plant survival and mortality during drought: why do some plants survive while others succumb to drought?, New Phytol., 178, 719–739, 2008.

Mooney, S. J., Pridmore, T. P., Helliwell, J., and Bennett, M. J.: Developing X-ray Computed Tomography to non-invasively image 3-D root systems architecture in soil, Plant. Soil, 352, 1–22, 2012.

Neumann, R. B. and Cardon, Z.: The magnitude of hydraulic redistribution by plant roots: a review and synthesis of empirical and modeling studies, New Phytol., 194, 337–352, 2012.

North, G. B. and Peterson, C. A.: Water flow in roots: Structural regulatory features, in: Vascular transport in plants, edited by: Holbrook, M. N. and Zwieniecki, M. A., Elevier Academic Press, Burlington, MA, USA, 131–156, 2005.

Oswald, S. E., Menon, M., Carmina, A., Vontobel, P., Lehmann, E., and Schulin, R.: Quantitative Imaging of Infiltration, Root Growth, and Root Water Uptake via Neutron Radiography, Vadose Zone J., 7, 1035–1047, 2008.

Pagès, L., Vercambre, G., Drouet, J.-L., Lecompte, F., Collet, C., and Le Bot, J.: Root Typ: a generic model to depict and analyse the root system architecture, Plant. Soil, 258, 103–119, 2004.

Pockman, W. T. and Sperry , J. S.: Vulnerability to xylem cavitation and the distribution of sonoran desert vegetation, Am. J. Botany, 87, 1287–1299, 2000.

Prieto, I., Armas, C., and Pugnaire, F. I.: Water release through plant roots: new insights into its consequences at the plant and ecosystem level, New Phytol., 193, 830–841, 2012.

Roose, T. and Fowler, A. C.: A model for water uptake by plant roots, J. Theoretical Biol., 228, 155–171, 2004.

Ryel, R. J., Caldwell, M. M., Yoder, C. K., Or, D., and Leffler, A. J.: Hydraulic redistribution in a stand of Artemisia tridentata: evaluation of benefits to transpiration assessed with a simulation model, Oecologia, 130, 173–184, 2002.

Schneider, C. L., Attinger, S., Delfs, J.-O., and Hildebrandt, A.: Implementing small scale processes at the soil-plant interface – the role of root architectures for calculating root water uptake profiles, Hydrol. Earth Syst. Sci., 14, 279–289, doi:10.5194/hess-14-279-2010, 2010.

Seneviratne, S. I., Corti, T., Davin, E. L., Hirschi, M., Jaeger, E. B., Lehner, I., Orlowsky, B., and Teuling, A. J.: Investigating soil moisture–climate interactions in a changing climate: A review, Earth-Sci. Rev., 99, 125–161, 2010.

Shukla, J. and Mintz, Y.: Influence of Land-Surface Evapotranspiration on the Earth's Climate, Science, 215, 1498–1501, 1982.

Šmůnek, J. and Hopmans, J. W.: Modeling compensated root water and nutrient uptake, Ecol. Model., 220, 505–521, doi:10.1016/j.ecolmodel.2008.11.004, 2008.

Siqueira, M., Katul, G., and Porporato, A.: Onset of water stress, hysteresis in plant conductance, and hydraulic lift: Scaling soil water dynamics from millimeters to meters, Water Resour. Res., 44, W01432, doi:10.1029/2007WR006094, 2008.

Steudle, E. and Peterson, C. A.: How does water get through roots?, J. Experiment. Botany, 49, 775–788, 1998.

Steudle, E.: Water uptake by plant roots: an integration of views, Plant. Soil, 226, 45–56, 2000.

Steudle, E.: The cohesion-tension mechanism and the acquisition of water by plant roots, Annu. Rev. Plant Physiol. Plant Molecular Biol., 52, 847–875, 2001.

Tuzet, A., Perrier, A., and Leuning, R.: A coupled model of stomatal conductance, photosynthesis and transpiration, Plant. Cell Environ., 26, 1097–1116, 2003.

Tyree, M. T. and Sperry , J. S.: Vulnerability of xylem to cavitation and embolism, Annu. Rev. Plant Biol., 40, 19–36, 1989.

Van Den Honert, T. H.: Water Transport in plants as a catenary process, Discussions of the Faraday Society, 3, 146–153, 1948.

Zarebanadkouki, M., Kim, Y. X., and Carminati, A.: Where do roots take up water? Neutron radiography of water flow into the roots of transpiring plants growing in soil, New Phytol., 199, 1034–1044, doi:10.1111/nph.12330, 2013.

Zwieniecki, M. A., Thompson, M. V., and Holbrook, N. M.: Understanding the Hydraulics of Porous Pipes: Tradeoffs Between Water Uptake and Root Length Utilization, J. Plant Growth Regul., 21, 315–323, 2003.

Appendix A: The functional form of effort and its dependence on boundary conditions

Any water potential ψ_w (m or 9810 J m^{-3}) describes the specific Gibbs free energy of water (Edlefsen and Anderson, 1948, article 62), comparable to the chemical potential. Differential changes in Gibbs free energy ΔG (J) in a system under consideration over a short period of time Δt (s) are therefore

$$\Delta G = \psi_\mathrm{w} \cdot \Delta V_\mathrm{w}, \tag{A1}$$

where ΔV_w (m^3) refers to the change of water volume in the system. When the system is closed and the change of energy is caused by a water flow Q_w (m^3 s^{-1}) over the boundary of the system, the above equation becomes

$$\Delta G = \psi_\mathrm{w} \cdot Q_\mathrm{w} \cdot \Delta t. \tag{A2}$$

Applying these equations to the coupled plant–root system in a closed container, where the only water flow out of the system is by root water uptake, we can therefore state that the change in Gibbs free energy of the system from a starting point t_0 (s) up to a time t (s) under consideration is

$$G(t) = \int_{\tau=t_0}^{t} \psi_\mathrm{C}(\tau) \cdot Q(\tau) d\tau, \tag{A3}$$

where $\psi_\mathrm{C}(\tau)$ (m) refers to the water potential at the root collar at time τ (s).

As the change of Gibbs free energy to go from state A to state B of a closed system equals the mechanical work to go from A to B (neglecting the work of expansion, Edlefsen and Anderson, 1948, article 21, 62), $G(t)$ is equivalent to the work required for root water uptake. We can define a normalized measure, $w(t)$ (J m^{-3}), which evaluates average work required per unit of water transpired between t_0 and t:

$$w(t) = \frac{G(t)}{\int_{\tau=t_0}^{t} Q(\tau)d\tau} = \frac{\int_{\tau=t_0}^{t} \psi_\mathrm{C}(\tau) \cdot Q(\tau)d\tau}{\int_{\tau=t_0}^{t} Q(\tau)d\tau}. \tag{A4}$$

This means that under arbitrary boundary conditions, effort can be understood as a flow-weighted average xylem water potential at the root collar.

Under a drying scenario, root water uptake causes soil water potential to decrease monotonically. Thus, at a unique time $\tilde{t}$(s) plant water stress occurs. Effort at time $\tilde{t}$ will in this case be denoted by $\tilde{w} = w(\tilde{t})$. Under a time constant transpiration rate $Q(\tau) = Q$, effort $\tilde{w} = w(\tilde{t})$ can be calculated as a temporal average xylem water potential at the root collar:

$$\tilde{w} = w(\tilde{t}) = \frac{\int_{\tau=0}^{\tilde{t}} Q(\tau) \cdot \psi_C(\tau)d\tau}{\int_{\tau=0}^{\tilde{t}} Q(\tau)d\tau} = \frac{Q \cdot \int_{\tau=0}^{\tilde{t}} \psi_C(\tau)}{Q \cdot \tilde{t}}$$

$$= \frac{\int_{\tau=0}^{\tilde{t}} \psi_C(\tau)d\tau}{\tilde{t}} = \bar{\psi}_C(\tilde{t}). \tag{A5}$$

In contrast to water yield, effort increases under water stress. However, this increase is small, as will be shown in the following.

In order to calculate effort at a time $t > \tilde{t}$, we use the general definition of effort and split the integrals in the enumerator and and denominator at the occurence of water stress $\tilde{t}$

$$w(t) = \frac{\int_{\tau=0}^{t} Q(\tau) \cdot \psi_C(\tau)d\tau}{\int_{\tau=0}^{t} Q(\tau)d\tau}$$

$$= \frac{\int_{\tau=0}^{\tilde{t}} Q(\tau) \cdot \psi_C(\tau)d\tau + \int_{\tau=\tilde{t}}^{t} Q(\tau) \cdot \psi_C(\tau)d\tau}{\int_{\tau=0}^{\tilde{t}} Q(\tau)d\tau + \int_{\tau=\tilde{t}}^{t} Q(\tau)d\tau}. \tag{A6}$$

We can now insert the flux boundary condition $Q(\tau) = Q$ for times $\tau = 0 \ldots \tilde{t}$ and the potential boundary condition $\psi(\tau) = \psi_\mathrm{crit}$ for times $\tau = \tilde{t} \ldots t$. We obtain

$$w(t) = \frac{\int_{\tau=0}^{\tilde{t}} Q(\tau) \cdot \psi_C(\tau)d\tau + \int_{\tau=\tilde{t}}^{t} Q(\tau) \cdot \psi_C(\tau)d\tau}{\int_{\tau=0}^{\tilde{t}} Q(\tau)d\tau + \int_{\tau=\tilde{t}}^{t} Q(\tau)d\tau}$$

$$= \frac{Q \cdot \int_{\tau=0}^{\tilde{t}} \psi_C(\tau)d\tau + \psi_\mathrm{crit} \cdot \int_{\tau=\tilde{t}}^{t} Q(\tau)d\tau}{Q \cdot \tilde{t} + \int_{\tau=\tilde{t}}^{t} Q(\tau)d\tau}. \tag{A7}$$

If we transform the integrals in the stress periods by replacing $\tau = \tilde{t} \ldots t$ by $\tau = 0 \ldots \Delta t$ ($\Delta t = t - \tilde{t}$ is the time since the occurrence of water stress), effort can be expressed as

$$w(t) = w(\tilde{t} + \Delta t)$$

$$= \frac{Q \cdot \int_{\tau=0}^{\tilde{t}} \psi_C(\tau)d\tau + \psi_{crit} \cdot \int_{\tau=0}^{\Delta t} Q(\tilde{t} + \tau)d\tau}{Q \cdot \tilde{t} + \int_{\tau=0}^{\Delta t} Q(\tilde{t} + \tau)d\tau}. \tag{A8}$$

By defining $E_U := Q \cdot \int_{\tau=0}^{\tilde{t}} \psi_C(\tau) d\tau = \text{const.}$, $V_U = Q \cdot \tilde{t} = \text{const.}$, and $V_s(\Delta t) = \int_{\tau=0}^{\Delta t} Q(\tilde{t} + \tau) d\tau$, effort can be expressed as

$$w(t) = w(\tilde{t} + \Delta t) = \frac{E_U + \psi_{crit} \cdot V_S(\Delta t)}{V_U + V_S(\Delta t)} = w(V_s(\Delta t)). \qquad (A9)$$

E_U (J) is the (time-independent) energy that was necessary to take up water under unstressed conditions, it is also the enumerator of $\tilde{w}$; V_U (m^3) is the (time-independent) amount of water that was extracted before the onset of water stress, it also is the denominator of $\tilde{w}$; and V_s (m^3) is the amount of water that was extracted after the onset of water stress. V_s depends on the duration Δt of water stress.

Using a first-order Taylor approximation of w around $\tilde{t}$ yields

$$w(t) = w(\tilde{t} + \Delta t) = \tilde{w} + (\psi_{crit} - \tilde{w}) \cdot \frac{V_s(\Delta t)}{V_u}. \qquad (A10)$$

For $\Delta t = 0$ ($t = \tilde{t}$, the onset of water stress) this approximation gives the correct value $\tilde{w}$ of effort. For $\Delta t > 0$, effort increases linearly with the amount of water V_s extracted under water stress. But as root water uptake rates of stressed plants decrease quickly in a drying soil, effort increases very slowly with time.

2

Regional co-variability of spatial and temporal soil moisture–precipitation coupling in North Africa

Irina Y. Petrova[1,2], **Chiel C. van Heerwaarden**[3], **Cathy Hohenegger**[1], **and Françoise Guichard**[4]

[1]Department of the Land in the Earth System, Max Planck Institute for Meteorology, Hamburg, Germany
[2]Laboratory of Hydrology and Water Management, Ghent University, Ghent, Belgium
[3]Meteorology and Air Quality Group, Wageningen University, Wageningen, the Netherlands
[4]CNRM (UMR 3589 CNRS and Météo-France), Toulouse, France

Correspondence: Irina Y. Petrova (irina.petrova@ugent.be)

Abstract. The magnitude and sign of soil moisture–precipitation coupling (SMPC) is investigated using a probability-based approach and 10 years of daily microwave satellite data across North Africa at a 1° horizontal scale. Specifically, the co-existence and co-variability of spatial (i.e. using soil moisture gradients) and temporal (i.e. using soil moisture anomaly) soil moisture effects on afternoon rainfall is explored. The analysis shows that in the semi-arid environment of the Sahel, the negative spatial and the negative temporal coupling relationships do not only co-exist, but are also dependent on one another. Hence, if afternoon rain falls over temporally drier soils, it is likely to be surrounded by a wetter environment. Two regions are identified as SMPC "hot spots". These are the south-western part of the domain (7–15° N, 10° W–7° E), with the most robust negative SMPC signal, and the South Sudanese region (5–13° N, 24–34° E). The sign and significance of the coupling in the latter region is found to be largely modulated by the presence of wetlands and is susceptible to the number of long-lived propagating convective systems. The presence of wetlands and an irrigated land area is found to account for about 30 % of strong and significant spatial SMPC in the North African domain.

This study provides the first insight into regional variability of SMPC in North Africa, and supports the potential relevance of mechanisms associated with enhanced sensible heat flux and mesoscale variability in surface soil moisture for deep convection development.

1 Introduction

Soil moisture can affect the state of the lower atmosphere through its impact on evapotranspiration and surface energy flux partitioning (e.g. Eltahir, 1998; Klüpfel et al., 2011). Especially in the "hot spots" of soil moisture–precipitation coupling (SMPC), like the semi-arid Sahel (Koster et al., 2004; Taylor et al., 2012; Miralles et al., 2012), soil moisture exerts strong control on evapotranspiration (e.g. Timouk et al., 2009; Dirmeyer, 2011; Lohou et al., 2014), influencing the development of the daytime planetary boundary layer (BL), and hence convective initiation and precipitation variability. Most of the physical understanding of how soil moisture could alter BL properties and affect development of convection comes from 1-D to 3-D model analyses (Seneviratne et al., 2010; Nicholson, 2015). Observational evidence of the SMPC largely relies on the measurements of recent field campaigns (like African HAPEX and AMMA: Goutorbe et al., 1994; Redelsperger et al., 2006), and hence is often limited to a short spatio-temporal scale. Such observational analyses present unique evidence of environmental conditions preceding convection development (e.g. Lothon et al., 2011) and can be further used as a testing ground to evaluate and improve the physical parametrizations of models (e.g. Couvreux et al., 2013). Both observational and modelling studies agree reasonably well on the effect of soil moisture availability and heterogeneity on the lower atmospheric stability (e.g. Kohler et al., 2010) and convective initiation at the mesoscale (e.g. Taylor, 2010; Birch et al., 2013).

However, the impact of soil moisture on convective precipitation remains more uncertain. At the mesoscale, there is a disagreement in the sign of the SMPC between observations and models which use parametrizations of deep convection (Hohenegger et al., 2009; Taylor et al., 2012, 2013). Recent satellite-based analysis has demonstrated that the choice of soil moisture parameter itself (temporal anomaly or spatial gradient) and related differences in physical mechanisms have a direct effect on the resulting sign of the coupling. Guillod et al. (2015) found a positive temporal (i.e. using the soil moisture anomaly) but negative spatial (i.e. using the spatial soil moisture gradient) SMPC over most of the globe at a 5° horizontal scale. Our study addresses the question of co-existence of spatial and temporal SMPC on a finer $1° \times 1°$ horizontal grid. Specifically, we use 10-year satellite records of daily soil moisture from the AMSR-E and 3-hourly TMPA precipitation product to investigate spatio-temporal co-variability of observed coupling relationships in the region of North Africa.

Both modelling and observational studies reported the possibility of negative as well as positive SMPC (Nicholson, 2015). Spatial gradients in soil moisture can affect BL state and convection initiation through thermally induced mesoscale circulations on 10 to 100 km scales (Taylor and Ellis, 2006; Taylor et al., 2007; Rochetin et al., 2017). In association with this mechanism and under favourable thermodynamic conditions, convection is likely to be initiated over spatially drier soils, indicating a negative SMPC (Taylor et al., 2011; Garcia-Carreras et al., 2011). However, whether the further development and propagation of moist convection will occur over drier or wetter soils remains less clear. The modelling study of Froidevaux et al. (2014) suggested that negative SMPC is possible under very weak surface wind conditions, and is associated with stationarity of convective systems once initiated. The opposite sign is expected under a stronger horizontal advection, which will support propagation of the developing moist convection downwind, i.e. from dry to wet soils, and its further amplification over wetter areas. Another important factor is related to the life cycle of mesoscale convective systems (MCSs) and thus their organization in space and time (Mathon et al., 2002). Small-scale convective systems are expected to be particularly sensitive to surface moisture variability and will propagate preferentially towards spatially drier soil, bounded by wetter surroundings (Taylor and Ellis, 2006). Alternatively, larger organized systems have been found to evolve towards wetter soils – areas of increased latent heat flux, convective available potential energy (CAPE) and moist static energy (MSE) (Taylor and Lebel, 1998; Clark et al., 2003). Hence, these systems are expected to be more sensitive to soil moisture availability.

The impact of temporal anomalies of soil moisture on the atmospheric BL and moist convection is largely governed by thermodynamical processes, and may likewise result in a coupling of both signs. Wet soils are expected to lead to an increase of boundary layer MSE or similarly equivalent potential temperature, through a decreased boundary layer height and subsequent less vigorous entrainment (e.g. Eltahir, 1998; Alonge et al., 2007). The enhanced MSE over wet soils is favourable for convective rainfall formation. Dry soils, on the contrary, are associated with a reduced MSE and thus provide lower potential for convection development and may even suppress the existing MCS (Gantner and Kalthoff, 2010) or deviate its propagation direction (Wolters, 2010). However, modelling and observational evidence indicates that both dry and wet soils can favour moist convection, depending on the morning stratification of the lower atmosphere (Findell and Eltahir, 2003a, b) into which the daytime convective BL is growing (Ek and Mahrt, 1994; van Heerwaarden and Guerau de Arellano, 2008; Couvreux et al., 2012).

The relevance of mesoscale spatial heterogeneity of soil moisture in favouring moist convection over wet and dry temporal soil moisture anomalies was demonstrated by e.g. Clark et al. (2003) and Taylor and Ellis (2006) respectively. However, until recently no attempts had been made to directly compare the temporal and spatial aspects. The first comparison of the spatial and temporal effects of soil moisture on precipitation was presented by the study of Guillod et al. (2015) (hereafter, G15) at a 5° horizontal scale. Applying the probability-based approach of Taylor et al. (2012) (hereafter, T12) to 10 years of global satellite-based soil moisture and precipitation data, they demonstrated that a negative spatial (rain over spatially drier soils) and a positive temporal (rain over temporally wetter soils) SMPC dominate over most of the globe and are not mutually exclusive of one another. G15 suggested that the two effects might be interconnected through spatial coupling mechanisms, in which adjacent precipitation would provide required moisture to enhance convection development over spatially but not temporally drier soil. Using multiple data sets, G15 showed that the signal is robust across different input data sets. However, in a few regions, including the Sahel in Africa, an opposite temporal relationship was revealed: spatially and temporally negative coupling was found to co-exist in contrast to the global relationship.

In this study, we further explore spatial and temporal SMPC as well as their co-existence in the North African region using the T12 method on a finer $1° \times 1°$ horizontal grid. Furthermore, we provide insight into the regional co-variability of the spatial and temporal effects on afternoon rainfall. The analysis is realized following two main steps.

1. *Identification of the factors that influence the magnitude and variability of the spatial SMPC measure.* By doing so we address the following question: which physical processes likely underlie the observed spatial SMPC relationship, if any?

2. *Analysis of the link between the spatial and temporal effects of soil moisture on precipitation.* This part addresses the two following questions: are spatial and tem-

poral negative coupling relationships independent, and if not, how do they inter-relate?

We reproduce and apply the probability-based approach of T12 to 10 years of daily AMSR-E soil moisture and 3-hourly TMPA precipitation records. In contrast to the previous studies, we estimate the temporal and spatial coupling effects at a finer 1° horizontal scale, which reveals previously hidden smaller scale effects of land cover features on the SMPC relationship.

The first part of the study (Sect. 4) includes an analysis of the regional variability and robustness of the observed spatial SMPC at varying horizontal scales i.e. from the original 5 to 2.5° and 1°. Identification of the factors relevant for the observed spatial SMPC distribution includes a sensitivity analysis of the spatial coupling measure to the presence of soil moisture parameter extremes (Sect. 4.3) and to the MCS life cycle (Sect. 4.4). The link between the temporal and spatial SMPC is assessed using correlation analysis in Sect. 5.1. The summary and discussion in Sect. 6 is delivered in two parts. Section 6.1 provides main conclusions on the potential physical mechanisms which likely underlie the observed coupling relationships. Section 6.2 reviews the reasons behind the opposite sign of the temporal coupling identified in the North African region as compared to the dominantly positive relationship identified in G15 over the globe. The paper concludes with more general discussion of the SMPC hot spots and conclusions in Sect. 7.

2 Domain and data

2.1 Study domain

We focus our analyses on the North African region (5–20° N, 20° W–40° E) (Fig. 1, dashed rectangle) during the summer period (JJAS). This region has been repeatedly pointed out as a hot spot of land–atmosphere interactions (Dirmeyer, 2011; Miralles et al., 2012; Gallego-Elvira and Taylor, 2016), and one of the most vulnerable regions with respect to climate change (Dirmeyer et al., 2012; Dirmeyer and Wang, 2014). A major feature affecting the Sahelian climate is the West African monsoon (Janicot and Thorncroft, 2008), which is associated with high precipitation variability (Nicholson, 2013). During the monsoon, soil moisture fluctuations are strongly influenced by precipitation at a large range of spatial and temporal scales. Atmospheric and surface fields display strong meridional gradients between 10 and 20° N (Fig. 1, zonal plot), shaped by the migration of the summer time rain belt, also referred to as the intertropical convergence zone (ITCZ). Wind convergence at the surface is observed further north, around 18–20° N, along the intertropical discontinuity (ITD), where the cool and moist monsoon flow meets hot and dry Saharan air. Associated with the meridional heat gradient, the monsoon circulation and related large-scale structures like the African easterly jet (AEJ), as well as synoptic

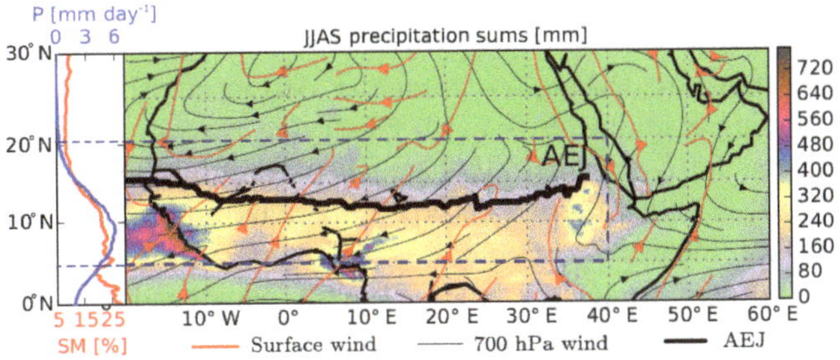

Figure 1. JJAS TMPA precipitation (shading), mean surface (red streamline) and 700 hPa (black streamline) ERA-Interim (1979–present, Dee et al., 2011) wind climatology averaged over the 2002–2011 period. The black thick line shows the mean location of the African easterly jet (AEJ). The inset plot to the left shows zonal means of daily AMSR-E soil moisture (red) and TMPA precipitation (blue) climatology calculated over the study domain (dashed rectangle, 5–20° N, 20° W–40° E).

disturbances like the African easterly waves (AEWs), largely modulate convection activity over the region (Duvel, 1990; Mohr and Thorncroft, 2006). Additionally, evidence supporting a significant role of the surface state in the triggering of deep precipitating convection is steadily growing (Nicholson, 2015).

The conditions in mid-July to August may be less favourable for a strong surface influence on convection. Compared to the drier early and late monsoon months of June and September, the wetter period – from July through August – is characterized by a typically lower level of free convection (LFC) (Guichard et al., 2009; Taylor et al., 2011) and less pronounced spatial contrast between fluxes due to more dense vegetation (Kohler et al., 2010; Lohou et al., 2014). In our study, the role of the monsoon dynamics is not addressed to preserve maximum sample size for the sake of statistical significance.

We intentionally extend our analysis further eastwards. Despite the inherent zonal symmetry of surface and atmospheric parameters (as in precipitation in Fig. 1), considerable differences exist in rainfall and large-scale circulation regimes between east and west. Distinct orography, intensity of surface and upper level jets and wave disturbances are likely to bring dissimilarities in the sensitivity of convection to the surface state between the two regions. The eastern African domain can also remotely influence convection in the western part of the region via the genesis of westward-propagating AEWs (Laing and Carbone, 2008) and long-lived MCSs (Laing et al., 2012). However, notably few studies have investigated land–atmosphere interactions in the eastern Sahel.

2.2 AMSR-E soil moisture

Soil moisture (SM) data from the Advanced Microwave Scanning Radiometer – Earth Observing System (AMSR-E, June 2002–October 2011) are analysed in this study. The AMSR-E unit is carried on board the polar orbiting AQUA satellite, measuring brightness temperatures in 12 channels, at 6 different frequencies (6.9–89 GHz) (Njoku and Jackson, 2003). Soil moisture derived from the lowest C-band frequency of 6.9 GHz is used here, as lower frequencies experience less signal contamination from vegetation and surface roughness, and are able to receive emission information from deeper soil layers (still just a few centimetres; Owe et al., 2008). The AQUA orbit is sun-synchronous, with typically one overpass per pixel per day at either 13:30 or 01:30 local solar time (LST). In order to capture the surface moisture state shortly before afternoon convection activity, only data of ascending day orbit, i.e. 13:30 LST, are used here. It is important to note that the day overpass is prone to higher biases compared to the night overpass because of the greater temperature differences between the surface and canopy involved in the physics algorithm (Njoku and Jackson, 2003).

We utilize the Level 3 estimates of AMSR-E soil moisture derived with the Land Parameter Model (LPRM; Owe et al., 2008) for JJAS 2002–2011. The product is available at a 0.25° × 0.25° spatial resolution. The LPRM is not valid for dense vegetation and water bodies. Therefore pixels with an optical depth > 0.8 are excluded. Water body and soil moisture quality masks were adopted from the materials of T12. Accordingly, pixels containing more than 5 % water are excluded, using water body classification of the 1 km Global Land Cover 2000 data set (available online at http://forobs.jrc.ec.europa.eu/products/glc2000/products.php, last access: September 2017). Application of the soil moisture quality mask, based on the correlation analysis between precipitation and soil moisture data sets, is intended to reduce the number of pixels covered with wetlands (for details, see the supplement in T12).

Many days (40–50 %) do not contain soil moisture information due to satellite revisit times. Over a given longitude per day the number of overpasses below 40° N does not exceed one, with occasionally daily sampling or sampling every third day (see Fig. 1 in Njoku and Jackson, 2003). The latter significantly reduces the size of the collected rainfall event sample available for the analyses. The AMSR-E instrument is chosen because it documents a relatively long period and performs better than ASCAT (Dorigo et al., 2010) over sparsely vegetated and deserted areas. The AMSR-E product also proved to be accurate at the precipitation event scale in capturing rain-related soil moisture variability and timing, when compared with in situ data in the Sahel (Gruhier and Rosnay, 2008).

2.3 TMPA-v7 precipitation

The Tropical Rainfall Measuring Mission (TRMM) Multi-satellite Precipitation Analysis (TMPA) represents a partial-global coverage (50° S–50° N) product of combined precipitation estimates. Three-hourly precipitation time series of the TMPA product (1998–2015, Huffman et al., 2007) at 0.25° × 0.25° horizontal resolution are used to estimate locations of afternoon convective precipitation events in the study.

The TMPA algorithm (Huffman et al., 2007) involves the following steps: (1) the merging of multiple independent passive microwave sensors, (2) their inter-calibration to the TRMM Combined Instrument (TCI) precipitation estimates, (3) further blending with preliminary calibrated infrared products from geostationary satellites and finally, (4) the scaling of the estimates to match monthly accumulated Global Precipitation Climatology Center (GPCC) rain gauge data.

In this study we utilize the product version 7 (TRMM-3B42), which includes several modifications to the algorithm and additional satellite data (Huffman and Bolvin, 2014). Consistent with the soil moisture record, only 10 years (2002–2011) of JJAS precipitation data are used. To ensure similar solar forcing on the surface and boundary layer, the 3 h precipitation time series for the present application are adjusted to LST (based on longitude) by taking the closest 3 h UTC time step. It is important to note that any 3 h TMPA value does not refer directly to its nominal hour, but it represents the average of the "best" overpass data within a 3-hourly window, centred around the nominal hour, i.e. ±90 min range. Variable time of the TMPA best data average is not expected to significantly affect our SMPC results.

3 Methods

3.1 Description of statistical framework

The SMPC in this study is referred to as the relationship between the afternoon convective rainfall and the antecedent soil moisture conditions. Using the method of T12, we examine first whether afternoon rainfall is more likely over soils that are untypically (relative to its climatology) drier or wetter than their surroundings. Next, following the definition of G15, we assess whether afternoon rainfall is more likely on days when soils are untypically drier or wetter than their temporal mean. Subsequently, the probability of convective rainfall events to occur over spatially drier or wetter soils being higher than expected is referred to as *spatial SMPC*, while the likelihood of convective rain events to occur over temporally wetter or drier soils being higher than expected is described by *temporal SMPC*. The following paragraph describes criteria which are used to define a convective precip-

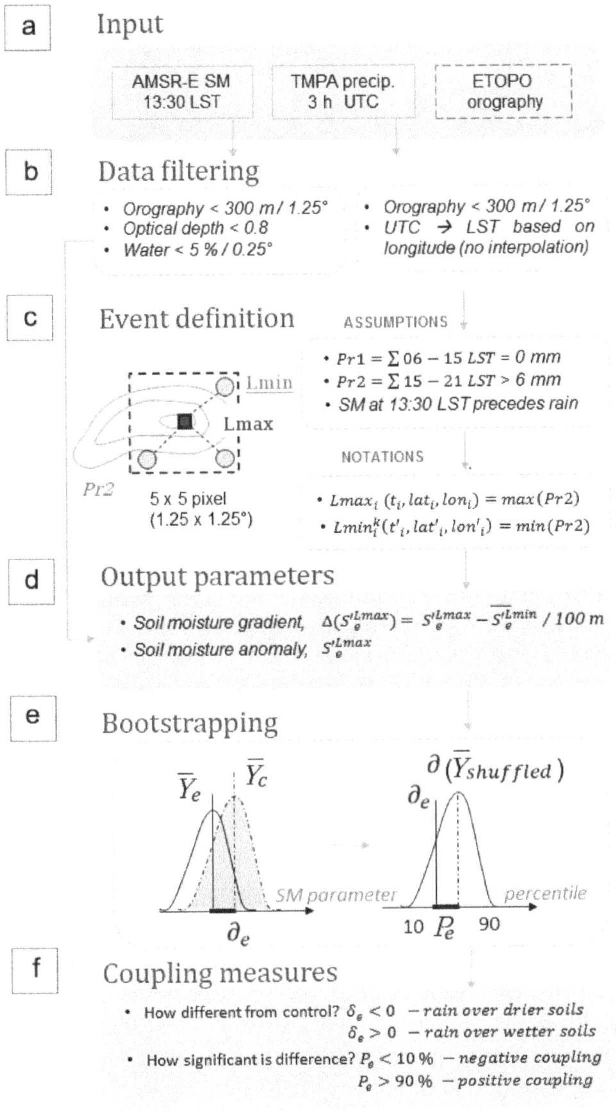

Figure 2. Schematic of data post-processing and statistical framework protocol implemented in the study.

itation event and evaluate soil moisture statistics antecedent to every event. The framework algorithm implemented in this study largely follows the method of T12 and is summarized in Fig. 2.

3.1.1 Definition of convective rainfall event (Fig. 2c)

We define a convective event location, $L_{\max}$, as the location where accumulated afternoon precipitation between 15:00 and 21:00 LST exceeds a threshold of 6 mm. Then, locations of afternoon accumulated precipitation minima, $L_{\min}$, are identified within a 5 × 5 pixel box (1.25° × 1.25°)[1] centred at $L_{\max}$ (Fig. 2c). The choice of a later accumulation time than in T12 (i.e. 15:00–21:00 LST instead of 12:00–21:00 LST) ensures that the soil moisture measurement at 13:30 LST precedes precipitation without introducing additional filters. The afternoon accumulated rainfall threshold that is twice larger than in T12 yields qualitatively similar results, though it leads to a slightly higher mean SMPC significance over the domain. According to additional sensitivity tests, the choice of higher threshold values in the method mostly influences the number of significant grid boxes linked to a reduction in the event sample size, yet it does not qualitatively affect the dominant preference of the afternoon rainfall over specific soil moisture conditions (Petrova, 2017).

The following set of assumptions is used to improve the accuracy of the convective event sample. If one of the conditions is not fulfilled, an event is excluded from further calculation.

1. Accumulated precipitation in the preceding hours (06:00–15:00 LST) in the entire 1.25° × 1.25° box must be zero.
2. The elevation height difference within the event box must not exceed 300 m. This is done to minimize the effect of orographic uplifting on the rainfall variability. The resulting distribution of the orography mask is shown in Fig. 3a.
3. The number of identified $L_{\min}$ locations within one box must be three or more (for averaging reasons). In that case, all $L_{\min}$ locations will have the same afternoon accumulated precipitation value, which will most often be zero.
4. If boxes overlap, the event with larger afternoon accumulated precipitation value is retained.

3.1.2 Soil moisture statistics in event locations (Fig. 2d–e)

Once events are identified, the soil moisture anomaly, S', measured prior to the precipitation event (at 13:30 LST) at $L_{\max}$, $\overline{L_{\min}}$ or any combination of the two is stored and analysed. $S'^{\overline{L_{\min}}}$ represents an average value of S' measured in every identified $L_{\min}$ location within a 1.25° × 1.25° event box. S' has its climatological mean subtracted, calculated as linear averaging within a 21-day moving window centred on the particular day and across the entire multi-year record. To exclude the contribution of a rain event from the anomaly values, the year of the event is excluded from the climatological mean calculation. To estimate whether soils were anomalously dry or wet in the location of maximum rain ($L_{\max}$) compared to the expected range (climatology) for that location, we store the pre-event soil moisture anomaly at $L_{\max}$,

[1]Following T12, a box size of 1.25° × 1.25° is selected as minimum possible size to resolve soil moisture variability around the centre of the box, taking into account the 50 km footprint of the AMSR-E soil moisture.

i.e. Y_e: $S_e^{\prime L_{max}}$ (Fig. 2d). To investigate whether soils in the location of maximum rain (L_{max}) were drier than in the neighbour region(s) where it rained less or did not rain (L_{min}), we calculate the pre-event soil moisture gradient between $\overline{L_{max}}$ and $\overline{L_{min}}$ scaled per 100 m, i.e. Y_e: $\Delta(S_e^{\prime L_{max}}) = S_e^{\prime L_{max}} - S_e^{\prime L_{min}}$ with the dimension of $m^3\,m^{-3}\,100\,m^{-1}$, where Δ stands for the gradient.

For each Y_e parameter we define the control sample Y_c, represented by an array of corresponding Y values measured in the same L_{max} and L_{min} event locations in the same calender month, but on the non-event years. The measure of coupling is then quantified by the magnitude of the difference between mean statistics of the event and control samples, $\delta_e = \text{mean}(Y_e) - \text{mean}(Y_c)$, and the measure of δ_e significance (Fig. 2e). Significance is represented by a percentile, P_e, of the observed δ_e in a bootstrapped sample of δ values that is observed by chance. For that, Y_e and Y_c are pooled together and resampled without replacement 5000 times.

3.1.3 Definition of temporal and spatial SMPC (Fig. 2f)

Parameters of δ_e and P_e calculated for the soil moisture gradients $\Delta(S_e^{\prime L_{max}})$ prior to the event quantify the preference of rain to occur over soils drier ($\delta_e < 0$, $P_e \leq 10\,\%$) or wetter ($\delta_e > 0$, $P_e \geq 90\,\%$) than their 1.25° × 1.25° environment, and are referred to as negative or positive spatial SMPC respectively. The same parameters estimated for the temporal soil moisture anomaly $S_e^{\prime L_{max}}$ instead specify the expressed preference of rain to occur over soils drier or wetter than their temporal mean, i.e. negative or positive temporal SMPC, accordingly.

In this study, estimation of δ_e and its significance P_e for the spatial and temporal coupling is realized over 5° × 5°, 2.5° × 2.5° and 1° × 1° boxes. Aggregation of event statistics at a higher resolution than used in the global studies of T12 and G15 results in a smaller event sample size per grid box, yet allows a reduction of the potential influence of meridional or zonal gradient in the parameter statistics; i.e. it makes the spread in underlying surface and atmospheric moisture conditions across the box latitudes smaller (Sect. 4.2). The latter is valuable for the interpretation of obtained statistics in terms of land cover and atmospheric state. Hence, most of the study focuses on the smallest 1° × 1° spatial grid.

3.2 Statistics of convective events

Application of the algorithm to the 10 years of JJAS AMSR-E soil moisture and TMPA precipitation time series yields 10 131 afternoon rainfall events. The distribution of identified events over the domain at 5° × 5° and 1° × 1° grid is shown in Fig. 3b and c respectively. The signature of orography and large-scale dynamic effects on event occurrence becomes more evident at the finer event-aggregation scale, thus giving an advantage to the highest horizontal resolution. Figure 3c shows that most events occur between 10 and 18° N,

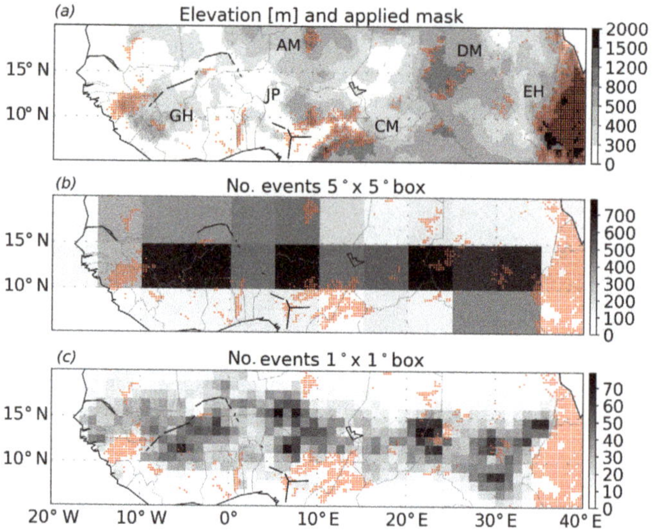

Figure 3. (a) Elevation map based on 1 arcmin Global Relief Model data ETOPO1 (Amante and Eakins, 2009) (grey shading) and the orography mask used in the study (red dots). The main orographic features of the region are the AM – Air Mountains, DM – Darfur Mountains, EH – Ethiopian Highlands, CM – Cameroon Mountains, JF – Jos Plateau and GH – Guinea Highlands. **(b–c)** Number of events in every **(b)** 5° × 5° and **(c)** 1° × 1° box (gray shading) and the applied orography mask at a 0.25° horizontal resolution (red dots).

and the occurrence maxima are zonally aligned. Two maxima are found over the central Sahel, covering the area between 10° W and 15° E – aligned with the mean position of the AEJ core (Fig. 1b). Another two maxima are evident at about 22 and 30° E. Overall, the obtained distribution of identified rain events at 1° × 1° grid resolution is consistent with the observed distribution of intense MCSs over the region (Mathon and Laurent, 2001).

4 Results of spatial SMPC analysis

4.1 SMPC at a 5° horizontal scale: comparison with previous studies

We start our assessment by investigating the spatial soil moisture–precipitation coupling relationship. In agreement with the global-scale studies of T12 and G15, we find a dominantly negative spatial SMPC in the North African domain on the 5° × 5° grid, i.e. a strong preference for convective rainfall events to occur over spatially drier soils (Fig. 4a). The majority of the 5° × 5° boxes (72 %) have percentile values P_e lower than 10 %, implying a significant negative difference in the mean magnitude of soil moisture gradients $\Delta(S_e^{\prime L_{max}})$ prior to the events relative to their typical (non-event) state. No significant positive difference between event and non-event conditions is found at the 5° scale (Table 1).

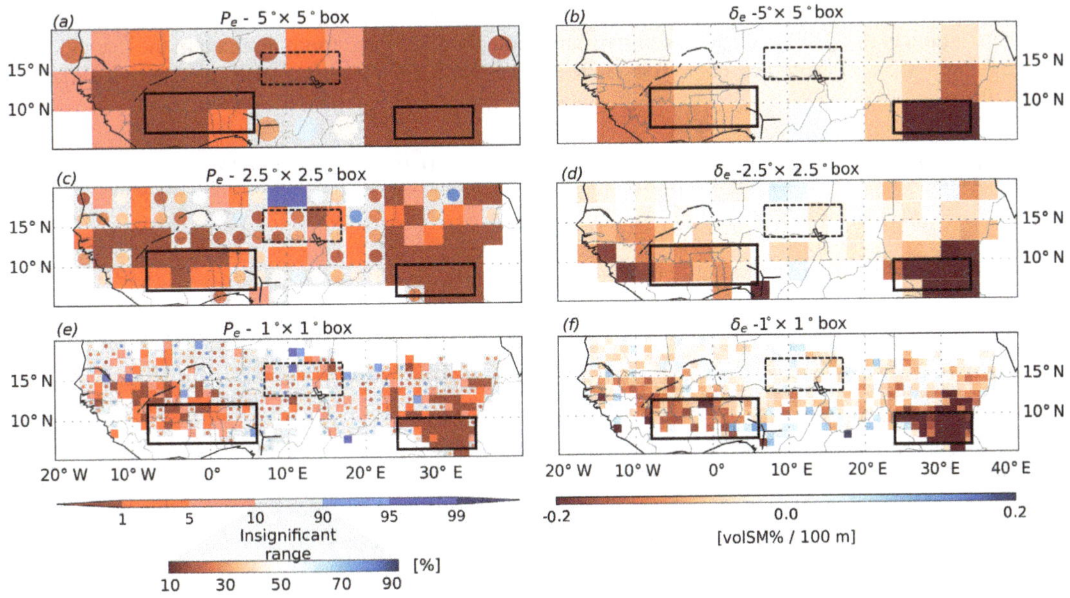

Figure 4. Distribution of percentiles P_e **(a, c, e)** of the observed δ_e difference **(b, d, f)**, estimated over **(a–b)** $5° \times 5°$, **(c–d)** $2.5° \times 2.5°$ and **(e–f)** $1° \times 1°$ boxes. Percentiles $< 10\,\%$ indicate significant negative coupling, i.e rain over spatially drier soils, and percentiles $> 90\,\%$ – significant positive coupling, i.e. rain over spatially wetter soils. The percentile values lying outside the significance range (i.e. between 10 and 90 % percentiles) are illustrated by circles. Black rectangles on the maps indicate featured domains selected for an in-depth analysis.

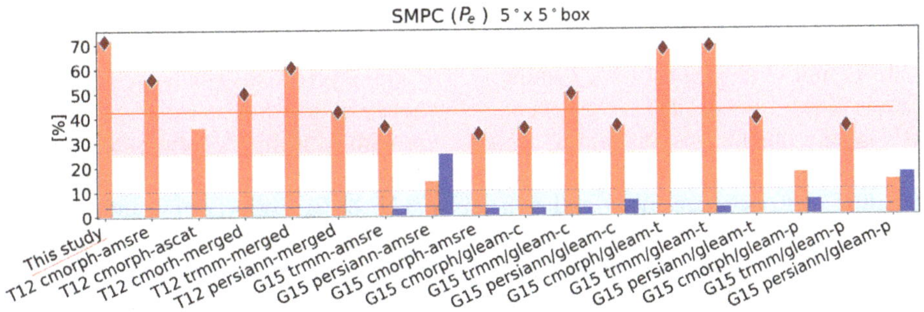

Figure 5. Percentage of $5° \times 5°$ grid boxes with significantly negative ($P_e < 10\,\%$, in red) and positive ($P_e > 90\,\%$, in blue) spatial SMPC over the North African domain in this study and previous studies of T12 and G15. Different data set combinations used in T12 and G15 are listed. Mean and SD of the negative (positive) SMPC fractions across the experiments are shown as a red (blue) solid line and shading accordingly. Following visual inspection, the experiments, in which a significant negative SMPC relationship exists in the western region of the North African domain, are marked with a rhomb.

Figure 5 further compares the percentage of the domain area with significant negative and positive coupling identified in our study and in the previous studies of T12 and G15 (see also Table A1 in the Appendix). The differences arise due to disparities in the data sets and methodologies. The weakest negative and the strongest positive coupling signals correspond to the estimates based on the PERSIANN (Precipitation Estimation from Remotely Sensed Information using Artificial Neural Networks; Hsu et al., 1997) data set from G15. This is possibly linked to the lower consistency between the PERSIANN precipitation and soil moisture variability in time (T12). On average, results based on different data set combinations summarized in Fig. 5 agree that afternoon precipitation occurs more often than expected over spatially drier soils in 42 % of the $5° \times 5°$ boxes, compared to only 4 % with a preference over spatially wetter soils (red and blue lines in Fig. 5 respectively).

The variability of spatial SMPC patterns among different data set combinations has shown to be quite strong over the globe and was not analysed further in G15. We find, however, that in the North African domain, areas of significant negative spatial coupling are fairly consistent. One of the most robust negative spatial SMPC signals is found in the south-western part of the domain (Fig. 4a, b). A total of 14 out of 18 data set combinations summarized in Fig. 5, including this study, locate the cluster of the lowest percentiles

Table 1. General statistics of the average event number and percentile P_e and δ_e difference estimated at various scales for three soil moisture parameters: soil moisture gradient $\Delta(S_e^{\prime L_{max}})$, temporal soil moisture anomaly $S_e^{\prime L_{max}}$ and (not presented in the methodology) soil moisture variance over the 1.25° box $\sigma S_e^{1.25}$. Percentiles $P_e < 10\,\%$ ($> 90\,\%$) indicate a significant negative (positive) δ_e difference, and hence a negative (positive) SMPC relationship.

Parameter	Scale	$\overline{\text{Num}_e}$	$P < 10$, (%)	$P > 90$, (%)	$\delta_{ev} < 0$, (%)	$\delta_{ev} > 0$, (%)
$\Delta(S_e^{\prime L_{max}})$:	5 × 5°	309	72	0	92	2.8
	2.5 × 2.5°	84	42	1.4	73	5.6
	1 × 1°	17	21	1.3	43	14
$S_e^{\prime L_{max}}$:	1 × 1°	–	67	0.8	92	8
$\sigma S_e^{1.25}$:	1 × 1°	–	33	3	78	22

roughly between 5–15° N and 10° W–10° E (Fig. 5, rhombs). This region occupies a relatively vast and flat area, associated with a reduced orographic forcing on convection development compared to the east, and a regional minimum in cold cloud occurrence (Laing and Carbone, 2008). The effect of large-scale structures like AEWs and the AEJ on convection, on the contrary, is expected to be stronger in the western Sahel than further east. However, this does not exclude and may even favour higher sensitivity of convection triggering to soil moisture heterogeneities (Gantner and Kalthoff, 2010; Adler et al., 2011). Overall, the identified negative spatial SMPC relationship in the region is consistent with the recent observation-based (Taylor and Ellis, 2006; Taylor, 2010; Lothon et al., 2011) and model-based (e.g. Gantner and Kalthoff, 2010; Garcia-Carreras et al., 2011; Birch et al., 2013; Taylor et al., 2013) studies in the western Sahel.

Another cluster of the lowest percentiles and the largest differences in soil moisture state between event and non-event days δ_e is identified in the south-east of the domain (Fig. 4a, b). The proximity to the Ethiopian Highlands and the presence of extensive seasonally flooded regions in this area makes it generally difficult to isolate the effect of the surface state on convection. This also possibly led to less agreement in the spatial SMPC estimates identified in our study, G15 analyses and T12 analyses (not shown). Unlike in the western Sahel, no accurate estimates of the SMPC exist in this eastern region.

4.2 Robustness of the negative SMPC at finer 2.5 and 1° scales

To identify factors and potential physical mechanisms that influence the magnitude and variability of the SMPC we reduce the event-aggregation scale to a finer 2.5° × 2.5° and 1° × 1° horizontal grid[2]. In particular, aggregation of the convective rainfall events and corresponding soil moisture statistics over the smallest 1° × 1° grid boxes reveal more details on the effects of land surface conditions on the SMPC.

[2]The SMPC statistic for the finer scales is calculated if at least eight events in a box are present.

The percentile maps obtained for the finer scales are presented in Fig. 4c and e. Despite the reduction in the number of significant δ_e values, largely due to the decreased number of events in every box, negative spatial SMPC relationships remain dominant at the finer scales, and exhibit a similar spatial pattern as on the 5° × 5° grid. The featured regions of significant negative coupling now scale down to the territories of Burkina Faso, Benin, parts of Côte d'Ivoire, Ghana and Mali (7–15° N, 10° W–7° E) in the west, as well as South Sudan (5–13° N, 24–34° E) in the east (Fig. 4e). In total, 42 % (21 %) of the boxes reveal significant negative difference δ_e for the 2.5° (1°) scale, versus an initial 72 % at the 5° × 5° grid (Table 1).

The overall distribution of the difference δ_e does not change at the finer scales (Fig. 4d, f). However, multiple pixels with a positive δ_e emerge. For example, a small region enclosed between the Cameroon mountains and Jos Plateau (7° N, 8° E; Fig. 4f) now indicates a higher likelihood of rainfall to occur over spatially wetter soils. The relationship, though non-significant, is plausible. This area includes part of the Niger River valley and represents a prominent location of intense convection and a local maximum of the cold cloud occurrence, linked to the initiation of convection at the lee side of the high terrain (Laing and Carbone, 2008). The potential link between the land surface characteristics and SMPC parameter is explored in more detail in the following section. In total, 14 % of the 1° × 1° boxes reveal a positive δ_e shift, compared to less than 6 and 3 % for coarser 2.5° × 2.5° and 5° × 5° grids respectively.

4.3 Evidence of a "wetland breeze" mechanism in the SMPC statistics

In areas like the Cameroon Mountains, where orography or floodplains have an effect on deep convection development, persistent wet and dry surface moisture patterns may pre-exist or develop, and therefore, lead to the occurrence of spatial soil moisture gradients that are stronger than usual. In the SMPC statistics, such gradients occur as extremes in a given distribution of soil moisture gradients $\Delta(S_e^{\prime L_{max}})$ within a 1° × 1° box. Here, we define a gradient $\Delta(S_e^{\prime L_{max}})$

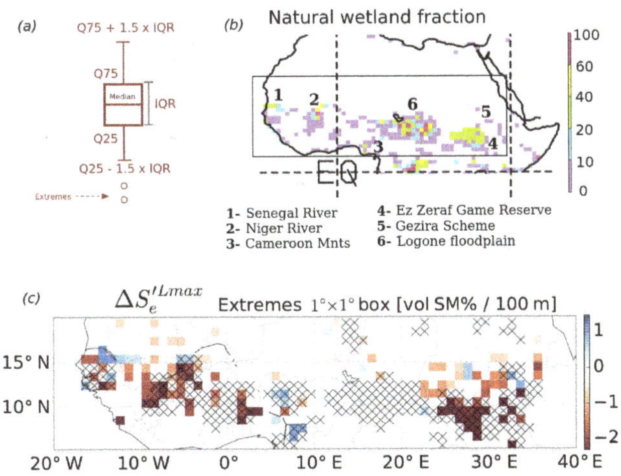

Figure 6. (a) Schematic box plot illustrating the quantile range (Q_{25}–$1.5 \times$ IQR, $Q_{75} + 1.5 \times$ IQR) used to identify extreme $\Delta(S_e^{\prime L_{\max}})$ values. Here, Q_{75} and Q_{25} are the third and first quartiles respectively, and the interquartile range (IQR) is the difference between them. **(b)** Natural wetland fraction from Matthews and Fung (1987) on a $1° \times 1°$ grid (adopted from Fig. 3 in Prigent et al., 2007). Names of major wetland and river locations are marked with a number. **(c)** Distribution of soil moisture gradient $\Delta(S_e^{\prime L_{\max}})$ extremes in the corresponding event sample of a $1° \times 1°$ box (colour). $\Delta(S_e^{\prime L_{\max}})$ is considered to be an extreme if it lies outside the (Q_{25}–$1.5 \times$ IQR, $Q_{75} + 1.5 \times$ IQR) range. If more than one extreme is found in one grid box, then its average value is shown. Black crosses indicate boxes containing $L_{\min}$ locations, in which climatology of daily soil moisture drying rates does not vary much with different absolute soil moisture values. This relationship is equivalent to low soil moisture variability in time and is representative of a wet (flooded) locations. For detailed algorithms, the reader is referred to Fig A1. The distribution of identified potentially flooded locations is consistent with the natural wetland fraction **(b)**.

to be an extreme if it lies outside the following range: (Q_{25}–$1.5 \times$ IQR) to ($Q_{75} + 1.5 \times$ IQR), where Q_{75} and Q_{25} are the third and first quartiles respectively, and the interquartile range (IQR) is the difference between them (Fig. 6a).

The distribution and magnitude of the extreme soil moisture gradients identified in the domain are shown in Fig. 6 (color shading). We find that 28 % of all valid $1° \times 1°$ boxes contain extreme $\Delta(S_e^{\prime L_{\max}})$. Most of the extreme soil moisture gradients are located in the regions of significant negative coupling in the west and east. In these areas, extreme $\Delta(S_e^{\prime L_{\max}})$ values lead to an overestimation of the $\Delta(S_e^{\prime L_{\max}})$ sample mean and therefore, δ_e magnitude. As a result, extreme gradients appear to predefine SMPC significance. Removal of extremes leads to a decrease in the number of boxes with significant negative spatial coupling by 30 %. However, in most cases the sign of the coupling remains unchanged (not shown).

Further analysis shows that extremes tend to cluster around major rivers and wetland areas in the east and west (Fig. 6b, c). Regional distribution of extreme $\Delta(S_e^{\prime L_{\max}})$ is consistent with the distribution of natural wetland fraction, presented earlier by Matthews and Fung (1987) and shown in Fig. 6b, and more recent estimates of inundation regions identified by Prigent et al. (2007). Strong positive soil moisture gradients are found around the Senegal River close to the coast, and on the lee side of the Cameroon Mountains. Strong negative soil moisture gradients are more numerous and seen all along the western flow of the Niger River, downwind of the permanent wetlands of Ez Zeraf Game Reserve and irrigated lands of the Gezira Scheme in Sudan. The scatter of the extremes in the east is likely related to the recurrent floods of the White Nile River. The reason for the absence of extremes around the Logone floodplains is obscure.

To verify the link between the location of extreme soil moisture gradients and flooded areas, we additionally identify $1° \times 1°$ boxes with the events, of which $L_{\min}$ pixels were likely covered with a wetland or a floodplain, and hence were wetter than the neighbouring $L_{\max}$ location (Fig. 6c, crosses). This is done by calculating a linear regression function between the climatology of 1-day soil moisture drying rate (i.e. difference in soil moisture on the event day 0 and on day -1) and the climatology of initial soil moisture value (i.e. soil moisture on day -1)[3] (Appendix Fig. A1). We consider the $L_{\min}$ location to be likely flooded, if (a) all initial soil moisture values are high, and (b) climatological values of drying rate are always small and do not vary much with the initial soil moisture (regression slope is close to zero). For details the reader is referred to Appendix A and Fig. A1. From Fig. 6c it is seen that the distribution of grid boxes with the potentially flooded $L_{\min}$ locations conforms with the distribution of the natural wetland fraction in Fig. 6b and includes the majority of extreme $\Delta(S_e^{\prime L_{\max}})$ locations. This result supports the presence of the link between the afternoon rainfall maximum and flooded areas.

The identified sensitivity of the afternoon rainfall to the strong negative soil moisture gradients around water bodies is in agreement with the results of the observational-based study of Taylor (2010). Analysing 24 years of Meteosat brightness temperatures over the Inner Niger Delta, he found that convection was initiated more often over and to the east of the wetland in the morning hours. However, later in the day, mesoscale convective systems tended to develop and propagate away from the wet areas towards drier soils, suggesting the formation of deep convection and afternoon precipitation over negative soil moisture gradients. Similarly, observed by Alter et al. (2015), enhancement of rain to the east of irrigated land at 14° N, 33° E and its suppression over the Gezira Scheme itself is consistent with the location of negative (positive) extreme soil moisture gradients to the west (east) of the irrigated region (Fig. 6c). All the above supports the consistency of the observed afternoon

[3]The climatologies are calculated for every $L_{\min}$ location, for the same month as the event but for the non-event years.

rainfall intensification over the drier soils adjacent to flooded areas with a wetland breeze mechanism. The ability of the method to capture these effects is also noteworthy.

4.4 Effect of propagation of deep convective events on the SMPC statistics in eastern and western domains

Another physical effect that may influence the SMPC relationship is related to the propagation and evolution of mesoscale convective systems (not accounted for by the current algorithm). Previous studies indicate that an opposite SMPC relationship might be expected at early versus late stages of MCS development (Taylor and Lebel, 1998; Taylor and Ellis, 2006; Taylor et al., 2010; Alonge et al., 2007; Clark et al., 2003; Gantner and Kalthoff, 2010). In this respect, a distinct strength or even sign of the spatial SMPC measure may result from separation of the rainfall events into those formed by weaker and smaller MCSs – mostly found in the early afternoon – or by long-lived and organized MCSs – dominant during late afternoon hours (Mathon and Laurent, 2001). Differences in the SMPC response to the MCS life cycle are also expected to exist between the two regions of significant negative coupling, in the east and west. To characterize these differences, we analyse precipitation diurnal cycles averaged over event days in the east and west first (Fig. 7), and then estimate sensitivity of the spatial SMPC to varying rainfall accumulation times (Fig. 8).

The Hovmöller diagram of rainfall averaged over 1000 event days in the western domain (black rectangle in Fig. 4) shows that intensification of the moist convection in the region is generally concentrated around main orographical features (Fig. 7a, c). The peak in precipitation occurs at similar times across the domain, and thereby does not reveal the expressed signature of the system propagation. Most of the MCSs are therefore expected to be shorter lived and smaller, as their dissipation locations would be found close to their initiation (Mathon and Laurent, 2001).

In the east, on the contrary, the strong south-western propagation component of moist convection dominates the zonal progression of the most intense rainfall during the diurnal cycle averaged over 754 event days (Fig. 7d). A large number of MCSs initiate at the lee side of the Ethiopian Highlands and propagate westward, undergoing cycles of regeneration and growing into mature and organized MCSs (Laing and Carbone, 2008; Laing et al., 2012). The emergence of an absolute rain rate maximum downwind of the permanent wetlands of the Ez Zeraf Game Reserve (at 30° E) during afternoon hours supports an influence of the flooded areas on moist convection intensification in the region (Fig. 7d, f). Consistent with the results of Taylor (2010) obtained for the Niger Inland Delta, the presence of wetlands in the eastern domain is expected to increase the number of organized and long-lived propagating MCSs in the late afternoon, originating from either locally triggered MCSs, i.e. formed at the dry land–wetland boundary, or from re-intensified pre-existing westward propagating systems. We may therefore expect a greater sample of long-lived and organized propagating MCSs to be found in the late afternoon hours in the Eastern than in the western domain. Accordingly, the response of the SMPC statistics to propagating MCSs is expected to be stronger in the east compared to the west. Figure 8, which shows the change of the SMPC parameter between different rainfall accumulation time periods, confirms this hypothesis. For this assessment an additional area in the north is considered (dashed rectangle in Fig. 4), as representative of a region where large-scale atmospheric and surface conditions differ from those of the east and west domains.

From Fig. 8 it is seen that the earlier rainfall accumulation time periods, i.e. 12:00–18:00 UTC in the east and 15:00–21:00 UTC in the west and north, result in the strongest negative δ_e difference, and hence spatial SMPC relationship in all three domains. No positive δ_e values are found for these time periods, and the fraction of negative soil moisture gradients preceding rainfall events is relatively high, i.e. 62, 57 and 55 % for east, west and north accordingly. Later accumulation times lead to a decrease in the magnitude and significance of the coupling parameter δ_e, and an increase in its spatial variability across the domains. These changes are associated with an increase in the amount of the positive soil moisture gradients in the regions (not shown).

Despite the similarities, differences in the SMPC response exist between the domains. In the east, the spatial SMPC shows the strongest sensitivity to the rainfall accumulation time and switches the sign to a positive one for the 18:00–24:00 UTC period. In accordance with Fig. 7d, the 18:00–24:00 UTC period reflects the afternoon progression of the mature MCS formed during early afternoon hours at the Ethiopian Highlands and around wetlands. According to Taylor and Lebel (1998) and Taylor et al. (2010), large and organized MCSs are expected to be more efficient in developing over wetter soils, associated with a well expressed BL moisture anomaly and higher MSE and CAPE and, at the same time, might get suppressed over drier surfaces (Clark et al., 2003). These observations are consistent with the increase in fraction of positive $\Delta(S_e'^{L_{max}})$ in all the domains towards late afternoon hours and the strongest SMPC response in the east identified here.

Additional analysis reveals that the majority of large and negative soil moisture gradients in all domains are linked to the rainfall events that are identified during the first afternoon time step (i.e 12:00 and 15:00 UTC for the east and west respectively) (not shown), and are therefore likely linked to weaker MCSs at the early stage of their development. The smaller and less organized MCSs have shown to be more sensitive to the thermally induced surface convergence zones and are likely to develop over spatially drier soils, adjacent to the strong gradients (e.g, Gantner and Kalthoff, 2010). This knowledge is consistent with the strongest negative δ_e difference identified here and hence with the prominent negative

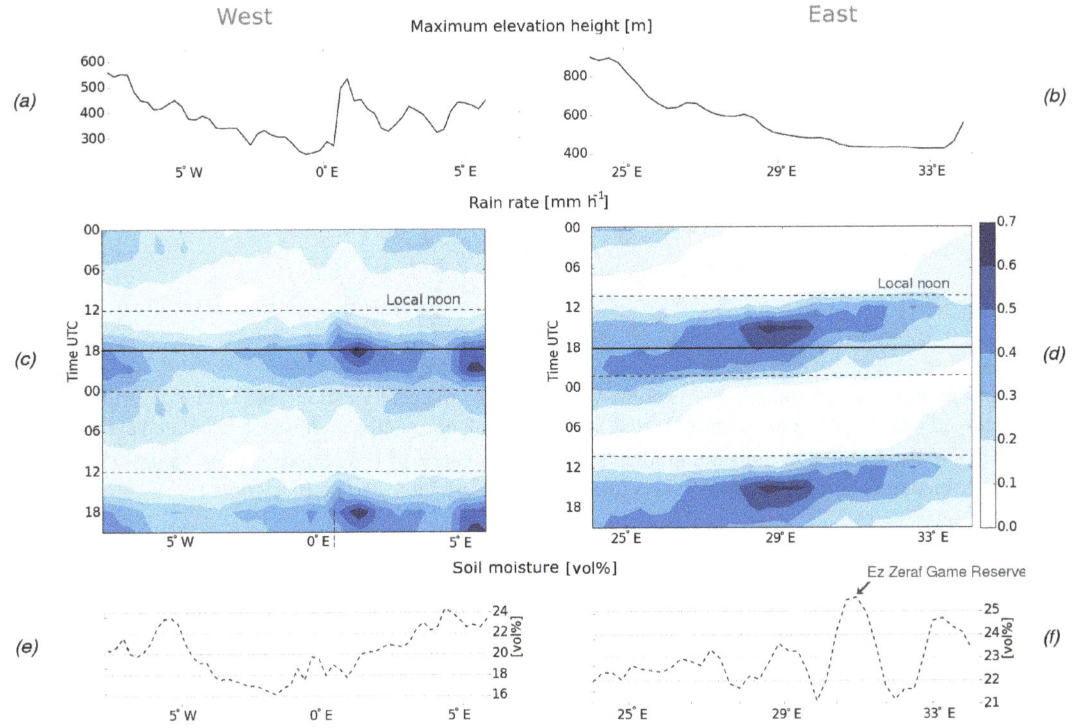

Figure 7. (a, b) Longitudinal cross-sections of maximum elevation height in the western and eastern domains respectively, **(c, d)** diurnal cycles of the rain rate averaged over event days and domain latitudes and **(e, f)** longitudinal cross section of soil moisture averaged over domain latitudes. Location of the Ez Zeraf Game Reserve permanent wetlands is marked by an arrow. All the times are given in UTC. Note the UTC+2 h difference to LST in the east.

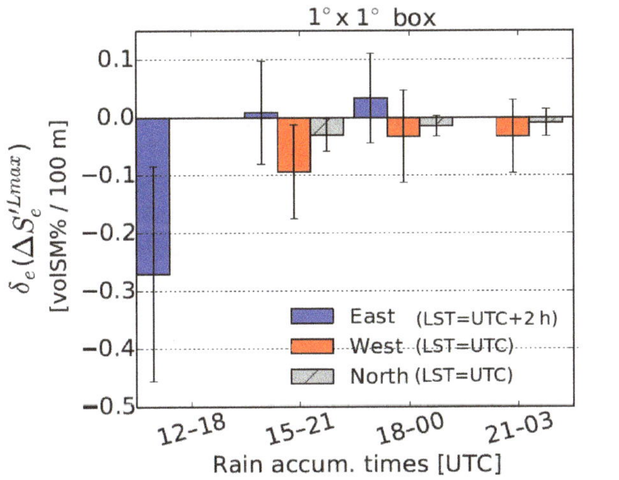

Figure 8. Value of the spatial coupling measure δ_e calculated for various afternoon rainfall accumulation times, and averaged over selected domains, i.e east (6–10° N, 24–34° E), west (7–12° N, 8° W–6° E) and north (14–17° N, 7–14° E). Locations of the domains are shown in Fig. 4. Error bars indicate one SD of δ_e values in every domain. Note that all times are indicated in UTC.

SMPC relationship observed during early afternoon times in all three domains.

5 Results of temporal SMPC analysis

5.1 Co-variability of the spatial and temporal SMPC

To further explore the consistency of the identified negative spatial SMPC with the physical effects we analyse the temporal SMPC relationship. By analogy to the spatial SMPC, we compute the soil moisture anomaly $S_e^{\prime L_{\max}}$ prior to the event and its difference δ_e to the typical state. Analysis of $S_e^{\prime L_{\max}}$ and its δ_e indicates a strong preference for rainfall events to occur over soils that are drier than their temporal mean (Fig. 9a) and drier than usual (Fig. 9b). The percentile values P_e lower than 10 % are found in 67 % of the studied $1° \times 1°$ boxes (Table 1). This implies that a temporally negative SMPC dominates over the domain, which reaffirms the co-existence of the negative spatial and temporal coupling identified by G15, but at a finer 1° horizontal scale.

The question remains as to whether the two coupling relationships are independent of one another. To answer this question we calculate the Spearman rank correlation coefficient[4] event-wise between the soil moisture anomaly $S_e^{\prime L_{\max}}$ and soil moisture gradients $\Delta(S_e^{\prime L_{\max}})$ in every $1° \times 1°$ box.

[4]Spearman correlation is a measure of the monotonic relationship. Therefore, a zero or low correlation value does not imply a zero relationship between two variables.

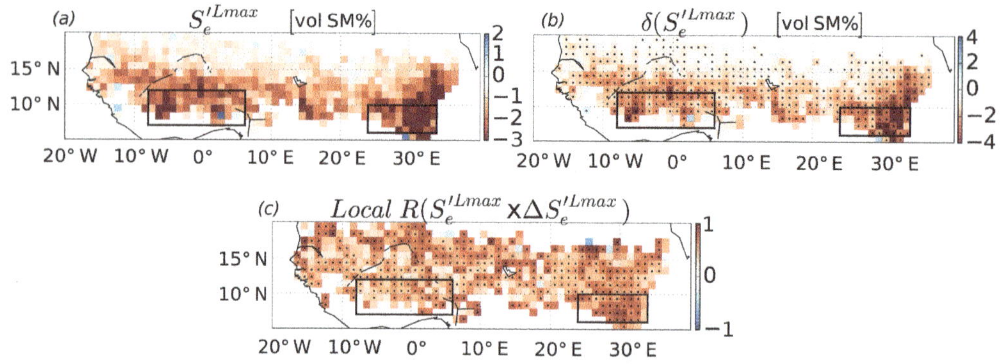

Figure 9. Distribution of the **(a)** pre-event soil moisture anomaly value $S_e'^{L\max}$ in event locations and **(b)** its departure from the typical non-event conditions, δ_e, averaged over $1° \times 1°$ boxes; and **(c)** local Spearman rank correlation coefficient calculated event-wise between the soil moisture anomaly $S_e'^{L\max}$ and the spatial soil moisture gradients $\Delta S_e'^{L\max}$ in every $1°$ box. Significant δ_e values with percentiles P_e below 10 % (above 90 %) and correlation coefficients with p values lower than 0.05 are indicated by black dots.

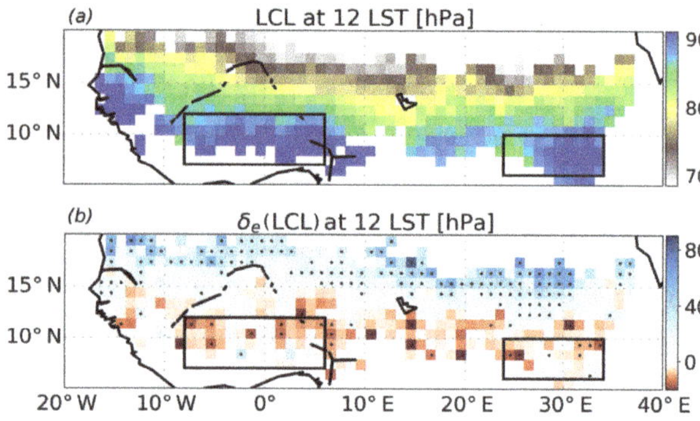

Figure 10. (a) Lifting condensation level (LCL) value derived from the 6-hourly ERA-Interim temperature, specific humidity profile and surface pressure data on event days at 12:00 LST and averaged over the $1° \times 1°$ box. **(b)** Corresponding δ_e difference of the mean LCL prior to the events relative to their typical state (climatology). The dot indicates significant δ_e values with percentiles P_e below 10 % (above 90 %). The positive (negative) δ_e values indicate an LCL that is lower (higher) than usual (relative to the ground level).

The correlation map in Fig. 9c shows that a high and significant correlation exists between $S_e'^{L\max}$ and $\Delta(S_e'^{L\max})$ anywhere in the domain. The mean correlation of 0.47 over the domain supports the existence of a relatively strong and positive monotonic relationship between the magnitude of the spatial soil moisture gradient and the soil moisture anomaly measured in $L_{\max}$. For comparison, the mean correlation estimated between soil moisture gradients and the mean soil moisture anomaly over the 1.25° event box is small (0.13). All the above suggests that in the North African region the spatial and temporal SMPC relationships, as defined by the current framework, are not independent of each other.

The strong and positive correlation (*in time*) identified between the soil moisture anomalies and gradients also yields a regional co-variability of the SMPC patterns. The *spatial* correlation between the two coupling distributions is high (0.64). The largest magnitudes of both $S_e'^{L\max}$ and $\Delta(S_e'^{L\max})$ parameters and their corresponding δ_e measures are found in the southern part of the domain. These regions are generally characterized as the areas of higher BL moisture and rainfall frequency, and therefore higher variability of soil moisture in time and space.

Mechanistically, the presence of the temporally negative SMPC in the areas of the highest BL moisture in the domain (or lowest lifting condensation level (LCL); Fig. 10a) is consistent with a higher relevance of mechanisms associated with the BL growth for convection initialization in regions of higher CAPE and lower convective inhibition (CIN) (Klüpfel et al., 2011; Gantner and Kalthoff, 2010; Adler et al., 2011). In this way, larger negative deviations of the soil moisture amount from its climatological mean, i.e. δ_e, would lead to thermals that are stronger than usual, which can more easily overcome CIN and release CAPE (Klüpfel et al., 2011).

Moreover, in combination with strong negative spatial gradients, these strong thermals can initiate breeze-like circulations, creating more favourable conditions for bringing the BL up to the LFC, especially over the southern regions, where BL moisture is in abundance. The relevance of drier surface conditions for moist convection development on event days over the wetter latitudes is supported by the observed significant increase of the LCL height (decrease in pressure) in the south on event days compared to the typical state (Fig. 10b, red shading). The higher LCL is associated with a decrease of BL relative and specific humidity (not shown) and supports the relevance of drier surface conditions for convection intensification as opposed to variations in BL water vapour amount prior to the events.

A different picture is observed over the drier latitudes of northern Sahel at the Sahara margin. There, the pre-event LCLs are found to be significantly lower (higher pressure) compared to its typical (climatological) state. This suggests a higher amount of water vapor in the BL (not shown) on the event days over the dry regions (Fig. 10b, blue shading).

This result is consistent with the previously reported decisive role of low-level moisture on MCS evolution in the drier Sahelian regions (Klüpfel et al., 2012). At these latitudes the northward excursion of moist monsoon air has been shown to favour convective activity (Barthe et al., 2010; Cuesta et al., 2010).

Considering also the relatively large number of dry days (10 days on average) preceding rain events in the north, it is less likely that underlying surface heterogeneity caused by a previous rainfall could have an influence on convection development on the event day. In the case study of Klüpfel et al. (2012) an MCS was initiated due to the arrival of the cold pool and convergence zone emanated by a remote convective system hundreds of kilometres away. Similar mechanisms may play a role in moist convection development in northern Sahel.

6 Summary and discussion

6.1 Role of thermally induced circulations in moist convection development

The dominant negative spatial SMPC relationship observed over the North African region agrees on the sign of the SMPC suggested by previous case studies and modelling studies (e.g. Taylor and Ellis, 2006; Birch et al., 2013; Froidevaux et al., 2014), and reconciles a number of physical effects. Following the sensitivity analyses, the two main factors are identified, which directly influence the magnitude and significance of the negative spatial SMPC relationship. These are the time of the afternoon rainfall accumulation and the extreme soil moisture gradients. Our results show that the observed relationship between the spatial SMPC measure and the time of rainfall accumulation conforms well with the varying sensitivity of rainfall to the underlying soil moisture conditions for different stages of MCS development (Taylor and Lebel, 1998; Clark et al., 2003; Taylor and Ellis, 2006; Alonge et al., 2007; Taylor et al., 2010). This further suggests potentially higher relevance of drier soils and soil moisture heterogeneity for rainfall when the rain systems are smaller and at an early stage of their development, consistent with results of e.g. Gantner and Kalthoff (2010). All the above also emphasizes the importance of considering the MCS properties as an additional factor in the analysis of the SMPC (Ford et al., 2015).

Extreme soil moisture gradients are found to predefine the significance of the negative spatial coupling in 30 % of the domain grid boxes. Concurrently, the extremes tend to cluster in the direct vicinity of major flooded areas and the irrigated land. The identified sensitivity of the afternoon rainfall to the strong soil moisture gradients adjacent to wetland areas shows consistency with the role of a wetland breeze mechanism in convection intensification over spatially drier soils. If this is true, our results demonstrate for the first time that the wetland breeze mechanism could be a systematic feature in the North African region, with further implications for the rainfall predictability.

This observed sensitivity of the SMPC measure to the flooded areas and to the MCS life cycle complies well with the potential role of thermally induced circulations in afternoon rainfall development over strong negative soil moisture gradients. Moreover, the identified preference of rainfall to occur simultaneously over temporally drier soils (negative temporal SMPC) and strong negative soil moisture gradients (negative spatial SMPC) might be considered as the most effective combination to initiate thermal circulations. This would imply a presence of a higher buoyancy and moisture flux at the event location, and hence a higher probability of convection development. Schematic representation of the deep convection initiation by the breeze-like circulations formed under the co-existence of the two SMPC effects is illustrated in Fig. 11.

6.2 Role of rainfall persistence

In the context of this study, the drying of the soil prior to the rainfall events might be considered as the primary process that underlies the magnitude of both SMPC relationships, and helps to explain the opposite sign of the temporal coupling identified in the North African region as compared to the temperate latitudes and wet climates (G15).

Consistent with the observed 2- to 4-day periodicity of rainfall in Western Africa (Laing et al., 2012; Taylor and Lebel, 1998), 2 to 3 dry days (rain < 1 mm) on average are found to precede each convective event day over southern latitudes, suggesting a strong drying of the upper soil layer in the event locations prior to the rain. The number of dry days reaches 10 over the dry and deserted regions in the north. Following the analysis of Schwendike et al. (2010), an almost complete recovery of the pre-rainfall surface moisture conditions may be expected in 2–3 days following the rainfall. Schematically, this typical variability of rainfall and soil moisture might be illustrated as a sequence of daily rain events separated by the periods of drying (Fig. 12a). From the figure it is seen that prior to the rain events the soil dries out, and soil moisture reaches a certain minimum value $S_{\min}$. The climatology value S_{clim} of soil moisture in the same location, however, is expected to be higher than any $S_{\min}$ in most of the cases, as it includes all dry and wet event days. Hence, when subtracted from the climatological value, a soil moisture value measured prior to the event will very likely yield a negative anomaly $-\ S_{\mathrm{e}}^{\prime L_{\max}}$, especially when averaged over many events. Therefore, a negative correlation between the soil moisture anomaly and rainfall might be expected. Though discussed in the framework of North Africa, similar behaviour might be expected in other water-limited regions of the world.

A different situation might occur in the wet temperate latitudes, where the variability of rainfall is to a large extent

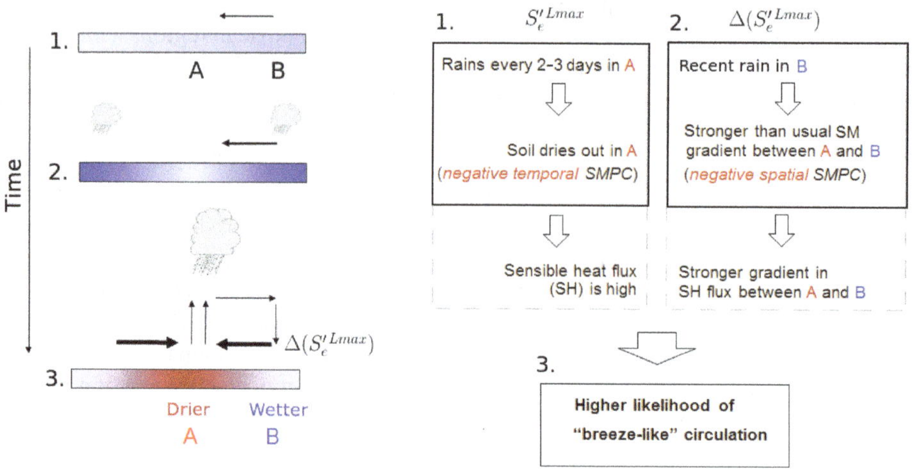

Figure 11. Conceptual diagram, illustrating intensification of moist convection by the initiated "breeze-like" circulations under favourable conditions of co-existing negative spatial and negative temporal SMPC effects. On the one hand, typically observed 2- to 4-day periodicity of rainfall in western Africa leads to a strong drying of the upper soil layer in the location A prior to the rain, and therefore increases sensible heat and buoyancy flux locally. Simultaneously, recent rainfall in B produces wet soils – a potential moisture supply area for the location A. Strong spatial gradients in soil moisture between locations A and B together with a relatively strong buoyancy flux in A can favour formation of thermally induced circulations under benign wind conditions. Considering a relatively weak mean surface wind of 2–3 $\mathrm{m\,s^{-1}}$ observed prior to the rainfall events over southern latitudes, the mesoscale circulations are likely to be initiated.

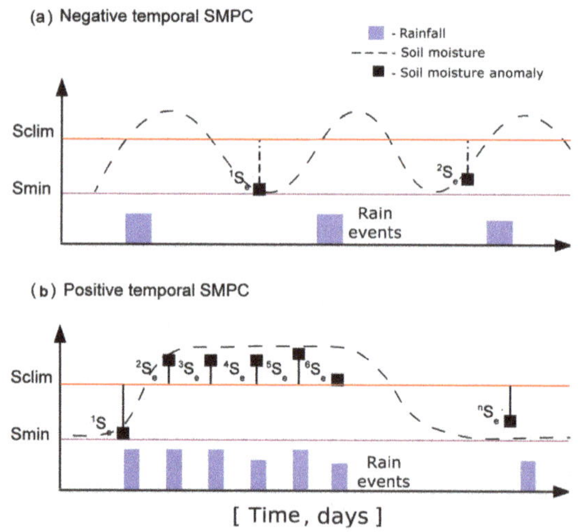

Figure 12. Conceptual diagrams of the relationship between daily rainfall occurrence and the surface moisture variability associated to it, in time, as representative of **(a)** West Africa and temporally negative SMPC and **(b)** central and northern Europe and positive temporal SMPC.

linked to fluctuations between the passage of a cyclone and a blocking situation (Schär et al., 1999). Such a behaviour might be illustrated as a multi-day sequence of rain events, associated with precipitation persistence as defined by the persistence in the weather regimes (Fig. 12b; see also Fig. 2 in Hohenegger et al., 2009). During these periods soil moisture increases and remains relatively high. Hence, a higher fraction of events might be expected to occur over soils that are wetter than usual, resulting in a positive soil moisture anomaly $S_e'^{L\max}$ prior to the event. The above relationship is consistent with the positive temporal SMPC, identified in G15. It is important to note, however, that the rainfall persistence may not be solely atmospherically driven, but may also reflect effects of the land surface (Salvucci et al., 2002; Guillod et al., 2014; Seneviratne et al., 2010).

The modulation of the SMPC sign depending on the large-scale weather regime was studied e.g. by Boé (2012) over France. The analysis showed that the synoptic blocking situations generally associated with drier conditions lead to a negative SMPC, while a positive correlation of rainfall to drier soil conditions was observed in a wet weather regime. Similarly, the most pronounced effect of negative soil moisture gradients on convection initiation over Europe and a higher correlation of the gradients to land surface temperatures was observed for the period with less antecedent rainfall (Taylor, 2015).

7 Conclusions

In this study, we revisit the negative spatial and negative temporal SMPC relationships identified earlier by T12 and G15 on the 5° × 5° horizontal grid. We use the probability-based approach of T12 and 10 years of satellite-based soil moisture and precipitation data (i) to identify the potential link of the observed statistical relationships to the physical mechanisms and (ii) to study the regional co-variability of the SMPC effects.

We find that the negative spatial coupling dominates over the region of North Africa. The result is independent of the choice of the observational data sets and is robust with respect to the event aggregation (spatial) scale. Compared to the coarser 5° × 5° grid previously used, the appeal of the finest considered 1° × 1° event-aggregation scale is that it reveals links to the wetland areas and rivers which can not be captured at the coarser scale.

The co-variability analysis of the two SMPC relationships indicates that spatial and temporal effects of soil moisture on afternoon precipitation in the North African region do not only co-exist but are also dependent on one another. The latter suggests that if rain falls over temporally drier soils, it is likely to be surrounded by a wetter environment. This combination is consistent with the relevance of processes associated with the dominance of the sensible heat flux and boundary layer growth on convection initiation, and supports the role of mesoscale variability in surface soil moisture for deep convection development.

The identified negative sign of the temporal coupling in the semi-arid conditions of the Sahelian environment is not unexpected. The drying of the soil for several days prior to the rainfall events is likely to underlay the preference of rain to occur over temporally drier soils. This additionally may play a role in the opposite sign of the temporal coupling in the North African region as compared to the positive relationship identified in wetter climates by G15. For the same reason, the predictability potential of the temporal effect on rainfall in the North African region is expected to be lower compared to the spatial effect.

Analysis of the spatial SMPC measure and factors which can influence its magnitude and variability in particular reveals two hot spot regions, where predictability skill of spatial soil moisture variability on rainfall might be higher. These are the western African domain (7–15° N, 10° W–7° E) and South Sudan in the east (5–13° N, 24–34° E). In the Western domain, the negative spatial SMPC signal is indicated to be more robust. In the east, the spatial coupling is found to be largely modulated by the presence of wetlands and is susceptible to the number of longer lived propagating MCSs. The analysis of the BL moisture conditions (here, LCL) preceding the rainfall events further supports the potential relevance of spatially and temporally drier soils for convection development in the south of the domain, where BL moisture is in abundance. In the drier northern latitudes the variability of BL moisture, associated with intrusions of moisture from the south, seems to be more important.

Following our analysis, a number of potential improvements to the framework might be summarized. Apparent non-local effects of water bodies, which are originally excluded by the method, hint towards the potential gaps in the filtering procedure and emphasize the potential role of moist convection evolution and propagation that are neglected by the method. The presence of wetland regions themselves, as we have shown, complicates the interpretation of the SMPC relationships. The uncertainty estimates of the soil moisture parameter derived over the recursively flooded regions are still missing. In the future, dynamical wetland products like the Global Inundation Extent from Multi-Satellites (GIEMS, Prigent et al., 2012, 2007) may be used to better isolate the effect of water bodies on moist convection development.

Notwithstanding these limitations, the present study demonstrates the ability of probability-based methods to identify characteristic features of physical effects. Considering continuous increase in the availability, time span and quality of satellite data, development of similar statistical methods should be valuable. The region of the strong negative SMPC identified in the east would highly benefit from more modelling and observational analysis in the eastern Sahel. The knowledge on the regional variability of the SMPC presented here can be further explored in drought and climate change research and observational campaigns and used for the validation of global climate models.

Data availability. The observational datasets used in this analysis are publically available and can be downloaded as follows:

- AMSR-E soil moisture (version 2), available at https://hydro1.gesdisc.eosdis.nasa.gov/data/s4pa/WAOB/LPRM_AMSRE_A_SOILM3.002/ (last access: September 2017);
- TRMM3B42 3-hourly precipitation, available at https://pmm.nasa.gov/data-access/downloads/trmm (last access: September 2017);
- ERA-Interim 6-hourly surface pressure, temperature and specific humidity profile data, available at http://apps.ecmwf.int/datasets/data/interim-full-daily/levtype=sfc/ (last access: September 2017).

Competing interests. The authors declare that they have no conflict of interest.

Acknowledgements. The authors would like to thank the Max Planck Institute for Meteorology (MPI-M) and the International Max Planck Research School (IMPRS) for providing facilities, material and scientific support which made publication of this paper possible. The authors would like to acknowledge Christopher Taylor, Benoît Guillod and Alexander Mahura for helpful comments on the study, and Stephan Kern for data support. We also thank George Huffman and Robert Parinussa for their clarifications related to TMPA and AMSR-E data products respectively.

Edited by: Pierre Gentine

References

Adler, B., Kalthoff, N., and Gantner, L.: The impact of soil moisture inhomogeneities on the modification of a mesoscale convective system: An idealised model study, Atmos. Res., 101, 354–372, 2011.

Alonge, C. J., Mohr, K. I., and Tao, W.-K.: Numerical studies of wet versus dry soil regimes in the West African Sahel, J. Hydrometeorol., 8, 102–116, 2007.

Alter, R., Im, E., and Eltahir, E.: Rainfall consistently enhanced around the Gezira Scheme in East Africa due to irrigation, Nat. Geosci., 8, 763–767, 2015.

Amante, C. and Eakins, B.: ETOPO1 1 arc-minute global relief model: procedures, data sources and analysis, available at: https://www.ngdc.noaa.gov/mgg/global/relief/ETOPO1/docs/ETOPO1.pdf (last access: September 2017) 2009.

Barthe, C., Asencio, N., Lafore, J.-P., Chong, M., Campistron, B., and Cazenave, F.: Multi-scale analysis of the 25–27 July 2006 convective period over Niamey: Comparison between Doppler radar observations and simulations, Q. J. Roy. Meteor. Soc., 136, 190–208, 2010.

Birch, C. E., Parker, D. J., O'Leary, A., Marsham, J. H., Taylor, C. M., Harris, P. P., and Lister, G. M. S.: Impact of soil moisture and convectively generated waves on the initiation of a West African mesoscale convective system, Q. J. Roy. Meteor. Soc., 139, 1712–1730, 2013.

Boé, J.: Modulation of soil moisture–precipitation interactions over France by large scale circulation, Clim. Dynam., 40, 875–892, 2012.

Clark, D. B., Taylor, C. M., Thorpe, A. J., Harding, R. J., and Nicholls, M. E.: The influence of spatial variability of boundary-layer moisture on tropical continental squall lines, Q. J. Roy. Meteor. Soc., 129, 1101–1121, 2003.

Couvreux, F., Rio, C., Guichard, F., Lothon, M., Canut, G., Bouniol, D., and A., G.: Initiation of daytime local convection in a semi-arid region analyzed with Large-Eddy Simulations and AMMA observations, Q. J. Roy. Meteor. Soc., 138, 56–71, 2012.

Couvreux, F., Guichard, F., Gounou, a., Bouniol, D., Peyrillé, P., and Köhler, M.: Modelling of the thermodynamical diurnal cycle in the lower atmosphere: A Joint evaluation of four contrasted regimes in the tropics over land, Bound.-Lay. Meteorol., 150, 185–214, https://doi.org/10.1007/s10546-013-9862-6, 2013.

Cuesta, J., Lavaysse, C., Flamant, C., Mimouni, M., and Knippertz, P.: Northward bursts of the West African monsoon leading to rainfall over the Hoggar Massif, Algeria, Q. J. Roy. Meteor. Soc., 136, 174–189, 2010.

Dee, D. P., Uppala, S. M., Simmons, A. J., Berrisford, P., Poli, P., Kobayashi, S., Andrae, U., Balmaseda, M. A., Balsamo, G., Bauer, P., Bechtold, P., Beljaars, A. C. M., van de Berg, L., Bidlot, J., Bormann, N., Delsol, C., Dragani, R., Fuentes, M., Geer, A. J., Haimberger, L., Healy, S. B., Hersbach, H., Hólm, E. V., Isaksen, L., Kå llberg, P., Köhler, M., Matricardi, M., McNally, A. P., Monge-Sanz, B. M., Morcrette, J.-J., Park, B.-K., Peubey, C., de Rosnay, P., Tavolato, C., Thépaut, J.-N., and Vitart, F.: The ERA-Interim reanalysis: configuration and performance of the data assimilation system, Q. J. Roy. Meteor. Soc., 137, 553–597, 2011.

Dirmeyer, P.: The terrestrial segment of soil moisture-climate coupling, Geophys. Res. Lett., 38, L16702, https://doi.org/10.1029/2011GL048268, 2011.

Dirmeyer, P. and Wang, Z.: Intensified land surface control on boundary layer growth in a changing climate, Geophys. Res. Lett., 45, 1290–1294, 2014.

Dirmeyer, P., Cash, B. A., Kinter, J. L., Stan, C., Jung, T., Marx, L., Towers, P., Wedi, N., Adams, J. M., Altshuler, E. L., Huang, B., Jin, E. K., and Manganello, J.: Evidence for enhanced land–atmosphere feedback in a warming climate, J. Hydrometeorol., 13, 981–995, 2012.

Dorigo, W. A., Scipal, K., Parinussa, R. M., Liu, Y. Y., Wagner, W., de Jeu, R. A. M., and Naeimi, V.: Error characterisation of global active and passive microwave soil moisture datasets, Hydrol. Earth Syst. Sci., 14, 2605–2616, https://doi.org/10.5194/hess-14-2605-2010, 2010.

Duvel, J.: Convection over tropical Africa and the Atlantic Ocean during northern summer. Part II: Modulation by easterly waves, Mon. Weather Rev., 118, 1885–1868, 1990.

Ek, M. and Mahrt, L.: Daytime evolution of relative humidity at the boundary layer top, American Meteorological Society, 122, 2709–2721, 1994.

Eltahir, E. A. B.: A soil moisture-rainfall feedback mechanism: 1. Theory and observations, Water Resour. Res., 34, 765–776, 1998.

Findell, K. and Eltahir, E.: Atmospheric controls on soil moisture-boundary layer interactions. Part II: Feedbacks within the continental United States, J. Hydrometeorol., 4, 570–583, 2003a.

Findell, K. and Eltahir, E.: Atmospheric controls on soil moisture-boundary layer interactions. Part I: Framework development, J. Hydrometeorol., 4, 552–569, 2003b.

Ford, T. W., Rapp, A. D., Quiring, S. M., and Blake, J.: Soil moisture–precipitation coupling: observations from the Oklahoma Mesonet and underlying physical mechanisms, Hydrol. Earth Syst. Sci., 19, 3617–3631, https://doi.org/10.5194/hess-19-3617-2015, 2015.

Froidevaux, P., Schlemmer, L., Schmidli, J., Langhans, W., and Schär, C.: Influence of the background wind on the local soil moisture-precipitation feedback, J. Atmos. Sci., 17, 782–799, 2014.

Gallego-Elvira, B. and Taylor, C.: Global observational diagnosis of soil moisture control on the land surface energy balance, Geophys. Res. Lett., 43, 2623–2631, 2016.

Gantner, L. and Kalthoff, N.: Sensitivity of a modelled life cycle of a mesoscale convective system to soil conditions over West Africa, Q. J. Roy. Meteor. Soc., 136, 471–482, 2010.

Garcia-Carreras, L., Parker, D. J., and Marsham, J. H.: What is the mechanism for the modification of convective cloud distributions by land surface–induced flows?, J. Atmos. Sci., 68, 619–634, 2011.

Goutorbe, J.-P., Lebel, T., Tinga, A., Bessemoulin, P., Brouwer, J., Dolman, A. J., Engman, E. T., Gash, J. H. C., Hoepffner, M., Kabat, P., Kerr, Y. H., Monteny, B., Prince, S., Said, F., Sellers, P., and Wallace, J. S.: HAPEX-Sahel: a large-scale study of land-atmosphere interactions in the semi-arid tropics, Ann. Geophys., 12, 53–64, https://doi.org/10.1007/s00585-994-0053-0, 1994.

Gruhier, C. and Rosnay, P. D.: Evaluation of AMSRE soil moisture product based on ground measurements over temperate and semiarid regions, Geophys. Res. Lett., 35, 2–7, 2008.

Guichard, F., Kergoat, L., Mougin, E., and Timouk, F.: Surface thermodynamics and radiative budget in the Sahelian Gourma: Seasonal and diurnal cycles, J. Hydrol., 375, 161–177, 2009.

Guillod, B. P., Orlowsky, B., Miralles, D. G., Teuling, A. J., and Seneviratne, S. I.: Reconciling spatial and temporal soil moisture effects on afternoon rainfall, Nat. Commun., 6, 1–6, 2015.

Guillod, B. P., Orlowsky, B., Miralles, D., Teuling, A. J., Blanken, P. D., Buchmann, N., Ciais, P., Ek, M., Findell, K. L., Gentine, P., Lintner, B. R., Scott, R. L., Van den Hurk, B., and I. Seneviratne, S.: Land-surface controls on afternoon precipitation diagnosed from observational data: uncertainties and confounding factors, Atmos. Chem. Phys., 14, 8343–8367, https://doi.org/10.5194/acp-14-8343-2014, 2014.

Hohenegger, C., Brockhaus, P., Bretherton, C. S., and Schär, C.: The soil moisture–precipitation feedback in simulations with explicit and parameterized convection, J. Climate, 22, 5003–5020, 2009.

Hsu, K.-l., Gao, X., Sorooshian, S., and Gupta, H. V.: Precipitation estimation from remotely sensed information using artificial neural networks, J. Appl. Meteorol., 36, 1176–1190, 1997.

Huffman, G. J. and Bolvin, D. T.: TRMM and other data precipitation data set documentation, Tech. Rep. May, Mesoscale Atmospheric Processes Laboratory, NASA Goddard Space Flight Center, available at: https://pmm.nasa.gov/sites/default/files/document_files/3B42_3B43_doc_V7.pdf (last access: September 2017), 2014.

Huffman, G. J., Bolvin, D. T., Nelkin, E. J., Wolff, D. B., Adler, R. F., Gu, G., Hong, Y., Bowman, K. P., and Stocker, E. F.: The TRMM multisatellite precipitation analysis (TMPA): quasi-global, multiyear, combined-sensor precipitation estimates at fine scales, J. Hydrometeorol., 8, 38–55, 2007.

Janicot, S., Thorncroft, C. D., Ali, A., Asencio, N., Berry, G., Bock, O., Bourles, B., Caniaux, G., Chauvin, F., Deme, A., Kergoat, L., Lafore, J.-P., Lavaysse, C., Lebel, T., Marticorena, B., Mounier, F., Nedelec, P., Redelsperger, J.-L., Ravegnani, F., Reeves, C. E., Roca, R., de Rosnay, P., Schlager, H., Sultan, B., Tomasini, M., Ulanovsky, A., and ACMAD forecasters team: Large-scale overview of the summer monsoon over West Africa during the AMMA field experiment in 2006, Ann. Geophys., 26, 2569–2595, https://doi.org/10.5194/angeo-26-2569-2008, 2008.

Klüpfel, V., Kalthoff, N., Gantner, L., and Kottmeier, C.: Evaluation of soil moisture ensemble runs to estimate precipitation variability in convection-permitting model simulations for West Africa, Atmos. Res., 101, 178–193, 2011.

Klüpfel, V., Kalthoff, N., Gantner, L., and Taylor, C. M.: Convergence zones and their impact on the initiation of a mesoscale convective system in West Africa, Q. J. Roy. Meteor. Soc., 138, 950–963, 2012.

Kohler, M., Kalthoff, N., and Kottmeier, C.: The impact of soil moisture modifications on CBL characteristics in West Africa: A case-study from the AMMA campaign, Q. J. Roy. Meteor. Soc., 136, 442–455, 2010.

Koster, R. D., Dirmeyer, P. A., Guo, Z., Bonan, G., Chan, E., Cox, P., Gordon, C. T., Kanae, S., Kowalczyk, E., Lawrence, D., Liu, P., Lu, C.-H., Malyshev, S., McAvaney, B., Mitchell, K., Mocko, D., Oki, T., Oleson, K., Pitman, A., Sud, Y. C., Taylor, C. M., Verseghy, D., Vasic, R., Xue, Y., and Yamada, T.: Regions of strong coupling between soil moisture and precipitation, Science, 305, 1138–40, 2004.

Laing, A. and Carbone, R.: The propagation and diurnal cycles of deep convection in northern tropical Africa, Q. J. Roy. Meteor. Soc., 109, 93–109, 2008.

Laing, A. G., Trier, S. B., and Davis, C. A.: Numerical simulation of episodes of organized convection in tropical Northern Africa, Mon. Weather Rev., 140, 2874–2886, 2012.

Lohou, F., Kergoat, L., Guichard, F., Boone, A., Cappelaere, B., Cohard, J.-M., Demarty, J., Galle, S., Grippa, M., Peugeot, C., Ramier, D., Taylor, C. M., and Timouk, F.: Surface response to rain events throughout the West African monsoon, Atmos. Chem. Phys., 14, 3883–3898, https://doi.org/10.5194/acp-14-3883-2014, 2014.

Lothon, M., Campistron, B., Chong, M., Couvreux, F., Guichard, F., Rio, C., and Williams, E.: Life cycle of a mesoscale circular gust front observed by a C-band Doppler radar in West Africa, Mon. Weather Rev., 139, 1370–1388, 2011.

Mathon, V. and Laurent, H.: Life cycle of Sahelian mesoscale convective cloud systems, Q. J. Roy. Meteor. Soc., 127, 377–406, 2001.

Mathon, V., Laurent, H., and Lebel, T.: Mesoscale convective system rainfall in the Sahel, J. Appl. Meteorol., 41, 1081–1092, 2002.

Matthews, E. and Fung, I.: Methane emission from natural wetlands: Global distribution, area, and environmental characteristics of sources, Global Biogeochem. Cy., 1, 61–86, 1987.

Miralles, D. G., van den Berg, M. J., Teuling, A. J., and de Jeu, R. A. M.: Soil moisture-temperature coupling: A multiscale observational analysis, Geophys. Res. Lett., 39, L21707, https://doi.org/10.1029/2012GL053703, 2012.

Mohr, K. and Thorncroft, C.: Intense convective systems in West Africa and their relationship to the African easterly jet, Q. J. Roy. Meteor. Soc., 132, 163–176, 2006.

Nicholson, S. E.: The West African Sahel: A review of recent studies on the rainfall regime and its interannual variability, ISRN Meteorology, 2013, 1–32, 2013.

Nicholson, S. E.: Evolution and current state of our understanding of the role played in the climate system by land surface processes in semi-arid regions, Global Planet. Change, 133, 201–222, 2015.

Njoku, E. and Jackson, T.: Soil moisture retrieval from AMSR-E, IEEE T. Geosci. Remote, 41, 215–229, 2003.

Owe, M., de Jeu, R., and Holmes, T.: Multisensor historical climatology of satellite-derived global land surface moisture, J. Geophys. Res., 113, F01002, https://doi.org/10.1029/2007JF000769, 2008.

Petrova, I.: Understanding soil moisture – precipitation coupling on mesoscales using observations over North Africa, Reports on Earth System Science/Max Planck Institute for Meteorology, 191, 2017.

Prigent, C., Papa, F., Aires, F., Rossow, W. B., and Matthews, E.: Global inundation dynamics inferred from multiple satellite observations, 1993–2000, J. Geophys. Res.-Atmos., 112, D12107, https://doi.org/10.1029/2006JD007847, 2007.

Prigent, C., Papa, F., Aires, F., Jimenez, C., Rossow, W. B., and Matthews, E.: Changes in land surface water dynamics since the 1990s and relation to population pressure, Geophys. Res. Lett., 39, L08403, https://doi.org/10.1029/2012GL051276, 2012.

Redelsperger, J.-L., Thorncroft, C. D., Diedhiou, A., Lebel, T., Parker, D. J., and Polcher, J.: African Monsoon Multidisciplinary Analysis: An international research project and field campaign, B. Am. Meteorol. Soc., 87, 1739–1746, 2006.

Rochetin, N., Couvreux, F., and Guichard, F.: Morphology of breeze circulations induced by surface flux heterogeneities and their impact on convection initiation, Q. J. Roy. Meteor. Soc., 143, 463–478, 2017.

Salvucci, G. D., Saleem, J. A., and Kaufmann, R.: Investigating soil moisture feedbacks on precipitation with tests of Granger causality, Adv. Water Resour., 25, 1305–1312, 2002.

Schär, C., Lüthi, D., and Beyerle, U.: The soil-precipitation feedback: A process study with a regional climate model, J. Climate, 12, 722–741, 1999.

Schwendike, J., Kalthoff, N., and Kohler, M.: The impact of mesoscale convective systems on the surface and boundary-layer structure in West Africa: Case-studies from the AMMA campaign 2006, Q. J. Roy. Meteor. Soc., 136, 566–582, 2010.

Seneviratne, S. I., Corti, T., Davin, E. L., Hirschi, M., Jaeger, E. B., Lehner, I., Orlowsky, B., and Teuling, A. J.: Investigating soil moisture-climate interactions in a changing climate: A review, Earth-Sci. Rev., 99, 125–161, 2010.

Taylor, C. M.: Feedbacks on convection from an African wetland, Geophys. Res. Lett., 37, L05406, https://doi.org/10.1029/2009GL041652, 2010.

Taylor, C. M.: Detecting soil moisture impacts on convective initiation in Europe, Geophys. Res. Lett., 42, 4631–4638, 2015.

Taylor, C. M. and Ellis, R. J.: Satellite detection of soil moisture impacts on convection at the mesoscale, Geophys. Res. Lett., 33, L03404, https://doi.org/10.1029/2005GL025252, 2006.

Taylor, C. M. and Lebel, T.: Observational evidence of persistent convective-scale rainfall patterns, Mon. Weather Rev., 126, 1597–1607, 1998.

Taylor, C. M., Parker, D. J., and Harris, P. P.: An observational case study of mesoscale atmospheric circulations induced by soil moisture, Geophys. Res. Lett., 34, L15801, https://doi.org/10.1029/2007GL030572, 2007.

Taylor, C. M., Harris, P. P., and Parker, D. J.: Impact of soil moisture on the development of a Sahelian mesoscale convective system: a case-study from the AMMA Special Observing Period, Q. J. Roy. Meteor. Soc., 136, 456–470, 2010.

Taylor, C. M., Gounou, A., Guichard, F., Harris, P. P., Ellis, R. J., Couvreux, F., and De Kauwe, M.: Frequency of Sahelian storm initiation enhanced over mesoscale soil-moisture patterns, Nat. Geosci., 4, 430–433, 2011.

Taylor, C. M., de Jeu, R. A. M., Guichard, F., Harris, P. P., and Dorigo, W. A.: Afternoon rain more likely over drier soils, Nature, 489, 423–426, 2012.

Taylor, C. M., Birch, C. E., Parker, D. J., Dixon, N., Guichard, F., Nikulin, G., and Lister, G. M. S.: Modeling soil moisture-precipitation feedback in the Sahel : Importance of spatial scale versus convective parameterization, Geophys. Res. Lett., 40, 6213–6218, 2013.

Timouk, F., Kergoat, L., Mougin, E., Lloyd, C., Ceschia, E., Cohard, J.-M., de Rosnay, P., Hiernaux, P., Demarez, V., and Taylor, C.: Response of sensible heat flux to water regime and vegetation development in a central Sahelian landscape, J. Hydrol., 375, 178–189, 2009.

van Heerwaarden, C. C. and Guerau de Arellano, J. V.: Relative humidity as an indicator for cloud formation over heterogeneous land surfaces, J. Atmos. Sci., 65, 3263–3277, 2008.

Wolters, D.: Effects of soil moisture gradients on the path and the intensity of a West African squall line, Q. J. Roy. Meteor. Soc., 136, 2162–2175, 2010.

Appendix A: Additional material relating to Sect. 4.3

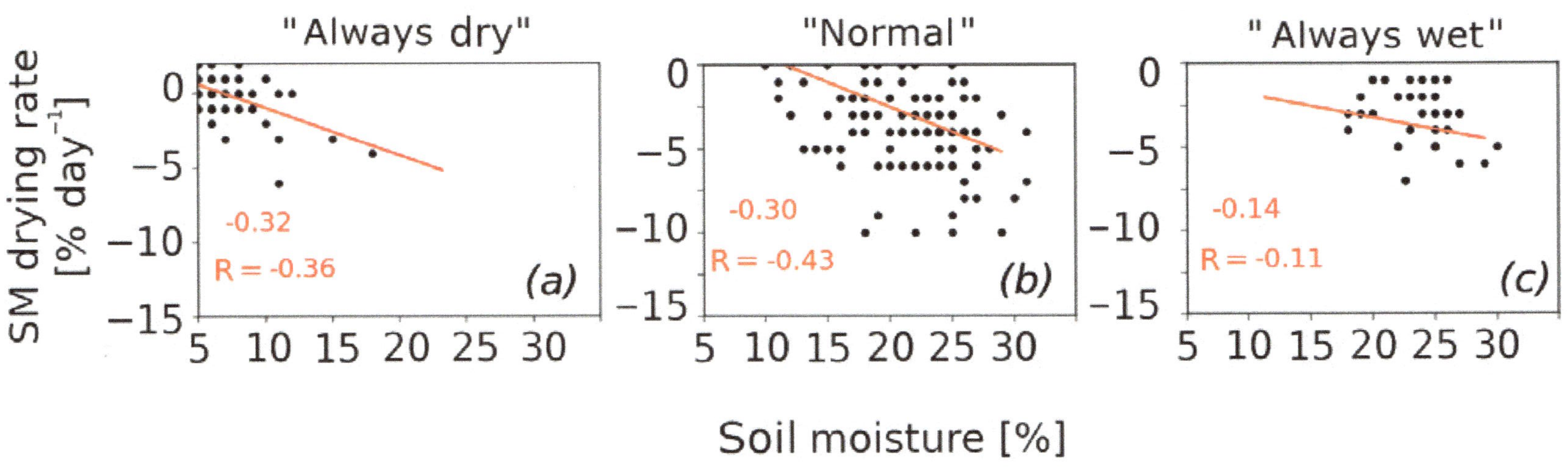

Figure A1. Examples of three possible relationships between cli- matology of soil moisture drying rate and absolute soil moisture in L_{min} locations. The red numbers indicate values of a slope and Spearman correlation respectively.

A1 Method used for identification of potentially flooded locations

The following paragraph describes the methodology used to identify events in which L_{min} locations are likely flooded. Following this methodology, a climatology of soil moisture drying rates is computed in every L_{min} location first. Climatology is calculated from values measured in the same L_{min} location, in the same month as the event but during the non-event years. Soil moisture drying rate is computed as the difference between soil moisture at 13:30 LST on the event day 0 and the previous day −1. Days with non-zero precipitation between two soil moisture measurements are excluded.

As a next step, we compute a climatology of soil moisture values on the day −1 to estimate a potential range of soil moisture conditions in every L_{min} location. Finally, for every L_{min} location we identify a linear regression function which fits best into a scatter plot relationship between drying rates and initial soil moisture values. Based on the slope of the linear regression and a climatological range of drying rates we stratify L_{min} locations as being potentially "always dry", "normal" or "always wet" (Fig. A1). Locations which are potentially always dry may be representative of rocky sand areas and hence will most often show low soil moisture values and drying rates close to zero. The cases of soil being always wet should indicate mostly high soil moisture values, but at the same time small drying rates, on the condition that a water supply is present. The "normal" case is expected to show a clear relationship between the drying rate and initial soil moisture content. The higher the soil moisture is, the larger the drying rate is expected to be.

Because the identification procedure requires selection of thresholds, the distribution of the potentially flooded locations will slightly vary depending on the selected threshold of the regression slope, minimum drying rate value or minimum absolute soil moisture. For the calculation of the potential wetland locations presented in Fig. 6c we chose two thresholds: the slope had to be larger than −0.15 and the soil moisture values had to be larger than 20 %.

Table A1. Percentage of 5° × 5° grid boxes with significantly negative ($P_e < 10\,\%$) and positive ($P_e > 90\,\%$) spatial SMPC over the North African domain in this study and previous studies of T12 and G15. Different data set combinations used in T12 and G15 are listed. "Merged" represents an integrated product composed of grid boxes in which either the AMSR-E or the ASCAT soil moisture data set was identified to be the best following a quality-control check. Following visual inspection, the experiments in which a significant negative SMPC relationship exists in the western region of the Sahelian domain are indicated with a * sign. For more details on various data set combinations, the reader is referred to the original papers.

No.	Experiment	Frac. $P < 10\,\%$ (%)	Frac. $P > 90\,\%$ (%)
1*	This study (trmm-amsre)	72	0
2*	T12 cmorph-amsre	56	0
3	T12 cmorph-ascat	36	0
4*	T12 cmorph-merged	50	0
5*	T12 trmm-merged	61	0
6*	T12 persiann-merged	42	0
7*	G15 trmm-amsre	36	3
8	G15 persiann-amsre	14	25
9*	G15 cmorph-amsre	33	3
10*	G15 cmorph-gleam(c)	36	3
11*	G15 trmm-gleam(c)	50	3
12*	G15 persiann-gleam(c)	36	6
13*	G15 cmorph-gleam(t)	67	0
14*	G15 trmm-gleam(t)	69	3
15*	G15 persiann-gleam(t)	39	0
16	G15 cmorph-gleam(p)	17	6
17*	G15 trmm-gleam(p)	36	0
18	G15 persiann-gleam(p)	14	17
	Mean	42	4
	SD	10	1

3

Effects of a deep-rooted crop and soil amended with charcoal on spatial and temporal runoff patterns in a degrading tropical highland watershed

Haimanote K. Bayabil[1], **Tigist Y. Tebebu**[1], **Cathelijne R. Stoof**[1,2], **and Tammo S. Steenhuis**[1,3]

[1]Department of Biological and Environmental Engineering, Cornell University, Ithaca, New York, USA
[2]Soil Geography and Landscape Group, Wageningen University, the Netherlands
[3]School of Civil and Water Resources Engineering, Bahir Dar University, Bahir Dar, Ethiopia

Correspondence to: Tammo S. Steenhuis (tss1@cornell.edu)

Abstract. Placement and hence performance of many soil and water conservation structures in tropical highlands has proven to be challenging due to uncertainty of the actual location of runoff-generating areas in the landscape. This is the case especially in the (sub-)humid areas of the Ethiopian highlands, resulting in limited success of such conservation measures. To improve understanding of the effect of land use on spatial and temporal runoff patterns in the Ethiopian highlands, we monitored runoff volumes from 24 runoff plots constructed in the 113 ha Anjeni watershed, where historical data of rainfall and stream discharge were available. In addition, we assessed the effectiveness of charcoal amendment of the soil and crop rooting depth in reducing runoff, and we compared the effect of lupine (a deep-rooted crop) to that of barley. We also measured daily rainfall, surface runoff, and root zone moisture contents during the monsoon seasons of 2012 and 2013 (with all plots being tilled in 2012, but only barley plots tilled in 2013). In addition, we analyzed long-term surface runoff from four plots, and outlet discharge data from the research site (1989–1993) were analyzed and compared with our observations. Results showed that the degrees of soil degradation and soil disturbance (tillage) were significant factors affecting plot-scale runoff responses. As expected, runoff was greater from more degraded soils. Overall, under the commonly applied lupine cropping practice, runoff was higher than under the commonly applied barley cropping practice. In particular, considerable difference was observed during smaller rainfall events (approximately <20 mm) in 2013, when lupine plots (non-tilled) had greater runoff than barley plots (tilled). Charcoal tended to decrease runoff, but results were not significant.

1 Introduction

Performance of many soil and water conservation structures in the tropical highlands has proven to be challenging due to uncertainty of their optimal placement. Ideally, the location of conservation structures should be directly related to where runoff is generated in the landscape. Evaluating the effectiveness of landscape modifications is especially timely in the Ethiopian highlands where the Ethiopian government is implementing land management practices to both increase rainwater productivity in the degrading landscape and increase the life of hydroelectric power plants such as the Grand Ethiopian Renaissance Dam on the Blue Nile near Sudan (Chen and Swain, 2014; Dagnew et al., 2015). The ultimate goal of these actions is to increase prosperity and ensure food security for the rapidly increasing population (Hurni, 1988; Nyssen et al., 2009). Most areas in the Ethiopian highlands receive high amounts of annual precipitation, aiding soil leaching and promoting land degradation; however, water scarcity is common for 8–9 months every year (Amsalu and Graaff, 2006; Bewket and Sterk, 2005; Biazin et al., 2011; Hugo et al., 2002). Rainfall distribution is variable not only spatially, but also temporally (Biazin et al., 2011; Bitew et al., 2009; McHugh et al., 2007). To counteract this problem of periodic water scarcity, soil and water conservation practices are ubiquitous in the Ethiopian highlands. However, surprisingly, most non-traditional soil and water conservation practices are ineffective because their placement neither addresses drivers of runoff nor considers spatial and temporal variation of runoff in a landscape.

Planning of effective soil and water management measures requires knowledge of dominant runoff-generating mechanisms and their controlling factors (e.g., land use, topography) (Orchard et al., 2013). There are two mechanisms of surface runoff generation: (1) Hortonian overland flow or infiltration excess surface runoff that occurs when rainfall intensity exceeds the infiltration capacity of the soil, and (2) saturation excess surface runoff that occurs when the (perched) water table rises, saturating the whole soil profile. However, there is still a lack of agreement regarding the nature of the runoff initiation mechanisms and their controlling factors in the Ethiopian highlands. Previous studies highlighted saturation excess as the dominant runoff mechanism (Bayabil et al., 2010; Steenhuis et al., 2009; Tilahun et al., 2013, 2014). A field study by Bayabil et al. (2010) found that in the Maybar watershed, with highly conductive soils, saturation excess runoff was mainly driven by topography, with water channeling through the hillsides as interflow, saturating the lower-lying fields. This is in line with findings from the Debra Mawi watershed in the northern Ethiopian highlands, where saturated lower-lying fields contributed most of the surface runoff (Tilahun et al., 2013). This strong evidence for saturation excess runoff being the driver of overland flow in the Ethiopian highlands is in contrast to findings from Bewket and Sterk (2005) and Taddese (2001). A study by Bewket and Sterk (2005) in the Chemoga watershed located in the Blue Nile basin, like the Anjeni and Debre Mawi watersheds, found that infiltration excess runoff mechanism was dominant mainly based on analysis of the hydrograph at the outlet focusing on land use change. Land use is important because it affects soil infiltration capacity. For example, several studies reported that land use change from natural vegetation to agricultural lands increased overland flow during the rainy monsoon phase and reduced baseflow during the dry phase in this region (Bewket and Sterk, 2005; Feoli et al., 2002; Taddese, 2001; Zeleke, 2000). In other countries as well, clearing of forests resulted in decreased infiltration rates and lower percolation to the sub-soils (Hanson et al., 2004; Mendoza and Steenhuis, 2002; Nyberg et al., 2012; Shougrakpam et al., 2010). Identification of the dominant runoff mechanism in relation to not only topography but also land use in the Ethiopian highlands is therefore essential for the development of effective soil and water conservation methods in this region.

On degraded fields with poor soil infiltration capacity, management practices should aim at improving infiltration rates. This can be done by restoring the soil macropore network by improving soil organic carbon pools, or by disturbing the soil profile either physically (tillage) or biologically (using deep-rooted crops). Deep-rooted crops can penetrate through the soil profile and thereby increase soil conductivity (Angers and Caron, 1998; Cresswell and Kirkegaard, 1995; Lesturgez et al., 2004; Meek et al., 1992). Moreover, upon decomposition of these roots, channels and biopores are created that could provide a network of macropores with considerable vertical and lateral conductivity (Yunusa and Newton, 2003).

Another solution for improving soil physical and hydraulic properties is by increasing the organic carbon pool through the addition of biochar or charcoal (Abel et al., 2013; Asai et al., 2009; Bayabil et al., 2015; Glaser et al., 2002; Kameyama et al., 2010; Karhu et al., 2011; Laird et al., 2010). Biochar and charcoal incorporation have been reported to improve soil bulk density (Abel et al., 2013; Laird et al., 2010), porosity (Abel et al., 2013; Atkinson et al., 2010), and hydraulic conductivity (Asai et al., 2009). Although both biochar and charcoal soil amendments can be effective in improving soil hydraulic properties, Bayabil et al. (2015) argued that charcoal is a more viable solution for rural Africa because it is widely produced in most rural areas of Africa (Lehmann et al., 2006) and is therefore more accessible to smallholder farmers than biochar. The analysis above shows that deep-rooted crops and additions of charcoal could ameliorate soil and water losses in a degrading landscape. However, field research on the effectiveness of these two management practices in a tropical highland setting with monsoon rainfalls, to our knowledge, does not exist.

The objective of this study was, therefore, to investigate spatial and temporal rainfall–runoff relationships in the Ethiopian highlands by investigating the effects of soil degradation status and landscape position. For this, soil degradation status was experimentally changed by adding biochar and growing a deep-rooted lupine crop.

The research was carried out in the Anjeni watershed in the Ethiopian highlands in 2012 and 2013. Twenty-four runoff plots were established along three transects going upslope in sets of three in each landscape position. Each set of three plots had one plot in which lupine was planted and two plots with barley – of which one was amended with charcoal.

2 Materials and methods

2.1 Study site

The Anjeni watershed is situated in the northwestern part of Ethiopia (Fig. 1), and was selected because of the availability of historical discharge records at the outlet and from runoff plots inside the watershed. The watershed has a drainage area of 113 ha and is one of the experimental watersheds established under the Soil Conservation and Research Program (SCRP) of the Ministry of Agriculture of Ethiopia in collaboration with the Swiss Agency for Development and Cooperation (SDC; Hurni et al., 2005). Its gauging station is located at 10°40′ N, 37°31′ E.

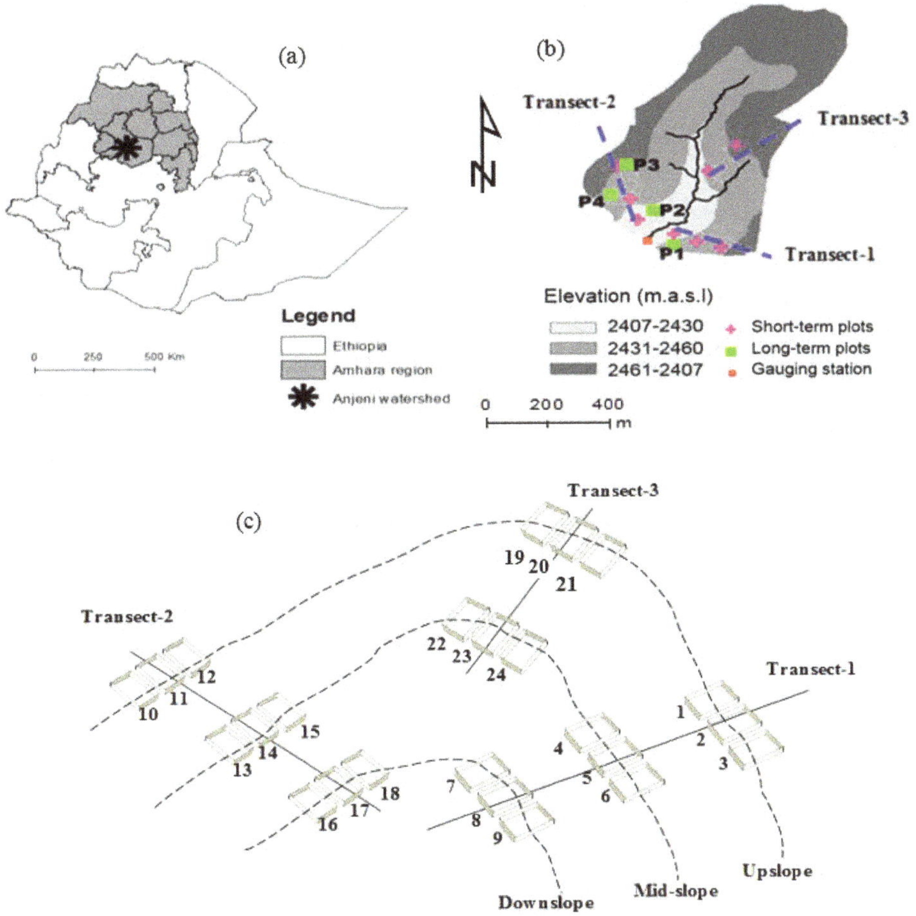

Figure 1. Location of the Anjeni watershed in the Amhara region in Ethiopia **(a)**, with the location of downslope transects and runoff plots indicated in **(b)** and **(c)** (not to scale). Dashed lines in **(c)** are elevation contours. Three treatments were applied: barley without soil amendment (control) was grown on plots 2, 4, 7, 12, 15, 18, 21, and 23; barley with charcoal amendment was grown on plots 1, 6, 8, 11, 13, 17, 20, and 22; and lupine without soil amendment was grown on plots 3, 5, 9, 10, 14, 16, 19, and 24. Soil and spatial attributes of plots are presented in Table 1.

The watershed has a unimodal rainy season that lasts from mid-May to mid-October, with a mean annual rainfall of 1690 mm yr^{-1}. The topography of Anjeni is typical of Tertiary volcanic landscapes: it has been deeply incised by streams, resulting in the current diversity of landforms (SCRP, 2000) with elevation between 2407 and 2507 m a.s.l. (Herweg and Ludi, 1999). The soils of Anjeni have been developed from the basaltic Trapp series of Tertiary volcanic eruptions and are similar to most parts of central Ethiopia, with the dominant soils being Alisols (41.5 ha), Nitisols (23.8 ha), Cambisols (18.9 ha) and Regosols (10 ha) covering more than 80 % of the watershed (Fig. S1 in the Supplement; SCRP, 2000; Zeleke, 2000). The deep Alisols cover the bottom part of the watershed; moderately deep Nitisols cover the mid-transitional, gently sloping parts of the watershed, while the shallow Regosols and Leptosols cover the high, steepest part of the watershed (Zeleke, 2000). Fields are intensively cultivated for crop production and a large proportion of the watershed is degraded (SCRP, 2000). In 1986, graded *fanya-juu* structures were installed for soil and water conservation, resulting in terraces across the landscape (SCRP, 2000).

2.2 Experimental setup

We studied the effect of land use and soil management on runoff patterns using 24 runoff plots installed across the watershed, accounting for spatial variability in soil degradation status and slope position (Table 1, Fig. 1). We also assessed effects of charcoal amendment and crop rooting depth along three transect locations. The 24 plots were positioned in groups of three along three transects perpendicular to the slope (Fig. 2). Soil degradation varied between transects: Transects 1 and 2 are located in the southeastern and southwestern parts of the watershed (Fig. 1b), and have deep soils, while Transect 3, located between Transects 1 and 2, is characterized by shallow and degraded soils. Transects 1 and 3 are steep (with slopes approximately 14.5 and 15.6 %, re-

Figure 2. Groups of three runoff plots setup in the downslope position along Transect 2. Water storage tanks are positioned below the plots, on the downslope side of the terrace edge. Dark brown lines above runoff plots are traditional conservation practices (drainage ditches) constructed by farmers to channel out excess water from fields.

spectively), while Transect 2 has a moderate slope (11.8 %). Effects of landscape position were assessed by placing plots at different slope positions: at downslope, mid-slope, and upslope positions along Transects 1 and 2; and in the two upper positions along Transect 3 (Fig. 1c). A randomized block experimental design type using transects as blocking factors was used during installation of plots, with the effect of charcoal and a deep-rooting crop assessed in every landscape position.

At the start of the 2012 growing season (June), all plots were plowed and, in each landscape position (Fig. 1), two plots were seeded with barley. Effects of charcoal amendment were assessed by amending one of the barley plots with charcoal during plowing, the non-amended barley plot serving as a control treatment. Effects of crop rooting depth were assessed by seeding the third plot at each transect location with the deep-rooted lupine (*Lupineus* albus L.) crop, with the non-amended barley plot again serving as a control treatment. Barley and lupine crops were assigned randomly to plots, and the same crop was maintained on each plot for 2 years (2012 and 2013). These crops were chosen as they are widely grown throughout the Ethiopian highlands. Farmers grow lupine as intercrop with cereals (e.g., barley and wheat) or as the sole crop on marginal lands without additional farm inputs. Barley has a fibrous root system, while lupine has a deep-rooted system and is widely grown on marginal lands for its leguminous nature. Because of their contrasting root architecture, lupine and barley are expected to have contrasting effects on soil hydraulic properties.

2.3 Agronomic practices on plots

Barley, one of the predominantly grown crops in the watershed (SCRP, 2000), was grown following local farmers' cultural practices, and thus barley plots were tilled in both 2012 and 2013. While lupine seed beds are typically not tilled, tillage was done in 2012 because the plots were originally designated to be sown with alfalfa, another deep-rooted crop though one that is always tilled. When the alfalfa did not establish itself successfully, lupine was sown on the tilled soil shortly after. The next growing season, in 2013, only barley plots were tilled and seeded, while lupine seeds were seeded on untilled plots (the more common practice in the area). Also in line with farmer practices, all barley plots were fertilized with 100 kg ha^{-1} di-ammonium phosphate (DAP; 46 % nitrogen, 23 % phosphorous, and 21 % potassium) during seeding, and 100 kg ha^{-1} of urea (100 % nitrogen) 1 month after sowing. Lupine plots were not fertilized. Both fertilization and tillage are different for lupine and barley treatments during the 2-year study period (2012 and 2013). To distinguish crop effects (barley and lupine crops grown under common practices) from tillage effects, we therefore analyzed data from the 2-year study separately.

On charcoal-amended barley plots, charcoal was applied at a fixed rate of 12 ton / ha during tillage in 2012 and 2013. Charcoal (prepared from *Eucalyptus camaladulensis* biomass in a way similar to that described by Bayabil et al., 2015) was manually crushed to obtain a relatively uniform particle size (2 mm diameter) and then manually mixed with the top 20 cm of the soil.

2.4 Plot installation and data collection

While crop and charcoal treatments were applied to 9 m^2 (3 m wide, 3 m long) areas, runoff was only measured on 4.5 m^2 plots (1.5 m wide, 3 m long) inside these areas, to allow for auxiliary measurements (e.g., soil moisture content) to be taken adjacent to instead of inside the runoff plots and thereby avoid trampling and soil disturbance inside the plots. For this, runoff plot boundaries were installed 0.75 m inside the seeded area from both sides. As illustrated in Fig. 2, all runoff plots were constructed at the level bottom ends of terraces. The plot boundaries consisted of 50 cm high metal sheets of which 25 cm were below ground and 25 cm were above ground, and the lower plot boundaries were reinforced with concrete. A 5 cm diameter PVC pipe carried surface runoff into primary collection tanks (76 L volume). When the primary tanks were full, excess water flowed through divisor slots directing one-tenth (10 %) of the excess flow into secondary tanks (76 L volume). The tanks were made from barrels cut in half and were covered on the top to minimize evaporation and prevent rainfall entry.

All plots were monitored manually for runoff on a daily basis during the monsoon season (from 29 June to 4 October 2012 and from 25 June to 8 October 2013). When runoff occurred, the depth of water in the two tanks was measured and then the water was drained out through valves fitted at the bottom of the tanks. Daily rainfall totals were measured using a manual rain gauge installed at the weather station situated in the watershed (see Fig. 1b, "Weather station"). In

Table 1. Spatial attributes and soil properties of plots.

Transect	Position	Plots	Elevation (m a.s.l.)	Slope (%)	Sand	Silt (%)	Clay	OM[1]	BD[2] ($g\,cm^{-3}$)	D[3] (m)
One	Upslope	1, 2, 3	2438	3.0	24.8	35.4	39.8	2.2	1.1	1.15
	Mid-slope	4, 5, 6	2431	2.5	31.7	28.0	40.3	2.1	1.1	1.22
	Downslope	7, 8, 9	2411	1.5	23.6	36.7	39.6	2.2	1.1	> 1.3
Two	Upslope	10, 11, 12	2461	2.5	23.8	32.2	44.0	2.1	1.1	0.84
	Mid-slope	13, 14, 15	2426	2.0	17.8	39.0	43.2	2.4	1.2	1.09
	Downslope	16, 17, 18	2415	1.0	24.7	36.3	39.0	2.4	1.3	> 1.3
Three	Upslope	19, 20, 21	2455	3.0	21.0	37.7	41.4	1.3	1.4	0.33
	Mid-slope	22, 23, 24	2438	2.0	30.6	37.4	32.0	1.4	1.3	0.72

[1] OM: organic matter; [2] BD: bulk density; and [3] D: soil depth.

addition, during the 2013 growing period, soil moisture content, θ ($g\,g^{-1}$), was measured gravimetrically by taking bulk soil samples from the top 20 cm depth at 10-day intervals. To prevent disturbance, samples were taken inside the seeded area but just outside each runoff plot.

2.5 Long-term plot runoff and river discharge data

In addition to runoff data from the 24 newly installed plots, we obtained long-term data from the Amhara Regional Agricultural Research Institute (ARARI). The data consist of runoff from four long-term 3 m^2 plots (3 m length, 1 m width; Fig. 1b, "Permanent plots") and discharge at the outlet of the watershed (Fig. 1b, "Gauging station"). To place our newly installed plot-scale runoff observations into a broader and longer-term context, we compared our data with historical plot-scale runoff data available in the watershed for the years 1989 through 1993. These data were measured on the four 3 m^2 plots that had slopes of 12, 16, 22, and 28 %. The 16 % sloped plot was on grassland, while the other three plots were cultivated with food crops (e.g., barley and wheat; SCRP, 2000). Discharge was measured continuously since 1984 (2 years before the installation of the *fanya juu* conservation structures) as part of the ongoing hydrological and erosion monitoring activities (SCRP, 2000), and we used discharge data for the 2012 and 2013 monsoon seasons to compare our plot-scale observations with watershed-scale patterns. Rainfall data obtained from the watershed (Fig. 1b, "Weather station") were available for the same period.

2.6 Data quality control and aggregation

To make sure that peaks of daily rainfall and runoff coincided, we checked all daily data visually and by calculating the daily runoff coefficients (R_{coef}) using Eq. (1):

$$R_{coef} = \frac{\text{runoff}}{\text{rainfall}}, \quad (1)$$

where runoff is daily runoff ($mm\,day^{-1}$) and rainfall is daily rainfall amount (mm).

Plot-scale rainfall–runoff data (Figs. S2–S4 in the Supplement) showed that there were 214 events (spread over 11 days in 2012 and 32 days in 2013) out of 5232 events in total (i.e., 4.1 % of the total) where daily runoff was greater than the rainfall amount recorded on the same day (i.e., $R_{coef} > 1$). In some cases, large rainfall events were visible that did not produce runoff on the same day, but for which peak runoff appeared on the following day. In other cases, there was more runoff than rainfall without delays (see spikes of blue, green, and red lines in Figs. S2–S4 in the Supplement). Runoff in excess of rainfall can be caused by rainfall and runoff measurement periods that do not coincide. Here, rainfall was measured at 08:00 every day. The first of the 24 runoff plots was also measured at 08:00, but emptying the barrels and scooping out the sediment is time consuming, causing the last plot to be emptied around noon. Rainfall and runoff periods therefore did not exactly coincide, which likely caused problems on days on which rainfall occurred between 08:00 and 12:00 (noon). Other potential causes of runoff exceeding rainfall are high spatial variation in rainfall that is not picked up by our single rain gauge, and interflow from outside the plot entering the plot during large rainstorms.

To reduce the impact of delayed peak runoff, we therefore decided to aggregate rainfall and runoff data over a 3-day period, resolving most of the high runoff coefficients. However, 47 events (2.6 % of total) observed on the 24 plots and recorded on 11 observation days spread over the 2-year study period were left with $R_{coef} > 1$ (Fig. S5 in the Supplement). Further data aggregation, even on a weekly interval, did not resolve the issue. One of the options to deal with such outlier data points would be to exclude observations from data analysis. However, to avoid bias between treatments and spatial locations, all observations from those 11 days would need to be discarded for all (24) plots, which would result in discarding 264 observations. Losing this many observations (14.9 % of a total of 1777 3-day observations) would considerably re-

Table 2. Summary of total runoff during the year ("Total runoff"), average 3-day runoff in mm ("Mean runoff") and standard error (SE*) in mm average of the plots in the transect for 2012 and 2013. The annual precipitation in 2012 and 2013 was 1036 and 1528 mm, respectively.

Plot	Rainfall/runoff (mm)	2012 Transect			2013 Transect		
		1	2	3	1	2	3
Control	Total runoff	438	484	598	300	367	482
	Mean runoff	13	15	18	9	10	14
	SE	2.8	2.5	3	1.7	2.7	2.5
Charcoal	Total runoff	365	468	513	282	365	397
	Mean runoff	11	14	15	8	10	11
	SE	2.2	2.5	2.6	2.1	3	2.3
Lupine	Total runoff	509	495	583	732	670	747
	Mean runoff	16	15	18	21	19	21
	SE	2.8	2.4	2.5	3.3	3.4	3.1
Rainfall	Total rainfall		1036			1528	
	Mean rainfall		32			43	
	SE		0.5			0.8	

* SE: standard error of the mean.

duce the power of our analysis. Thus, to achieve a balance between the number of runoff events remaining for analysis and the objective of analyzing large runoff events, the remaining high runoff events ($R_{\text{coef}} > 1$) after data aggregation on 3-day intervals were therefore assigned a maximum value that equals the 3-day rainfall amount – resulting in a runoff coefficient of 1. As such, adjusted 3-day aggregate runoff data were used for all statistical data analyses in this paper.

In addition to this analysis of runoff coefficients, to assess the differences in soil water storage between plots, the SCS curve number was fitted to 3-day rainfall and 3-day adjusted runoff data for each treatment type and cropping year using Eq. (2). The SCS equation was effectively used in predicting rainfall–runoff relationships in the Ethiopian highlands (Tilahun, 2012) and for different regions in the USA and Australia (Steenhuis et al., 1995). The SCS curve number equation (Rallison, 1980) is represented as

$$Q = \frac{(P_e - 0.2S)^2}{(P_e + 0.8S)}, \qquad (2)$$

where Q is 3-day runoff (mm), P_e is 3-day rainfall (mm), and S (mm) is potential maximum soil storage.

2.7 Statistical analysis

Data analysis aimed at detecting differences in runoff response between land uses and spatial locations (transects and elevation ranges) during the 2-year study period. Statistical data analysis was performed using R (R Development Core Team, 2014). To determine the effect of charcoal amendment and deep-rooted lupine as well as spatial location with different soil degradation levels (transects) and slope position, a linear mixed effect model was fitted using the nlme package in R. In this model, crop type, slope position, and transect were used as fixed factors, and individual plots as random factors. For fixed factors with significant effects, post hoc mean comparison tests were performed using the lsmeans package in R to identify group pairs with significant difference.

3 Results and discussion

3.1 Plot-scale rainfall–runoff response and effect of charcoal amendment and deep-rooted lupine

The adjusted runoff depths during the monsoon seasons of 2012 and 2013 are shown in Fig. 3 for all eight groups of plots along the three transects. In 2013, runoff response from lupine plots was considerably greater than barley plots, while in 2012, runoff tended to be more or less similar for all treatments. In addition, a summary of observed rainfall data recorded is presented in Table 2. Average monthly rainfall in 2012 was similar to the 5-year average (based on 1989–1993 observations; Fig. S6 in the Supplement), while in 2013 it exceeded the 5-year average.

As discussed in the "Methods" section, runoff exceeding rainfall (i.e., $R_{\text{coef}} > 1$), as shown in Figs. S2–S4 and S5 in the Supplement, is not expected and worrisome. We therefore checked historical long-term data (1989–1993) from four permanent plots (3 m length, 1 m width) measured by the well-trained technicians at the experimental station, and found the same "problem" that in many cases there was more runoff than rainfall (Fig. 4a). This indicates that our daily observations with $R_{\text{coef}} > 1$ (Fig. 4b) are real and not caused by measurement errors. This phenomenon of runoff exceeding rainfall has not been reported often for temperate climates, and it is therefore likely that rainfall in monsoon climates is more variable over short distances than rains in temperate climates. Studies found that rainfall in the Ethiopian highlands significantly varies in space (Bewket and Conway, 2007; Bitew et al., 2009). Bitew et al. (2009) observed an up to 424 % coefficient of variation of daily rainfall between rain gauges. These authors further noted that in areas with complex topography (like the Anjeni watershed), extrapolation of point rainfall observations to larger scales could be less accurate.

3.2 Plot runoff and outlet discharge

All plots on degraded soils along Transect 3 produced significantly greater runoff than plots along the other two transects with relatively deeper soils (Fig. S7 in the Supplement). While we expected slope position to affect runoff, results from the linear mixed effects model showed that plot-scale runoff responses between slope positions were not significant. Because of this, 2012 and 2013 runoff responses of barley (both control and charcoal amended) and deep-rooted lupine were grouped by transect and then compared. Statistical test results showed that, for all transects, lupine plots produced significantly more runoff than both the control and charcoal-amended barley plots. Charcoal amendment, on the other hand, caused no significant effects (Fig. 5). The cumulative runoff for the lupine plots followed the cumulative runoff for the outlet more than the barley plots, particularly in 2013 (Fig. 6).

Comparison of plot-scale cumulative runoff (colored lines, Fig. 6) and cumulative river discharge observed at the watershed outlet (black line, Fig. 6) with cumulative rainfall indicated that approximately 100 mm of cumulative rainfall was needed before runoff was initiated from all plots. In general, during the start of the monsoon season (until 500 mm cumulative rainfall in Fig. 6), plot-scale runoff response generally exceeded watershed-scale discharge response. In agreement with this, Mutema et al. (2015), a study in South Africa, observed a significant reduction in unit runoff as plot size increased from micro-plots ($1\,m^2$) to plots ($10\,m^2$) and subsequently to mico-catchments ($0.23\,km^2$). Similarly, van de Giesen et al. (2011) reported that runoff decreased with increases in plot size. Nevertheless, as the rainy season progressed, starting from the middle of August and at approximately 500 mm cumulative rainfall, watershed-scale discharge started to exceed plot-scale runoff depths (with the exception of the lupine plots in 2013; see below). The difference between plot-scale runoff and outlet discharge during the onset of the monsoon season indicates that detention storage at a watershed scale occurs, while the difference between the plot and watershed scale later in the monsoon season is caused by baseflow at the watershed outlet. This is consistent with previous observations by Tilahun et al. (2013, 2014) and

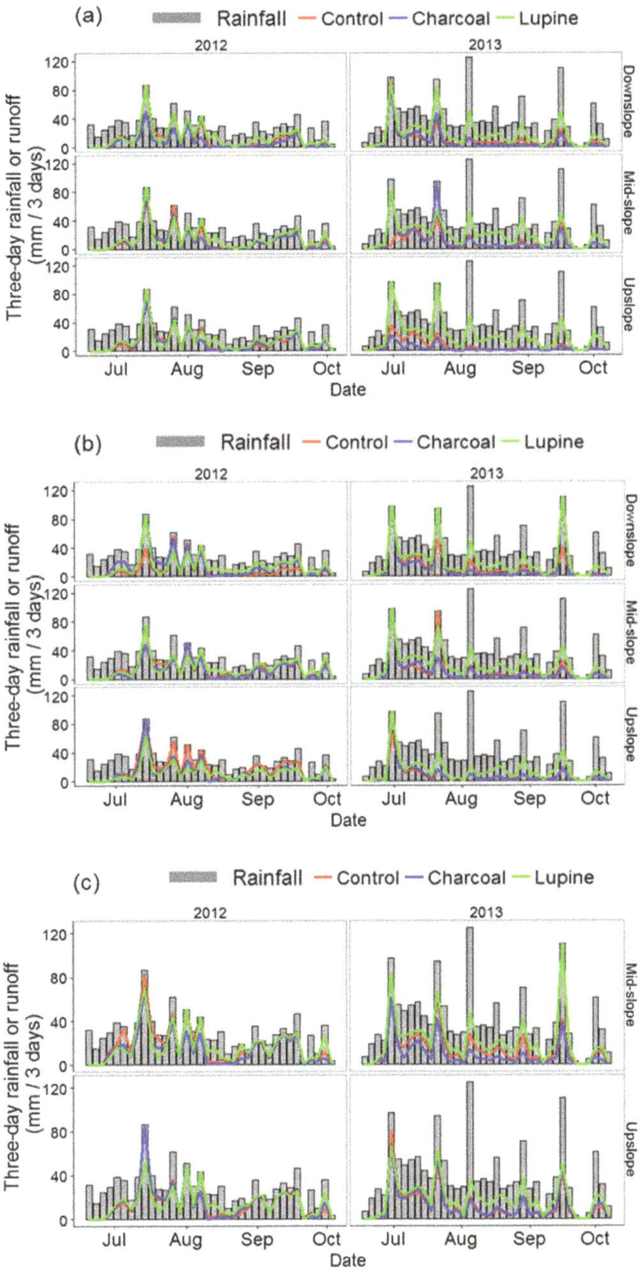

Figure 3. Three-day rainfall and adjusted 3-day runoff depths (aggregated over 3 days) from individual plots in different slope positions along Transect 1 **(a)**, Transect 2 **(b)**, and Transect 3 **(c)**.

Bayabil et al. (2010), who observed that initially, the runoff from the hillsides infiltrates to a lower slope position as interflow, while later in the season these bottom lands start to contribute both subsurface flow and surface runoff.

A considerable difference in the runoff response of barley and lupine plots was observed between the monsoon seasons of 2012 and 2013. In 2012, runoff tended to be more or less similar for all treatments, whereas in 2013 runoff from barley and lupine plots began to deviate after approx-

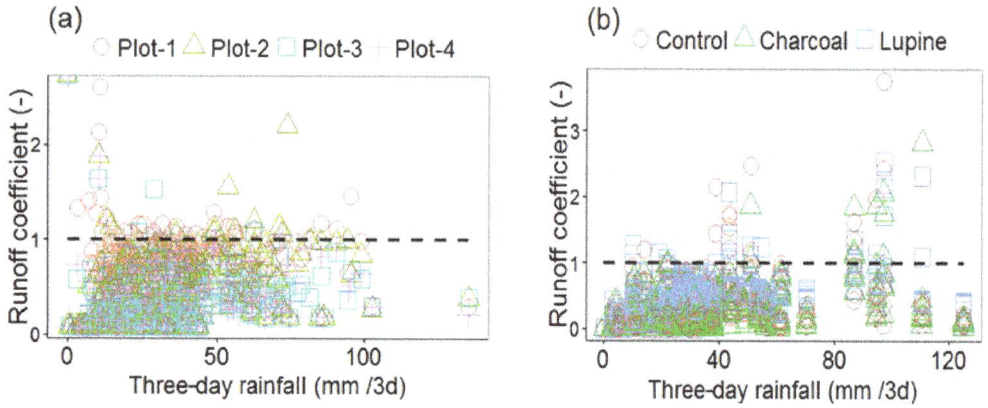

Figure 4. Runoff coefficients computed from observations from long-term monitoring plots (3 m length, 1 m width) **(a)** and plots in 2012 and 2013 **(b)**. The black-dashed horizontal line represents $R_{\text{coef}} = 1$.

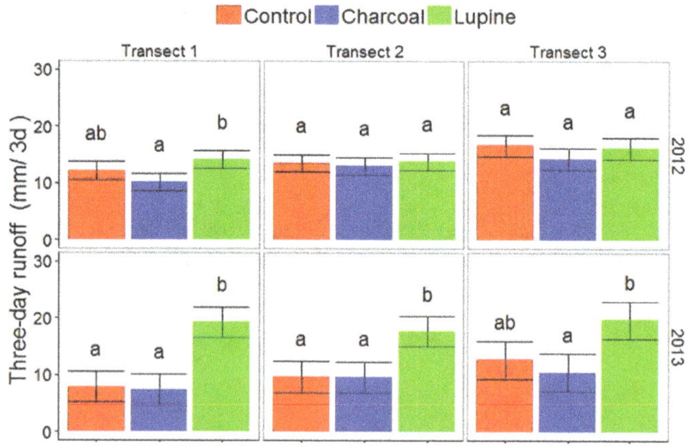

Figure 5. Effect of charcoal amendment and deep-rooted lupine crop on plot-scale runoff (3-day total) for each transect and year. Treatments not sharing the same letter within an individual transect for a given year are significantly different at $p < 0.05$.

imately 250 mm cumulative rainfall (Fig. 6). In agreement with this, a closer look at the plots (Fig. 3) clearly shows that for most of the high rainfall amounts, there is little difference in runoff response between the barley and lupine plots. Only for smaller rain events (approximately <20 mm) and during the start of the 2013 rainy season (around 1 July) did runoff from lupine plots exceed that of barley plots. It is interesting that this is the case for all three transects in 2013 but does not occur in 2012. The only management difference between these 2 years is that lupine was tilled in 2012 but not in 2013. This implies that tillage resulted in relatively greater soil water storage for lupine plots, and that the difference in rainfall–runoff response between these crop treatments in 2013 may be ascribed to the fact that barley plots were tilled and lupine plots were not. Soil water storage estimated by fitting the SCS-CN equation (Rallison, 1980) confirmed smaller storage for lupine than for barley (Fig. 7). This would mean that there is very little infiltration in the lupine plots other than to refill the water abstracted by the lupine for evapotranspiration.

These findings indicate that both soil degradation status (soil depth) and disturbance (tillage) are important factors affecting rainfall–runoff relationships in the landscape. In addition to tillage activities, inherent differences in plant root morphology (e.g., length and density) between the barley and lupine could likely be another factor. Most of the root masses of barley are located at shallow depths in the upper part of the soil profile (Lugg et al., 1988) and thereby take water from the top soil, whereas lupine roots grow deeper (Fig. S8 in the Supplement) than barley and extract water from deeper depths (French and Buirchell, 2005). These differences in root water uptake are somewhat visible in slightly greater, albeit not significant, root zone moisture readings (measured from the top 20 cm) observed for lupine plots beginning in August 2013 (Fig. S9 in the Supplement).

It is important to note that the fact that lupine did not decrease runoff during this study period does not imply that it would not reduce runoff in the long term. When the roots of lupine decompose, it is likely that biopores and channels would be created (as reported by Meek et al., 1992, and Lesturgez et al., 2004) and that the resulting high vertical and lateral continuity improves the network of macropores (Yunusa and Newton, 2003), which would result in reduced surface runoff and associated erosion.

4 Conclusion

We set out to investigate the factors that control runoff initiation by investigating the effects of soil degradation status, landscape position, and different land uses (barley with and without charcoal, and deep-rooted lupine crop) on spatial and temporal rainfall–runoff relationships. We observed and analyzed the discharge of 24 runoff plots installed in groups of three in three transects over a 2-year period. Each group con-

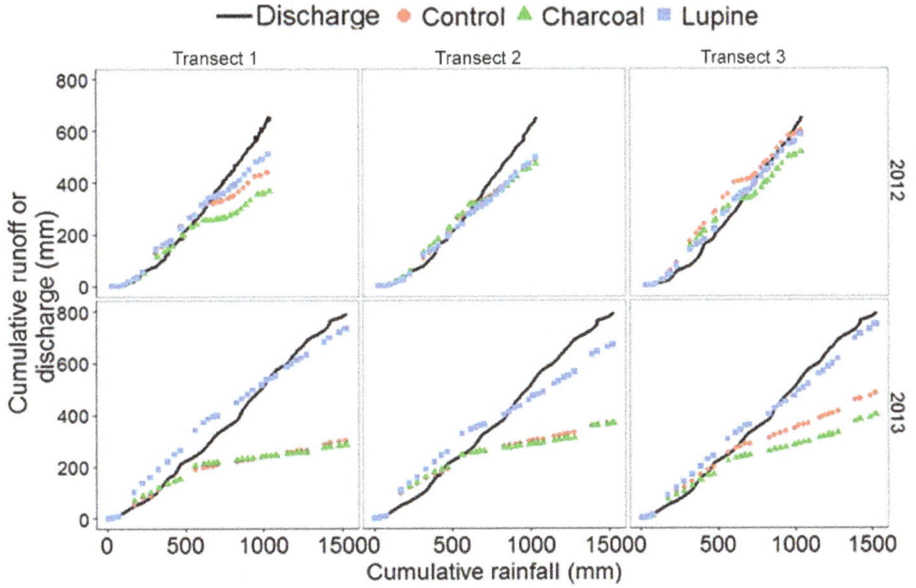

Figure 6. Cumulative rainfall vs. cumulative runoff (from control, charcoal, and lupine plots along three transects) and discharge at the watershed outlet, for 2012 and 2013.

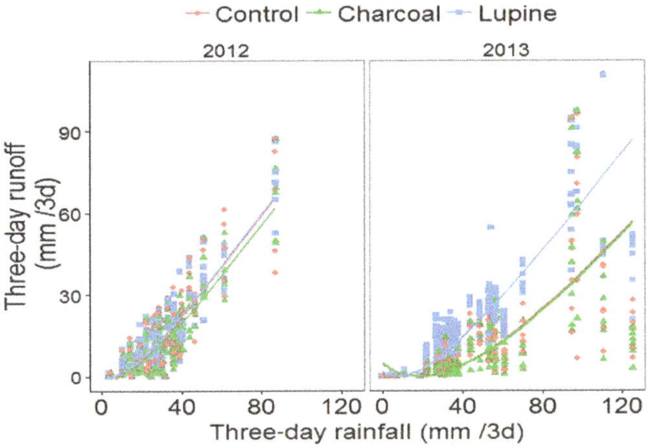

Figure 7. Effect of charcoal amendment and deep-rooted lupine on 3-day soil water storage: 3-day rainfall vs. 3-day runoff with SCS-CN fitted lines fitted using the SCS-CN equation (Rallison, 1980). Fitted 3-day storages values in 2012 were 22, 26, and 21 mm for control, charcoal, and lupine, respectively, and, in 2013, 93, 94, and 40 mm for control, charcoal, and lupine, respectively.

sisted of plots grown with lupine with no amendment, barley with no amendment, and barley with a charcoal amendment. Monsoonal rains are highly variable even over short distances, and in several cases there was more runoff from the plot than rainfall at the rain gauge. In general, we found that, first, watershed detention storage increased during the first half of the rainy phase, and plot-scale runoff depths exceeded those at watershed scale. The opposite was true later on in the rainy phase due to the occurrence of baseflow at the watershed outlet. Second, under the commonly applied cropping practices (tillage for barley, no tillage for lupine), runoff was greater for lupine than barley. Especially during small rainfall events (approximately <20 mm) in 2013, runoff from non-tilled lupine plots exceeded that of tilled barley plots. Charcoal amendment tended to decrease runoff, but results were not significant. Third, plot-scale rainfall–runoff relationships are greatly affected by root-zone soil water storage capacity, which is directly affected by a range of factors including soil degradation and the amount of water that can percolate to deeper soil layers, tillage practices and fertilization (that were different for lupine and barley treatments), and root morphology of crops (e.g., root length and density).

In the short term, the decreased soil water storage for lupine compared to barley crops in this region implies that lupine has a smaller rainfall threshold for runoff initiation. In the long term, however, lupine may have the potential to actually reduce runoff by improving infiltration rates through the creation of bio-pores once its large taproot decomposes (Fig. S8 in the Supplement). The long-term impact of lupine cropping on runoff processes therefore requires further investigation. Understanding the drivers of hardpan formation and permeability is essential for the development of management approaches that can effectively tackle hardpan occurrence and its hydrologic impacts, in order to ultimately reverse the land degradation trend and reduce erosion.

Our findings are in agreement with other studies that show that rainfall–runoff relationships at a small plot scale are different than at the outlet (e.g., Han et al., 2012; Stoof et al.,

2012), and that better use of green water (rainfall) for smallholder agriculture systems in the Ethiopian highlands could be achieved by decreasing runoff by increasing the storage of water in the root zone. However, more research has to be done on how best to achieve the latter.

Acknowledgements. This study was funded by the N. Borlaug Leadership Enhancement in Agriculture Program (LEAP) in cooperation with IWMI's East Africa office, and by a grant from USAID in cooperation with the Higher Education for Development (HED) office by Cornell's Bradfield Award Committee. The authors would like to thank Hans Hurni for his foresight and efforts in establishing the SCRP watershed sites in the 1980s. We also thank Molla Addisu, Debre Markos University, for his help in obtaining some of the materials used in the field. Birhanu Mehiretu (field technician in the Anjeni watershed) was very helpful during the research and field data collection.

Edited by: G. Jewitt

References

Abel, S., Peters, A., Trinks, S., Schonsky, H., Facklam, M., and Wessolek, G.: Impact of biochar and hydrochar addition on water retention and water repellency of sandy soil, Geoderma, 202–203, 183–191. doi:10.1016/j.geoderma.2013.03.003, 2013.

Amsalu, A. and Graaff, J.: Farmers' Views of Soil Erosion Problems and their Conservation Knowledge at Beressa Watershed, Central Highlands of Ethiopia, Agric. Hum. Values, 23, 99–108, doi:10.1007/s10460-005-5872-4, 2006.

Angers, D. A. and Caron, J.: Plant-induced changes in soil structure: Processes and feedbacks, Biogeochemistry, 42, 55–72, 1998.

Asai, H., Samson, B. K., Stephan, H. M., Songyikhangsuthor, K., Homma, K., Kiyono, Y., Inoue, Y., Shiraiwa, T., and Horie, T.: Biochar amendment techniques for upland rice production in Northern Laos: Soil physical properties, leaf SPAD and grain yield, Field Crops Res., 111, 81–84, 2009.

Atkinson, C. J., Fitzgerald, J. D., and Hipps, N. A.: Potential mechanisms for achieving agricultural benefits from biochar application to temperate soils: a review, Plant Soil, 337, 1–18, 2010.

Bayabil, H. K., Tilahun, S. A., Collick, A. S., Yitaferu, B., and Steenhuis, T. S.: Are Runoff Processes Ecologically or Topographically Driven in the (Sub) Humid Ethiopian Highlands? The Case of the Maybar Watershed, Ecohydrology, 3, 457–466, 2010.

Bayabil, H. K., Stoof, C. R., Lehmann, J. C., Yitaferu, B., and Steenhuis, T. S.: Assessing the potential of biochar and charcoal to improve soil hydraulic properties in the humid Ethiopian Highlands: The Anjeni watershed, Geoderma, 243–244, 115–123, doi:10.1016/j.geoderma.2014.12.015, 2015.

Bewket, W. and Conway, D.: A note on the temporal and spatial variability of rainfall in the drought-prone Amhara region of Ethiopia, Int. J. Climatol., 27, 1467–1477. doi:10.1002/joc.1481, 2007.

Bewket, W. and Sterk, G.: Dynamics in land cover and its effect on stream flow in the Chemoga watershed, Blue Nile basin, Ethiopia, Hydrol. Process., 19, 445–458, doi:10.1002/hyp.5542, 2005.

Biazin, B., Stroosnijder, L., Temesgen, M., AbdulKedir, A., and Sterk, G.: The effect of long-term Maresha ploughing on soil physical properties in the Central Rift Valley of Ethiopia, Soil Tillage Res., 111, 115–122, doi:10.1016/j.still.2010.09.006, 2011.

Bitew, M. M., Gebremichael, M., Hirpa, F. A., Gebrewubet, Y. M., Seleshi, Y., and Girma, Y.: On the local-scale spatial variability of daily summer rainfall in the humid and complex terrain of the Blue Nile: observational evidence, Hydrol. Process., 3670–3674. doi:10.1002/hyp.7468, 2009.

Chen, H. and Swain, A.: The Grand Ethiopian Renaissance Dam: Evaluating Its Sustainability Standard and Geopolitical Significance, Energy Dev. Front., 3, 11–19, 2014.

Cresswell, H. and Kirkegaard, J.: Soil Amelioration by Plant Roots – the Process adn the Evidence, Aust. J. Soil Res., 33, 221–239, 1995.

Dagnew, D. C., Guzman, C. D., Zegeye, A. D., Tibebu, T. Y., Getaneh, M., Abate, S., Zemale, F. A., Ayana, E. K., Tilahun, S. A., and Steenhuis, T. S.: Impact of conservation practices on runoff and soil loss in the sub-humid Ethiopian Highlands: The Debre Mawi watershed, J. Hydrol. Hydromech., 63, 210–219, doi:10.1515/johh-2015-0021, 2015.

Feoli, E., Gallizia, L., and Woldu, Z.: Processes of Environmental Degradation and Opportunities for Rehabilitation in Adwa, Northern Ethiopia, Landsc. Ecol., 17, 315–325, 2002.

French, R. J. and Buirchell, B. J.: Lupin: the largest grain legume crop in Western Australia, its adaptation and improvement through plant breeding, Aust. J. Agric. Res., 56, 1169, doi:10.1071/AR05088, 2005.

Glaser, B., Lehmann, J., and Zech, W.: Ameliorating physical and chemical properties of highly weathered soils in the tropics with charcoal – a review, Biol. Fertil. Soils, 35, 219–230, doi:10.1007/s00374-002-0466-4, 2002.

Han, S., Xu, D., and Wang, S.: Runoff formation from experimental plot, field, to small catchment scales in agricultural North Huaihe River Plain, China, Hydrol. Earth Syst. Sci., 16, 3115–3125, doi:10.5194/hess-16-3115-2012, 2012.

Hanson, D. L., Steenhuis, T. S., Walter, M. F., and Boll, J.: Effects of soil degradation and management practices on the surface water dynamics in the Talgua River Watershed in Honduras, Land Degrad. Dev., 15, 367–381, doi:10.1002/ldr.603, 2004.

Herweg, K. and Ludi, E.: The performance of selected soil and water conservation measures – case studies from Ethiopia and Eritrea, Catena, 36, 99–114, 1999.

Hugo, L., Johann, B., Juergen, G., Hiremagalur, G., Mohammad, J., Victor, M., John, M., Martin, O., and Mohamed, S.: Linking Natural Resources, Agriculture and Human Health: Case Studies from East Africa, LEISA Mag., supplement, 17–20, 2002.

Hurni, H.: Principles of Soil Conservation for Cultivated Land. Soil Technol. 1, 101–116, 1988.

Hurni, H., Tato, K., and Zeleke, G.: The Implications of Changes in Population, Land Use, and Land Management for Surface Runoff in the Upper Nile Basin Area of Ethiopia, Mt. Res. Dev., 25, 147–154, 2005.

Kameyama, K., Miyamoto, T., and Shinogi, Y.: Increases in available water content of soils by applying bagasse-charcoals, in: Proceedings of the 19th World Congress of Soil Science: Soil Solutions for a Changing World, Brisbane, Australia, 1–6 August 2010, 105–108, 2010.

Karhu, K., Mattila, T., Bergström, I., and Regina, K.: Biochar addition to agricultural soil increased CH_4 uptake and water holding capacity – Results from a short-term pilot field study, Agric. Ecosyst. Environ., 140, 309–313, doi:10.1016/j.agee.2010.12.005, 2011.

Laird, D. A., Fleming, P., Davis, D. D., Horton, R., Wang, B., and Karlen, D. L.: Impact of biochar amendments on the quality of a typical Midwestern agricultural soil, Geoderma, 158, 443–449, doi:10.1016/j.geoderma.2010.05.013, 2010.

Lehmann, J., Gaunt, J., and Rondon, M.: Bio-charhar sequestration in terrestrial ecosystems – a review, Mitig. Adapt. Strat. Global Change, 11, 403–427, doi:10.1007/s11027-005-9006-5, 2006.

Lesturgez, G., Poss, R., Hartmann, C., Bourdon, E., Noble, A., and Ratana-Anupap, S.: Roots of Stylosanthes hamata create macropores in the compact layer of a sandy soil, Plant Soil, 260, 101–109, 2004.

Lugg, D. G., Tubaileh, A. S., Kallsen, C. E., and Sammis, T. W.: Irrigation effects on rooting patterns of spring barley, Irrig. Sci., 9, 27–43, 1988.

McHugh, O. V., Steenhuis, T. S., Berihun, A., and Fernandes, E. C. M.: Performance of in situ rainwater conservation tillage techniques on dry spell mitigation and erosion control in the drought-prone North Wello zone of the Ethiopian highlands, Soil Tillage Res., 97, 19–36, doi:10.1016/j.still.2007.08.002, 2007.

Meek, B. D., Rechel, E. R., Carter, L. M., DeTar, W. R., and Urie, A. L.: Infiltration rate of a sandy loam soil: effects of traffic, tillage, and plant roots, Soil Sci. Soc. Am. J. 56, 908–913, 1992.

Mendoza, G. and Steenhuis, T. S.: Determination of Hydraulic Behavior of Hillsides with a Hillslope Infiltrometer, Soil Sci. Soc. Am. J., 66, 1501–1504, 2002.

Mutema, M., Jewitt, G., Chivenge, P., Kusangaya, S., and Chaplot, V.: Spatial scale impact on daily surface water and sediment fluxes in Thukela river, South Africa, Phys. Chem. Earth, 1–10, doi:10.1016/j.pce.2015.10.001, in press, 2015.

Nyberg, G., Bargués Tobella, A., Kinyangi, J., and Ilstedt, U.: Soil property changes over a 120-yr chronosequence from forest to agriculture in western Kenya, Hydrol. Earth Syst. Sci., 16, 2085–2094, doi:10.5194/hess-16-2085-2012, 2012.

Nyssen, J., Poesen, J., Haile, M., Moeyersons, J., Deckers, J., and Hurni, H.: Effects of land use and land cover on sheet and rill erosion rates in the Tigray highlands, Ethiopia, Z. Geomorphol., 53, 171–197, doi:10.1127/0372-8854/2009/0053-0171, 2009.

Orchard, C. M., Lorentz, S. A., Jewitt, G. P. W., and Chaplot, V. A. M.: Spatial and temporal variations of overland flow during rainfall events and in relation to catchment conditions, Hydrol. Process., 27, 2325–2338, doi:10.1002/hyp.9217, 2013.

Rallison, R. E.: Origin and Evolution of the SCS Runoff Equation, in: Proceedings of Symposium on Watershed Management, Am. Soc. Civil Eng., Boise, ID, USA, 21–23 July 1980, 912–924, 1980.

R Core Team: R: A language and environment for statistical computing, R Foundation for Statistical Computing, Vienna, Austria, available at: https://www.R-project.org/, last access: 11 January 2015, 2014.

SCRP: Area of Anjeni, Gojam, Ethiopia: long-term monitoring of the agricultural environment 1984–1994, Soil erosion and conservation database, Soil Conservation Research Programme (SCRP), Centre for Development and Environment and the Ministry of Agriculture, Ethiopia, University of Berne, Berne, Switzerland, 89 pp., 2000.

Shougrakpam, S., Sarkar, R., and Dutta, S.: An experimental investigation to characterise soil macroporosity under different land use and land covers of northeast India, J. Earth Syst. Sci., 119, 655–674, 2010.

Steenhuis, T. S., Winchell, M., Rossing, J., Zollweg, J., and Walter, M.: SCS Runoff Equation Revisited For Variable-Source Runoff Areas, J. Irrig. Drain. Eng., 121, 234–238, 1995.

Steenhuis, T. S., Collick, A. S., Easton, Z. M., Leggesse, E. S., Bayabil, H. K., White, E. D., Awulachew, S. B., Adgo, E., and Ahmed, A. A.: Predicting discharge and sediment for the Abay (Blue Nile) with a simple model, Hydrol. Process., 3728–3737, doi:10.1002/hyp.7513, 2009.

Stoof, C. R., Vervoort, R. W., Iwema, J., van den Elsen, E., Ferreira, A. J. D., and Ritsema, C. J.: Hydrological response of a small catchment burned by experimental fire, Hydrol. Earth Syst. Sci., 16, 267–285. doi:10.5194/hess-16-267-2012, 2012.

Taddese, G.: Land Degradation: A Challenge to Ethiopia, Environ. Manage., 27, 815–824, doi:10.1007/s002670010190, 2001.

Tilahun, S. A.: Observations and modeling of erosion from spatially and temporally distributed sources in the (semi) humid Ethiopian highlands, PhD thesis, Cornell University, Ithaca, New York, available at: http://soilandwater.bee.cornell.edu/publications/Seifu2012PhDthesis.pdf,last access: 6 November 2014, 2012.

Tilahun, S. A., Guzman, C. D., Zegeye, A. D., Engda, T. A., Collick, A. S., Rimmer, A., and Steenhuis, T. S.: An efficient semi-distributed hillslope erosion model for the subhumid Ethiopian Highlands, Hydrol. Earth Syst. Sci., 17, 1051–1063, doi:10.5194/hess-17-1051-2013, 2013.

Tilahun, S. A., Guzman, C. D., Zegeye, A. D., Dagnew, D. C., Collick, A. S., Yitaferu, B., and Steenhuis, T. S.: Distributed discharge and sediment concentration predictions in the sub-humid Ethiopian highlands: the Debre Mawi watershed, Hydrol. Process., 1817–1828, doi:10.1002/hyp.10298, 2014.

van de Giesen, N., Stomph, T.-J., Ajayi, A. E., and Bagayoko, F.: Scale effects in Hortonian surface runoff on agricultural slopes in West Africa: Field data and models, Agric. Ecosyst. Environ., 142, 95–101, doi:10.1016/j.agee.2010.06.006, 2011.

Yunusa, I. A. and Newton, P.: Plants for amelioration of subsoil constraints and hydrological control: the primer-plant concept, Plant. Soil, 257, 261–281, 2003.

Zeleke, G.: Landscape Dynamics and Soil Erosion Process Modeling in the North-Western Ethiopian Highlands, African Studies Series A 16, Geographica Bernensia, Berne, Switzerland, 2000.

4

Towards improved parameterization of a macroscale hydrologic model in a discontinuous permafrost boreal forest ecosystem

Abraham Endalamaw[1], W. Robert Bolton[1], Jessica M. Young-Robertson[2], Don Morton[3], Larry Hinzman[4], and Bart Nijssen[5]

[1]International Arctic Research Center, University of Alaska Fairbanks, Fairbanks, AK, USA
[2]School of Natural Resources and Extension, University of Alaska Fairbanks, Fairbanks, AK, USA
[3]Boreal Scientific Computing LLC, Fairbanks, AK, USA
[4]University of Alaska Fairbanks, Fairbanks, AK, USA
[5]Civil and Environmental Engineering, University of Washington, Seattle, WA, USA

Correspondence to: Abraham Endalamaw (amendalamaw@alaska.edu)

Abstract. Modeling hydrological processes in the Alaskan sub-arctic is challenging because of the extreme spatial heterogeneity in soil properties and vegetation communities. Nevertheless, modeling and predicting hydrological processes is critical in this region due to its vulnerability to the effects of climate change. Coarse-spatial-resolution datasets used in land surface modeling pose a new challenge in simulating the spatially distributed and basin-integrated processes since these datasets do not adequately represent the small-scale hydrological, thermal, and ecological heterogeneity. The goal of this study is to improve the prediction capacity of mesoscale to large-scale hydrological models by introducing a small-scale parameterization scheme, which better represents the spatial heterogeneity of soil properties and vegetation cover in the Alaskan sub-arctic. The small-scale parameterization schemes are derived from observations and a sub-grid parameterization method in the two contrasting sub-basins of the Caribou Poker Creek Research Watershed (CPCRW) in Interior Alaska: one nearly permafrost-free (LowP) sub-basin and one permafrost-dominated (HighP) sub-basin. The sub-grid parameterization method used in the small-scale parameterization scheme is derived from the watershed topography. We found that observed soil thermal and hydraulic properties – including the distribution of permafrost and vegetation cover heterogeneity – are better represented in the sub-grid parameterization method than the coarse-resolution datasets. Parameters derived from the coarse-resolution datasets and from the sub-grid parameterization method are implemented into the variable infiltration capacity (VIC) mesoscale hydrological model to simulate runoff, evapotranspiration (ET), and soil moisture in the two sub-basins of the CPCRW. Simulated hydrographs based on the small-scale parameterization capture most of the peak and low flows, with similar accuracy in both sub-basins, compared to simulated hydrographs based on the coarse-resolution datasets. On average, the small-scale parameterization scheme improves the total runoff simulation by up to 50 % in the LowP sub-basin and by up to 10 % in the HighP sub-basin from the large-scale parameterization. This study shows that the proposed sub-grid parameterization method can be used to improve the performance of mesoscale hydrological models in the Alaskan sub-arctic watersheds.

1 Introduction

The sub-arctic region of Interior Alaska lies in the transition zone between the warm temperate region to the south and the cold arctic region to the north. This region is underlain by discontinuous permafrost and is very sensitive to climate warming (Hinzman et al., 2006). Along with climate and geology, topography is a significant factor that controls the distribution of permafrost in Interior Alaska, as it has a strong control on the amount and intensity of solar radiation received at the land surface (Viereck et al., 1983; Morrissey and Strong, 1986). A difference in solar radiation between north- and south-facing slopes supports the existence of permafrost on north-facing slopes and valley bottoms but not on south-facing slopes (Slaughter and Kane, 1979; Slaughter et al., 1983; Hinzman et al., 2006). Subsequently, permafrost-underlain and permafrost-free areas in this region display contrasting watershed characteristics and hydrological responses. The presence or absence of permafrost is the primary factor that creates a complex landscape mosaic of sharp spatial boundaries of contrasting vegetation cover and soil hydraulic and thermal properties, moisture dynamics, and water pathways (Bolton et al., 2000; Hinzman et al., 2002; Bolton, 2006; Petrone et al., 2006; White et al., 2008; Jones and Rinehart, 2010). Due to these small-scale complexities associated with permafrost distribution, simulating large-scale hydrological processes remains a challenge in the Interior Alaskan boreal forest.

The hydraulic conductivity of permafrost-affected soils is several orders of magnitude less than that of the overlying organic layers and the nearby permafrost-free soil (Rieger et al., 1972; Burt and Williams, 1976; Kane and Stein, 1983; Woo, 1986; Ping et al., 2005; Zhang et al., 2010a). The ice-rich permafrost is an impermeable layer at the permafrost surface that limits the hydraulic flow to the active layer – the thin, seasonally thawed soil layer above permafrost (Romanovsky and Osterkamp, 1995; Romanovsky et al., 2003). Streamflow in permafrost-dominated watersheds has been described as "flashy", responding rapidly to precipitation and snowmelt with storm hydrographs displaying a sharp rise and prolonged recession (Bolton et al., 2000; Quinton and Carey, 2008) and relatively low baseflow between precipitation events (Kane, 1980; Kane et al., 1981; Kane and Stein, 1983; Slaughter et al., 1983; Hinzman et al., 2002; Bolton, 2006; Petrone et al., 2006, 2007). On the other hand, in watersheds with no or relatively low areal permafrost extent, the soil hydraulic conductivity and infiltration capacity are much higher, resulting in a slower streamflow response to precipitation and snowmelt, relatively higher baseflow between storm events, and greater residence time of water in catchments (Bolton et al., 2000; Carey and Woo, 2001).

In the Alaskan sub-arctic, vegetation type, density, physiological, and structural properties such as leaf area index (LAI) and stomatal conductance display a strong variation between permafrost-dominated and permafrost-free soils (Viereck et al., 1983; Viereck and Van Cleve, 1984). These variations can lead to significant variation in the partitioning of precipitation and snowmelt into runoff and evapotranspiration (ET), and change in soil water content between permafrost-dominated and permafrost-free soils (Hinzman et al., 2002, 2006; Bolton et al., 2006; Cable et al., 2014; Young-Robertson et al., 2016). Permafrost-affected soils typically support coniferous vegetation that is shallowly rooted, has low physiological activity, is tolerant of cold and wet soils, and is able to survive a short growing season (Viereck et al., 1983; Viereck and Van Cleve, 1984; Morrissey and Strong, 1986; Mölders, 2011). In contrast, the well-drained, relatively warm, permafrost-free soils support deciduous vegetation that is much larger, has higher LAI and stomatal conductance, and has a deeper rooting network (Cable et al., 2014; Young-Robertson et al., 2016). This difference in vegetation between permafrost-dominated and permafrost-free soil can further influence streamflow responses (Naito and Cairns, 2011) due to the large differences in the rates of and controls on ET, particularly transpiration, canopy interception (Baldocchi et al., 2000; Ewers et al., 2005; Andreadis et al., 2009), and vegetation water storage (Young-Robertson et al., 2016). In Interior Alaska, deciduous trees have significantly higher transpiration rates and vegetation water storage compared to coniferous trees (Ewers et al., 2005; Yuan et al., 2009; Cable et al., 2014; Young-Robertson et al., 2016), which limits the availability of water for runoff in the permafrost-free watersheds.

Plot-scale and hill-slope studies have documented the differences in and relationships between soil thermal and hydrologic properties, and ecosystem vegetation composition in high-latitude cold regions (Dingman, 1973; Woo, 1976; Kirkby, 1978; Kane et al., 1981; Woo and Steer, 1983). However, larger-scale land surface parameterizations and the data products used in land surface, hydrological, and climate models do not adequately represent the complex sub-arctic watersheds with significant spatial variability in soil and vegetation dynamics. Hence, hydrological modeling using these coarse-resolution datasets cannot produce accurate estimates of the spatially variable and basin-integrated watershed responses. This is primarily because coarse-resolution datasets smoothed the non-linear small-scale watershed heterogeneities that control hydrological responses of the sub-arctic watersheds. Until the small-scale hydrologic processes, soil properties, and vegetation distributions are well represented, accurate large-scale hydrologic simulation and modeling remains extremely challenging (Walsh et al., 2005).

Several techniques, including parameter regionalization (Parajka et al., 2005; Pokhrel et al., 2008; Samaniego et al., 2010; Kumar et al., 2013), are proposed to address the limitation of large-scale hydrological simulation due to sub-grid variabilities of soil and vegetation cover properties. Standard regionalization (SR; Pokhrel et al., 2008; Troy et al., 2008) and multiscale parameter regionalization (MPR; Samaniego et al., 2010; Kumar et al., 2013) are widely used to parameterize the basin hydrological predictors. The latter approach (MPR) takes into account sub-grid heterogeneity of hydrological processes by implementing the prior predictors (the relationships between elevation, slope, vegetation characteristics, soil properties, etc.) into model parameterization (Samaniego et al., 2010; Kumar et al., 2013). However, the first approach (SR) lacks explicit representation of basin characteristics and primarily depends on calibration of subsets of the region (Troy et al., 2008). In this study, we proposed a sub-grid parameterization approach of soil property and vegetation characteristic parameterization of a mesoscale

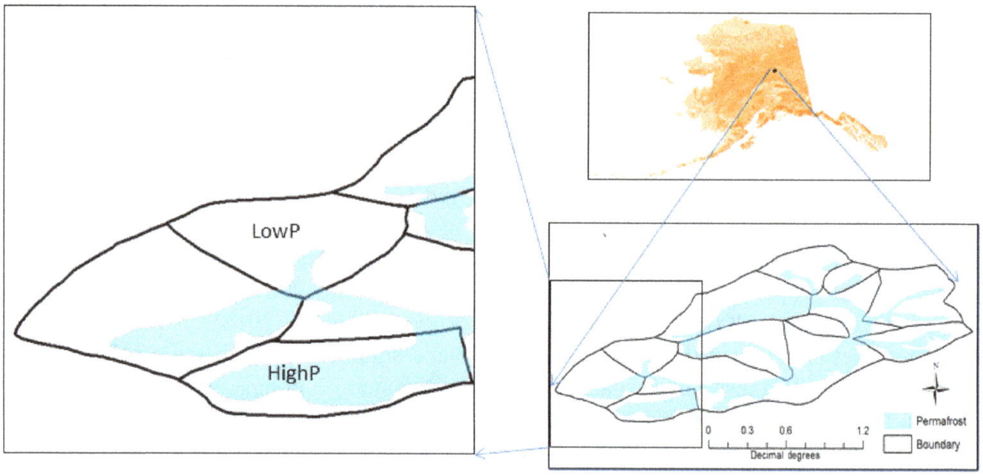

Figure 1. Location and permafrost extent (light blue shading) in the Caribou Poker Creek Research Watershed, Alaska, and the two sub-basins of interest (low permafrost or LowP sub-basin and high permafrost or HighP sub-basin). The permafrost map of CPCRW was produced by a small-scale observation across the watershed (Rieger et al., 1972; Yoshikawa et al., 2002).

hydrological model. The sub-grid parameterization method proposed in this study closely fits the MPR approach. In both approaches, basin hydrological predictors are explicitly taken into account, and parameters are transferable to scales and locations other than a particular area of similar basin characteristics.

The primary objective of this study is to improve the prediction capability of hydrological models in Interior Alaska's boreal forest by implementing a small-scale parameterization scheme, which represents the spatial heterogeneity of soil properties and vegetation cover, into a mesoscale hydrological model. The study is conducted in two small sub-basins of the CPCRW (LowP and HighP sub-basins of areas 5.2 and 5.7 km^2, respectively) that are representative of the region. This study transfers the plot- and hill-slope-scale knowledge into a mesoscale distributed hydrological model (the variable infiltration capacity (VIC) model at a spatial resolution of approximately 1 km) so that its application can be extended to large basins in the region. A sub-grid parameterization method – on which the small-scale parameterization scheme is based – is derived from a high-resolution digital elevation model (DEM). Streamflow, ET and soil moisture simulations, based on the small-scale parameterization method and the coarse-resolution datasets, are presented and investigated in both sub-basins.

2 Methods

2.1 Study area

Located approximately 50 km northwest of Fairbanks, Alaska, the Caribou Poker Creek Research Watershed (CPCRW; centered at 65°10′ N and 147°30′ W; basin area ~ 101 km^2) is within the zone of discontinuous permafrost and in the North American boreal forest region (Fig. 1). CPCRW is also within the Yukon-Tanana Uplands of the Northern Plateaus physiographic province (Wahrhaftig, 1965) with elevations ranging from 187 to 834 m. CPCRW has a long-term record of ecological, meteorological, and hydrological data (Bolton et al., 2000; Knudson and Hinzman, 2000; Hinzman et al., 2002, 2003; Bolton, 2006). The climate of CPCRW is characterized as continental with large diurnal and annual temperature variations, low annual precipitation, low cloud cover, and low humidity (Haugen et al., 1982). The average annual air temperature range at CPCRW is very large (July mean temperature of 14.7 °C, January mean temperature of −18.4 °C) with a mean annual temperature of −2.1 °C (Fig. 2, Table 1). The mean annual precipitation is 417 mm, of which two-thirds occurs as rainfall (Bolton, 2006).

Seven soil series have been identified in CPCRW (Rieger et al., 1972). We group the seven series into two general categories: permafrost-dominated soils that are poorly drained with a thick organic layer, and permafrost-free soils that are well-drained with a shallow organic layer (Rieger et al., 1972). Permafrost in CPCRW is generally found along north-facing slopes and valley bottoms where the solar input is very low (Rieger et al., 1972; Haugen et al., 1982). Soils free of permafrost are generally found on south- to southwest-facing slopes.

The vegetation distribution in CPCRW displays a strong relationship with permafrost distribution. Coniferous vegetation that consists primarily of black spruce (*Picea mariana*) is generally found in areas underlain by permafrost. Feather moss (*Hylocomium* spp.), tussock tundra (mostly *Carex aquatilis*), and sphagnum mosses (*Sphagnum* sp.) are also found along the valley bottoms. Deciduous vegetation consists primarily of aspen (*Populus tremuloides*), birch (*Betula papyrifera*), alder (*Alnus crispa*), and sporadic patches of white spruce (*Picea glauca*), and is found on the well-drained, south-facing permafrost-free soils (Haugen et al., 1982).

There are several sub-basins in the CPCRW that differ in permafrost coverage and vegetation composition. The nearly permafrost-free or LowP (5.2 km^2) and permafrost-dominated or HighP (5.7 km^2) sub-basins (Fig. 1) are selected for this study for their contrasting permafrost extents and vegetation cover compositions (Table 3). Unlike vegetation cover, permafrost extent, and solar input, local climate does not vary between the two sub-basins (Fig. 2, Table 1) due to the local understory convective mixing of the bulk atmosphere (Hinzman et al., 2006). The coniferous/deciduous vegetation composition, derived from Haugen et al. (1982), is approximately 30/70 % for the LowP sub-basin and 95/5 % for the HighP sub-basin (Fig. 3c, Table 3). The permafrost coverage is different between the two sub-basins, with approximately 2 and 53 % in the LowP and HighP sub-basins, respectively (Table 1; Rieger et al., 1972; Yoshikawa et al., 2002).

2.2 Sub-grid parameterization method

The sub-grid parameterization method is a simple method used to produce high-resolution soil property and vegetation cover maps for accurate representation of the watershed characteristics into the hydrological model. A fine-resolution DEM is primarily used to parameterize the sub-grid heterogeneity of the landscape. In addition, the relationships between vegetation, permafrost, slope, and aspect (Viereck et al., 1983; Morrissey and Strong, 1986; Hinzman et al., 2006) are included in the sub-grid parameterization method.

First, aspect of the landscape is calculated using the 30 m DEM (Aster, 2009) of the watershed with the ArcGIS aspect spatial analyst toolbox. The aspect toolbox uses elevation values of eight surrounding gird cells to calculate the gradient and aspect of each grid cell in the model domain (Burrough et al., 1998) resulting in an aspect map with nine classes (Fig. 3a) that are aggregated into two aspect classes: permafrost-underlain and permafrost-free aspects. The following assumptions are made in the classification of each aspect into permafrost-dominated and permafrost-free aspects:

1. North-facing slopes and valley bottoms are underlain by permafrost, and south-facing slopes are permafrost free (Rieger et al., 1972; Hinzman et al., 2002; Yoshikawa et al., 2002).

2. Coniferous vegetation, primarily black spruce trees, are found along poorly drained north-facing slopes and valley bottoms (permafrost-underlain soil; Haugen et al., 1982).

3. Deciduous vegetation, primarily birch and aspen trees, are found on the well-drained, south-facing soils (permafrost-free soil; Haugen et al., 1982).

4. The hydraulic conductivity of frozen soil is 2 orders of magnitude less than the same soil in unfrozen conditions (Burt and Williams, 1976; Kane and Stein, 1983; Woo, 1986).

The small-scale observed (Rieger et al., 1972) and modeled (Yoshikawa et al., 2002) permafrost maps are used as a reference during the grouping of the nine aspect classes into permafrost-underlain and permafrost-free aspects. Finally, north, northeast, northwest, southwest, and flat aspects are classified as permafrost-underlain aspects with coniferous vegetation cover. South, southeast, east, and west are classified as permafrost-free aspects and with a deciduous vegetation cover. The vegetation cover and soil hydraulic property maps obtained from the sub-grid parameterization method are shown in Figs. 3b and 4c, respectively. Since each grid cell has a unique soil property and the spatial scale or length of the observed permafrost map is on the order of few meters, a threshold by which a grid cell can be assumed 100 % permafrost or permafrost-free soil is required. After several rounds of testing, 0.5 grid cell fraction threshold is assumed to classify a grid cell as permafrost-underlain or permafrost-free soil. We used the small-scale observed (Rieger et al., 1972) permafrost map in the determination of this threshold value. As a result, a grid cell is assigned with permafrost soil property when the fraction of permafrost is greater than 0.5 (Fig. 4c).

2.3 VIC model description

The VIC model (Liang et al., 1994, 1996; Nijssen et al., 1997) is a mesoscale process-based distributed hydrological model that represents vegetation heterogeneity, multiple soil layers, variable infiltration, and non-linear baseflow. The version of the VIC model used in this study, VIC 4.1.2, contains several explicit formulations for snow accumulation and ablation, frozen soil, and evapotranspiration and solves the full energy balance or water balance at sub-daily or daily time steps to simulate the energy and water fluxes for individual grid cells. A separate routing model (Duband et al., 1993; Lohmann et al., 1996, 1998a, b) is used to collect the runoff and baseflow simulations from each grid cell and simulate streamflow at the outlet of each watershed. VIC has been successfully used to simulate major hydrological and thermal processes at different spatial and temporal scales, such as basin streamflow (Abdulla et al., 1996; Abdulla and Lettenmaier, 1997a, b; Schnorbus et al., 2011; Bennett et al., 2012, 2015), evaporation and transpiration, canopy interception, soil moisture (Robock et al., 2003; Andreadis et al., 2005; Slater et al., 2007; Wu et al., 2007; Meng and Quiring, 2008; Billah and Goodall, 2012; Wu et al., 2015), surface and ground heat fluxes, ground temperature, and snowpack energy balance, including ablation and accumulation processes (Cherkauer and Lettenmaier, 1999; Cherkauer et al., 2003; Andreadis et al., 2009).

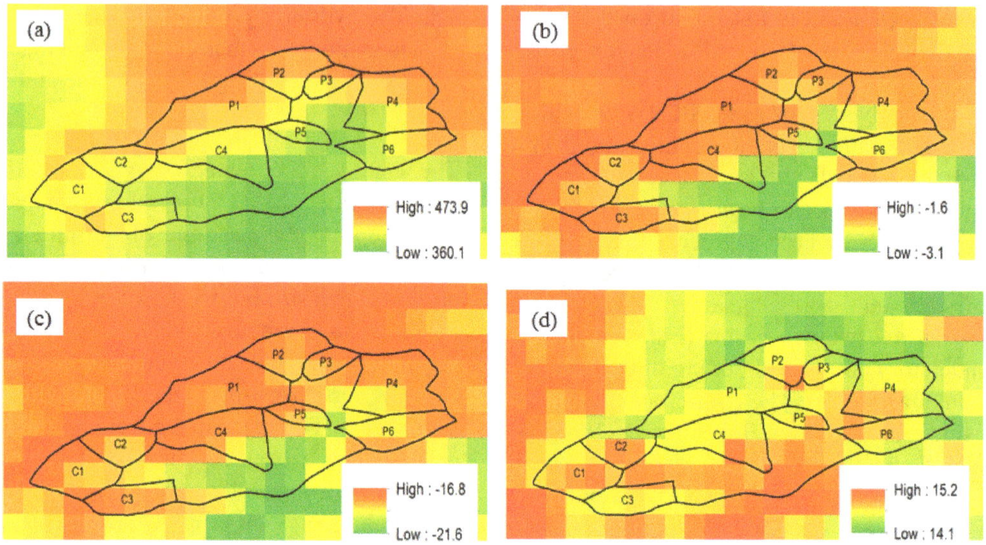

Figure 2. Climatology of the Caribou Poker Creek Research Watershed, Alaska, from 1970 to 2010: **(a)** mean annual precipitation (mm), **(b)** mean annual air temperature (°C), **(c)** mean January air temperature (°C), and **(d)** mean July air temperature (°C). These climate datasets are extracted from the Alaska-Pacific River Forecast Center (APRFC; Bennett, 2014). Only US National Climate Data Center cooperative daily climate stations were used to generate the gridded fields of 800 m resolution using an inverse distance weighting function. Note that these data are not used to force the model in this study. In this study, data from four very close observation stations are used.

Table 1. Mean (1970–2012) climatology of the Caribou Poker Creek Research Watershed (CPCRW), Alaska, and the sub-basins (LowP, HighP): MAP (mean annual precipitation, mm), MAT (mean annual air temperature, °C), MIN (mean minimum air temperature, °C), MAX (mean maximum air temperature, °C), MJan (mean January air temperature, °C), and MJuly (mean July air temperature, °C).

Basin/sub-basin	Permafrost extent (%)	MAP	MAT	MIN	MAX	MJan	MJuly
CPCRW	30	416.7	−2.1	−7.2	3.0	−18.4	14.7
LowP	2	421.5	−1.9	−6.7	2.9	−17.8	14.7
HighP	53	408.2	−2.1	−7.3	3.0	−18.5	14.8

In this study, a fully coupled water–energy balance mode of the VIC model, with 1/64 degree grid resolution (approximately 1 km), is used in the LowP and HighP sub-basins of the CPCRW. We utilize this higher model resolution in order to explicitly represent and simulate the contrasting hydrological responses of the valley bottoms and hill slopes that co-exist in a very short spatial distance. The frozen soil algorithm is activated for permafrost grid cells (Cherkauer and Lettenmaier, 1999; Cherkauer et al., 2003; Bowling et al., 2008) in order to solve the ground heat flux and surface energy balance. The frozen soil algorithm solves the thermal/heat flux equation through the soil column to calculate the frozen soil penetration and the ice content of each soil layer (Cherkauer and Lettenmaier, 1999). The frozen soil algorithm also solves the effect of frozen soil on moisture transport (Cherkauer and Lettenmaier, 1999).

VIC was forced with daily or sub-daily minimum and maximum air temperatures, precipitation, and wind speed. The remaining forcing data including incoming shortwave radiation, longwave radiation, atmospheric pressure, and relative humidity are estimated based on the daily temperature range and precipitation as calculated by Mountain Microclimate Simulation Model (MTCLIM; Thornton and Running, 1999). MTCLIM is implemented within the VIC model (Bohn et al., 2013). The radiation estimate by the VIC model indicates an average of 15–20 $\mathrm{W m^{-2}}$ higher shortwave radiation in the LowP sub-basin than the HighP sub-basin. In this study, we generate the daily gridded minimum and maximum temperatures, precipitation, and wind speed data from four meteorological stations located within the CPCRW (http://www.lter.uaf.edu/data.cfm) using the inverse distance weighting (IDW) interpolation method.

In addition to forcing, VIC requires parameters that describe soil properties, vegetation distribution and characteristics, and topographic information. In this study, a three-layer soil column is used with a top layer with a depth of 0.1 m and variable depth for the second and third layers based on model calibration. Soil properties are uniform within a grid cell, but these properties are allowed to vary for each layer in the grid cell. VIC requires several soil hydraulic and thermal property parameters. Due to a lack of high-spatial-resolution soil data in the region, most of the soil data – such as soil textural classes, saturated hydraulic conductivity, bulk density, and porosity – were extracted directly from the 5 arcmin (approximately 9 km) Food and Agriculture Organization digital soil map of the world (FAO, 1998). Other soil hydraulic properties – including field capacity wilting point, residual soil moisture content, water retention, and bubbling pressure – are estimated by the Brooks and Corey formulation (Rawls and Brakensiek, 1985; Saxton et al., 1986) based on soil texture classes, porosity, and bulk density information. All soil parameters are resampled to 1/64 degree using bilinear interpolation. Selected soil property values are shown in Table 2.

VIC uses a mosaic representation of vegetation coverage and subdivides each grid cell's land cover into a specified number of tiles. Each tile represents the fraction of the cell covered by a particular land cover type (coniferous, evergreen forest, grassland, etc.). The coarse-resolution vegetation cover composition data used in this study are obtained from the University of Alaska Fairbanks (UAF), Scenarios Network for Alaska and Arctic Planning (SNAP) 1 km × 1 km land cover map originating from the North American Land Change Monitoring System (NALCMS) 2005 dataset (http://www.snap.uaf.edu/data.php). Rooting depth data are extracted from the International Satellite Land Surface Climatology Project (ISLSCP) Initiative II Ecosystem Rooting Depths (Schenk and Jackson, 2009). Tree height, trunk ratio, displacement, and roughness of the primary vegetation classes are derived from field measurements by Young-Robertson et al. (2016). The remaining vegetation parameter values including LAI for each vegetation class in the region are derived from Myneni et al. (1997), Hansen et al. (2000), and Nijssen et al. (2001a, b).

2.4 VIC model parameterization methods

Based on sources of vegetation cover and soil hydraulic property data, we developed two parameterization methods: large-scale parameterization (large-scale, hereafter) and small-scale parameterization (small-scale, hereafter) schemes. The two schemes differ in vegetation cover, saturated hydraulic conductivity, and organic layer thickness. The remaining model inputs and parameters including meteorological forcing, vegetation characteristics, and soil properties are the same in both schemes.

2.4.1 Large-scale parameterization

The large-scale parameterization scheme is derived from the coarse-resolution SNAP vegetation cover (Fig. 3d) and FAO soil property (Fig. 4a) datasets as described in the previous section (Table 4).

2.4.2 Small-scale parameterization

The small-scale parameterization scheme consists of two small-scale parameterization sub-schemes: one derived from the sub-grid parameterization method (aspect parameterization, hereafter) and one derived from a permafrost map (permafrost parameterization, hereafter). In both small-scale parameterization schemes, the top layer of the permafrost cell is also assumed to be the organic layer (Ping et al., 2005; Zhang et al., 2010b) while mineral soil, as obtained from FAO dataset, is assumed for permafrost-free grid cells.

Aspect parameterization

The aspect parameterization is based on vegetation cover (Fig. 3b) and soil property (Fig. 4c) products of the sub-grid parameterization method (Table 4). Saturated hydraulic conductivity and organic layer thickness in the aspect parameterization depend on whether the grid cell is permafrost underlain or permafrost free (Fig. 4c).

Permafrost parameterization

The permafrost parameterization is primarily based on the small-scale observed (Rieger et al., 1972) and modeled (Yoshikawa et al., 2002) permafrost maps. In the permafrost parameterization, we assume the proportion of coniferous and deciduous vegetation in each grid cell is the same as the fraction of permafrost-underlain and permafrost-free soil, respectively (Table 4). Like the aspect parameterization, we classified a grid cell as containing permafrost when the fraction of permafrost in the grid cell is greater than 0.5 (Fig. 4b), as described in Sect. 2.2.

2.5 Calibration and validation

Model calibration was performed by comparing streamflow simulation with observation for the period of 2001–2005 at LowP and HighP sub-basin outlets of CPCRW. The large-scale parameterization scheme – FAO soil (Fig. 4a) and SNAP vegetation cover (Fig. 3d) datasets – is implemented during the calibration process. The multi-objective complex evolution (MOCOM) automated calibration approach developed by Yapo et al. (1998) was used to match the simulated and observed streamflow. MOCOM is a multi-criteria calibration approach based on random parameter sampling strategy to optimize several user-defined criteria (Yapo et al., 1998; Wagener et al., 2001). As suggested by Liang et al. (1994), the model was calibrated using baseflow generation parameters including maximum velocity of the baseflow (D_{smax}), infiltration parameter (bi), fraction of D_{smax} where non-linear baseflow begins (D_s), maximum soil moisture for non-linear baseflow to occur (W_s), thickness of the second soil layer (D2), and thickness of the third soil layer (D3). Table 2 shows final calibration values of the user-defined parameters for both sub-basins.

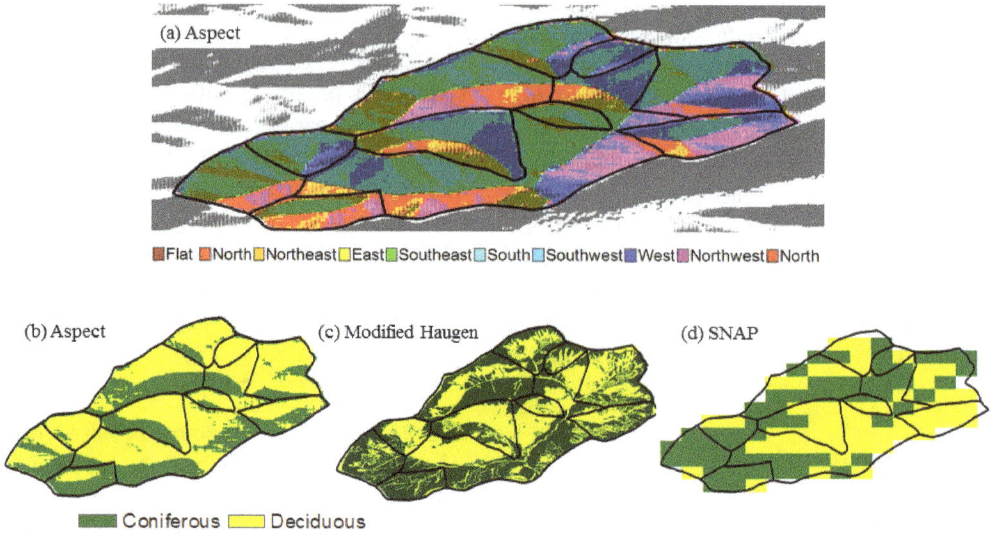

Figure 3. (a) Aspect map – derived from the DEM – of the CPCRW. Coniferous and deciduous vegetation composition at the CPCRW as derived from **(b)** model-based aspect or topography of the watershed, **(c)** an observation-based vegetation coverage map modified from Haugen et al. (1982), and **(d)** a large-scale-based Scenario Network for Alaska and Arctic Planning (SNAP) vegetation coverage map is shown. Note that panel **(b)** is prepared from the aspect map **(a)** and the sub-grid parameterization method discussed in Sect. 2.2.

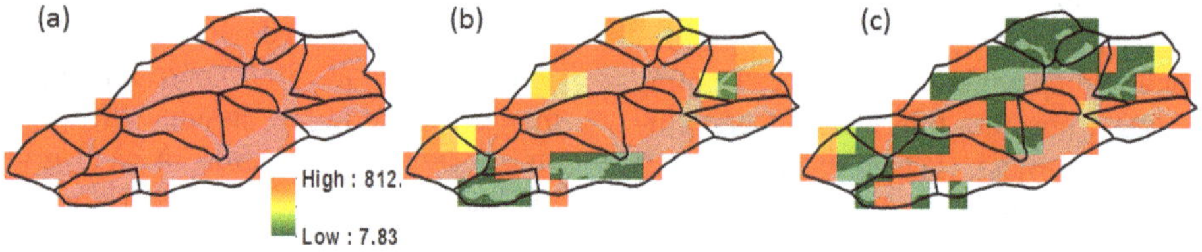

Figure 4. Saturated hydrologic conductivity (mm day^{-1}) as derived from **(a)** a large-scale FAO soil dataset, **(b)** a permafrost map of Rieger et al. (1972), and **(c)** an aspect map of CPCRW is shown. Panels **(b)** and **(c)** are small-scale parameterizations obtained from altering large-scale values according to the presence or absence of permafrost. Grid cells that are classified as permafrost soil are assigned hydraulic conductivity 2 orders of magnitude less than the value obtained from the large-scale (FAO) dataset. FAO soils map do not indicate any of the permafrost-underlain soil hydraulic properties which are known to be present in the CPCRW.

After the lumped sub-basin baseflow generation parameter values are derived by calibration, validation of the model with the large-scale and small-scale parameterization methods was conducted by comparing the observed and simulated runoff at the outlets of the LowP and HighP sub-basins from 2005 to 2008. Calibration results obtained from large-scale parameterization are directly implemented into small-scale parameterization during validation in order to examine how processes change with the new parameterizations without recalibration. We utilized two verification statistics: correlation coefficient (R^2; Eq. 1) and Nash–Sutcliffe efficiency (NSE; Eq. 2).

$$R^2 = \left(\frac{N\left(\sum_{i=1}^{N} S_i . Q_i\right) - \left(\sum_{i=1}^{N} S_i\right) . \left(\sum_{i=1}^{N} Q_i\right)}{\left[\left(N\sum_{i=1}^{N} S_i^2 - \left(\sum_{i=1}^{N} S_i\right)^2\right)\left(N\sum_{i=1}^{N} Q_i^2 - \left(\sum_{i=1}^{N} Q_i\right)^2\right)\right]^{0.5}} \right)^2 \quad (1)$$

$$\mathrm{NSE} = 1 - \left(\frac{\sum_{i=1}^{N} (S_i - Q_i)^2}{\sum_{i=1}^{N} (\overline{Q} - Q_i)^2} \right), \quad (2)$$

where N is equal to the number of data points (i.e., daily streamflow realizations), i is the time step (days), S is the simulated streamflow (mm day^{-1}), and Q is the observed streamflow (mm day^{-1}).

R^2 (Eq. 1) describes the degree of linear correlation of simulated and observed runoff or goodness of fit. The value of R^2 ranges from 0.0 to 1.0, and larger values indicate better fit between simulation and observation. NSE (Eq. 2) describes the relative magnitude of simulated runoff variances compared to variance in observed streamflow. NSE is also an indicator of model fit in terms of a scatter plot of simulated versus observed streamflow values, wherein a slope near the 1 : 1 line indicates a better fit. The value of NSE ranges from 1 (perfect fit) to $-\infty$. While values larger than

Table 2. Selected vegetation and soil parameters, their values and estimating methods or sources: runoff influence (RI) – runoff increases when parameter values increase (+) and runoff decreases when parameter values increase (−), coniferous vegetation (conif), deciduous vegetation (decid), permafrost-underlain soil (PF), permafrost-free soil (NPF), parameter values specifically for the LowP sub-basin (LowP), and parameter values specifically for the HighP sub-basin (HighP).

Parameters	Value ranges	RI	Sources
Vegetation parameters			
Leaf area index: LAI	3.8–4.2 (conif) and 0.11–6.0 (decid)	(−)	(Hansen et al., 2000; Nijssen et al., 2001a, b)
Roughness length: Z_0	1.23 (conif) and 2.24 (decid)		
Displacement length: d_0	6.7 (conif) and 13.4 (decid)		
Minimum stomata resistance: r_{min} ($s\,m^{-1}$)	130 (conif) and 150 (decid)		
Architectural resistance: r_{arc} ($s\,m^{-1}$)	60 (conif) and 60 (decid)		
Canopy albedo: α (fraction)	0.11 (conif) and 0.09–0.13 (decid)		
Trunk ratio: (fraction)	0.1 (conif) and 0.4 (decid)		This study
Depth of the two root zone: r_depth_1, r_depth_2 (m)	0.1, 0.5 (conif) and 0.1, 1.0 (decid)		Schenk and Jackson (2009)
Fraction of root in the two-root zone: r_fract_1, r_fract_2 (fraction)	0.5, 0.5 (conif and decid)		
Soil parameters			
Porosity: ρ ($m^3\,m^{-3}$)	43.9–66.5		FAO (1998)
Bulk density: ($kg\,m^{-3}$)	∼ 1200 (PF and NPF)		
Exponent used in the estimation of unsaturated hydraulic conductivity: expt	∼ 10.84 (LowP and HighP)		
Exponent used in baseflow curve: c	2 (LowP and HighP)		
Saturated hydraulic conductivity: K_s ($mm\,day^{-1}$)	∼ 8 (PF) and ∼ 800 (NPF)		FAO (1998); this study
Dumping depth: (m)	4 (LowP) and 2 (HighP)		
Calibration parameters			
Variable infiltration curve parameter: b	0.11 (LowP), 0.31 (HighP)	(+)	This study
Maximum velocity of the baseflow: D_{smax} ($mm\,day^{-1}$)	2.16 (LowP), 2.86 (HighP)	(−)	
Fraction of baseflow, D_{smax}, where non-linear baseflow begins: D_s (fraction)	0.17 (LowP), 0.45 (HighP)	(−)	
Fraction of maximum soil moisture where non-linear baseflow begins: W_s (fraction)	0.79 (LowP), 0.74 (HighP)	(−)	
Three soil-layer thicknesses: d_1, d_2, and d_3 (m)	0.1, 0.37, and 0.75 (LowP); 0.1, 0.35, and 0.47 (HighP)	(−)	

0.0 can be considered as acceptable levels of model performance (Krause et al., 2005; Schaefli and Gupta, 2007), values approaching 1.0 are more preferred depending on the study area. NSE uses of the mean observed value as a reference (Schaefli and Gupta, 2007). Hence, factors that affect the mean value of observed streamflow will have a stronger effect on the values NSE. In Interior Alaska, a lower value of NSE can be acceptable due to the large uncertainties of mean observed streamflow, which results from aufeis-related measurement errors at beginning of the snowmelt runoff season (Bolton, 2006). NSE values below zero indicates that the mean observed streamflow is better predictor than the simulated runoff (Krause et al., 2005).

Percent difference (PD) is also used to indicate and visualize differences in simulated runoff from each parameterization scenario:

$$\mathrm{PD} = \frac{(Q_\mathrm{R} - Q_i)}{0.5(Q_\mathrm{R} + Q_i)} \times 100\,\%, \tag{3}$$

where PD is the percent difference, Q_R is reference observed streamflow, and Q_i is simulated runoff. The range of PD (Eq. 3) falls between −100 and 100 % with 0 being the perfect match between observed and simulated streamflow.

Large negative PD in the time series indicates model overprediction. On the other hand, large positive PD represents model underprediction. PD values closer to 0 indicate a better model fit.

In addition to the three measures of success mentioned above, we compared the simulated and observed hydrographs to determine whether our sub-grid parameterization method can reproduce the strong relative contrast in runoff behavior between the two catchments.

3 Results

3.1 Comparisons of vegetation cover in each parameterization scenario

Figure 3b, c, and d show the three vegetation cover representations of CPCRW derived from the sub-grid parameterization method (aspect), permafrost, and SNAP, respectively. The comparison of vegetation cover focuses on the coniferous and deciduous vegetation composition in the LowP and HighP sub-basins. Table 3 summarizes the proportion of coniferous and deciduous vegetation in the different parameterization schemes. The small-scale observed vegetation cover from Haugen et al. (1982; Fig. 3c) is used as a reference vegetation cover to evaluate the other vegetation cover scenarios.

The SNAP vegetation cover map represents the HighP sub-basin well but overestimates the coniferous proportion of the LowP sub-basin. It also shifts the dominant vegetation type from deciduous to coniferous for the LowP sub-basin (30 % in small-scale vegetation to 63 % from SNAP; see Table 3). Vegetation cover from the sub-grid parameterization method (aspect) captures the dominant vegetation type for both sub-basins (Table 3). Coniferous and deciduous distribution in the permafrost map parameterization is generally similar to permafrost-underlain and permafrost-free proportions. A 3/97 and 53/47 proportion of coniferous/deciduous vegetation is obtained from parameterization based on the permafrost map in the LowP and HighP sub-basins, respectively (Table 3)

3.2 Calibration

Figure 5a and b show daily simulated versus observed streamflow for the calibration period of 2001–2005 in the LowP and HighP sub-basins, respectively. Table 5 summarizes the coefficient of determination (R^2) and NSE during the calibration of two sub-basins. Calibration results show that runoff simulation in the HighP sub-basin (Fig. 5b) is in better agreement than in the LowP sub-basin with the observed streamflow (Fig. 5a). Although the R^2 for the LowP (0.51) is greater than for the HighP (0.48), the NSE of the LowP is less than that of the HighP (0.17 for the LowP and 0.38 for the HighP sub-basin) indicating the peak flows are systematically underestimated in the LowP than in the HighP sub-basin (Table 5). The simulated streamflow for both sub-basins is believed to be satisfactory given the inadequate soil and vegetation parameter representation from the coarse-resolution datasets. Moreover, streamflow measurement during the early snowmelt period is mostly missing or not accurate most of the time due to the icing problem (Bolton, 2006); this contributes to the lower calibration values in this region. Figure 5c and d show the observed streamflow differences between LowP and HighP sub-basins for the calibration period of 2001–2005. The notable discharge contrast between the two sub-basins is shown by the flow duration curve in Fig. 5d. The peak flow in the HighP sub-basin is about 5 times higher than the peak flow in the LowP sub-basin. Streamflow in the LowP sub-basin has lower peak flow, higher baseflow, and longer runoff response to precipitation and snowmelt than HighP. Final calibration parameter values suggest that the HighP sub-basin requires parameters that favor direct runoff while the LowP sub-basin requires parameters that favor more baseflow and infiltration (Table 2).

3.3 Hydrological fluxes under different parameterization methods

In order to test the improvement of VIC simulations in the small-scale parameterization over the coarser-resolution data products, three simulations were completed: one using large-scale parameterization and two using small-scale parameterizations (aspect and permafrost). Streamflow, ET, and soil moisture were simulated in the LowP and HighP sub-basins from 2006 to 2008. Table 4 summarizes how the soil property and vegetation cover are parameterized in the three parameterization schemes.

3.3.1 Streamflow

The simulated hydrograph changes with vegetation cover and soil properties in both sub-basins. Figure 6a and c show the simulated versus observed streamflow of each parameterization scheme in the LowP and HighP sub-basins of CPCRW. The percent difference (PD; Eq. 3) of each simulation from observation is also shown in Fig. 6b and d. Table 5 compares the three streamflow simulations against observations in both sub-basins. Streamflow simulation based on large-scale parameterization yields the lowest performance with a R^2/NSE of 0.34/0.17 and 0.48/0.42 for the LowP and HighP sub-basins, respectively (Table 5). The mean annual streamflow simulation in both sub-basins is largely underestimated under the large-scale parameterization scheme (by 45 to 68 and 27 to 52 % of the total annual runoff in the LowP and HighP sub-basins). The underestimation is higher in the LowP sub-basin. Further, the simulated spring peak flow tends to occur earlier than observed peak for both sub-basins (Fig. 6).

The simulated streamflow hydrograph obtained from aspect parameterization – the small-scale parameterization scheme which is primarily derived from the sub-grid param-

Table 3. Estimated percent of the sub-basins (LowP and HighP) covered by coniferous and deciduous vegetation, and percent of the sub-basins underlain by permafrost. Values were obtained by applying different parameterization methods: SNAP vegetation cover (SNAP), a modified Haugen et al. (1982) vegetation cover map (modified Haugen), aspect-derived vegetation cover (aspect), the CPCRW permafrost map (permafrost), and large-scale FAO digital soil map of the world (FAO soil dataset). NA indicates that a given method was not utilized to obtain values for either the vegetation or permafrost distribution.

Parameterization	Vegetation, % landscape distribution				Permafrost, % landscape distribution	
	Coniferous		Deciduous			
Method	LowP	HighP	LowP	HighP	LowP	HighP
SNAP	63	93	37	7	NA	NA
Modified Haugen	30	95	70	5	NA	NA
Aspect	13	78	87	22	10	42
Permafrost map	3	53	97	47	3	83
FAO soil dataset	NA	NA	NA	NA	0	0

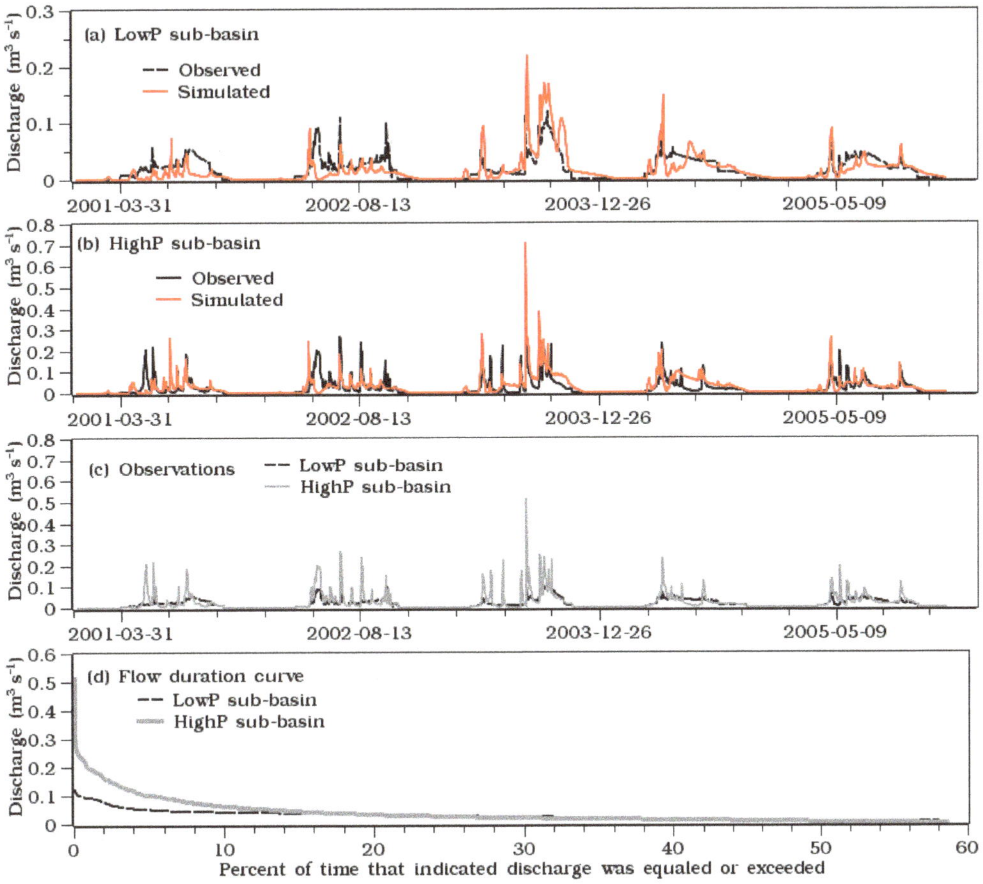

Figure 5. Observed versus simulated streamflow during the calibration period of 2001–2005: **(a)** at the LowP sub-basin and **(b)** at the HighP sub-basin of the CPCRW. The discharge contrasts between HighP and LowP sub-basins are shown in panels **(c)** and **(d)**. Panel **(c)** denotes the streamflow hydrograph and **(d)** denotes the flow duration curve in both sub-basins.

eterization method – is closest to the observed streamflow hydrograph for both sub-basins (Fig. 6a and c) with an R^2 /NSE of 0.51/0.48 and 0.53/0.52 for the LowP and HighP sub-basins, respectively (Table 5). Throughout the simulation period, the PD is also close to zero under the aspect parameterization compared to the other two parameterizations (Fig. 6b and d). Both peak and low flows in both sub-basins are well simulated in the small-scale parameterization scheme (Fig. 6). Mean annual PD values are also close to zero for streamflow simulations using the aspect parameter-

Table 4. Definition of the parameterization scenarios with respect to corresponding vegetation cover and soil property parameters.

Scenarios	Vegetation cover	Soil hydraulic property
Aspect	Small-scale landscape model (aspect)-based vegetation cover map (Fig. 3b)	Aspect-based soil property (Fig. 4c)
Permafrost	Permafrost distribution map (Fig. 1); permafrost areas are assumed coniferous while the remainder is deciduous	Soil hydraulic property derived from permafrost distribution map (Fig. 4b)
Large scale	SNAP vegetation cover (Fig. 3d)	FAO soil dataset (Fig. 4a)

ization (−7.6 to −14.4 and −2.2 to −8.3 % in the LowP and HighP sub-basins) compared to simulation under the large-scale parameterization (−17.7 to −25.4 and −11.5 to −15.9 % in the LowP and HighP sub-basins.).

3.3.2 Evapotranspiration

Unlike streamflow simulations, ET simulations generally do not display significant differences between the LowP and HighP sub-basins in all parameterizations. However, on an annual scale, there is a notable difference in the simulated ET between parameterization schemes in both sub-basins. The difference in simulated mean annual ET between LowP and HighP sub-basins under aspect parameterization (356.5 and 335.2 mm in the LowP and HighP sub-basins) is almost half of the values of ET simulation when permafrost (396.6 and 352.2 mm in the LowP and HighP sub-basins) and large-scale parameterization (402.8 and 362.8 mm in the LowP and HighP sub-basins) are used.

For further comparison between the LowP and HighP sub-basins (Fig. 7), ET and streamflow simulations under the aspect parameterization are selected due to the better streamflow simulation performance under aspect parameterization (Fig. 6). Generally, ET is higher than runoff most of the time in both sub-basins. The HighP sub-basin, however, displays more runoff than ET during snowmelt, late summer/fall, and large storm events (Fig. 7b).

3.4 Soil moisture

Variation in soil properties and vegetation cover is generally found to be sensitive to VIC's soil moisture simulation. We also note that the rates of snowmelt, melt/refreeze, and soil moisture increase/decrease between the parameterization schemes are notably different between the two sub-basins (Fig. 8). Unfrozen soil moisture content over the winter does not show any significant variation between the two sub-basins (Fig. 8). However, the change in storage in the LowP sub-basin is higher than in the HighP sub-basin. Additionally, the fluctuations of soil moisture and change in storage with the snowmelt and rainfall events are higher in the LowP sub-basin as compared to the fluctuation in the HighP sub-basin (Fig. 8). These results are consistent with previous observations (Young-Robertson et al., 2016) and modeling (Bolton, 2006) studies in the Interior Alaskan boreal forest ecosystem. These results indicate that soil moisture in the LowP sub-basin is more sensitive to climate forcing and land surface representation as compared to the wet system of the HighP sub-basin.

Table 5. Coefficient of determination (R^2) and Nash–Sutcliffe efficiency (NSE) values for the LowP and HighP sub-basins for calibration and validation periods.

	LowP sub-basin		HighP sub-basin	
	R^2	NSE	R^2	NSE
Calibration	0.51	0.17	0.48	0.38
Validation				
Large-scale parameterization	0.34	0.17	0.48	0.42
Permafrost parameterization	0.43	0.3	0.51	0.48
Aspect parameterization	0.51	0.48	0.53	0.52

4 Discussion

One of the important findings from our small-scale subgrid parameterization method, as shown by Figs. 3 and 4, is that an acceptable spatial vegetation cover and soil property representation of the Interior Alaskan sub-arctic can be produced without any direct measurements or remote sensing methods. Large-scale vegetation cover and soil property products that are typically used in mesoscale hydrological modeling do not accurately represent spatial heterogeneities of the Interior Alaskan boreal forest ecosystem. This is because of the fact that coarse-resolution data products include regions outside the watershed area that influence the average value of the grid cell, rather than capturing the spatial heterogeneity of ecosystem properties within the watershed boundaries. This study indicates that previously documented relationships among soil hydraulic and thermal properties, vegetation cover, topography, slope, and aspect (Viereck et al., 1983; Viereck and Van Cleve, 1984; Morrissey and Strong, 1986; Hinzman et al., 2006; Mölders, 2011) can be used to formulate a methodology by which local to regional scale

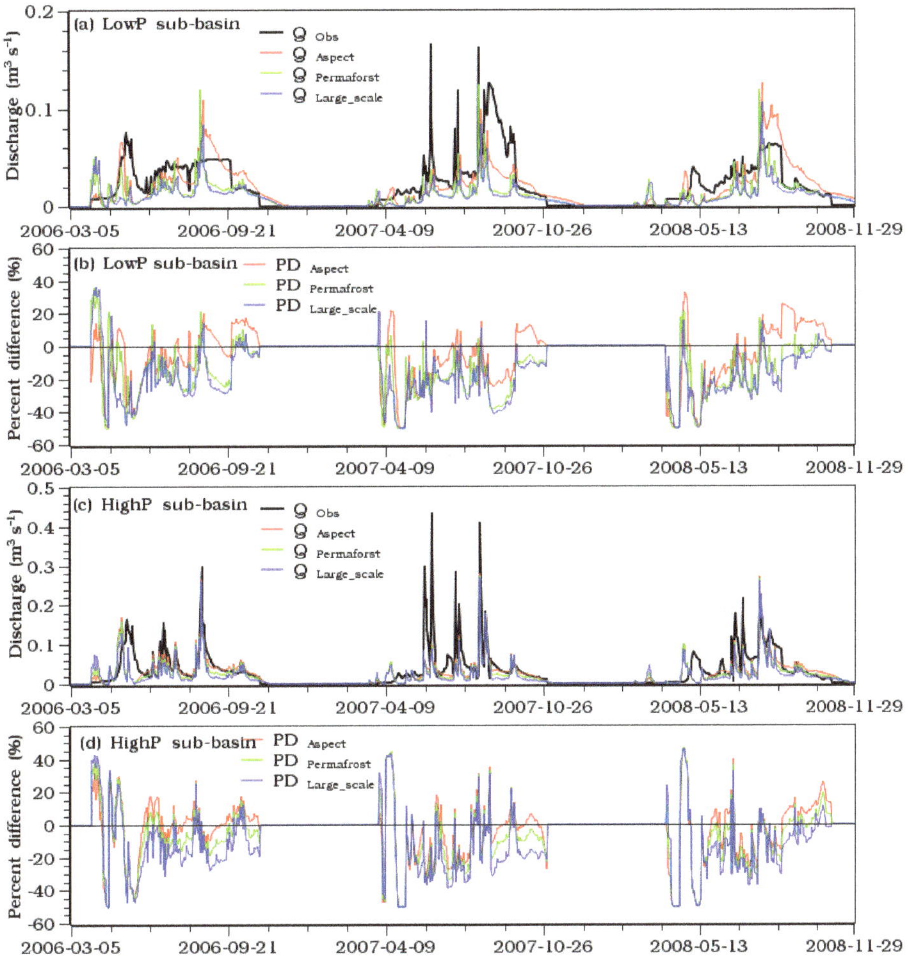

Figure 6. Comparison of streamflow simulations – obtained from different parameterizations – with observations **(a)** in the LowP sub-basin and **(c)** in the HighP sub-basin. Panels **(b)** and **(d)** show the percent difference (PD) for each runoff simulation from the observed streamflow in the LowP and HighP sub-basins, respectively.

landscape properties can be incorporated into mesoscale hydrological models.

The soil hydraulic properties obtained from the large-scale FAO soil data do not reflect any variation between permafrost-affected and permafrost-free soils (Fig. 4a). However, many surface and sub-surface soil hydraulic properties including the hydraulic conductivity and thickness of the organic layer show a large difference between permafrost-affected and permafrost-free soil (Rieger et al., 1972; Burt and Williams, 1976; Kane and Stein, 1983; Woo, 1986; Ping et al., 2005; Zhang et al., 2010a). Therefore, hydrological modeling using such data creates large uncertainty that cannot be easily corrected with model calibration, as the hydrology and landscape properties including the vegetation composition (Viereck et al., 1983; Viereck and Van Cleve, 1984; Morrissey and Strong, 1986; Hinzman et al., 2006; Mölders, 2011) are primarily controlled by the presence or absence of permafrost (Kane, 1980; Kane et al., 1981; Kane and Stein, 1983; Slaughter et al., 1983; Hinzman et al., 2002; Bolton, 2006; Petrone et al., 2006, 2007).

Our multi-objective model calibration indicates that watersheds with different permafrost proportions display consistent and contrasting parameter values between nearly permafrost-free and permafrost-dominated watersheds (Fig. 5, Table 2). As reported by previous studies (Bolton et al., 2000; Bolton, 2006; Hinzman et al., 2002), simulated runoff in permafrost-free watersheds is best fitted with observations for parameter values that favor high infiltration and baseflow and a longer recession period (Fig. 5a, Table 2). On the other hand, the best fit between simulated and observed streamflow in the permafrost-dominated watersheds (Fig. 5b) is achieved only when baseflow parameters that favor more direct runoff and inhibit infiltration (Table 2) are introduced into the model. During model calibration, especially for the heavily parameterized distributed models, it is important to consider these variabilities in the baseflow parameters of the discontinuous permafrost watersheds. It

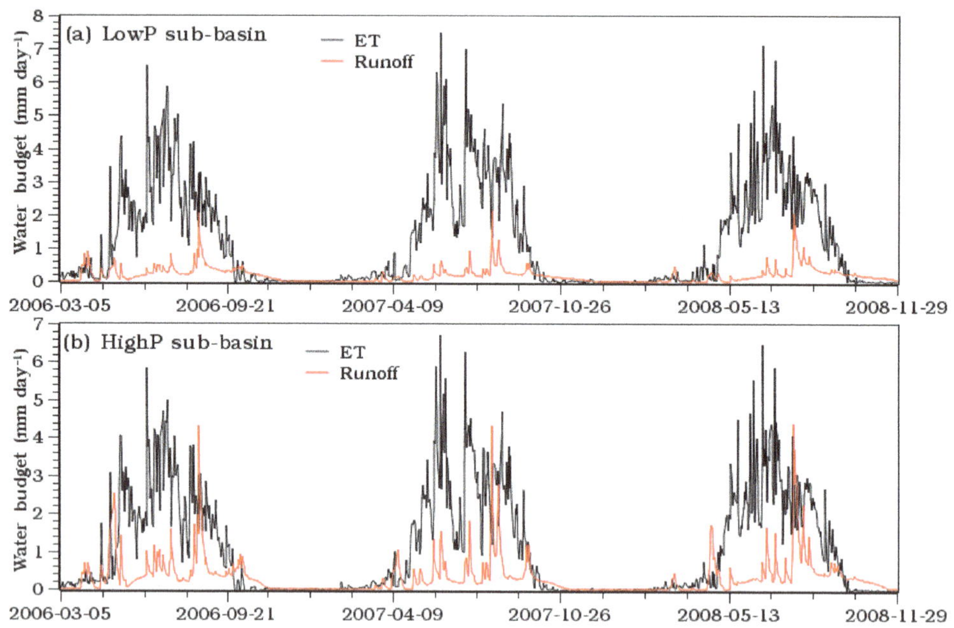

Figure 7. Comparison of evapotranspiration (ET) and runoff simulations – obtained from parameterization based on aspect parameterization **(a)** in the LowP sub-basin and **(b)** in the HighP sub-basin.

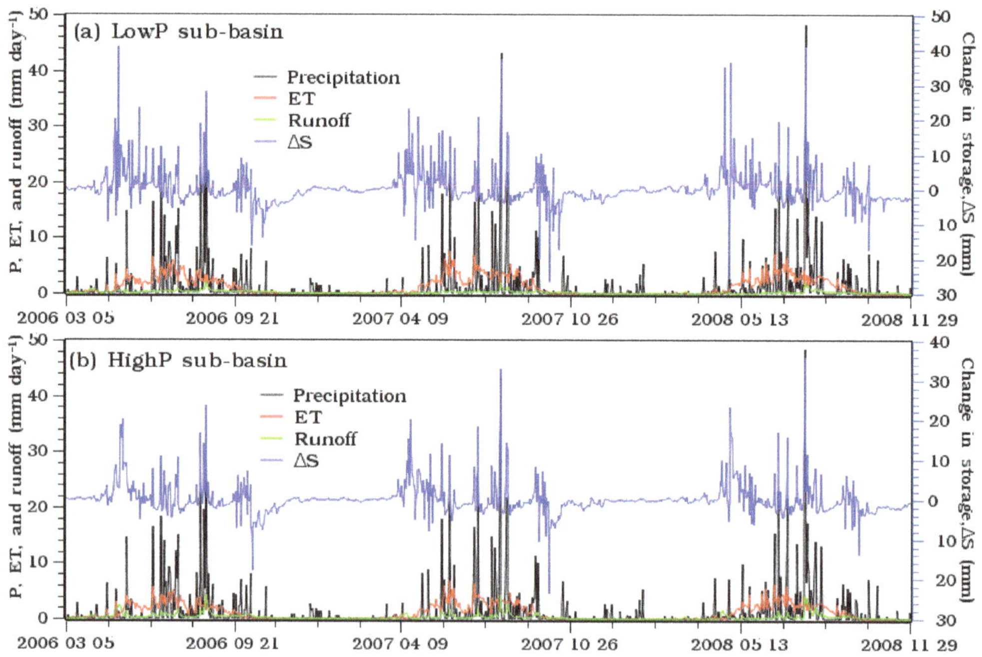

Figure 8. Partitioning of precipitation or snowmelt (P, precipitation) into evapotranspiration (ET), runoff, and change in soil moisture storage (ΔS) in the **(a)** LowP and **(b)** HighP sub-basins. All simulations are under aspect parameterization.

could save a considerable amount of time and resources, especially for big watersheds and limited computational facilities.

Based on the three model evaluation criteria (Eqs. 1–3) and the comparison of runoff behaviors between simulation and observation, streamflow simulations in both the permafrost-free/LowP (Fig. 6a and b) and permafrost-dominated/HighP (Fig. 6c and d) methods perform better when the small-scale sub-grid soil hydraulic and thermal properties (Fig. 4c) and vegetation cover (Fig. 3b) heterogeneity are introduced into the VIC model soil property and vegetation cover parameterizations. Comparison of the improvement between the LowP

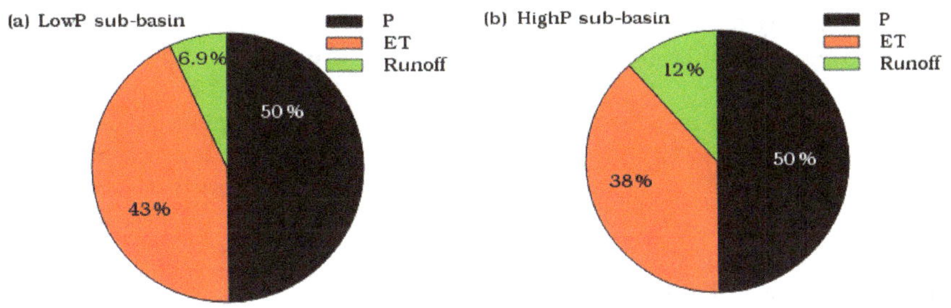

Figure 9. VIC-model-simulated (based on aspect parameterization) mean annual (2006–2008) percentage of water balance components, *P* (black), ET (red), and runoff (green) in the **(a)** LowP and **(b)** HighP sub-basins. The change in storage component is very small compared with the other components; hence, it is not included.

and HighP sub-basins indicates that the strongest streamflow simulation improvement (Tables 3 and 5) was observed in sub-basins that are highly unrepresented by the large-scale data products – the LowP sub-basin (Figs. 3, 4, and 6a and b). In general, our study shows that most of the peak and low flows in both sub-basins are captured well with the small-scale parameterization scheme as compared to the direct use of coarse-resolution land surface data products (Fig. 6, Table 5), except the 2007 spring flooding event (Fig. 6). The exception for the 2007 spring runoff peak can be explained by the high rainfall and mixed precipitation in the beginning of the 2006–2007 winter (nadp.isws.illinois.edu?/data?/sites?/siteDetails.aspx??id=AK01?&net?=NTN). Rainfall at the onset of winter generally freezes on the surface soil layer over the winter and limits snowmelt from infiltrating during the spring snowmelt period to generate flooding.

Although we did not detect significant differences in the ET simulations between parameterization methods, we found that sub-basin average ET in the LowP sub-basin is higher than in the HighP sub-basin, except at the beginning of snowmelt when deciduous trees of the LowP are storing snowmelt water but not yet transpiring (Young-Robertson et al., 2016; Figs. 7 and 9). Unlike other snow-dominated regions in middle latitudes, where ET tends to have a strong relationship with precipitation and a change in storage (Lohmann et al., 1998b, a), ET in this region is more strongly related to temperature than precipitation or change in soil moisture storage. This apparent decoupling between ET and precipitation is likely because the deciduous trees in this region are utilizing the snowmelt water stored in their trunks rather than being directly tied to rainfall (Young-Robertson et al., 2016). This implies that temperature is the limiting factor for ET in this region. This may have important implications for the general warming trend on a broader spectrum, as in the case of permafrost degradation (Romanovsky and Osterkamp, 1995, 2000; Romanovsky et al., 2002; O'Donnell et al., 2009).

Several studies have documented that VIC soil moisture simulation is strongly sensitive to how vegetation cover (Ford and Quiring, 2013; Tesemma et al., 2015) and soil properties (Liang et al., 1996; Lohmann et al., 1998b; Lee et al., 2011; Billah and Goodall, 2012; Ford and Quiring, 2013) are represented. This study shows that vegetation cover and soil property representation are not only sensitive to the soil moisture content simulations, but they also have a strong influence on the rate of snowmelt, and snowmelt and refreeze processes in the Alaskan sub-arctic environment. Small-scale parameterization schemes are the best at capturing the expected soil moisture pattern in both the LowP and HighP sub-basins. This includes high soil moisture in the permafrost-dominated sub-basin due to lower hydraulic conductivity of the soil and the faster winter freeze-up of the soil column in the permafrost-dominated sub-basin than in the permafrost-free one.

We also made an effort to understand how the annual water balance is partitioned into runoff, ET, and change in soil moisture in the permafrost-free/LowP and permafrost-dominated/HighP sub-basins. Both sub-basins do not differ in the percentage of each component between the large-scale and small-scale parameterizations. As shown in the previous studies (Hinzman et al., 2002; Bolton, 2006; Young-Robertson et al., 2016), our results indicate that in both sub-basins ET dominates the annual water balances (Figs. 8 and 9), while the change in storage is the lowest in both sub-basins. There are, however, variations in the percentage and pattern of the water balance components between LowP and HighP sub-basins. The runoff from LowP sub-basin is lower than that from the HighP sub-basin due to the high tree water storage and transpiration (Cable et al., 2014; Young-Robertson et al., 2016) of the deciduous trees and higher infiltration of the permafrost-free soil of the LowP sub-basin than HighP sub-basin. The mean annual change in storage in the LowP sub-basin is about 35 % higher than in the HighP sub-basin. From this result, we can conclude that the soil thermal and hydraulic properties dictate the partitioning of water into the different processes in this region.

This is the first study that introduces vegetation and soil property heterogeneity from high-resolution topographic information into the mesoscale hydrological model. The results need to be verified for regional-scale basins to determine the transferability of our findings to similar areas where the land surface data products do not adequately represent the spatial heterogeneity for accurate hydrological simulations.

5 Conclusions

This study indicates that coarse-resolution soil and vegetation data products – data that are used extensively in land surface modeling in the middle and lower latitudes – do not adequately represent the North American boreal forest discontinuous permafrost ecosystems. Hydrological modeling with coarse-resolution products cannot adequately simulate important small-scale processes. This is because small-scale permafrost distribution and ecosystem composition primarily control the land surface processes in this region. This indicates the need for landscape modeling or sub-grid parameterization that can produce these small-scale features and incorporate them into land surface models. The strong correlative relationship between topography, vegetation, and permafrost distribution (Burt and Williams, 1976; Haugen et al., 1982; Viereck et al., 1983; Hinzman et al., 1991, 1998) in this region can be used to produce a sub-grid parameterization method that represents the small-scale soil property and vegetation cover heterogeneity for distributed hydrological models.

This study also shows that our sub-grid parameterization method – based primarily on this strong relationship between permafrost, topography, and vegetation composition – produced a better representation of permafrost and vegetation cover than large-scale soil and vegetation cover products. Hydrological simulations – including basin-integrated and spatially variable runoff, evapotranspiration, and soil moisture dynamics – using a small-scale parameterization scheme derived from a sub-grid parameterization method are a notable improvement over parameterizations based on coarse-resolution data products. Finally, in our effort to demonstrate methodologies that can improve hydrological modeling through a small-scale parameterization method, we intend to implement and test results from this pilot study into a large river basin in the next phase of the research.

Competing interests. The authors declare that they have no conflict of interest.

Special issue statement. This article is part of the special issue "Observations and modeling of land surface water and energy exchanges across scales: special issue in Honor of Eric F. Wood". It is a result of the Symposium in Honor of Eric F. Wood: Observations and Modeling across Scales, Princeton, New Jersey, USA, 2–3 June 2016.

Acknowledgements. The financial support for this work was provided by the grant from the Department of Energy (SciDAC grant no. DE-SC0006913). This work was also made possible through financial support from the Alaskan Climate Science Center, funded by the Cooperative Agreement G10AC00588 from the United States Geological Survey. Its contents are solely the responsibility of the author and do not necessarily represent the official view of the USGS.

Edited by: Albert van Dijk

References

Abdulla, F. A. and Lettenmaier, D. P.: Application of Regional Parameter Estimation Schemes to Simulate the Water Balance of a Large Continental River, J. Hydrol., 197, 258–285, 1997a.

Abdulla, F. A. and Lettenmaier, D. P.: Development of Regional Parameter Estimation Equations for a Macroscale Hydrologic Model, J. Hydrol., 197, 230–257, 1997b.

Abdulla, F. A., Lettenmaier, D. P., Wood, E. F., and Smith, J. A.: Application of a Macroscale Hydrologic Model to Estimate the Water Balance of the Arkansas-Red River Basin, J. Geophys. Res.-Atmos., 101, 7449–7459, 1996.

Andreadis, K. M., Clark, E. A., Wood, A. W., Hamlet, A. F., and Lettenmaier, D. P.: Twentieth-Century Drought in the Conterminous United States, J. Hydrometeorol., 6, 985–1001, 2005.

Andreadis, K. M., Storck, P., and Lettenmaier, D. P.: Modeling Snow Accumulation and Ablation Processes in Forested Environments, Water Resour. Res., 45, W05429, https://doi.org/10.1029/2008WR007042, 2009.

Aster, G. V. T.: Aster Global Dem Validation Summary Report, Technical Report No. 1, Meti/Ersdac, Nasa/Lpdaac, Usgs/Eros, 2009.

Baldocchi, D., Kelliher, F. M., Black, T. A., and Jarvis, P.: Climate and Vegetation Controls on Boreal Zone Energy Exchange, Glob. Change Biol., 6, 69–83, 2000.

Bennett, K. E.: Changes in Extreme Hydroclimate Events in Interior Alaskan Boreal Forest Watersheds, PhD dissertation, University of Alaska, Fairbanks, 2014.

Bennett, K. E., Werner, A. T., and Schnorbus, M.: Uncertainties in Hydrologic and Climate Change Impact Analyses in Headwater Basins of British Columbia, J. Climate, 25, 5711–5730, 2012.

Bennett, K. E., Cannon, A. J., and Hinzman, L. D.: Historical Trends and Extremes in Boreal Alaska River Basins, J. Hydrol., 527, 590–607, 2015.

Billah, M. M. and Goodall, J. L.: Applying Drought Analysis in the Variable Infiltration Capacity (VIC) Model for South Carolina, South Carolina Water Resources Conference, 2012.

Bohn, T. J., Livneh, B., Oyler, J. W., Running, S. W., Nijssen, B., and Lettenmaier, D. P.: Global Evaluation of Mtclim and Related Algorithms for Forcing of Ecological and Hydrological Models, Agr. Forest Meteorol., 176, 38–49, 2013.

Bolton, W. R.: Dynamic Modeling of the Hydrologic Processes in Areas of Discontinuous Permafrost, PhD Dissertation, University of Alaska, Fairbanks, University of Alaska Fairbanks, 2006.

Bolton, W. R., Hinzman, L. D., and Yoshikawa, K.: Stream Flow Studies in a Watershed Underlain by Discontinuous Permafrost,

Paper presented at the American Water Resources Association Proceedings on Water Resources in Extreme Environments, 1–3, 2000.

Bolton, W. R., Hinzman, L. D., Jones, K. C. P., Jeremy, B., and Adams, P. C.: Watershed Hydrology and Chemistry in the Alaskan Boreal Forest the Central Role of Permafrost Alaska's Changing Boreal Forest, Oxford University Press, 269 pp., 2006.

Bowling, L. C., Cherkauer, K. A., and Adam, J. C.: Current Capabilities in Soil Thermal Representations within a Large Scale Hydrology Model, Paper presented at the 9th International Conference on Permafrost, 2008.

Burrough, P. A., Mcdonnell, R., Burrough, P. A., and Mcdonnell, R.: Principles of Geographical Information Systems, vol. 333, Oxford university press Oxford, 1998.

Burt, T. P. and Williams, P. J.: Hydraulic Conductivity in Frozen Soils, Earth Surf. Processes, 1, 349–360, 1976.

Cable, J. M., Ogle, K., Bolton, W. R., Bentley, L. P., Romanovsky, V., Iwata, H., Harazono, Y., and Welker, J.: Permafrost Thaw Affects Boreal Deciduous Plant Transpiration through Increased Soil Water, Deeper Thaw, and Warmer Soils, Ecohydrology, 7, 982–997, https://doi.org/10.1002/eco.1423, 2014.

Carey, S. K. and Woo, M. K.: Slope Runoff Processes and Flow Generation in a Subarctic, Subalpine Catchment, J. Hydrol., 253, 110–129, 2001.

Cherkauer, K. A. and Lettenmaier, D. P.: Hydrologic Effects of Frozen Soils in the Upper Mississippi River Basin, J. Geophys. Res.-Atmos., 104, 19599–19610, 1999.

Cherkauer, K. A., Bowling, L. C., and Lettenmaier, D. P.: Variable Infiltration Capacity Cold Land Process Model Updates, Global Planet. Change, 38, 151–159, 2003.

Dingman, S. L.: Effects of Permafrost on Stream Flow Characteristics in the Discontinuous Permafrost Zone of Central Alaska, Paper presented at the North American Contribution to Second International Conference of Permafrost, National Academy of Sciences, Washington, DC, 447–453, 1973.

Duband, D., Obled, C., and Rodriguez, J. Y.: Unit Hydrograph Revisited: An Alternate Iterative Approach to Uh and Effective Precipitation Identification, J. Hydrol., 150, 115–149, 1993.

Ewers, B. E., Gower, S. T., Bond-Lamberty, B., and Wang, C.: Effects of Stand Age and Tree Species Composition on Transpiration and Canopy Conductance of Boreal Forest Stands, Plant Cell Environ., 28, 660–678, 2005.

FAO: Digital Soil Map of the World and Derived Soil Properties, Land and Water Digital Media Series 1 Food and Agriculture Organization of the United States (FAO), Rome, Italy, 1998.

Ford, T. W. and Quiring, S. M.: Influence of Modis-Derived Dynamic Vegetation on VIC-Simulated Soil Moisture in Oklahoma, J. Hydrometeorol., 14, 1910–1921, 2013.

Hansen, M. C., Defries, R. S., Townshend, J. R. G., and Sohlberg, R.: Global Land Cover Classification at 1 Km Spatial Resolution Using a Classification Tree Approach, Int. J. Remote Sens., 21, 1331–1364, 2000.

Haugen, R. K., Slaughter, C. W., Howe, K. E., and Dingman, S. L.: Hydrology and Climatology of the Caribou-Poker Creeks Research Watershed, Alaska, US Army Corps of Engineers, Cold Regions Research and Engineering Laboratory Hanover, NH, 1982.

Hinzman, L. D., Kane, D. L., Gieck, R. E., and Everett, K. R.: Hydrologic and Thermal Properties of the Active Layer in the Alaskan Arctic, Cold Reg. Sci. Technol., 19, 95–110, 1991.

Hinzman, L. D., Goering, D. J., and Kane, D. L.: A Distributed Thermal Model for Calculating Soil Temperature Profiles and Depth of Thaw in Permafrost Regions, J. Geophys. Res.-Atmos., 103, 28975–28991, 1998.

Hinzman, L. D., Ishikawa, N., Yoshikawa, K., Bolton, W. R., and Petrone, K. C.: Hydrologic Studies in Caribou-Poker Creeks Research Watershed in Support of Long Term Ecological Research, Eurasian Journal of Forest Research, 5, 67–71, 2002.

Hinzman, L. D., Fukuda, M., Sandberg, D. V., Chapin, F. S., and Dash, D.: Frostfire: An Experimental Approach to Predicting the Climate Feedbacks from the Changing Boreal Fire Regime, J. Geophys. Res.-Atmos., 108, 8153, https://doi.org/10.1029/2001JD000415, 2003.

Hinzman, L. D., Viereck, L. A., Adams, P. C., Romanovsky, V. E., and Yoshikawa, K.: Climate and Permafrost Dynamics of the Alaskan Boreal Forest Alaska's Changing Boreal Forest, Oxford University Press, New York, 39–61, 2006.

Jones, J. B. and Rinehart, A. J.: The Long-Term Response of Stream Flow to Climatic Warming in Headwater Streams of Interior Alaska, Can. J. Forest Res., 40, 1210–1218, 2010.

Kane, D. L.: Snowmelt Infiltration into Seasonally Frozen Soils, Cold Reg. Sci. Technol., 3, 153–161, 1980.

Kane, D. L. and Stein, J.: Water Movement into Seasonally Frozen Soils, Water Resour. Res., 19, 1547–1557, 1983.

Kane, D. L., Bredthauer, S. R., and Stein, J.: Subarctic Snowmelt Runoff Generation, Paper presented at the Conference on The Northern Community, edited by: Vinson, T. S., ASCE, Seattle, Washington, 591–601, 1981.

Kirkby, M. J.: Hillslope Hydrology, Wiley-Interscience, New York, New York, 348 pp., 1978.

Knudson, J. A. and Hinzman, L. D.: Streamflow Modeling in an Alaskan Watershed Underlain by Permafrost, in: Water Resources in Extreme Environments, American Water Resources Association, 309–313, 2000.

Krause, P., Boyle, D. P., and Bäse, F.: Comparison of Different Efficiency Criteria for Hydrological Model Assessment, Adv. Geosci., 5, 89–97, 2005.

Kumar, R., Samaniego, L., and Attinger, S.: Implications of Distributed Hydrologic Model Parameterization on Water Fluxes at Multiple Scales and Locations, Water Resour. Res., 49, 360–379, 2013.

Lee, H., Seo, D.-J., and Koren, V.: Assimilation of Streamflow and in Situ Soil Moisture Data into Operational Distributed Hydrologic Models: Effects of Uncertainties in the Data and Initial Model Soil Moisture States, Adv. Water Resour., 34, 1597–1615, 2011.

Liang, X., Lettenmaier, D. P., Wood, E. F., and Burges, S. J.: A Simple Hydrologically Based Model of Land Surface Water and Energy Fluxes for General Circulation Models, J. Geophys. Res.-Atmos., 99, 14415–14428, 1994.

Liang, X., Wood, E. F., and Lettenmaier, D. P.: Surface Soil Moisture Parameterization of the VIC-2l Model: Evaluation and Modification, Global Planet. Change, 13, 195–206, 1996.

Lohmann, D., Nolte-Holube, R., and Raschke, E.: A Large-Scale Horizontal Routing Model to Be Coupled to Land Surface Parametrization Schemes, Tellus A, 48, 708–721, 1996.

Lohmann, D., Raschke, E., Nijssen, B., and Lettenmaier, D. P.: Regional Scale Hydrology: I. Formulation of the VIC-21 Model Coupled to a Routing Model, Hydrol. Sci. J., 43, 131–141, 1998a.

Lohmann, D., Raschke, E., Nijssen, B., and Lettenmaier, D. P.: Regional Scale Hydrology: II. Application of the VIC-21 Model to the Weser River, Germany, Hydrol. Sci. J., 43, 143–158, 1998b.

Meng, L. and Quiring, S. M.: A Comparison of Soil Moisture Models Using Soil Climate Analysis Network Observations, J. Hydrometeorol., 9, 641–659, 2008.

Mölders, N.: Land-Use and Land-Cover Changes: Impact on Climate and Air Quality, in: Vol. 44, Springer Science & Business Media, 2011.

Morrissey, L. A. and Strong, L. L.: Mapping Permafrost in the Boreal Forest with Thematic Mapper Satellite Data, Photogramm. Engi. Rem. S., 52, 1513–1520, 1986.

Myneni, R. B., Ramakrishna, R., Nemani, R., and Running, S. W.: Estimation of Global Leaf Area Index and Absorbed Par Using Radiative Transfer Models, IEEE T. Geosci. Remote, 35, 1380–1393, 1997.

Naito, A. T. and Cairns, D. M.: Patterns and Processes of Global Shrub Expansion, Prog. Phys. Geogr., 35, 423–442, 2011.

Nijssen, B., Lettenmaier, D. P., Liang, X., Wetzel, S. W., and Wood, E. F.: Streamflow Simulation for Continental-Scale River Basins, Water Resour. Res., 33, 711–724, 1997.

Nijssen, B., O'Donnell, G. M., Hamlet, A. F., and Lettenmaier, D. P.: Hydrologic Sensitivity of Global Rivers to Climate Change, Climatic Change, 50, 143–175, 2001a.

Nijssen, B., O'Donnell, G. M., Lettenmaier, D. P., Lohmann, D., and Wood, E. F.: Predicting the Discharge of Global Rivers, J. Climate, 14, 3307–3323, 2001b.

O'Donnell, J. A., Romanovsky, V. E., Harden, J. W., and Mcguire, A. D.: The Effect of Moisture Content on the Thermal Conductivity of Moss and Organic Soil Horizons from Black Spruce Ecosystems in Interior Alaska, Soil Sci., 174, 646–651, 2009.

Parajka, J., Merz, R., and Blöschl, G.: A comparison of regionalisation methods for catchment model parameters, Hydrol. Earth Syst. Sci., 9, 157–171, https://doi.org/10.5194/hess-9-157-2005, 2005.

Petrone, K. C., Jones, J. B., Hinzman, L. D., and Boone, R. D.: Seasonal Export of Carbon, Nitrogen, and Major Solutes from Alaskan Catchments with Discontinuous Permafrost, J. Geophys. Res.-Biogeo., 111, G02020, https://doi.org/10.1029/2005JG000055, 2006.

Petrone, K. C., Hinzman, L. D., Shibata, H., Jones, J. B., and Boone, R. D.: The Influence of Fire and Permafrost on Sub-Arctic Stream Chemistry During Storms, Hydrol. Process., 21, 423–434, 2007.

Ping, C., Michaelson, G., Packee, E., Stiles, C., Swanson, D., and Yoshikawa, K.: Soil Catena Sequences and Fire Ecology in the Boreal Forest of Alaska, Soil Sci. Soc. Am. J., 69, 1761–1772, 2005.

Pokhrel, P., Gupta, H. V., and Wagener, T.: A Spatial Regularization Approach to Parameter Estimation for a Distributed Watershed Model, Water Resour. Res., 44, W12419, https://doi.org/10.1029/2007WR006615, 2008.

Quinton, W. L. and Carey, S. K.: Towards an Energy-Based Runoff Generation Theory for Tundra Landscapes, Hydrol. Process., 22, 4649–4653, 2008.

Rawls, W. J. and Brakensiek, D. L.: Prediction of Soil Water Properties for Hydrologic Modeling, ASCE, 1985.

Rieger, S., Furbush, C. E., Schoephorster, D. B., Summerfield Jr., H., and Geiger, L. C.: Soils of the Caribou-Poker Creeks Research Watershed, Interior Alaska (No. Crrel-Tr-236), Cold regions research and engineering Lab Hanover NH, 1972.

Robock, A., Luo, L., Wood, E. F., Wen, F., Mitchell, K. E., Houser, P. R., Schaake, J. C., Lohmann, D., Cosgrove, B., and Sheffield, J.: Evaluation of the North American Land Data Assimilation System over the Southern Great Plains During the Warm Season, J. Geophys. Res.-Atmos., 108, 8846, https://doi.org/10.1029/2002JD003245, 2003.

Romanovsky, V. E. and Osterkamp, T. E.: Interannual Variations of the Thermal Regime of the Active Layer and near-Surface Permafrost in Northern Alaska, Permafrost Periglac., 6, 313–335, 1995.

Romanovsky, V. E. and Osterkamp, T. E.: Effects of Unfrozen Water on Heat and Mass Transport Processes in the Active Layer and Permafrost, Permafrost Periglac., 11, 219–239, 2000.

Romanovsky, V. E., Burgess, M., Smith, S., Yoshikawa, K., and Brown, J.: Permafrost Temperature Records: Indicators of Climate Change, EOS, Transactions American Geophysical Union, 83, 589–594, 2002.

Romanovsky, V. E., Sergueev, D. O., and Osterkamp, T. E.: Temporal Variations in the Active Layer and near-Surface Permafrost Temperatures at the Long-Term Observatories in Northern Alaska, Paper presented at the the 8th International Conference on Permafrost, 21–25 July, 2003, Zurich, Switzerland, vol. 2, edited by: Phillips, M., Springman, S. M., and Arenson, L. U., A. A. Balkema, Brookfield, Vt., 989–994, 2003.

Samaniego, L., Kumar, R., and Attinger, S.: Multiscale Parameter Regionalization of a Grid-Based Hydrologic Model at the Mesoscale, Water Resour. Res., 46, W05523, https://doi.org/10.1029/2008WR007327, 2010.

Saxton, K. E., Rawls, W. J., Romberger, J. S., and Papendick, R. I.: Estimating Generalized Soil-Water Characteristics from Texture, Soil Sci. Soc. Am. J., 50, 1031–1036, 1986.

Schaefli, B. and Gupta, H. V.: Do Nash Values Have Value?, Hydrol. Process., 21, 2075–2080, 2007.

Schenk, H. J. and Jackson, R. B.: Islscp Ii Ecosystem Rooting Depths. ISLSCP Initiative II Collection, Data set, Oak Ridge National Laboratory Distributed Active Archive Center, Oak Ridge, Tennessee, USA, 10, available: http://daac.ornl.gov/ (last access: 13 September 2013), 2009.

Schnorbus, M. A., Bennett, K. E., Werner, A. T., and Berland, A. J.: Hydrologic Impacts of Climate Change in the Peace, Campbell and Columbia Watersheds, British Columbia, Canada, Pacific Climate Impacts Consortium, University of Victoria, Victoria, BC, 157, 2011.

Slater, A. G., Bohn, T. J., Mccreight, J. L., Serreze, M. C., and Lettenmaier, D. P.: A Multimodel Simulation of Pan-Arctic Hydrology, J. Geophys. Res.-Biogeo., 112, G04S45, https://doi.org/10.1029/2006JG000303, 2007.

Slaughter, C. W. and Kane, D. L.: Hydrologic Role of Shallow Organic Soils in Cold Climates, Paper presented at the Canadian Hydrology Symposium: 79 Cold Climate Hydrology, Vancouver, B.C., 1979.

Slaughter, C. W., Hilgert, J. W., and Culp, E. H.: Summer Streamflow and Sediment Yield from Discontinuous-Permafrost Headwaters Catchments, Paper presented at the Permafrost, Fourth International Conference, Proceedings, 17–22 July 1983, Fairbanks, Alaska, National Academy Press, Washington, D.C., 1172–1177, 1983.

Tesemma, Z. K., Wei, Y., Peel, M. C., and Western, A. W.: Including the dynamic relationship between climatic variables and leaf area index in a hydrological model to improve streamflow prediction under a changing climate, Hydrol. Earth Syst. Sci., 19, 2821–2836, https://doi.org/10.5194/hess-19-2821-2015, 2015.

Thornton, P. E. and Running, S. W.: An Improved Algorithm for Estimating Incident Daily Solar Radiation from Measurements of Temperature, Humidity, and Precipitation, Agr. Forest Meteorol., 93, 211–228, 1999.

Troy, T. J., Wood, E. F., and Sheffield, J.: An Efficient Calibration Method for Continental-Scale Land Surface Modeling, Water Resour. Res., 44, W09411, https://doi.org/10.1029/2007WR006513, 2008.

Viereck, L. A. and Van Cleve, K.: Some Aspects of Vegetation and Temperature Relationships in the Alaska Taiga, Miscellaneous publication-University of Alaska, Agricultural and Forestry Experiment Station, USA, 1984.

Viereck, L. A., Dyrness, C. T., Cleve, K. V., and Foote, M. J.: Vegetation, Soils, and Forest Productivity in Selected Forest Types in Interior Alaska, Can. J. Forest Res., 13, 703–720, 1983.

Wagener, T., Boyle, D. P., Lees, M. J., Wheater, H. S., Gupta, H. V., and Sorooshian, S.: A framework for development and application of hydrological models, Hydrol. Earth Syst. Sci., 5, 13–26, https://doi.org/10.5194/hess-5-13-2001, 2001.

Wahrhaftig, C.: Physiographic Divisions of Alaska: US Geological Survey Professional Paper 482, US Geological Survey, 1965.

Walsh, J. E., Anisimov, O., Hagen, J. O. M., Jakobsson, T., Oerlemans, J., Prowse, T. D., Romanovsky, V., Savelieva, N., Serreze, M., and Shiklomanov, A.: Cryosphere and Hydrology, Arctic climate impact assessment, 183–242, 2005.

White, D., Autier, V., Yoshikawa, K., Jones, J., and Seelen, S.: Using Doc to Better Understand Local Hydrology in a Subarctic Watershed, Cold Reg. Sci. Technol., 51, 68–75, 2008.

Woo, M.-K.: Hydrology of a Small Canadian High Arctic Basin During the Snowmelt Period, Catena, 3, 155–168, 1976.

Woo, M.-K.: Permafrost Hydrology in North America 1, Atmosphere-Ocean, 24, 201–234, 1986.

Woo, M.-K. and Steer, P.: Slope Hydrology as Influenced by Thawing of the Active Layer, Resolute, Nwt, Can. J. Earth Sci., 20, 978–986, 1983.

Wu, Z., Lu, G., Wen, L., Lin, C. A., Zhang, J., and Yang, Y.: Thirty-Five Year (1971–2005) Simulation of Daily Soil Moisture Using the Variable Infiltration Capacity Model over China, Atmosphere-Ocean, 45, 37–45, 2007.

Wu, Z., Mao, Y., Lu, G., and Zhang, J.: Simulation of Soil Moisture for Typical Plain Region Using the Variable Infiltration Capacity Model, Proceedings of the International Association of Hydrological Sciences, 368, 215–220, 2015.

Yapo, P. O., Gupta, H. V., and Sorooshian, S.: Multi-Objective Global Optimization for Hydrologic Models, J. Hydrol., 204, 83–97, 1998.

Yoshikawa, K., Hinzman, L. D., and Gogineni, P.: Ground Temperature and Permafrost Mapping Using an Equivalent Latitude/Elevation Model, J. Glaciol. Geocryol., 24, 526–532, 2002.

Young-Robertson, J. M., Bolton, W. R., Bhatt, U. S., Cristobal, J., and Thoman, R.: Deciduous Trees Are a Large and Overlooked Sink for Snowmelt Water in the Boreal Forest, Sci. Rep., 6, 29504, https://doi.org/10.1038/srep29504, 2016.

Yuan, W., Luo, Y., Richardson, A. D., Oren, R. A. M., Luyssaert, S., Janssens, I. A., Ceulemans, R., Zhou, X., Gruenwald, T., and Aubinet, M.: Latitudinal Patterns of Magnitude and Interannual Variability in Net Ecosystem Exchange Regulated by Biological and Environmental Variables, Glob. Change Biol., 15, 2905–2920, 2009.

Zhang, Y., Carey, S. K., Quinton, W. L., Janowicz, J. R., Pomeroy, J. W., and Flerchinger, G. N.: Comparison of algorithms and parameterisations for infiltration into organic-covered permafrost soils, Hydrol. Earth Syst. Sci., 14, 729–750, https://doi.org/10.5194/hess-14-729-2010, 2010a.

Zhang, Y., Carey, S. K., Quinton, W. L., Janowicz, J. R., Pomeroy, J. W., and Flerchinger, G. N.: Comparison of algorithms and parameterisations for infiltration into organic-covered permafrost soils, Hydrol. Earth Syst. Sci., 14, 729–750, https://doi.org/10.5194/hess-14-729-2010, 2010b.

5

Calibration of a parsimonious distributed ecohydrological daily model in a data-scarce basin by exclusively using the spatio-temporal variation of NDVI

Guiomar Ruiz-Pérez[1,6], Julian Koch[2,3], Salvatore Manfreda[4], Kelly Caylor[5], and Félix Francés[6]

[1]Department of Crop Production Ecology, Swedish University of Agricultural Sciences, Uppsala, Sweden
[2]Department of Hydrology, Geological Survey of Denmark and Greenland, Copenhagen, Denmark
[3]Department of Geosciences and Natural Resources Management, University of Copenhagen, Copenhagen Denmark
[4]Department of European and Mediterranean Cultures, University of Basilicata, Matera, Italy
[5]Bren School of Environmental Science and Management, Department of Geography, UC Santa Barbara, Santa Barbara, California, USA
[6]Research Group of Hydrological and Environmental Modelling (GIHMA), Research Institute of Water and Environmental Engineering, Universitat Politècnica de València, Valencia, Spain

Correspondence to: Guiomar Ruiz-Pérez (Guiomar.ruiz.perez@slu.se)

Abstract. Ecohydrological modeling studies in developing countries, such as sub-Saharan Africa, often face the problem of extensive parametrical requirements and limited available data. Satellite remote sensing data may be able to fill this gap, but require novel methodologies to exploit their spatio-temporal information that could potentially be incorporated into model calibration and validation frameworks.

The present study tackles this problem by suggesting an automatic calibration procedure, based on the empirical orthogonal function, for distributed ecohydrological daily models. The procedure is tested with the support of remote sensing data in a data-scarce environment – the upper Ewaso Ngiro river basin in Kenya. In the present application, the TETIS-VEG model is calibrated using only NDVI (Normalized Difference Vegetation Index) data derived from MODIS. The results demonstrate that (1) satellite data of vegetation dynamics can be used to calibrate and validate ecohydrological models in water-controlled and data-scarce regions, (2) the model calibrated using only satellite data is able to reproduce both the spatio-temporal vegetation dynamics and the observed discharge at the outlet and (3) the proposed automatic calibration methodology works satisfactorily and it allows for a straightforward incorporation of spatio-temporal data into the calibration and validation framework of a model.

1 Introduction

Drylands cover 30 % of the Earth's land surface and 50 % of Africa (Franz et al., 2010). Projections of the IPCC (Intergovernmental Panel on Climate Change, 2007) indicate that the extent of these regions will likely increase in the coming decades. Dryland expansion would have a considerable additional impact on water resources, which should be taken into account by water management plans (Franz et al., 2010).

In water-controlled ecosystems, the vegetation assumes a critical role influencing all components of the hydrological cycle (Rodriguez-Iturbe et al., 2001; Manfreda and Caylor, 2013). For instance, actual evapotranspiration (aET) may account for more than 90 % of the annual precipitation in water-controlled areas (Zhang et al., 2016; Jasechko et al., 2013). Montaldo et al. (2005) affirmed that the use of constant LAI (leaf area index) values, commonly used in hydrological applications, produces large errors in land surface flux predictions. Given the strong control exerted on aET by the vegetation, reliable estimates of spatio-temporal variations of vegetation patterns are vital to obtain trustworthy predictions of available water resources (Andersen, 2008). In this sense, ecohydrological modeling becomes essential in order to include the vegetation dynamics as an additional state variable (Rodriguez-Iturbe et al., 2001).

Evidence of aET being a prevalent driver of hydrological records of streamflow and water table depth, i.e., available water resources, has been observed in many studies (Bond et al., 2002; Nyholm et al., 2003; Gribovszki et al., 2008; Conradt et al., 2013). Recently, Tsang et al. (2014) showed that adding a better evapotranspiration scheme in a widely used runoff model improves streamflow predictions. Conradt et al. (2013), who compared three different strategies for deriving sub-basin aET, affirmed that incorporating spatial variation of aET in a semi-distributed model increases its robustness. Conversely, Stisen et al. (2011) and others stressed that those improvements are not necessarily seen in the outlet hydrograph. However, it could also be interpreted in the inverse sense; good performances in terms of the outlet hydrograph do not necessarily mean more reliable estimates of aET.

The streamflow record is traditionally the only observation used for the calibration of hydrological models, but several studies demonstrated the limited capabilities of such an approach when models are validated at interior points of a river basin. Discharge represents an integrated catchment response, and hence provides only limited insight on the lumped behavior of a catchment (Stisen et al., 2011; Koch et al., 2016a; Michaud and Sorooshian, 1994; Reed et al., 2004; Smith et al., 2013). In that sense, Conradt et al. (2013) provided several examples for large simulation errors within the model domain and they mentioned, among others, the outcomes given by Feyen et al. (2008), Merz et al. (2009) and Smith et al. (2012). Moreover, Wi et al. (2015) also pointed out that caution is needed when using an outlet calibration approach for streamflow predictions under future climate conditions. This leads to the idea of using spatial state variables, and the new era of distributed (temporal and spatial) models emerges in order to balance the conceptual distributed nature of this kind of model (Stisen et al., 2011).

Traditional observation, which generally consists of point data with little spatial support, is effectively strengthened by remote sensing data, which offer the capacity to provide detailed spatial coverage and pattern information (Franssen et al., 2008; McCabe et al., 2008; Stisen et al., 2011). Additionally, satellite data have the great advantage of also being available in data-scarce areas. In this sense, the application of remotely sensed data represents an excellent source that provides information with a fairly good spatial and temporal resolution (Yang et al., 2012). In modeling, remote sensing data have been utilized in three different ways: (1) as forcing data (Xiao et al., 2004; Yuan et al., 2010; Samaniego et al., 2011; Stisen et al., 2011), (2) as a priori information of particular parameters (Winsemnius et al., 2008; Stisen et al., 2011) and (3) for model calibration and validation (see next section for an in-depth discussion of this point).

Satellite imagery provides not only temporal information but also valuable information on spatial patterns, which can facilitate a spatial-pattern-orientated model evaluation. As highlighted by Koch et al. (2015), spatial model evaluation is an active field of research not only in hydrology but also in other disciplines such as atmospheric sciences (Brown et al., 2011; Gilleland et al., 2010). However, up to now, there exists no formal guideline on how to assess the goodness of fit of the simulated spatial patterns and little information can be found about how to effectively utilize spatio-temporal data. Some authors, such as Conradt et al. (2013), Graf et al. (2014) and Koch et al. (2015, 2016b), focused on developing and testing metrics to be employed when spatio-temporal data are involved. For example, Koch et al. (2015) compared kappa statistics, fuzzy theory and empirical orthogonal function (EOF) analysis in an attempt towards a true spatial model evaluation of distributed models. But, besides these efforts, there are only a limited number of spatial validation studies that fully embrace the availability of satellite remote sensing data by means of true spatial performance metrics (Koch et al., 2016b). EOF analysis is a versatile methodology to investigate the spatio-temporal patterns of fluxes and states in the soil–vegetation–atmosphere continuum (Fang et al., 2015). As mentioned previously, Koch et al. (2015) carried out a validation of a distributed model using satellite-based land surface temperature data by means of an EOF analysis. With other statistical purposes, the EOF analysis was used by Graf et al. (2014), Kim and Barros (2002) and Liu (2003). A fine-scale study was carried out by Drewry and Albertson (2006) who used EOF analysis to associate spatial pattern in the errors of a canopy-atmosphere model with errors in the parameters. However, to our knowledge, EOF analysis has not been applied in model calibration yet.

In summary, the main objectives of this research are the following: (1) to incorporate spatio-temporal data into the calibration process by applying the EOF methodology as an objective function and (2) to exploit satellite data as a proxy of reliable estimates of vegetation dynamics for both the calibration and validation of an ecohydrological model. To address these key challenges, a distributed parsimonious ecohydrological model was applied in a water-controlled basin located in Kenya.

2 Satellite data and model calibration and validation

The applicability of remote sensing to calibrate and/or validate a model by exploiting information on spatial patterns still remains a challenging task. In order to better understand this issue in more detail, a bibliographic survey of the ISI Web of Knowledge science citation index was undertaken using the following word combinations in the topic search: (1) satellite calibration, (2) satellite implementation, (3) satellite ecohydrological modeling and (4) remote sensing ecohydrology. From the total number of publications obtained by this search, only those that incorporated satellite data to specifically model calibration were selected.

On the plot scale, Quevedo and Francés (2008) and Pasquato et al. (2015) calibrated and validated a parsimonious ecohydrological model using satellite information.

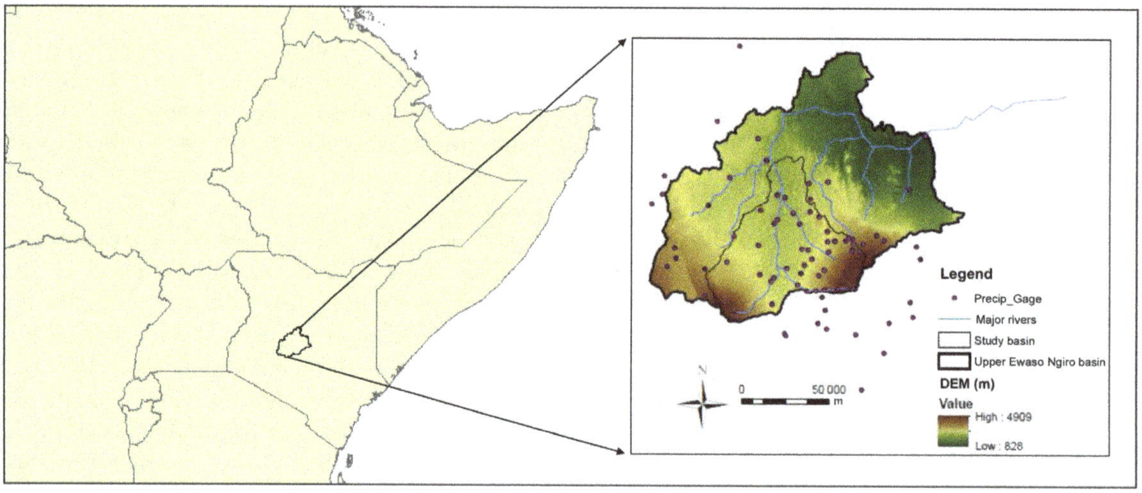

Figure 1. Map showing the general location of the upper Ewaso Ngiro river basin within the boundaries of the east sub-Saharan Africa region. The study sub-catchment (study basin) was selected because of the density of rainfall stations (points in dark red).

More recently, Ruiz-Pérez et al. (2016) discussed the applicability of satellite data during the calibration process comparing the results obtained by a parsimonious model calibrated using only satellite data against the results obtained by a complex model calibrated using field measurements on the pixel scale. Similarly, Quevedo and Francés (2008) and Pasquato et al. (2015) used time series of NDVI to validate a parsimonious ecohydrological model named HORAS. On the catchment scale, Immerzeel and Droogers (2008) used satellite-based evapotranspiration in combination with observed streamflow to calibrate the semi-distributed SWAT model. Zhang et al. (2009) concluded that multi-objective calibration of the SIMHYD model against streamflow and satellite-based aET produced better daily and monthly runoff compared to calibration with streamflow alone. More recently, Rientjes et al. (2013) calibrated a semi-distributed hydrological model using streamflow data and satellite-based aET. Regarding other satellite products, GRACE data, which can be used to detect variations in terrestrial moisture storage (e.g., Lettenmaier and Famiglietti, 2006), have been used to calibrate both global and regional-scale surface hydrology models, in combination with stream discharge data (e.g., Lo et al., 2010). Zhang et al. (2011) calibrated the AWRA-L model with streamflow, NOAA-AVHRR LAI and TRMM-MI (Tropical Rainfall Measuring Mission's, TRMM, Microwave Imager, TMI) soil moisture using a multi-objective calibration framework. Only few studies carried out the calibration exclusively against remote sensing data. For instance, Gutmann et al. (2010) calibrated landscape hydraulic properties in the Noah land surface model using only MODIS surface temperatures from 14 different sites and using observed flux data for model verification. Also, Velpuri et al. (2012) modeled Lake Turkana water levels using only satellite information. All the above-mentioned studies shared the same conclusion: including remote sensing data into the model calibration and/or validation improves the overall performance.

In general, from the total of reviewed publications, calibration using only satellite data was performed in the 47 % of cases, while a combination of satellite data and field measurements (especially streamflow at the outlet) was used in the remaining contributions. Similar results were obtained regarding the validation: 35.3 % of publications only adopted field measurements (especially historical streamflow), employing satellite data exclusively for the model calibration; 47 % used a combination of field measurements and satellite data; 11.8 % only used satellite data; and one publication without any specification. However, more interesting is how the different calibrations were carried out. In most of the cited examples, a sort of multi-objective calibration was used adopting only some points or pixels to calibrate the entire catchment. In other cases, lumped or semi-distributed models were implemented instead of fully distributed ones, considering aggregated values of the satellite data. In other words, the spatial heterogeneity of the basin is neglected and the full potential of satellite imagery, namely the information on spatial patterns, is not fully exploited. Therefore, a method able to make use of the potential of the spatio-temporal information contained in remote sensed data is highly desirable and a calibration scheme which relies solely on remote sensing data will be greatly beneficial in modeling of data-scarce catchments (Kunnath-Poovakka et al., 2016).

3 Study area and data

The upper Ewaso Ngiro basin is located in the Laikipia region of Kenya (Fig. 1). The basin is part of the Laikipia plateau, which lies between Mount Kenya (south east) and the Aberdare mountains (south west). The basin has a drainage area of 15 200 km^2, with the largest river being the Ewaso Ngiro. This region is characterized by distinct rainy and dry seasons. The first rainy season occurs from March to May, while the second rainy season occurs from October to December. Both air temperature and precipitation patterns are heavily influenced by elevation. A full description of the precipitation patterns in the region can be found in Franz (2007).

Soil texture ranges from sandy clay to clay soils (according to the 1980 UNESCO soil map). Although the most characteristic landscape is savanna, higher elevations are dominated by forests and a large piece of land has been converted to cropland (Franz, 2007). The remaining surfaces of the study region are classified as grassland, shrubland and wooded grassland (savanna ecosystems).

As meteorological data, we used the weather stations of the natural resource monitoring, modeling and management project (NRM3) of Nanyuki, Kenya (illustrated in Fig. 1). Daily precipitation and temperature from 1959 to 2003 were validated by Franz et al. (2010). Considering the available hydrological information, we selected a sub-basin with an area of about 4600 km^2 for the present study (Fig. 1). The selected catchment is equipped with a streamflow gauge at the outlet (operational from 1980 to 2002). The reference evapotranspiration (ET_0) was calculated using the Penman–Monteith equation with the simplifications proposed by Allen et al. (2006). This approach is extremely useful to describe the spatial distribution of solar radiation and to derive the ET_0 maps during any day of the year (Manfreda et al., 2013).

Regarding the satellite data, we adopted the Normalized Difference Vegetation Index (NDVI) included in the MOD13Q1 and MYD13Q1 products provided by NASA (NASA Land Processes Distributed Active Archive Center, LP DAAC). This satellite product is available from 2000 to present. The MOD13Q1 and MYD13Q1 data are provided every 16 days at 250 m of spatial resolution. The used NDVI products (MOD13Q1 and MYD13Q1) are in level 3, which means that they are not raw satellite data. NDVI indices are retrieved from daily, atmosphere-corrected, bidirectional surface reflectance.

Finally, based on previous experiences (Ruiz-Pérez et al., 2016; Pasquato et al., 2015) and in similar climatic conditions, we declined to use other products such as LAI or ET derived from MODIS because these kinds of products are produced by models. And, for example, Ruiz-Pérez et al. (2016) found large discrepancies between the LAI provided by satellite and the LAI measured in field. At this point, we had no information to determine the accuracy of these particular models and the spatial information used to implement them. In contrast, NDVI values are calculated by direct differences of spectrum bands, i.e., no models are involved and we therefore decided to use this latter product instead of satellite LAI and/or ET.

4 Model description: TETIS-VEG

The proposed model, called TETIS-VEG, is based on a distributed hydrological model called TETIS (Francés et al., 2007) coupled with a dynamic vegetation model. Both models have simplicity of model structure in common. The used equations are as simple as possible in order to reduce the number of parameters (Table 1). The sub-models are interconnected. The transpiration calculated in the hydrological sub-model depends on the LAI simulated by the dynamic vegetation model. At the same time, the simulated LAI depends on the water stress, which is calculated using the hydrological sub-model. The hydrological sub-model can be used on different timescales (from a few minutes up to daily time steps), while the vegetation dynamic sub-model has to be applied on a daily scale. Hence, the TETIS-VEG model must be used on a daily scale. Both sub-models can be used on a broad range of spatial scales. In this research, the resolution of the implemented model was 90 m × 90 m.

4.1 The hydrological sub-model: TETIS

TETIS's conceptual scheme consists of a series of connected reservoirs, each one representing different water storages in the soil column: (i) vegetation interception, (ii) first static soil layer (retained water by upper soil capillary forces, i.e., below field capacity plus water detention in surface puddles; evaporation and transpiration can occur), (iii) second static soil layer (retained water in deeper soil by capillary forces; only transpiration can occur), (iv) surface (for overland runoff), (v) gravitational soil layer (upper soil water content above field capacity for interflow) and (vi) aquifer (for river baseflow). Vertical connections between reservoirs describe the precipitation, evapotranspiration, infiltration and percolation processes. The horizontal flows describe the three different hydrological responses that give the discharge at the catchment outlet: overland runoff, interflow and baseflow. A more detailed description of the TETIS model can be found in Francés et al. (2007) and GIMHA (2014).

The TETIS model uses a split structure for the effective parameter value at each cell (Francés and Benito, 1995; Francés et al., 2007). The effective parameter is calculated using a correction factor multiplied by the estimated value of the parameter in each cell using all the available information (land cover map, soil type map, DEM, depth of roots and soil layer, etc.) and expert's knowledge. Hence, the effective parameter in each grid cell is computed as the product of two terms: (1) a common correction factor for each type of parameter that takes into account the model, information and input errors and the temporal and spatial scale effects; and (2) the a priori estimated value at each cell. For a given parameter, the a priori and effective values are different from cell to cell, while the correction factor is common for all cells (and different from map to map).

Table 1. Summary of the initial values, the search range and the final value of the parameters or correction factors of both sub-models (hydrological and dynamic vegetation sub-models) as well as the units and the reviewed references.

Model	Correction factor or parameter[a]	Units		Initial value	Search range	Final value	References
Hydrological sub-model	FC1-Maximum static storage	[–]		1.00	[0.5,2.5]	1.80	[1]
	FC2-Evapotranspiration	[–]		0.70	[0.7, 1.2]	1.05	[1]
	FC3-Infiltration	[–]		0.20	[0.01, 2]	0.12	[1]
	FC4-Slope velocity	[–]		1.00	[0.1, 1.2]	1.00	[1]
	FC5-Percolation	[–]		0.08	[0.001, 2]	0.05	[1]
	FC6-Interflow	[–]		140.00	[0.001, 100000]	150.12	[1]
	FC7-Deep percolation	[–]		0.06	[0.001, 2]	0.04	[1]
	FC8-Connected aquifer	[–]		20.00	[0.001, 100000]	16.82	[1]
	FC9-Flow velocity	[–]		1.00	[0.2, 1.2]	1.00	[1]
Vegetation sub-model	Specific Leaf Storages	mm	Tree	0.50	[0.5, 3]	0.43	[2], [3], [4]
			Shrub	2.00	[0.5, 3]	2.00	
			Grass	2.00	[0.5, 3]	2.00	
	Light use efficiency (LUE)	$kg\,m^{-2}\,MJ$	Tree	1.50	[1.2, 2.5]	1.14	[5], [6]
			Shrub	1.50	[1.2, 2.5]	1.14	
			Grass	1.50	[1.2, 2.5]	1.71	
	Coverage factor	[–]	b	0.80	[0.1, 1.0]	0.90	[3], [4]
	Distribution of roots	[–]	Tree	0.30	[0.0, 1.0]	0.10	[3], [4], [7]
			Shrub	0.5	[0.0, 1.0]	0.20	
			Grass	0.7	[0.0,1.0]	0.34	
	Maximum LAI	$m^2\,m^{-2}$	Tree	2.50	[0.5, 3.5]	3.10	[5], [8], [9], [10]
			Shrub	2.00	[0.5, 3.5]	2.00	
			Grass	1.00	[0.5, 3.5]	1.50	
	Light extinction coefficient	[–]	All	0.50	[0.4, 0.6]	0.52	[11]
	Specific leaf area (SLA)	$m^2\,kg^{-1}$	Tree	4.00	[2.0, 5.0]	4.00	
			Shrub	6.00	[4.0, 20.0]	10.00	[5], [12]
			Grass	6.00	[6.0, 50.0]	30.00	
	Optimal temperature	°C	All	16	[10, 30]	18	[11]

[a] Regarding the hydrological sub-model, the table shows the value of the correction factors while for the vegetation sub-model, the table shows the parameter values.[b] The coverage factor depends on the location. The value in the table is the mean value. We used the reported information by Franz et al. (2007) and Caylor et al. (2006). References: [1] GIMHA Team (2014); [2] Van Dijk et al. (2011); [3] Franz et al. (2007); [4] Caylor et al. (2006); [5] TRY Database (www.try-db.org); [6] Yuan et al. (2007); [7] Le Roux et al. (1995); [8] Pasquato et al. (2015); [9] Ceballos and Ruiz de la Torre (1979); [10] López-Serrano et al. (2000); [11] Ruiz-Pérez et al. (2016); [12] Castro de Costa et al. (2014).

With the split-parameter structure, only nine correction factors are calibrated. Each one related to one of these estimated parameter maps: maximum static storage, reference evapotranspiration, infiltration capacity, hillslope velocity, percolation capacity, horizontal saturated conductivity for interflow, horizontal saturated conductivity for aquifer and river channel velocity.

4.2 The dynamic vegetation sub-model: LUE model

The proposed dynamic vegetation sub-model is based on the concept of light use efficiency (LUE; Medlyn, 1998) and calculates the leaf biomass (Bl) according to the Eq. (1). The LUE is the proportionality between plant biomass production by terrestrial vegetation and absorbed photosynthetically active radiation (APAR) in optimal conditions. However, the LUE can be strongly affected by stress conditions. The key factors contributing to the variation of this efficiency are: soil moisture content, air temperature (Landsberg and Waring, 1997; Sims et al., 2006) and nutrient levels (Gamon et al., 1997; Ollinger et al., 2008). Since this model is designed to be used in water-controlled areas, the nutrient levels are not considered.

$$\frac{\mathrm{d}B_\mathrm{l}}{\mathrm{d}t} = (\mathrm{LUE} \times \varepsilon \times \mathrm{PAR} \times \mathrm{fPAR} - \mathrm{Re}) \times \varphi_l\,(B_\mathrm{l}) - k_\mathrm{l} \times B_\mathrm{l}, \tag{1}$$

where B_l is the leaf biomass, LUE is the above-mentioned light use efficiency, ε is the water stress factor, PAR is the photosynthetically active radiation, fPAR is the fraction of photosynthetically active radiation, Re is the respiration, φ_l (B_l) is the fractional leaf allocation and k_l is the leaf natural decay factor to reproduce the senescence. The water stress factor depends on the amount of water contained in the two static reservoirs and it is calculated according to Porporato et al. (2001). Basically, the stress factor is equal to 1 (maximum stress) if the water storage is less than the water storage at wilting point; it is equal to 0 (minimum stress) if the water storage is higher than the water storage at critical point (plants start the stomatal closure); and, it varies from 0 to 1 using a potential function which depends on the wilting point, the critical point and an exponent set to equal 2. Then, this stress multiplies the LUE index, reducing the efficiency when its value is lower than 1 (non-optimal conditions).

The LAI is simulated through the product between the leaf biomass, the specific leaf area (SLA) and the vegetation fractional cover. Later, the LAI is used to calculate the transpiration in the hydrological sub-model according to Eq. (2).

$$T_i = (\mathrm{ET}_0 - \mathrm{EI}) \times \min(1, \mathrm{LAI}) \times \zeta \times Z_i \times \mathrm{fc}, \quad (2)$$

where T_i is the transpiration from the i soil layer, ET_0 is the reference evapotranspiration, EI is the evaporation of the intercepted water, LAI is simulated by the model, Z_i is the percentage of roots in the i soil layer and fc is the coverage factor.

Therefore, the LUE model has eight parameters to be calibrated: (1) Specific leaf storage (the maximum interception storage is calculated as the product between the specific leaf storage and the LAI simulated by the model), (2) the LUE index (explained above), (3) the coverage factor, (4) the distribution of roots between the first and the second static storage layers, (5) the maximum LAI sustainable by the system (the simulated LAI is limited by a maximum), (6) the light extinction coefficient k, (7) the SLA and (8) the optimal temperature (the stress factor also depends on the temperature).

5 Methodology

One of the main objectives of this research was to explore the potential of satellite remotely sensed data for model calibration. The TETIS-VEG model was calibrated purely against MODIS NDVI according to the following three steps: (1) a manual calibration in order to obtain a first approximation of model parameters, (2) an automatic calibration based on the combined use of EOFs and a genetic algorithm in order to refine model parameterization and (3) a model validation carried out with both remote sensed data and traditional data (such as streamflow measurements). Since the meteorological data (precipitation and temperature) were available from 1960 to 2003 and the MODIS NDVI was available from 2000 to present, we decided to use the year 2003 as the calibration period and the period from 2000 to 2002 for validation. In order to avoid the effect of the initial conditions, we used one year as warming-up period (the year 2002 and 1999 for model calibration and validation, respectively).

For these purposes, we adopted the NDVI as a descriptor of the state of the vegetation assuming that LAI and NDVI are intimately related. Studies on various vegetation types, e.g., agroecosystems (Cohen et al., 2003), grasslands (Friedl et al., 1994), shrublands (Law and Waring, 1994), conifer forests (Chen and Cihlar, 1996) and broadleaf forests (Frassnacht et al., 1997) have led to the general conclusion that the spectral vegetation indices such as NDVI have considerable sensitivities to LAI. The relationship between LAI and NDVI can be considered linear for low values, while it becomes nonlinear for the higher values of NDVI due to the greenness saturation (e.g., Turner et al., 1999). In this case study, the maximum LAI values are around 2.0–2.5, according to Franz (2007), and are lower than the greenness saturation threshold. Therefore, the relationship between the observed NDVI and the simulated LAI is expected to be linear.

5.1 Empirical orthogonal function method (EOF)

The EOF method is generally used to analyze the spatio-temporal variability of a single variable, but a comparison between different variables can also be performed using coupled EOF techniques (Björnssson and Venegas, 1997). The method decomposes a dataset in a time series and spatial patterns. Furthermore, the method allows for estimating a measure of the "importance" of each spatial pattern. We refer to the spatial patterns as the EOFs (in literature, they are also referred to as principal components), and to the time variation as loadings (in literature, there are several terms: expansion coefficient time series, expansion coefficients, EOF time series, principal component time series, etc.).

The EOF method is essentially a linear algebra methodology based on matrix transformation. The first step is thus the conversion of the spatio-temporal data to be analyzed into a matrix. Basically, we construct a matrix (**F**) in which each column is the temporal variation of the data in a particular cell while each row represents the cells values during a particular time step. Usually, the second step is to compute the anomalies of the analyzed data, which was not needed in this study because we used normalized data (for reasons that will be explained below).

The next step of the applied EOF method consists of the calculation of the spatial **F** covariance matrix (**R**) according to Eq. (3). Then, the eigenvalue problem is solved by Eq. (4).

$$\mathbf{R} = \mathbf{F}^T \times \mathbf{F}, \quad (3)$$

$$\mathbf{R} \times \mathbf{C} = \mathbf{C} \times \mathbf{\Lambda}. \quad (4)$$

$\mathbf{\Lambda}$ is a diagonal matrix containing the eigenvalues λ_i of **R**. The $\boldsymbol{c}_i$ column vectors of **C** are the eigenvectors of **R** corresponding to the i-respective eigenvalues. Each of these eigenvectors can be regarded as a map which denote the EOFs (or principal spatial patterns). In what follows, we always assume that the eigenvectors are ordered according to the value of the eigenvalues. Thus, EOF1 is the eigenvector associated with the biggest eigenvalue. The fraction of the total variance in **R** explained by EOF_i is found by dividing the λ_i by the sum of all the other eigenvalues. The time evolution of an EOF_i ($\boldsymbol{a}_i$) is calculated according to Eq. (5). The components of these time vectors are referred to as loadings in this paper.

$$a_i = \mathbf{F} \times \mathrm{EOF}_i \tag{5}$$

Using the spatial covariance calculated according to Eq. (3), the EOF technique provides three different results: the main patterns or EOFs, their time evolution whose components are called loadings and the portion of spatial variance explained by each EOF, which is calculated by dividing each λ by the trace of $\mathbf{\Lambda}$.

5.2 Manual calibration

The manual calibration was done with a dual purpose. First, we wanted to test the applicability of the proposed TETIS-VEG model in the study basin. Second, we wanted to obtain a first approximation for the parameters and, at the same time, constrain the automatic calibration. This manual calibration consisted of the usual ad hoc method (manual adjustment of parameter values) considering the Pearson correlation coefficient between the simulated LAI and the observed NDVI in a total of 32 different points inside the basin. These points were selected within homogeneous areas defined according to the main spatial patterns of the observed NDVI (EOFs) and the available maps of land cover, soil texture, DEM, slope and soil depth.

In this case, the EOF analysis was used to identify the main spatial patterns of the observed NDVI. Once the main spatial patterns were identified, we combined our own human perception with the confusion matrices between the main spatial patterns and the spatial maps of model parameterization. Confusion matrices are widely applied for map comparison in distributed modeling comparing actual to predicted values for each specific category defined previously (García-Arias et al., 2016; Bennett et al., 2013; Van Vliet et al., 2013 among many others). Generally, the rows in the matrix represent the values predicted by the model, whereas the columns represent the actual values. By its nature, the confusion matrix is an overall measure for similarity between two categorized maps. However, the comparison of numerical maps is feasible if they are categorized previously. While most of the spatial maps of the basin characteristics (land use, soil type, etc.) were categorical, the main patterns obtained by the EOF analysis were numerical. To build the confusion matrices, the main patterns were therefore discretized according to the number of river basin features (such as land cover map, soil type map, etc.) and based on the similitude between the corresponding histograms. Once the discretization was done, by a cell-by-cell comparison of the discretized NDVI main pattern maps obtained after the EOF analysis and the available spatial maps, the confusion matrices were built.

These confusion matrices allowed the calculation of the weighted kappa (κ) coefficient (Cohen, 1968). This coefficient, whose maximum value of 1 represents a perfect agreement, was employed to identify which spatial maps (land cover map, soil type map, DEM, etc.) were linked with the main patterns of the observed NDVI. Then, they were used in order to select the most appropriate points for the manual calibration.

5.3 Automatic calibration

The most innovative aspect of the proposed procedure was the direct use of the EOF analysis in the automatic calibration. As proposed by Koch et al. (2015), we decided to build one integral matrix by concatenating both the observed and predicted data: the matrix contained the normalized values of the NDVI provided by MODIS and the normalized values of the LAI simulated by the model. In this way, the upper part of this matrix contained the temporal variation of the normalized observed NDVI in all cells as columns, while the lower part contained the temporal variation of the normalized simulated LAI in all cells as columns. We decided to use the normalized values of the NDVI and LAI because, although they are correlated, they differ in range.

However, normalization implies that some spatial information is lost. In order to avoid these losses, we added two rows in the matrix $\mathbf{F}$: the first containing the difference between the temporal mean of the observed NDVI at a particular cell and the general mean using the complete NDVI dataset, and the second with the same content but referred to the simulated LAI. In this way, we included the spatial gradient of the observed NDVI and the spatial gradient of the simulated LAI. These two rows represent two additional maps included in the evaluation of the model performance. If they were similar, it would mean that the spatial gradient remains and is properly reproduced.

The number of pixels was 1 034 706. For the calibration period (year 2003), there were 44 NDVI maps (one every 8 days more or less). Hence, the built integral matrix's size was 90 rows ($44 + 44 + 2$ additional rows) $\times$ 1 034 706 columns. After the construction of this matrix, the EOF analysis was applied obtaining the following: the EOF maps for the matrix containing both NDVI and LAI, the portion of variance explained by each EOF map and the loadings of each EOF map. The combined EOF analysis yielded orthogonal EOF maps that explained the combined intervariability and intravariability of both datasets. For each time step, the loadings express how much the respective LAI and NDVI maps contribute to the direction of the corresponding EOF. Hence, if the observed NDVI and the simulated LAI were completely correlated, the temporal evolution of the EOF maps for both, NDVI and LAI, would be essentially equal.

The automatic model calibration was carried out by trying to minimize the differences between the loadings of simulated and observed data. The used objective function was based on that idea and it also took into account the portion of variance explained by each EOF in order to consider that the variance contribution decreases consecutively for the EOFs. The adopted error measure is described in the following equation:

$$\text{Error} = \sum_{i=1}^{k} w_i \times \sum_{i=1}^{t} \left|\text{load\_sim}_{i,j} - \text{load\_obs}_{i,j}\right|, \tag{6}$$

where Error is the objective function to minimize, w_i is the portion of variance explained by the EOF_i, $\text{load\_sim}_{i,j}$ is the loading of the EOF_i at time step j for the simulated data (in this particular case, the normalized LAI) and $\text{load\_obs}_{i,j}$ is the loading of the EOF_i at time step j for the observed data (in this particular case, the observed NDVI).

The calibration was performed using a genetic algorithm called Pyevolve (see http://pyevolve.sourceforge.net/). This algorithm needs a seed (initial values of the parameters) and a searching boundary of the parameters to be calibrated. We used the results obtained after the manual calibration explained above as seed and made sure that the searching boundaries were wide enough (Table 1).

In order to explore the outcomes of the proposed calibration framework, we additionally calculated both the temporal Pearson correlation coefficient between the NDVI provided by MODIS and the LAI simulated by the TETIS-VEG model in each cell and the spatial Pearson correlation at each time step. For the spatial and temporal correlation coefficients, we used the original values of both datasets (NDVI and LAI), not the normalized values as used by the EOF analysis. It is important to mention that the Pearson correlation coefficient between two datasets X and Y is positive if X and Y tend to be simultaneously greater than, or simultaneously less than, their respective means. Hence, the mean should be representative. For this reason, in the case of the spatial correlation coefficient, we distinguished between the main land covers whose means can be significantly different: tree, shrubs and grass.

5.4 Validation

The period selected for the model validation was the three years from 2000 to 2002. As during the calibration period (year 2003), there were data of precipitation, temperature and, also, NDVI provided by MODIS. To validate the model, we used the same performance indexes applied during the automatic calibration process. Keeping the parameter values obtained by the automatic calibration, we built the matrix concatenating the normalized value of the observed NDVI and the normalized value of the simulated LAI with two additional rows used to incorporate the spatial gradient of both datasets as explained above. We also plotted these two maps and compared them as we did during the model calibration. Using EOF techniques, we obtained the coupled EOF maps and their associated loadings and portion of variance explained by them. As during the calibration, we compared the deviation of the loadings for each EOF map and we calculated the Error function defined in Eq. (6). We also calculated the temporal and spatial Pearson correlation coefficients.

In addition to this, we explored the reliability of the calibrated model in reproducing streamflow. In fact, during the validation period, the observed discharge at the outlet point was available unlike during the calibration period. The reliability of the hydrological sub-model in reproducing the streamflow was an extremely challenging task considering that the entire modeling structure had been calibrated using only vegetation data from remote sensing along with physical information about the basin.

Furthermore, we included the Nash and Sutcliffe efficiency index (NS, Nash and Sutcliffe, 1970) and the bias (or volume) error (E) value between the observed and simulated discharges at the basin outlet in the model validation. We also decided to strengthen our discharge analysis by using the concept of flow duration curves (FDCs). FDCs are simple and powerful tools, commonly used in hydrology, to describe the runoff regime in a river basin that can be representative of the model's ability to reproduce the different components of the streamflow (e.g., Manfreda et al., 2005). In fact, FDCs represent the relationship between magnitude and frequency of streamflows, thus providing an important synthesis of the relevant hydrological processes occurring on the basin scale (Pumo et al., 2013). Actually, the shape of a flow-duration curve in its upper and lower regions is particularly significant in evaluating the stream and basin characteristics (Coopersmith et al., 2012). The shape of the curve in the high-flow region indicates the type of flood regime the basin is likely to have, whereas the shape of the low-flow region characterizes the ability of the basin to sustain low flows during dry seasons (Cheng et al., 2012). Hence, the FDC represents the full spectrum of variability in terms of their magnitudes (Wagener et al., 2013).

6 Results

6.1 Manual calibration

The main objective of this a priori manual calibration was the identification of the most appropriate points where the model could be tested. To accomplish that, we identified the main spatial patterns of the observed NDVI and then we compared the EOFs with the spatial features of the river basin (such as land cover map, DEM, soil type map, etc.).

Using our own perception, we identified a certain relationship between EOF_1 (which explained the 61.5 % of the observed NDVI spatial variance) and the land-use map. This potential relationship was supported by the κ coefficient (described in Sect. 5) that assumed a value of 0.34. This is not a high value but it showed the existence of a relationship between the two maps. Regarding EOF_2 (which explained the 10.5 % of the observed NDVI spatial variance), no connections with the basin physical characteristics were found. It might contain a mix of several drivers and, therefore, it can't be directly linked to a single one. Conversely, EOF_3 showed

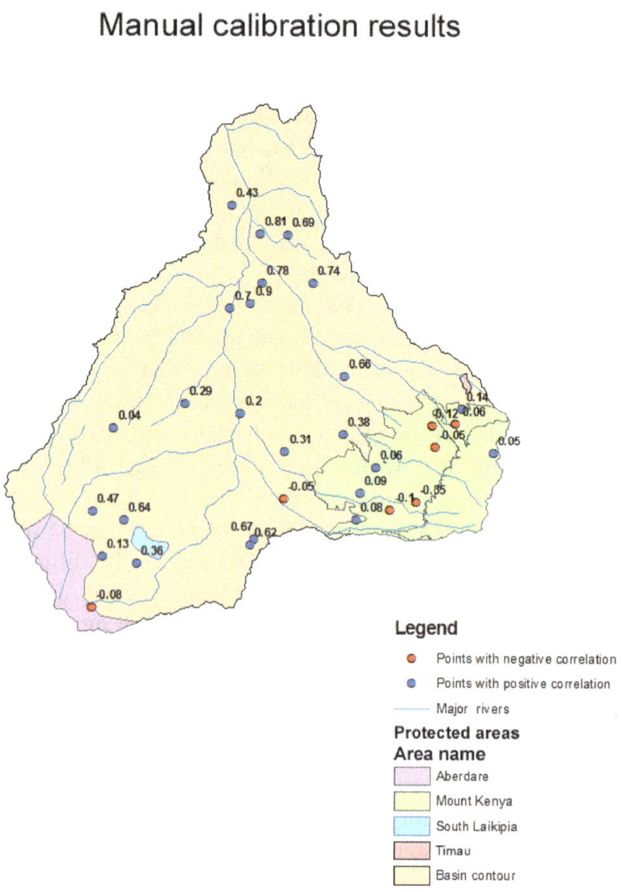

Figure 2. Location of the points where the manual calibration was carried out. The value of the Pearson correlation coefficient between the satellite NDVI and the simulated LAI appears together with the point used for the manual calibration of the model.

a good agreement with the soil texture map (the κ coefficient was 0.32). Therefore, we can state that the observed patterns of NDVI are strongly influenced by the spatial distribution of land cover and soil texture. In the following, we combined these two maps, extracted all possible combinations and randomly selected two points of each of these combinations obtaining 32 points covering all of the catchment area.

After conducting the manual calibration, the Pearson correlation coefficient between the observed NDVI and the simulated LAI was positive in 25 points of the 32 considered points. All points with negative correlations had in common the fact that they were located near to the Mount Kenya or Aberdare mountains (Fig. 2).

Finally, Table 1 shows the obtained set of parameters. This set was used as seed during the automatic calibration. It must be underlined that all parameters had values consistent with the reviewed literature (references embedded in Table 1).

6.2 Automatic calibration

The proposed automatic calibration is based on the assumption that the closer the loadings of the simulated values are to the loadings of the observed values, the higher the similarity between the spatial patterns is. Calibration produced a good agreement between the observed and simulated loadings of EOF_1 (Fig. 3; upper left panel) but small deviation between the observed and simulated loadings related to EOF_2 and EOF_3. The loadings of the remaining EOFs were completely scattered mainly due to their corresponding low contribution (low weight) in the objective function of the automatic calibration process (Eq. 6). It is needed to remark that EOF_1 explained more than 60 % of the dataset spatial variance, while EOF_2 and EOF_3 explained around 10 % each. The remaining EOFs explained less than 3 % each, but in any case they were considered during the calibration process (weighted by the portion of variance explained by each one).

We also used three additional metrics to evaluate the model performance: (1) the temporal Pearson correlation coefficient evaluated in each cell, (2) the spatial Pearson correlation distinguishing between trees, shrubs and grasses computed at any time and (3) comparison of the spatial gradient maps. First, the temporal Pearson correlation coefficient between the observed NDVI and the simulated LAI was higher than 0.4 (Fig. 4; left panel) in most of the catchment. The weakest correlations were obtained in the two highest areas of the basin near to the Mount Kenya and Aberdare mountains with zero to negative values.

The spatial Pearson correlation coefficients were calculated excluding the regions with negative temporal Pearson correlation coefficient. Although slightly worse than the results in terms of temporal correlation, the mean spatial correlations were higher than 0.45 for all main land covers: trees (mean = 0.58), shrubs (mean = 0.49) and grasses (mean = 0.55) (Fig. 5; upper panel). The best scores were obtained in cells classified as trees. In fact, the median was almost 0.60 and the variance was not high (standard deviation = 0.16). Conversely, the cells classified as grasses obtained the worst results with the lowest median and the highest variance (standard deviation = 0.18).

Figure 6 (upper panels) shows the comparison between the maps which represent, in each cell, the difference between the temporal mean and the general mean of the observed NDVI and the simulated LAI, respectively. No great differences were found by comparing both maps indicating the good spatial performance of the ecohydrological model, at least from the vegetation point of view.

6.3 Validation

Similarly to the calibration process, EOF_1 explained more than 60 % of the spatial variance, while EOF_2 and EOF_3 explained around 10 % and 5 %, respectively, for the validation period. The remaining EOF maps are not presented because

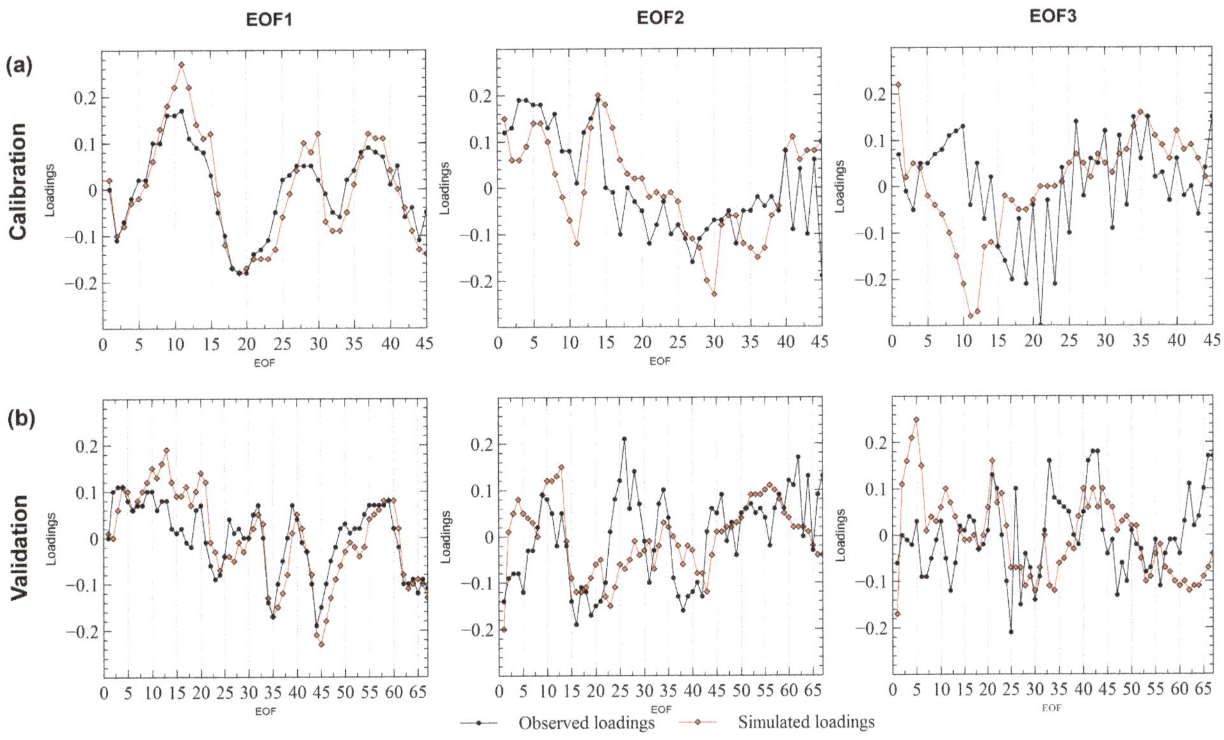

Figure 3. The three first EOFs during the calibration **(a)** and during the validation **(b)** are represented. The y axes reflect the unitless loadings of each EOF. The x axes reflect the time step.

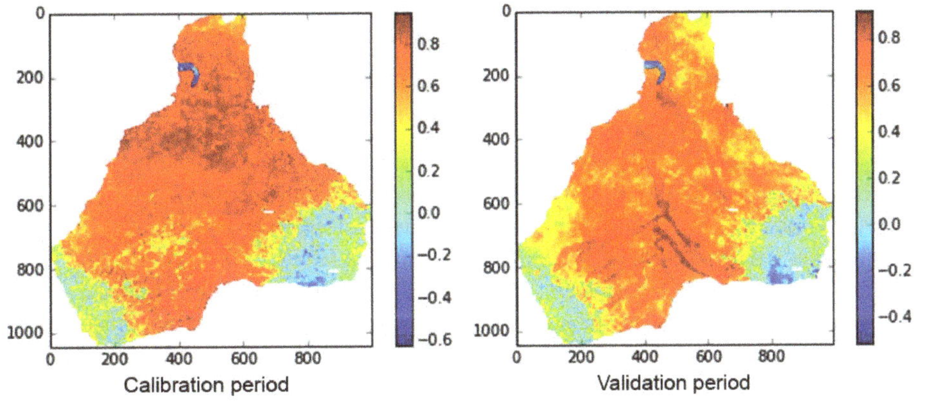

Figure 4. Temporal Pearson correlation coefficient between the NDVI provided by MODIS and the LAI simulated by the model during the calibration and validation periods. The two areas with negative values correspond to the Mount Kenya and Aberdare mountains.

none of them explained more than 3 %. The simulated and observed loadings of EOF_1 were almost equal, while the obtained results in relation to EOF_2 and EOF_3 were slightly worse (Fig. 3; lower row of panels). However, it is important to stress that both showed the same clear temporal dynamics. Indeed, the resulting "Error" for the validation period was 4.03, just slightly worse than the Error for the calibration period. It must be considered that the Error value was calculated considering all EOFs (Eq. 6).

The temporal Pearson correlation map between simulated LAI and NDVI showed the same pattern observed in the calibration period: the two areas located near Mount Kenya and the Aberdare mountains had a temporal correlation coefficient equal to zero or negative. However, in more than 80 % of the catchment, this coefficient was between 0.3 and 0.9 (Fig. 4; right panel).

Regarding the spatial Pearson correlation coefficient between simulated LAI and NDVI in the three main land covers, the results were not as good as the results obtained in terms of temporal correlation. Nevertheless, there were no negative spatial correlation coefficients at any time step. In the case of shrubs and grasses, both the mean and median

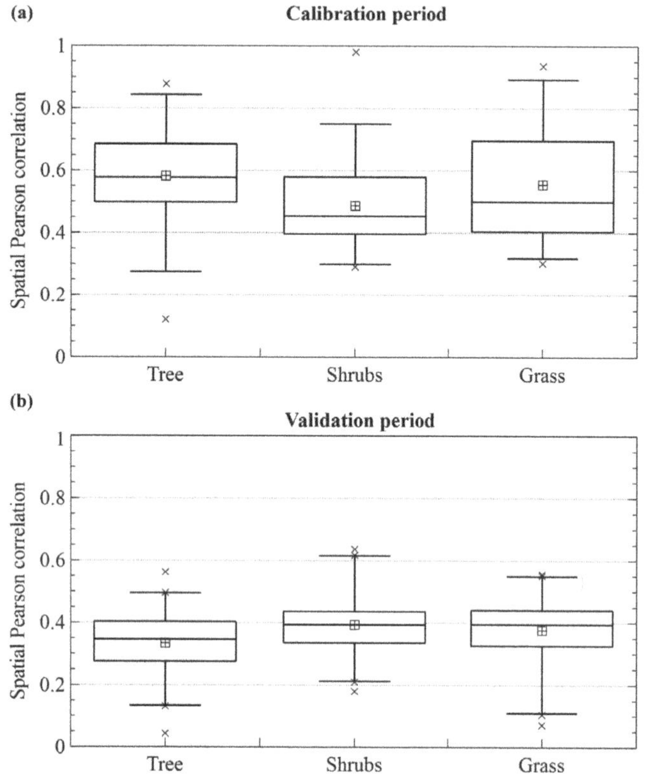

Figure 5. Spatial Pearson correlation coefficient during the calibration **(a)** and during the validation **(b)** distinguishing between the main land covers: tree, shrubs and grass. The whiskers were calculated according to the 98th percentile and the outliers were plotted as × symbol. The median is the line inside the boxplot and the mean is the square tile symbol.

were almost 0.4, while the corresponding value for the trees was around 0.35 (Fig. 5; lower panel). The variance obtained during the validation period was narrower than that obtained during the calibration period for the three land covers. The spatial pattern of LAI was, as for the calibration period, well captured by the model (Fig. 6; see the lower panels). The cells with high differences between their own temporal mean and the general mean were consistent in both maps.

Finally, since there was observed discharge at the basin outlet during the years 2000, 2001 and 2002, it was possible to compare the discharge simulated by the model with the observations. The volume error (E) was equal to −0.40, while the NS index was equal to 0.32. E is strongly affected by the results obtained at the beginning of the validation period, probably due to the absence of information regarding the initial conditions. Although we used a year as warming-up period, the simulations improved only after 2001. In fact, having calculated the performance indexes in each year, E decreased in magnitude from −0.88 in 2000 to only −0.17 during the year 2002 (Fig. 7). Regarding the NS index, the worst result was also obtained in the first year and it improved from a negative value in 2000 to 0.35 during the year 2002, as one should expect considering the visual comparison in Fig. 7. This trend is emphasized in the plot of the FDCs (Fig. 8), where the underestimation in the first 2 years is clearly highlighted. The upper panel (Fig. 8a) compares the FDC of observations and simulations within the whole period, while the lower panels (Fig. 8 b–d) compare the corresponding FDCs within the 2000, 2001 and 2002. In these plots, the simulation seems to closely interpret hydrological response in the year 2002.

7 Discussion

From the a priori manual calibration step to the model validation step, it was possible to identify a behavioral pattern which would also be observed during the following automatic calibration and validation steps: EOF_1 explains more than 60 % of the spatial variance, EOF_2 around 10 % and EOF_3 around 5 %, while the remaining EOFs could be considered negligible. The fact that EOF_1 and EOF_3 of the observed NDVI were related to the land cover and soil type maps, respectively, was consistent with our expectations that NDVI is a suitable proxy of vegetation dynamics.

After the automatic calibration, the model matched the observed loadings of EOF_1 and its accuracy was slightly poorer for EOF_2 and EOF_3. Thus EOF_1 captured the predominant pattern that was found in both the observed NDVI and the simulated LAI data. On one hand, the temporal variation of EOF_1 loadings seemed to be related to the two typical growing seasons in the catchment: the first one during March–May and the second one during October–December (Franz et al., 2010; Fig. 3). On the other hand, the loadings of EOF_2 and EOF_3 were not strongly connected with any feature. The loadings of the remaining EOFs were scattered, which implies that mainly measurement and model noise were covered by these EOFs.

The weakness of the proposed calibration methodology is that, although the associated weights to the loading deviation in Eq. (6) are needed, they are also misleading some spatial information. New ways to weigh the loading deviations must emerge in future research as proposed by Koch et al. (2015). In fact, due to the portion of variance explained by EOF_1, this first main pattern controlled the calibration process. In future applications, the proposed error index may be improved by focusing or excluding specific EOFs. A popular method for deciding which EOF to keep and which to discard is to use "selection rules". Basically, there are three classes of selection rules depending on whether they focus on the amount of variance explained by each EOF, the loadings or the EOF maps (Preisendorfer, 1988). Another option could be to rotate the EOFs as proposed by Bonaccorso et al. (2003). As each rotated EOF will not explain the same variance of the unrotated one, this approach would be an option to use different combinations of EOFs, which explain different amounts of variance in order to reduce the influence of EOF_1.

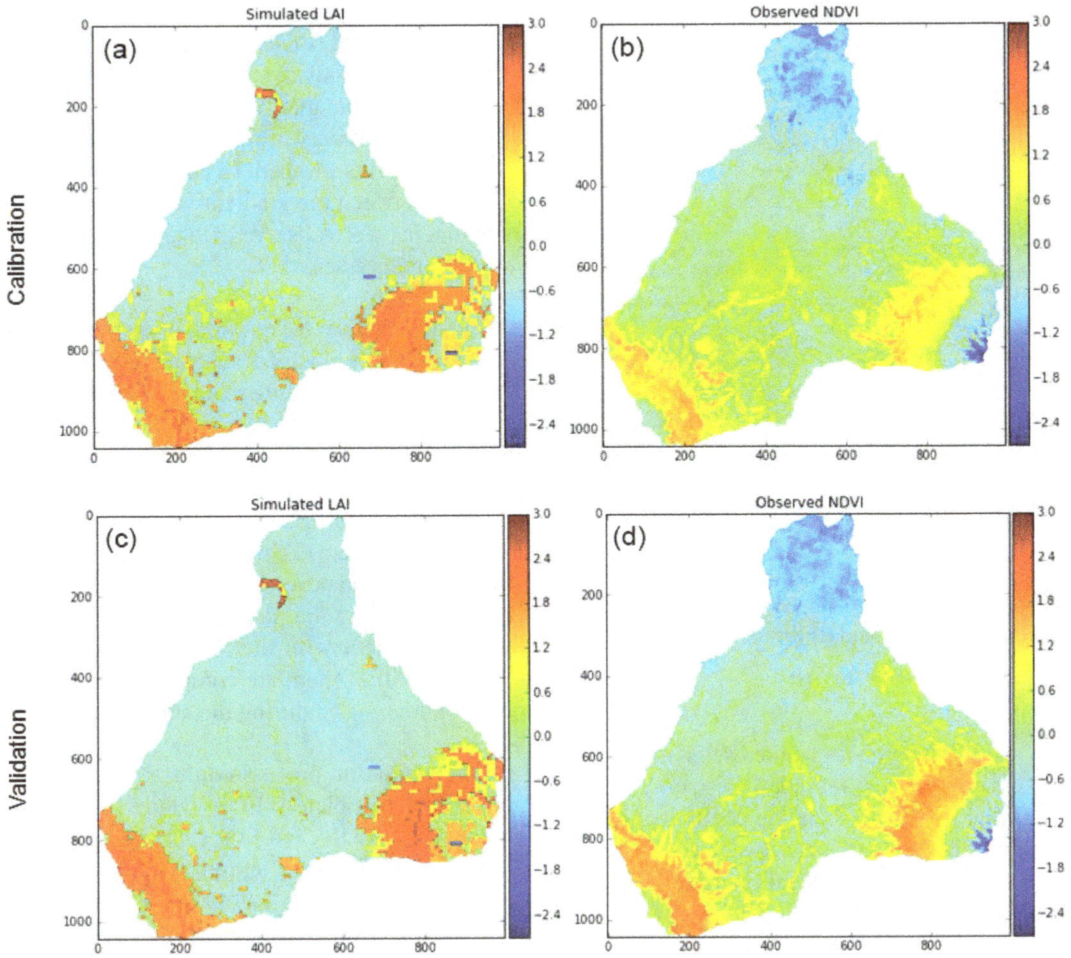

Figure 6. Comparison between the maps where each pixel color represents the difference between the temporal mean calculated in this particular pixel and the general mean calculated using all datasets of the simulated LAI **(a, c)** and observed NDVI **(b, d)** in both periods: calibration **(a, b)** and validation **(c, d)**. This difference is a measure of the spatial gradient of both variables (LAI and NDVI).

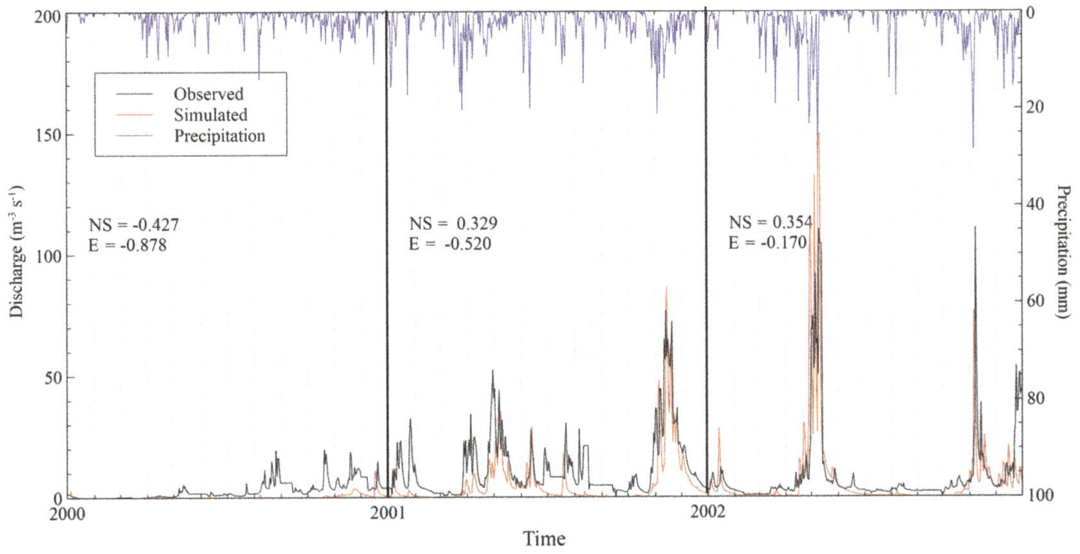

Figure 7. Time series of rainfall and observed and simulated daily discharge ($m^3\ s^{-1}$) during the validation period (2000, 2001 and 2002).

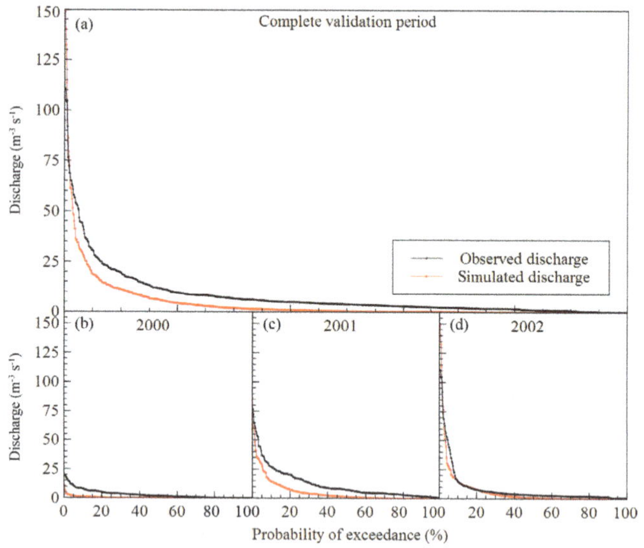

Figure 8. Observed (in black) and simulated (in red) flow duration curves (FDCs) for the whole validation period **(a)** and for the corresponding 3 years in isolation (**b**, **c** and **d**) for the years 2000, 2001 and 2002, respectively.

The automatic calibration process worked satisfactorily as shown by the additional metrics: temporal Pearson correlation coefficient, spatial Pearson correlation coefficient in the main land covers and the comparison between the gradient maps. In terms of spatial Pearson correlation coefficient, the weakest values were obtained in the higher portion of the basin near Mount Kenya and the Aberdare mountains, while the remaining cells within the study area showed a good agreement between observed NDVI and simulated LAI. This same behavior was also observed when calibrating manually. Two reasons could explain such results. First, the observed NDVI in some cells of those areas had a really bad quality as testified by the unrealistic oscillations of the NDVI from 0.8 to 0.1 (even zero) in just one week. These unrealistic oscillations could be produced by the presence of clouds over the area near to the mountains. The second reason is related to the conceptual limitation of the proposed model. The TETIS-VEG was designed to be used only in water-controlled areas. Franz (2007) analyzed the correlation between the fractional woody cover and the mean annual precipitation within the catchment and they were strongly correlated. However, two different slopes were observed. The transition point, which indicates when water availability had a smaller influence on the fractional woody cover, occurred approximately around 800 mm year^{-1}. Physically, the transition point is believed to be a good approximation of the transition from a water-controlled ecosystem to a nutrient-controlled ecosystem. This approach allowed us to define the higher areas within the study catchment (where Mount Kenya and the Aberdare mountains are included) as nitrogen limited ecosystems instead of water-controlled.

With the exemption of these two areas, a strong correlation between NDVI and LAI existed, i.e., the model captured the temporal dynamic of LAI. However, this did not necessarily mean that the magnitude of LAI was reasonable. This last point was proven by calculating the good performance in terms of spatial Pearson correlation and the comparison between the gradient maps. No differences and good agreements were observed along the main land covers: trees, shrubs and grasses.

Finally, there were four parameters in the automated calibration which changed substantially (in relative terms) in comparison to the values obtained during the manual calibration: the correction factor of the maximum static storage, the correction factor of the reference evapotranspiration, the factor related to the distribution of roots between the first and second static storage layers and the maximum LAI sustainable by the system (Table 1). These parameters directly affect the transpiration process and the amount of available water to be consumed by the plants. In any case, all obtained values were consistent with the reviewed literature (embedded in Table 1). All of them are completely included in the searching boundary used during the automatic calibration.

The validation process confirmed (1) the model was able to capture completely EOF_1 while the model performance worsened in EOF_2 and EOF_3, (2) the simulated LAI and the observed NDVI were temporally correlated in most of the catchment and (3) the spatial distribution of LAI was consistent as shown by the comparison between the gradient maps and the value of the spatial Pearson correlation coefficient at any time.

An additional interesting outcome provided by the validation was the comparison between simulated and observed hydrograph data at the outlet. The simulated stream flow was promising, but not completely convincing. This limitation is obvious since the model parameters were calibrated on NDVI data, i.e., the model was calibrated on vegetation dynamics. Therefore, a direct comparison between hydrographs should not be too exigent when no information about the parameters involved in the river flow routing or aquifer discharge was included in the calibration.

We strengthened our discharge analysis by using the concept of FDCs. By graphical comparison (Fig. 8), it could be observed that the model is able to reproduce the shape of the observed FDC, while some discrepancies were found in terms of magnitude. However, its performance improved considerably year by year. Since the FDC shape is an important indicator of the relevant hydrological processes occurring on the basin scale, this result pointed out the capability of the proposed model calibration methodology to reproduce the main hydrological behavior of the study basin.

8 Conclusions

The main two objectives of this research were (1) to explore if it is possible to calibrate and validate an ecohydrological model using only satellite information and (2) to incorporate spatio-temporal data about a model state variable into the automatic calibration process. In order to tackle these challenges, a parsimonious distributed ecohydrological model was calibrated by exclusively using NDVI data provided by MODIS. A methodology based on EOF analysis was proposed to carry out manual and automatic calibration of the model. Finally, the results were validated using satellite data referring to different periods and the observed discharge at the basin outlet, which was not used for calibration.

In general, the proposed model was able to properly reproduce the vegetation dynamics and the observed stream flow. The results highlight the usefulness of satellite data. It was possible to implement the hydrological and the vegetation components of the TETIS-VEG daily model using only NDVI data and also to validate the model with satisfactory results. Such outcomes are promising because they demonstrate that satellite data could be exploited in order to predict river discharge in ungauged basins. More specifically, we expect this result given the key role played by vegetation in water-controlled areas such as the upper Ewaso Ngiro river basin in Kenya, where having an appropriate description of vegetation and transpiration is critical for a correct description of the water balance on the local and basin scale.

The proposed automatic calibration was completely designed to incorporate spatio-temporal data in order to take maximum advantage of the available satellite data. After calibrating, the simulated vegetation patterns display good agreement with measured NDVI in most of the basin except for some portions at higher altitudes. This non-satisfactory result may be due to the bad quality of the NDVI data and/or the limitation of the vegetation sub-model (that was specifically designed for semiarid regions).

Model limitations along with poor data quality and resolution negatively affected the overall model performance, but the proposed procedure allowed us to exploit the amount of information available addressing the critical issue of identifying a procedure for the calibration of a distributed model. This allowed us to obtain a correct description of vegetation dynamics in space and time also providing, as a marginal benefit, a fairly good streamflow prediction. In this context, it was mandatory to adopt a daily time step in order to have a coherence with satellite NDVI data and also removing the need for a runoff propagation module in our model.

Finally, we should consider that the potential of the present study is due to the large availability of remote sensing information (not only satellite) concerning spatial state variables and more information will be available in the future. Many efforts are being made to improve the quality and quantity of remote sensing data (drones, better devices, etc.). Additionally, the scientific community must also be ready to exploit the enormous amount of information contained in this data (temporal, spatial and spatio-temporal). Therefore, we should identify the best way to use all of this new available information, not only for data assimilation but also for model calibration and validation.

Competing interests. The authors declare that they have no conflict of interest.

Acknowledgements. The research leading to these results has received funding from the Spanish Ministry of Economy and Competitiveness and FEDER funds, through the research projects ECOTETIS (CGL2011-28776-C02-014) and TETISMED (CGL2014-58127-C3-3-R). The collaboration between Universitat Politècnica de València, Università degli studi della Basilicata and Princeton University was funded by the Spanish Ministry of Economy and Competitiveness through the EEBB-I-15-10262 fellowship.

Edited by: Dimitri Solomatine

References

Allen, R. G., Pruitt, W. O., Wright, J. L., Howell, T. A., Ventura, F., Snyder, R., Itenfisu, D., Steduto, P., Berengena, J., Yrisarry, J. B., Smith, M., Pereira, L. S., Raes, D., Perrier, A., Alves, I., Walter, I., Elliott, R.: A recommendation on standardized surface resistance for hourly calculation of reference ET0 by the FAO56 Penman-Monteith method, Agr. Water Manage., 81, 1–22, https://doi.org/10.1016/j.agwat.2005.03.007, 2006.

Andersen, F. H.: Hydrological modeling in a semi-arid area using remote sensing data, Doctoral Thesis, Department of Geography and Geology, University of Copenhagen, Denmark, 2008.

Bjornsson, H. and Venegas, S. A.: A manual for EOF and SVD analyses of climate data, CCGCR Rep. 97-1, McGill University, Montréal, Canada, 52 pp., 1997.

Bonaccorso, B., Bordi, I., Cancelliere, A., Rossi, G., and Sutera, A.: Spatial variability of drought: an analysis of the SPI in Sicily, Water Resour. Manage., 17, 273–296, 2003.

Bond, B. J., Jones, J. A., Moore, G., Phillips, N., Post, D., and McDonnell, J. J.: The zone of vegetation influence on baseflow revealed by diel patterns of streamflow and vegetation water use in a headwater basin, Hydrol. Process., 16, 1671–1677, 2002.

Brown, B. G., Gilleland, E., and Ebert, E. E.: Forecasts of spatial fields, in: Forecast Verification, John Wiley, 95–117, 2011.

Caylor, K. K., D'Odorico P., and Rodriguez-Iturbe I.: On the ecohydrological organization of spatially heterogeneous semi-arid landscapes, Water Resour. Res., 42, W07424, https://doi.org/10.1029/2005WR004683, 2006.

Ceballos, Y. and Ruiz de la Torre, J.: Árboles y arbustos de la España peninsular, ETSI Montes Publications, Madrid, 1979.

Chen, J. M. and Cihlar, J.: Retrieving leaf area index of boreal conifer forests using Landsat TM images, Remote Sens. Environ., 55, 153–162, 1996.

Cheng, L., Yaeger, M., Viglione, A., Coopersmith, E., Ye, S., and Sivapalan, M.: Exploring the physical controls of regional patterns of flow duration curves – Part 1: Insights from statistical analyses, Hydrol. Earth Syst. Sci., 16, 4435–4446, https://doi.org/10.5194/hess-16-4435-2012, 2012.

Cohen, W. B., Maiersperger, T. K., Gower, S. T., and Turner, D. P.: An improved strategy for regression of biophysical variables and Landsat ETM+data, Remote Sens. Environ., 84, 561–571, 2003.

Cohen, J.: Weighted kappa: nominal scale agreement provision for scaled disagreement or partial credit, Psychol. Bull., 70, 213–220, https://doi.org/10.1037/h0026256, 1968.

Conradt, T., Wechsung, F., and Bronstert, A.: Three perceptions of the evapotranspiration landscape: comparing spatial patterns from a distributed hydrological model, remotely sensed surface temperatures, and sub-basin water balances, Hydrol. Earth Syst. Sci., 17, 2947–2966, https://doi.org/10.5194/hess-17-2947-2013, 2013.

Coopersmith, E., Yaeger, M. A., Ye, S., Cheng, L., and Sivapalan, M.: Exploring the physical controls of regional patterns of flow duration curves – Part 3: A catchment classification system based on regime curve indicators, Hydrol. Earth Syst. Sci., 16, 4467–4482, https://doi.org/10.5194/hess-16-4467-2012, 2012.

Drewry, D. T. and Albertson, J. D.: Diagnosing model error in canopy-atmosphere exchange using empirical orthogonal function analysis, Water Resour. Res., 42, W06421, https://doi.org/10.1029/2005WR004496, 2006.

Fang, Z., Bogena, H., Kollet, S., Koch, J., and Vereecken, H.: Spatio-temporal validation of long-term 3D hydrological simulations of a forested catchment using empirical orthogonal functions and wavelet coherence analysis, J. Hydrol., 529, 1754–1767, 2015.

Feyen, L., Kalas, M., and Vrugt, J. A.: Semi-distributed parameter optimization and uncertainty assessment for large-scale streamflow simulation using global optimization, Hydrolog. Sci. J., 53, 293–308, https://doi.org/10.1623/hysj.53.2.293, 2008.

Francés, F. and Benito, J.: La modelación distribuida con pocos parámetros de las crecidas, Ingeniería del Agua, 2, 7–24, 1995.

Francés, F., Vélez, J. I., and Vélez, J. J.: Split-parameter structure for the automatic calibration of distributed hydrological models, J. Hydrol., 332, 226–240, 2008.

Franssen, H. J. H, Brunner, P., Makobo, P., and Kinzelbach, W.: Equally likely inverse solutions to a groundwater flow problem including pattern information from remote sensing images, Water Resour. Res., 44, W01419, https://doi.org/10.1029/2007WR006097, 2008.

Franz, T. E.: Ecohydrology of the upper Ewaso Ngiro river basin, Kenia, Doctoral Thesis, Princeton University, Princeton, NJ, USA, 2007.

Franz, T. E., Caylor, K. K., Nordbotten, J. M., Rodríguez-Iturbe, I., and Celia, M. A.: An ecohydrological approach to predicting regional woody species distribution patterns in dryland ecosystems, Adv. Water Resour., 33, 215–230, 2010.

Frassnacht, K. S., Gower, S. T., MacKenzie, M. D., Nordheim, E. V., and Lillesand, T. M.: Estimating the leaf area index of north central Wisconsin forests using the Landsat Thematic Mapper, Remote Sens. Environ., 61, 229–245, 1997.

Friedl, M. A., Michaelsen, J., Davis, F. W., Walker, H., and Schimel, D. S.: Estimating grassland biomass and leaf area index using ground and satellite data, Int. J. Remote Sens., 15, 1401–1420, 1994.

Gamon, J. A., Serrano, L., and Surfus, J. S.: The photochemical reflectance index: an optical indicator of photosynthetic radiation use efficiency across species, functional types, and nutrient levels, Oecologia, 112, 492–501, 1997.

García-Arias, A. and Francés, F.: The RVDM: modelling impacts, evolution and competition processes to determine riparian vegetation dynamics, Ecohydrology, 9, 438–459, https://doi.org/10.1002/eco.1648, 2016.

Gigante, V., Iacobellis, V., Manfreda, S., Milella, P., and Portoghese, I.: Influences of Leaf Area Index estimations on water balance modeling in a Mediterranean semi-arid basin, Nat. Hazards Earth Syst. Sci., 9, 979–991, https://doi.org/10.5194/nhess-9-979-2009, 2009.

GIMHA team (Vélez, I., Vélez, J., Puricelli, M., Montoya, J. J., Camilo, J. C., Bussi, G., Medici, C., Orozco, I., Ruiz-Pérez, G.): Description of the distributed conceptual hydrological model TETIS v.8, Universitat Politècnica de València, 2014.

Gilleland, E., Ahijevych, D. A., Brown, B. G., and Ebert, E. E.: Verifying forecasts spatially, B. Am. Meteorol. Soc., 91, 1365–1373, 2010.

Graf, A., Bogena, H. R., Drüe, C., Hardelauf, H., Pütz, T., Heinemann, G., and Vereecken, H.: Spatiotemporal relations between water budget components and soil moisture in a forested tributary catchment, Water Resour. Res., 50, 4837–4857, https://doi.org/10.1002/2013WR014516, 2014.

Gribovszki, Z., Kalicz, P., Szilágyi, J., and Kucsara, M.: Riparian zone evapotranspiration estimation from diurnal groundwater level fluctuations, J. Hydrol., 349, 6–17, https://doi.org/10.1016/j.jhydrol.2007.10.049, 2008.

Gutmann, E. D. and Small, E. E.: A method for the determination of the hydraulic properties of soil from MODIS surface temperature for use in land-surface models, Water Resour. Res., 46, W06520, https://doi.org/10.1029/2009WR008203, 2010.

Immerzel, W. and Droogers, P.: Calibration of a distributed hydrological model based on satellite evapotranspiration, J. Hydrol., 349, 411–424, 2008.

Jasechko, S., Sharp, Z. D., Gibson, J. J., Birks, S. J., Yi, Y., and Fawcett, P. J.: Terrestrial water fluxes dominated by transpiration, Nature, 496, 347–350, https://doi.org/10.1038/nature11983, 2013.

Kim, G. and Barros, A. P.: Space-time characterization of soil moisture from passive microwave remotely sensed imagery and ancillary data, Remote Sens.Environ., 81, 393–403, 2002.

Koch, J., Jensen, K. H., and Stisen, S.: Toward a true spatial model evaluation in distributed hydrological modeling: Kappa statistics, Fuzzy theroy, and EOF-analysis benchmarked by the human perception and evaluated against a modeling case study, Water Resour. Res., 51, 1225–1246, https://doi.org/10.1002/2014WR016607, 2015.

Koch, J., Cornelissen, T., Fang, Z., Bogena, H., Diekkrüger, B., Kollet, S., and Stisen, S.: Inter-comparison of three distributed hydrological models with respect to seasonal variability of soil moisture patterns at a small forested catchment, J. Hydrol., 533, 234–249, 2016a.

Koch, J., Siemann, A., Stisen, S., and Sheffield, J.: Spatial validation of large-scale land surface models against monthly land surface temperature patterns using innovative performance metrics, J. Geophys. Res.-Atmos., 121, 5430–5452, https://doi.org/10.1002/2015JD024482, 2016b.

Kunnath-Poovakka, A., Ryu, D., Renzullo, L. J., and George, B.: The efficacy of calibrating hydrologic model using remotely sensed evapotranspiration and soil moisture for streamflow prediction, J. Hydrol., 535, 509–524, 2016.

Landsberg, J. J. and Waring, R. H.: A generalised model of forest productivity using simplified concepts of radiation-use efficiency, carbon balance and partiotioning, Forest Ecol. Manage., 95, 209–228, 1997.

Law, B. E. and Waring, R. H.: Remote sensing of leaf area index and radiation intercepted by understory vegetation, Ecol. Appl., 4, 272–279, 1994.

Le Roux, X., Bariac, T., and Mariotti, A.: Spatial partitioning of the soil water resource between grass and shrub components in a West African humid savanna, Oecologia, 104, 147–155, 1995.

Liu, Y.: Spatial patterns of soil moisture connected to monthly-seasonal precipitation variability in a monsoon region, J. Geophys. Res.-Atmos., 108, 8856, https://doi.org/10.1029/2002JD003124, 2003.

Lo, M. H., Famiglietti, J. S., Yeh, P. J., and Syed, T. H.: Improving parameter estimation and water table depth simulation in a land surface model using GRACE water storage and estimated base flow data, Water Resour. Res., 46, W05517, https://doi.org/10.1029/2009WR007855, 2010.

López-Serrano, F. R., Landete-Castillejos, T., Martínez-Millán, J., and Cerro-Barja, A.: LAI estimation of natural pine forest using a non-standard sampling technique, Agr. Forest Meteorol., 101, 95–111, https://doi.org/10.1016/S0168-1923(99)00171-9, 2000.

Manfreda, S. and Caylor, K. K.: On The Vulnerability of Water Limited Ecosystems to Climate Change, Water, 5, 819–833, 2013.

Manfreda, S., Pizzolla, T., and Caylor, K. K.: Modelling vegetation patterns in semiarid environments, Procedia Environ. Sci., 19, 168–177, 2013.

Manfreda, S., Fiorentino, M., and Iacobellis, V.: DREAM: a distributed model for runoff, evapotranspiration, and antecedent soil moisture simulation, Adv. Geosci., 2, 31–39, https://doi.org/10.5194/adgeo-2-31-2005, 2005.

McCabe, M. F., Wood, E. F., Wojcik, R., Pan, M., Sheffield, J., Gao, H., and Su, H.: Hydrological consistency using multi-sensor remote sensing data for water and energy cycle studies, Remote Sens. Environ., 112, 430–444, 2008.

Medlyn, B. E.: Physiological basis of the light use efficiency model, Tree Physiol., 18, 167–176, 1998.

Merz, R., Parajka, J., and Bloschl, G.: Scale effects in conceptual hydrological modeling, Water Resour. Res., 45, W09405, https://doi.org/10.1029/2009WR007872, 2009.

Michaud, J. and Sorooshian, S.: Comparison of simple versus complex distributed runoff models on a midsized semiarid watershed, Water Resour. Res., 30, 593–605, 1994.

Montaldo, N., Rondena, R., Albertson, J. D., and Mancini, M.: Parsimonious modeling of vegetation dynamics for ecohydrologic studies of water limited ecosystems, Water Resour. Res., 41, W10416, https://doi.org/10.1029/2005WR004094, 2005.

Nash, J. E. and Sutcliffe, J. V.: River flow forecasting through conceptual models part I – A discussion of principles, J. Hydrol., 10, 282–290, 1970.

Nyholm, T., Rasmussen, K. R., and Christensen, S.: Estimation of stream flow depletion and uncertainty from discharge measurements in a small alluvial stream, J. Hydrol., 274, 129–144, 2003.

Ollinger, S. V., Richardson, A. D., Martin, M. E., Hollinger, D. Y., Frolking, S., Reich, P. B., Plourde, L. C., Katul, G., Munger, J. W., Oren, R., Smith, M. L., Paw, U., Bolstad, K. T., Cook, P. V., Day, B., Martin, M. C., Monson, T. A., and Schmidt, R. K. H. P.: Canopynitrogen, carbon assimilation, and albedo in temperate and boreal forests: functional relations and potential climate feedbacks, P. Natl. Acad. Sci. USA, 105, 19335–19340, 2008.

Pasquato, M., Medici, M., Friend, A. D., and Francés, F.: Comparing two approaches for parsimonious vegetation modelling in semiarid regions using satellite data, Ecohydrology, 8, 1024–1036, https://doi.org/10.1002/eco.1559, 2015.

Porporato, A., Laio, F., Ridolfi, L., and Rodriguez-Iturbe, I.: Plants in water-controlled ecosystems: active role in hydrologic processes and response to water stress: III. Vegetation water stress, Adv. Water Resour., 24, 725–744, 2001.

Pumo, D., Noto, L. V., and Viola, F.: Ecohydrological modelling of flow duration curve in Mediterranean river basins, Adv. Water Resour., 52, 314–327, 2013.

Preisendorfer, R. W. and Mobbley, C. D.: Principal component analysis in meteorology and oceanography, Vol. 425, Amsterdam, Elsevier, 1988.

Quevedo, D. I. and Francés, F.: A conceptual dynamic vegetation-soil model for arid and semiarid zones, Hydrol. Earth Syst. Sci., 12, 1175–1187, https://doi.org/10.5194/hess-12-1175-2008, 2008.

Rientjes, T. H. M., Muthuwatta, L. P., Bos, M. G., Booij, M. J., and Bhatti, H. A.: Multi-variable calibration of a semi-distributed hydrological model using streamflow data and satellite-based evapotranspiration, J. Hydrol., 505, 276–290, 2013.

Reed, S., Koren, V., Smith, M., Zhang, Z., Moreda, F., Seo, D.-J., and DMIP participants: Overall distributed model intercomparison project results, J. Hydrol., 298, 27–60, 2004.

Rodriguez-Iturbe, I., Porporato, A., Laio, F., and Ridolfi, L.: Plants in water-controlled ecosystems: active role in hydrologic processes and response to water stress, I. Scope and general outline, Adv. Water Resour., 24, 695–705, 2001.

Ruiz-Pérez, G., González-Sanchis, M., Del Campo, A. D., and Francés, F.: Can a parsimonious model implemented with satellite data be used for modelling the vegetation dynamics and water cycle in water-controlled environments?, Ecol. Model., 324, 45–53, 2016.

Samaniego, L., Kumar, R., and Jackisch, C.: Predictions in a data-sparse region using a regionalized grid-based hydrologic model driven by remotely sensed data, Hydrol. Res., 42, 338–355, https://doi.org/10.2166/nh.2011.156, 2011.

Sims, D. A., Luo, H., Hastings, S., Oechel, W. C., Rahman, A. F., and Gamon, J. A.: Parallel adjustments in vegetation greenness and ecosystem CO_2 exchange in response to drought in a Southern California chaparral ecosystem, Remote Sens. Environ., 103, 289–303, 2006.

Smith, M. B., Koren, V., Reed, S., Zhang, Z., Zhang, Y., Moreda, F., Cui, Z., Mizukami, N., Anderson, E. A., and Cosgrove, B. A.: The distributed model intercomparison project – Phase 2: Motivation and design of the Oklahoma experiments, J. Hydrol., 418–419, 3–16, https://doi.org/10.1016/j.jhydrol.2011.08.055, 2012.

Smith, M. B., Koren, V., Zhang, Z., et al.: The Distributed Model Intercomparison project – Phase 2: Experiment design and summary results of the western basin experiments, J. Hydrol., 207, 300–329, https://doi.org/10.1016/j.jhydrol.2013.08.040, 2013.

Stisen, S., McCabe, M. F., Refsgaard, J. C., Lerer, S., and Butts, M. B.: Model parameter analysis using remotely sensed pattern information in a multi-constraint framework, J. Hydrol., 409, 337–349, 2011.

Tsang, Y. P., Hornberger, G., Kaplan, L. A., Newbold, J. D., and Aufdenkampe, A. K.: A variable source area for groundwater evapotranspiration: Impacts on modeling stream flow, Hydrol. Proc., 28, 2439–2450, 2014.

Turner, D. P., Cohen, W. B., Kennedy, R. E., Fassnacht, K. S., and Briggs, J. M.: Relationships between leaf area index and Landsat TM spectral vegetation indices across three temperate zone sites, Remote Sens. Environ., 70, 52–68, 1999.

van Dijk, A. I. J. M. and Renzullo, L. J.: Water resource monitoring systems and the role of satellite observations, Hydrol. Earth Syst. Sci., 15, 39–55, https://doi.org/10.5194/hess-15-39-2011, 2011.

Velpuri, N. M., Senay, G. B., and Asante, K. O.: A multi-source satellite data approach for modelling Lake Turkana water level: calibration and validation using satellite altimetry data, Hydrol. Earth Syst. Sci., 16, 1–18, https://doi.org/10.5194/hess-16-1-2012, 2012.

Wagener, T., Blöschl, G., Goodrich, D. C., Gupta, H., Sivapalan, M., Tachikawa, Y., and Weiler, M.: A synthesis framework for runoff prediction in ungauged basins, chap., 2, 11–28, 2013.

Wi, S., Yang, Y. C. E., Steinschneider, S., Khalil, A., and Brown, C. M.: Calibration approaches for distributed hydrologic models in poorly gaged basins: implication for streamflow projections under climate change, Hydrol. Earth Syst. Sci., 19, 857–876, https://doi.org/10.5194/hess-19-857-2015, 2015.

Winsemius, H. C., Savenije, H. H. G., and Bastiaanssen, W. G. M.: Constraining model parameters on remotely sensed evaporation: justification for distribution in ungauged basins?, Hydrol. Earth Syst. Sci., 12, 1403–1413, https://doi.org/10.5194/hess-12-1403-2008, 2008.

Xiao, X., Zhang, Q., Braswell, B., Urbanski, S., Boles, S., Wofsy, S., Moore, B., and Ojima, D.: Modeling gross primary production of temperate deciduous broadleaf forest using satellite images and climate data, Remote Sens. Environ., 91, 256–270, 2004.

Yang, Y., Shang, S., and Jiang, L.: Remote sensing temporal and spatial patterns of evapotranspiration and the responses to water management in a large irrigation district of North China, Agr. Forest Meteorol., 164, 112–122, 2012.

Yuan, W., Liu, S., Zhou, G., Zhou, G., Tieszen, L. L., Baldocchi, D., Bernhofer, C., Gholz, H., Goldstein, A. H., Goulden, M. L., Hollinger, D. Y., Hu, Y., Law, B. E., Stoy, P. C., Vesala, T., and Wofsy, S. C.: Deriving a light use efficiency model from eddy covariance flux data for predicting daily gross primary production across biomes, Agr. Forest Meteorol., 143, 189–207, https://doi.org/10.1016/j.agrformet.2006.12.001, 2007.

Yuan, W., Liu, S., Yu, G., Bonnefond, J.-M., Chen, J., Davis, K., Desai, A. R., Goldstein, A. H., Gianelle, D., Rossi, F., Suyker, A. E., and Verma, S. B.: Global estimates of evapotranspiration and gross primary production based on MODIS and global meteorology data, Remote Sens. Environ., 114, 1416–1431, 2010.

Zhang, Y. Q., Francis, H. S., Chiew, H. S., Zhang, L., and Li, H.: Use of remotely sensed actual evapotranspiration to improve rainfall-runoff modeling in southeast Australia, Am. Meteorol. Soc., 10, 969–980, 2009.

Zhang, Y. Q., Viney, N. R., Chiew, F. H. S., van Dijk, A. I. J. M., and Liu Y. Y.: Improving hydrological and vegetation modelling using regional model calibration schemes together with remote sensing data, 19th International Congress on Modelling and Simulation, Perth, Australia, 12–16 December, 2011.

Zhang, Y., Peña-Arancibia, J. L., McVicar, T. R., Chiew, F. H., Vaze, J., Liu, C., Lu, X., Zheng H., Wang, Y., Liu, Y., and Miralles, D. G.: Multi-decadal trends in global terrestrial evapotranspiration and its components, Sci. Rep.-UK, 6, 19124, https://doi.org/10.1038/srep19124, 2016.

6

Towards ecosystem accounting: a comprehensive approach to modelling multiple hydrological ecosystem services

C. Duku[1], H. Rathjens[2], S. J. Zwart[3], and L. Hein[1]

[1]Environmental Systems Analysis Group, Wageningen University, P.O. Box 47, 6700AA, Wageningen, the Netherlands
[2]Department of Earth, Atmospheric, and Planetary Sciences, Purdue University, 550 Stadium Mall Drive, West Lafayette, IN 47907-2093, USA
[3]Africa Rice Center, 2031 PB, Cotonou, Benin

Correspondence to: C. Duku (confidence.duku@wur.nl, confidence.duku@gmail.com)

Abstract. Ecosystem accounting is an emerging field that aims to provide a consistent approach to analysing environment–economy interactions. One of the specific features of ecosystem accounting is the distinction between the capacity and the flow of ecosystem services. Ecohydrological modelling to support ecosystem accounting requires considering among others physical and mathematical representation of ecohydrological processes, spatial heterogeneity of the ecosystem, temporal resolution, and required model accuracy. This study examines how a spatially explicit ecohydrological model can be used to analyse multiple hydrological ecosystem services in line with the ecosystem accounting framework. We use the Upper Ouémé watershed in Benin as a test case to demonstrate our approach. The Soil Water and Assessment Tool (SWAT), which has been configured with a grid-based landscape discretization and further enhanced to simulate water flow across the discretized landscape units, is used to simulate the ecohydrology of the Upper Ouémé watershed. Indicators consistent with the ecosystem accounting framework are used to map and quantify the capacities and the flows of multiple hydrological ecosystem services based on the model outputs. Biophysical ecosystem accounts are subsequently set up based on the spatial estimates of hydrological ecosystem services. In addition, we conduct trend analysis statistical tests on biophysical ecosystem accounts to identify trends in changes in the capacity of the watershed ecosystems to provide service flows. We show that the integration of hydrological ecosystem services into an ecosystem accounting framework provides relevant information on ecosystems and hydrological ecosystem services at appropriate scales suitable for decision-making.

1 Introduction

Ecosystem accounting provides a systematic framework to link ecosystems to economic activities (Boyd and Banzhaf, 2007; Maler et al., 2008; EC et al., 2013; Edens and Hein, 2013; Obst et al., 2013). Specifically, ecosystem accounting aims to integrate the concept of ecosystem services into a national accounting context as described in UN et al. (2009). There is increasing interest in ecosystem accounting as a new, comprehensive tool for environmental monitoring and management (Obst et al., 2013). The recently released System of Environmental-Economic Accounting (SEEA)–Experimental Ecosystem Accounting guideline (EC et al., 2013) provides guidelines for setting up both biophysical and monetary ecosystem accounts. Biophysical accounting for ecosystem services forms the basis for monetary accounting.

Ecosystem services are the contributions of ecosystems to human welfare (TEEB, 2010; EC et al., 2013). Hydrological ecosystem services, specifically, are the contributions to human benefits produced by the effects of terrestrial ecosystem components on freshwater as it moves through the landscape. Terrestrial ecosystem components directly modify different attributes (such as quantity, quality, location and timing) of various ecohydrological processes, resulting in augmentation or degradation of these processes (Brauman et al., 2007). Factors such as the presence of beneficiaries (Boyd and Banzhaf, 2007), spatiotemporal accessibility (Fisher et al., 2009), and management pressure (Schröter et al., 2014) then determine whether the ecohydrological processes constitute hydrological ecosystem services. Hydrological ecosystem services are diverse and can be broadly classified into five categories: improvement of extractive water supply, improvement of in-stream water supply, water damage mitigation, provision of water-related cultural services, and water-associated supporting services (Brauman et al., 2007). Production of these services underlies water and food security and the protection of human lives and properties.

Biophysical accounting for hydrological ecosystem services allows for the organization and analysis of biophysical data on these services at different spatial and temporal scales suitable for the development, monitoring and evaluation of public policy (EC et al., 2013). Biophysical accounting also allows for the distinction between the flow of hydrological ecosystem services and the capacity of watershed ecosystems to provide service flows (EC et al., 2013). Service flow is the contribution in space and time of an ecosystem to either a utility function (e.g. private household) or a production function (e.g. crop production) that leads to a human benefit, whereas service capacity is a reflection of ecosystem condition and extent at a point in time, and the resulting potential to provide service flows (EC et al., 2013; Edens and Hein, 2013). For hydrological ecosystem services, high service capacity areas and high service flow areas may occur in different points or areas in space (Fisher et al., 2009), making the need for their empirical distinction and separate spatial characterization crucial for land and watershed management.

Many approaches have been used for modelling, mapping and quantifying hydrological ecosystem services (e.g. Le Maitre et al., 2007; Naidoo et al., 2008; Liquete et al., 2011; Maes et al., 2012; Notter et al., 2012; Willaarts et al., 2012; Leh et al., 2013; Liu et al., 2013; Terrado et al., 2014, for an overview). For ecosystem accounting, however, key aspects requiring further research include the modelling of hydrological ecosystem services with adequate spatiotemporal detail and accuracy at aggregated scales, distinguishing between service capacity and service flow, and linking ecohydrological processes (and ecosystem components) to the supply of dependent hydrological ecosystem services. Addressing these issues requires the consideration of among others physical and mathematical representation of ecohydrological processes, spatial heterogeneity of ecosystems, temporal resolution, and required model accuracy (Guswa et al., 2014). Adequate representation of the spatial heterogeneity of biophysical environments in ecohydrological models is crucial in ecosystem accounting because spatial units form the basic focus of measurement similar to functions of economic units in national accounting (EC et al., 2013). In addition, if ecosystem accounting is to provide reliable information for the assessment of integrated policy responses at the landscape level, then physical and mathematical representation of model processes should be based on scientific consensus (Vigerstol and Aukema, 2011). Furthermore, model results should be accurate and model uncertainties should be understood and reported (Seppelt et al., 2011; Martínez-Harms and Balvanera, 2012). Finally, ecohydrological modelling for ecosystem accounting necessitates the use of continuous simulation watershed models that are able to capture short- and long-term temporal variability in ecohydrological processes.

Our objective is to present a spatially explicit modelling approach aligned with an ecosystem accounting framework to map and quantify the capacities and the flows of multiple hydrological ecosystem services. We use the Soil Water and Assessment Tool (SWAT), which has been configured with a grid-based landscape discretization and further enhanced to simulate water flow across the discretized landscape units, to simulate the watershed ecohydrology. The model is calibrated and validated, and indicators consistent with the ecosystem accounting framework are used to map and quantify the capacities and the flows of multiple hydrological ecosystem services based on the model outputs. Biophysical ecosystem accounts are subsequently set up based on the spatial estimates of hydrological ecosystem services. We use the Upper Ouémé watershed in Benin as a test case to demonstrate our approach. This case-study area was selected because of a relatively high data availability (Judex and Thamm, 2008; AMMA-CATCH, 2014). It is also a microcosm of rural sub-Saharan Africa, where large sections of the population depend on smallholder rainfed agriculture for their livelihood, where groundwater is the major source of drinking water, and where rapid population growth and increasing land use change are prevalent. The hydrological ecosystem services we model and account for are crop water supply, household water supply (groundwater supply and surface water supply), water purification, and soil erosion control. We select these four services because they are critical to food and water security for the population. Agriculture is the major source of income and livelihood in the watershed and is predominantly rainfed. Furthermore, groundwater is the major source of household water use (for both drinking and non-drinking purposes).

2 Description of the case-study area

The Upper Ouémé watershed as depicted in Fig. 1 is located in central Benin, covering an area of approximately 14 500 km^2. The natural vegetation is a mosaic of savannah woodland and small forest islands. The Upper Ouémé forest reserve is the major protected forest area in the watershed, with an approximate area of 2420 km^2. Smallholder rainfed agriculture is the major economic activity and is supported by climatic conditions that are characterized by a unimodal rainfall season from May to October of about 1250 mm per year.

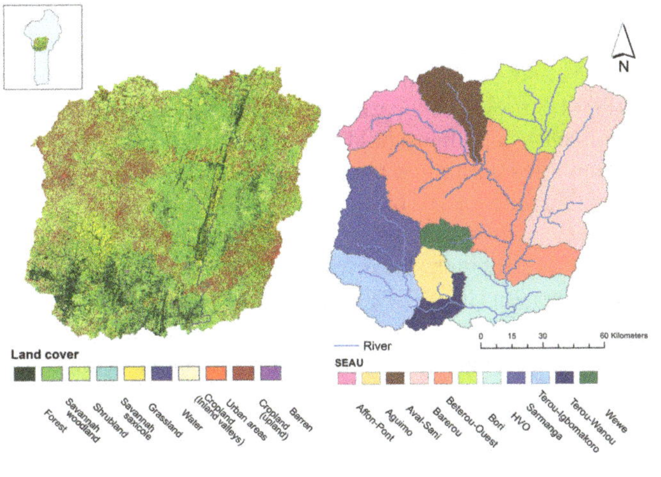

Figure 1. Land cover and subwatershed ecosystem accounting units (SEAUs) of the Upper Ouémé watershed. Land cover data adapted from Judex and Thamm (2008).

Maize, rice, yam, cassava and millet are some of the important food crops cultivated in this area, with cotton being the major cash crop. These crops are predominantly cultivated using rainfed agriculture. The irrigation sector is relatively poorly developed. Rice is mostly cultivated in inland valley lowlands due to their higher water availability, lower soil fragility and higher fertility compared to upland areas (Giertz et al., 2012; Rodenburg et al., 2014). Fertilizer use is increasing in the region and high fertilizer inputs are associated with crops such as maize, rice and cotton (Bossa et al., 2012). An estimated average of 100–250 $kg\,ha^{-1}$ of fertilizer (nitrogen, phosphorus and potassium NPK + urea) is applied to cotton, rice and maize (Bossa et al., 2012). With a population of about 400 000, there is low demographic density (28 inhabitants km^{-2}) in the watershed (Judex and Thamm, 2008). However, the population is growing rapidly (about 4 % yr^{-1}) due to migrants coming from different parts of the country and other neighbouring countries to farm. Rapid population growth has caused the expansion of agricultural areas and led to both deforestation and increasing scarcity of agricultural land (Judex and Thamm, 2008) accompanied by increasing soil degradation due to shortening of the fallow period (Giertz et al., 2012). It has been estimated that there will be nearly complete deforestation in some parts of the Upper Ouémé watershed assuming a 6 % yr^{-1} expansion of agricultural areas (Orekan, 2007).

Conversion of savannah woodland and forests for crop cultivation is mainly through slash and burn techniques. In addition, the population obtains about 90 % of its drinking water needs directly from ground-water, with about 5 % from small lakes, ponds and rivers collectively referred to in this study as surface water (Judex and Thamm, 2008).

3 Methods

3.1 Modelling framework

3.1.1 Model selection

In order to address modelling challenges regarding model process inclusion, spatial heterogeneity, physical and mathematical representation, temporal resolution, and model accuracy, we considered several watershed models and selected the SWAT model (Arnold et al., 1998) to be most appropriate for this study. The SWAT model has a comparative advantage in integrated assessment modelling of ecohydrological interactions that underpin hydrological ecosystem services provision (Vigerstol and Aukema, 2011). The SWAT model is a physically based, ecohydrological model that simulates the impact of land use and land management practices on water, sediments and agricultural chemicals in large complex watersheds (Neitsch et al., 2009). It is a continuous simulation watershed model operated at a daily time step. In the SWAT model, a watershed can be spatially discretized using three approaches. They are grid cells, representative hillslopes, and hydrologic response units (HRUs) (Arnold et al., 2013). The HRU-based discretization is the most popular and most geographic information system interfaces are set up to use this discretization (e.g. ArcSWAT). Each HRU is a lumped area within a subwatershed that is comprised of unique land cover, soil and management combinations (Neitsch et al., 2009). The hydrological cycle is divided into two phases. The first is the land phase which controls the amount of water, sediment, nutrient and pesticide loadings to the main channel in each subwatershed. Land phase processes include weather, hydrology (canopy storage, infiltration, evapotranspiration, surface runoff, lateral subsurface flow, return flow) plant growth, erosion, nutrients and management operations (Neitsch et al., 2009). Surface runoff, lateral flow and return flow from the land phase are then routed through the channel network of the watershed to the outlet in the second phase called the routing phase. This phase also includes processes such as sediment and nutrient routing (Neitsch et al., 2009).

3.1.2 Model modification

The SWAT model used in this study had two major modifications; the first one was a model process modification, whereas the second one was a modification of the spatial discretization scheme. The process modification involved the incorporation of a landscape routing sub-model that simulates surface water, lateral and groundwater flow interactions across discretized landscape units. This sub-model was developed and incorporated into the standard SWAT model by Volk et al. (2007) and Arnold et al. (2010). The modified model, the SWAT Landscape model, addresses an inherent weakness in the standard SWAT model. The standard SWAT model uses an HRU-based discretization and transported water, sediment, nutrient and pesticide loadings from upstream HRUs are routed directly into stream channels bypassing downstream HRUs (Gassman et al., 2007; Volk et al., 2007; Arnold et al., 2010; Bosch et al., 2010). Therefore, the impact of management of upstream HRUs on downstream HRUs cannot be sufficiently assessed. This weakness is a result of the lack of spatial interactions among different HRUs in the land phase of the hydrological cycle (Neitsch et al., 2009). The SWAT Landscape model addresses this weakness by using a constant flow separation ratio to partition landscape and channel flow in each HRU (Arnold et al., 2010). The channel flow portion is routed through the stream network, whereas the landscape flow portion is routed from upstream HRUs to downstream HRUs.

The second modification was a change from the HRU-based spatial discretization scheme of the standard SWAT model to a grid-based landscape discretization scheme. We set up the SWAT Landscape model with this grid-based landscape discretization using SWATgrid (Rathjens and Oppelt, 2012). The grid-based set-up of the SWAT Landscape model uses a modified topographic index to estimate spatially distributed proportions of landscape and channel flow (Rathjens et al., 2014), unlike the HRU-based set-up which uses a constant flow separation ratio. A new parameter called the drainage density factor controls the spatially distributed flow separation in the SWATgrid set-up (Rathjens et al., 2014). This parameter can be adjusted during calibration. For this study, the grid-based set-up of the SWAT Landscape model was used to delineate the watershed into spatially interacting grid cells. Flow paths were determined from the digital elevation model (DEM) and the TOPAZ digital landscape analysis tool (Garbrecht and Martz, 2000) and runoff from a grid cell flowed to one of eight adjacent cells (Rathjens et al., 2014). A detailed description of the two modifications can be found in Arnold et al. (2010) and Rathjens et al. (2014).

3.1.3 Model input data

A combination of spatial and non-spatial input data from a variety of sources was used to set up the model. The spatial input data are described in Table 1. A 30 m DEM was obtained from the National Aeronautics and Space Administration (NASA) ASTER Global Digital Elevation Map to generate stream networks and watershed configurations, and to estimate topographic parameters. Land cover and soil maps were obtained from the Integrated Approach to Efficient Management of Scarce Water Resources in West Africa (IMPETUS) project database (Judex and Thamm, 2008). The land cover map had been derived from classification of LANDSAT-7 ETM+ satellite images. Gridded daily precipitation data were obtained from the African Monsoon and Multidisciplinary Analysis–Coupling the Tropical Atmosphere and the Hydrological Cycle (AMMA-CATCH) database (AMMA-CATCH, 2014) and gridded temperature data were obtained from the Climate Research Unit (CRU) TS 3.21 database (Jones and Harris, 2013). Data on groundwater and surface water household consumption (including drinking and non-drinking purposes) were obtained from the IMPETUS project database. These had been derived from national census and household surveys in about 200 towns and communities within the watershed (INSAE, 2003; Hadjer et al., 2005; Judex and Thamm, 2008). For our study area, per capita groundwater consumption was 19 L per day per person and per capita surface water consumption was 14 L per day per person (INSAE, 2003; Hadjer et al., 2005).

3.1.4 Model configuration and performance evaluation

The initial model set-up was carried out with the ArcSWAT interface, which is based on an HRU configuration. This was essential for generating input data for the grid-based configuration. Simulations of the HRU-based SWAT model were conducted for the period 1999–2012. The first 2 years (1999 and 2000) served as a warm-up period for the model to assume realistic initial conditions. Potential evapotranspiration was modelled with the Hargreaves method (Hargreaves et al., 1985) and water transfers for households were modelled as constant extraction rates from shallow aquifers (groundwater extractions) and streams (surface water extractions). The Soil Conservation Service curve number approach was used to model surface runoff and the daily curve number value was calculated as a function of plant evapotranspiration (Neitsch et al., 2009). The HRU-based SWAT model was first calibrated and validated with streamflow data before calibration and validation of sediment and nitrogen loads. A split-time calibration and validation technique was carried out on the HRU-based model using the Sequential Uncertainty Fitting (SUFI-2) optimization algorithm of the SWAT Calibration and Uncertainty Program (Abbaspour et al., 2008). For calibration and validation of streamflow, we used daily observed streamflow data from 11 monitoring stations within the watershed. These stations had drainage areas of varying spatial scale to capture watershed-scale and subwatershed-scale ecohydrological processes. Calibration was mostly from 2001 to 2007 and validation was from 2008 to 2011. To evaluate transport of sediments and nutrients, the model was further calibrated with weekly measured sediment and organic nitrogen load data. Two years of data (2008–2009) were available from a single monitoring station, Beterou station. Sediment and organic nitrogen load data for 2008 were used for the calibration, whereas data for 2009 were used for the validation.

The calibrated and validated input parameter sets from the HRU-based set-up were transferred to the grid-based set-up of the SWAT Landscape model using the SWATgrid interface (Rathjens and Oppelt, 2012). Given the computational resources and time required to run a grid-based set-up of the SWAT Landscape model at a higher spatial resolution (e.g. 1 ha) for a relatively large watershed such as the Upper Ouémé (Arnold et al., 2010; Rathjens and Oppelt, 2012), we

Table 1. Description of spatial input data of the Upper Ouémé watershed for the SWAT Landscape model.

Data type	Description	Resolution	Source
Topography	ASTER digital elevation model (DEM)	30 m	NASA
Land use/land cover	Classified LANDSAT-7 ETM+ image	28.5 m	IMPETUS
Soil types	Soil map and associated parameters derived from geological maps and field surveys	1 : 200 000	IMPETUS
Precipitation	Gridded daily precipitation data (1999–2012)	25 km	AMMA-CATCH
Temperature	Gridded monthly average minimum and maximum temperatures (1999–2012)	50 km	CRU TS 3.21
Household water consumption	Groundwater and surface water extractions	(village level)	IMPETUS

resampled the DEM, soil and land cover data to a resolution of 500 m × 500 m. The resampling allowed for a balance between computational efficiency during model simulation and maintenance of accurate spatial representation of landscape patterns. Grid-based simulations of the SWAT Landscape model were conducted for the period 1999–2012. The first 2 years served as a model warm-up period. The grid-based set-up of the SWAT Landscape model was then calibrated manually by adjusting only the drainage density factor parameter. The full calibrated parameter values are listed in Table 3. Three quantitative statistics recommended by Moriasi et al. (2007) were selected to evaluate model performance: Nash–Sutcliffe efficiency (NSE), percent bias (PBIAS), and ratio of the root mean square error to the standard deviation of measured data (RSR). Nash–Sutcliffe efficiency is a normalized statistic that determines the relative magnitude of the residual variance compared to the measured data variance (Nash and Sutcliffe, 1970); PBIAS measures the average tendency of the simulated data to be larger or smaller than their observed counterparts (Gupta et al., 1999); RSR standardizes root mean squared error using the observations' standard deviation (Moriasi et al., 2007).

3.2 Spatial assessment of hydrological ecosystem services

For each hydrological ecosystem service, two appropriate indicators were selected to model service flow and service capacity. Computations were made for each grid cell enabling the model to reflect spatial differences in service flow and in service capacity. The selected hydrological ecosystem services and their service flow and service capacity indicators are shown in Table 2.

3.2.1 Crop water supply

An important hydrological ecosystem service input to crop production in rainfed agricultural systems is the provision of plant available water by ecohydrological processes that affect the soil water balance (Pattanayak and Kramer, 2001; IWMI, 2007; Zang et al., 2012). Crop water stress is a major limitation to crop production in rainfed agricultural systems (IWMI, 2007). We modelled service flow in croplands, which is referred to in this study as upland agricultural fields, and in inland valley lowlands (Rodenburg et al., 2014). Whereas inland valley lowlands in the study area are predominantly used for rice cultivation, the land cover input data did not differentiate the types of crops grown in upland agricultural fields. For our simulations we assumed that all upland agricultural fields were used for only maize cultivation (which is the most common crop in our study area in terms of extent of cultivated land area). For maize cultivation, the growing period (GP), i.e. the time period between crop establishment and harvesting, was 103 days, whereas the GP for rice cultivation was 123 days. For both maize and rice, crop establishment was in the month of June. Service flow of crop water supply was modelled as the total number of days during a growing period in which there was no water stress (i.e. days when the total plant water uptake was sufficient to meet maximum plant water demand). Service flow depends on the specific type of crop cultivated. This approach is based on the model output variable, daily water stress, and is a modification of Notter et al. (2012). For each day, the model used Eq. (1) to compute water stress for a given grid cell, j (Neitsch et al., 2009). After model simulation, service flow was computed using Eq. (2).

$$W_{\mathrm{strs},\,j} = 1 - T_{\mathrm{act},\,j}/T_{\mathrm{max},\,j}, \tag{1}$$

where W_{strs} is daily water stress, T_{act} is plant water uptake or actual transpiration (mm), and T_{max} is maximum plant water demand or maximum transpiration (mm):

$$S_{\mathrm{f},\,j} = N(d_1, d_2, \ldots, d_n | W_{\mathrm{strs}} = 0)_j, \tag{2}$$

where S_{f} is the service flow (days GP^{-1}), and N is the number of days d_1 to d_n when W_{strs} was zero.

Service capacity on the other hand was modelled as the total number of days in a year when the sum of actual evapotranspiration and the amount of residual moisture added to the soil profile equalled or exceeded potential evapotranspiration. For a given spatial unit, this gives an indication of the number of days when potentially there will be no crop water stress irrespective of the crop type to be cultivated. This approach has management relevance. Our approach was

Table 2. Overview of selected hydrological ecosystem services and associated service flow and service capacity indicators (GP is growing period).

Hydrological ecosystem service	Service flow indicator	Service capacity indicator
1. Crop water supply	Total number of days during the growing period in which there was no water stress (days GP^{-1})	Total number of days in a year when the sum of actual evapotranspiration and the amount of residual moisture added to the soil profile equalled or exceeded potential evapotranspiration (days yr^{-1})
2. Household water supply		
a. Groundwater supply	Amount of groundwater extracted ($m^3\,ha^{-1}\,yr^{-1}$)	Groundwater recharge ($m^3\,ha^{-1}\,yr^{-1}$)
b. Surface water supply	Amount of surface water extracted ($m^3\,ha^{-1}\,yr^{-1}$)	Water yield ($m^3\,ha^{-1}\,yr^{-1}$)
3. Water purification	Rate of denitrification ($kg\,ha^{-1}\,yr^{-1}$)	Denitrification efficiency (% denitrified)
4. Soil erosion control	Reduction in soil loss (metric $t\,ha^{-1}\,yr^{-1}$)	Maximum potential reduction in soil loss (metric $t\,ha^{-1}\,yr^{-1}$)

Table 3. Description of calibrated parameter values of the SWAT Landscape model. Superscript a indicates that the fitted values depended on the land cover type. Superscript b indicates that this parameter was used only in the calibration of the grid-based SWAT Landscape model. Subscript v_ indicates that the parameter value is replaced by the fitted value. Subscript r_ indicates the parameter value is multiplied by (1+ the fitted value).

Parameter name	Description	Fitted values
r_CN2	Initial SCS (Soil Conservation Service) runoff curve number for moisture condition II	(From −0.2 to −0.05)[a]
v_RCHRG_DP	Deep aquifer percolation fraction	0.2
v_GW_REVAP	Groundwater re-evaporation coefficient	0.18
v_GWQMN	Threshold depth of water in the shallow aquifer required for return flow to occur	1000
v_REVAPMN	Threshold depth of water in the shallow aquifer for re-evaporation or percolation to the deep aquifer to occur	500
v_SURLAG	Surface runoff lag coefficient	0.12
r_SOL_AWC	Available water capacity of the soil	0.1
v_ESCO	Soil evaporation compensation factor	(From 0.001 to 0.2)[a]
v_EPCO	Plant uptake compensation factor	(From 0.1 to 1)[a]
v_USLE_P	USLE (Universal Soil Loss Equation) support practice factor	0.13
v_USLE_C	Minimum value of the USLE C factor for water erosion applicable to the land cover	(From 0.038 to 0.45)[a]
v_NPERCO	Nitrate percolation coefficient	0.2
v_N_UPDIS	Nitrogen uptake distribution parameter	70
v_SDNCO	Denitrification threshold water content	1.1
v_CDN	Denitrification exponential rate coefficient	1.4
v_DD[b]	Drainage density factor which affects the flow separation ratio	7.5

based on the commonly used method (FAO, 1978, 1983) for determining the length of growing periods in rainfed agricultural systems. Unlike that method where moisture supply was based on precipitation, moisture supply in our approach was based on simulated spatiotemporal soil moisture dynamics. We used this approach for our study because at the local scale terrestrial ecosystem components have very little effect on precipitation attributes such as quantity, location, timing, etc. For a given day, the SWAT model used Eq. (3) to compute water balance. From the water balance components we computed the total available soil moisture and subsequently calculated potential water stress using Eq. (4). Service capacity of crop water supply was then computed using Eq. (5).

$$\Delta SW_i = \sum_{i=1}^{n} \left(R_{day} - Q_{surf} - ET_a - W_{seep} - Q_{gw} \right), \quad (3)$$

where ΔSW is the amount of residual moisture added to the soil profile on day i (mm); n is number of days in the year; R_{day} is the amount of precipitation (mm); Q_{surf} is the amount of surface runoff (mm); ET_a is actual evapotranspiration (mm); W_{seep} is percolation exiting the soil profile (mm); and Q_{gw} is return flow (mm).

$$W_{\text{pstrs}} = 1 - \left[(\Delta\text{SW} + \text{ET}_\text{a})/\text{ET}_\text{p}\right] \text{ if } \Delta\text{SW} > 0. \quad (4)$$

W_{pstrs} is potential daily water stress; ΔSW is the amount of residual moisture added to the soil profile (mm); ET_a is actual evapotranspiration (mm); ET_p is potential evapotranspiration (mm).

$$S_\text{c} = N\left(d_1, d_2, \ldots, d_n \mid W_{\text{pstrs}} \leq 0\right), \quad (5)$$

where S_c is service capacity (days yr^{-1}), and N is the number of days d_1 to d_n in a year when potentially there will be no water stress.

3.2.2 Household water supply

This hydrological ecosystem service refers to the amount of water extracted before treatment for household consumption (drinking and non-drinking purposes) (EC et al., 2013). This measurement boundary excluded other sources of water (e.g. tap water) where economic agents or inputs (e.g. water treatment facilities) were used to modify the state of the water resources before household consumption. We acknowledge that inflows to reservoirs of water distribution and processing facilities that deliver tap water can be considered as a hydrological ecosystem service. However, we excluded this from our study. This is because in our study area, the population obtain about 90 % of their drinking water needs from groundwater, with about 5 % from small lakes, ponds and rivers collectively referred to in this study as surface water (Judex and Thamm, 2008). A distinction was made between service capacity and service flow from groundwater, and service capacity and service flow from surface water.

To model service flow from groundwater and surface water, data on water consumption per capita, village population and water access for about 200 communities within the watershed were used. These data had been extracted from the 2002 national census (INSAE, 2003) and from household surveys in the study area (Hadjer et al., 2005). The data represented household water consumption (including drinking and non-drinking purposes) and lacked information on the actual points of extraction. Therefore, in modelling the service flow, we assumed that there is a positive spatial correlation between points of consumption and points of extraction. Furthermore, to estimate the village population from 2003 to 2012, we applied a 4 % yr^{-1} growth rate (Judex and Thamm, 2008). Water consumption per capita, however, was kept constant. A population density grid was created using the ArcGIS kernel density function (ESRI, 2012) and multiplied by water consumption per capita to estimate the amount of water consumed per grid cell. The amount consumed per grid cell then gives an indication of the amount extracted per grid cell.

The ecosystem's capacity to support groundwater extractions was modelled as groundwater recharge, which is the total amount of water entering the aquifers within a specified time step (e.g. month or year) (Arnold et al., 2013). The ecosystem's capacity to support surface water extractions, however, was modelled as the water yield. Water yield is the net amount of water contributed by a grid cell to the river network within a specified time step (Arnold et al., 2013). Both groundwater recharge and water yield are model output variables.

3.2.3 Water purification

In the Upper Ouémé watershed, fertilizer application is increasing and high fertilizer input is associated with crops such as maize, rice and cotton (Bossa et al., 2012). Increasing fertilizer application can lead to contamination of groundwater and surface water resources through nutrient leaching. This poses serious environmental and health risks to beneficiaries of these systems (Tilman et al., 2002; Wolfe and Patz, 2002). In our study area, groundwater provides over 90 % of the total household water consumption. Water purification is, therefore, an essential ecosystem service in the Upper Ouémé watershed that increases the quality of groundwater for human consumption as well as other purposes. One of the challenges in terms of quantifying hydrological ecosystem services is the identification of management relevant indicators that can be enhanced through management interventions to augment the service production. For this study, we used soil denitrification as an indicator of this hydrological ecosystem service. Soil denitrification controls the rate of nitrate leaching by determining the quantities (after plant uptake) of nitrate available for leaching into groundwater systems (Jahangir et al., 2012). For example, Kramer et al. (2006) observed that organic farming supports more active and efficient denitrifier communities, leading to a considerable reduction in nitrate leaching as compared to conventional farming. In this study, the SWAT Landscape model was used to simulate the complete nitrogen cycle and service flow was estimated directly as the rate of denitrification, a model output variable. We should emphasize that the SWAT Landscape model does not explicitly simulate microbial processes and dynamics, but rather that it simulates the ecohydrological conditions suitable for denitrification to occur (Boyer et al., 2006). The model, therefore, computes denitrification as a function of soil moisture content, soil temperature, presence of a carbon source and nitrate availability using Eqs. (6) and (7) (Neitsch et al., 2009).

$$N_{\text{dn}} = \text{NO}_3 \cdot \left(1 - \exp\left[-\beta_{\text{dn}} \cdot \gamma_{\text{tmp}} \cdot C_{\text{org}}\right]\right) \text{ if } \gamma_{\text{sw}} \geq \gamma_{\text{sw,thr}}, \quad (6)$$

$$N_{\text{dn}} = 0 \qquad \text{if } \gamma_{\text{sw}} < \gamma_{\text{sw,thr}}, \quad (7)$$

where N_{dn} is the amount of nitrogen lost through denitrification (kg ha^{-1}), NO_3 is the amount of nitrate in the soil (kg ha^{-1}), β_{dn} is the rate coefficient for denitrification, γ_{tmp}

is the nutrient cycling temperature factor, γ_{sw} is the nutrient cycling water factor, $\gamma_{sw,thr}$ is the threshold value of the nutrient cycling water factor for denitrification to occur, and C_{org} is the amount of organic carbon (%). The values of β_{dn} and $\gamma_{sw,thr}$ are user-defined values and were adjusted during calibration; β_{dn} was 1.4 and $\gamma_{sw,thr}$ was 1.1.

Service capacity was estimated as the denitrification efficiency, which in this study was computed using Eq. (8). When the ecohydrological conditions required for denitrification are present, the rate of denitrification (service flow) is determined by the amount of nitrate available in the soil. Unlike other land cover types (which only receive nitrogen or nitrates from wet deposition or from overland flow), cropland areas receive additional nitrogen or nitrates through fertilizer application. Therefore, for a given grid cell, denitrification efficiency determines the proportion of the total nitrate that is denitrified. As a measure of service capacity, denitrification efficiency gives an indication of the suitability of a spatial unit for denitrification.

$$\mathrm{DN}_{\mathrm{eff}} = (N_{\mathrm{dn}}/N_{\mathrm{total}}) \cdot 100, \tag{8}$$

where $\mathrm{DN}_{\mathrm{eff}}$ is the denitrification efficiency (%), N_{dn} is the amount of nitrogen lost through denitrification in the time step ($\mathrm{kg\,ha^{-1}}$), and N_{total} is the total amount of nitrogen available (e.g. through fertilizer application, wet deposition, etc.) in the time step ($\mathrm{kg\,ha^{-1}}$).

3.2.4 Soil erosion control

Controlling soil erosion in the watershed has numerous benefits including maintaining soil fertility, preventing river sedimentation, and downstream water quality. There are inherent physical soil and landscape properties such as soil erodibility and slope that affect soil erosion (Williams, 1975). However, we focussed on the role of vegetation cover in controlling soil erosion. Service flow was modelled as the actual reduction in soil loss produced by the existing vegetation cover, and was computed using Eq. (9).

$$\mathrm{SD}_{\mathrm{rtd}} = S_{\mathrm{yld,pot}} - S_{\mathrm{yld}}, \tag{9}$$

where $\mathrm{SD}_{\mathrm{rtd}}$ is the reduction in soil loss produced by the existing vegetation cover ($\mathrm{metric\,t\,ha^{-1}}$), $S_{\mathrm{yld,pot}}$ is the maximum potential soil loss in the absence of vegetation cover ($\mathrm{metric\,t\,ha^{-1}}$), and S_{yld} is the soil loss under prevailing vegetation cover and land management practices ($\mathrm{metric\,t\,ha^{-1}}$). Both $S_{\mathrm{yld,pot}}$ and S_{yld} were computed with the modified universal soil loss equation (Williams, 1975) incorporated into the SWAT Landscape model.

For the service capacity of soil erosion control, we used the maximum potential reduction in soil loss produced by the vegetation cover as an indicator. This maximum potential reduction in soil loss (maximum potential soil retention) can be said to be equal to the maximum potential soil loss. For example, for a specified spatial unit, if the maximum potential soil loss in the absence of the vegetation cover is estimated as 2 $\mathrm{metric\,t\,ha^{-1}\,yr^{-1}}$, then it indicates that the potential reduction in soil loss due to the vegetation cover cannot be greater than 2 $\mathrm{metric\,t\,ha^{-1}\,yr^{-1}}$. The maximum potential soil loss was modelled assuming there was no vegetation cover (e.g. Leh et al., 2013; Terrado et al., 2014).

3.3 Accounting for hydrological ecosystem services

Biophysical ecosystem accounts are the basis for monetary accounting and were set up in accordance with SEEA-Experimental Ecosystem Accounting guidelines (EC et al., 2013). We defined 11 subwatershed ecosystem accounting units (SEAUs) as the spatial scales of aggregation. We set up annual biophysical service capacity and service flow accounts for each SEAU. The 11 SEAUs were defined from a total of 44 subwatersheds based on the drainage areas of streamflow monitoring stations within the watershed. The monitoring stations are listed in Table 4. The 44 subwatersheds were delineated from the ASTER Global Digital Elevation Map as part of the initial model set-up with ArcSWAT. Some monitoring stations with smaller drainage areas were nested within those with larger drainage areas. In such cases the SEAU was defined as the drainage area of the nested monitoring station because we wanted to set up spatially disaggregated accounts. Large drainage areas of other monitoring stations had nested subwatersheds within them that were ungauged. In these cases also, the SEAU was defined as the nested subwatershed. For each SEAU, the spatial estimates of service capacity load per grid cell (500 m × 500 m) and service flow load per grid cell (500 m × 500 m) that had been computed in Sect. 3.2 were then aggregated.

A key motivation for ecosystem accounting is to provide information for tracking changes in ecosystems and linking those changes to economic and other human activities (EC et al., 2013). Trend analysis statistical tests were conducted on the total annual values (or total seasonal values for crop water supply) of service capacity accounts in each SEAU. Trend analysis determines whether the changes in service capacity over time are due to random variability or statistically significant and consistent changes. This was conducted using the non-parametric Mann–Kendall test for trend. The Mann–Kendall test for trend statistically determines whether there is a monotonic upward or downward trend of a variable over time. A trend was detected if temporal variation in service capacity was statistically significant at the 5 % significance level (P value < 0.05). If a trend was detected, the Mann–Kendall statistic and Sen's slope estimator were calculated.

The Mann–Kendall statistic is a measure of the strength and direction of a trend, whereas Sen's slope estimator is a measure of the magnitude of a trend.

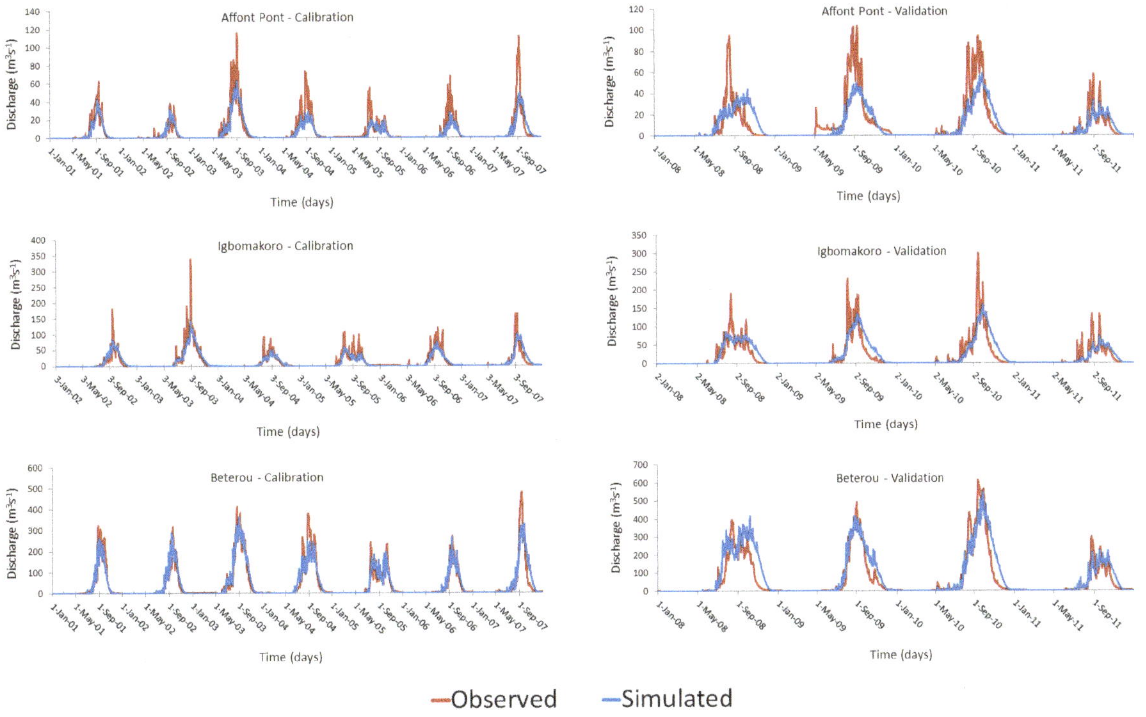

Figure 2. Comparing simulated and observed streamflow for three monitoring stations with varying drainage areas: Affont-Pont, 1172 km^2; Igbomakoro, 2309 km^2; Beterou, 10 046 km^2.

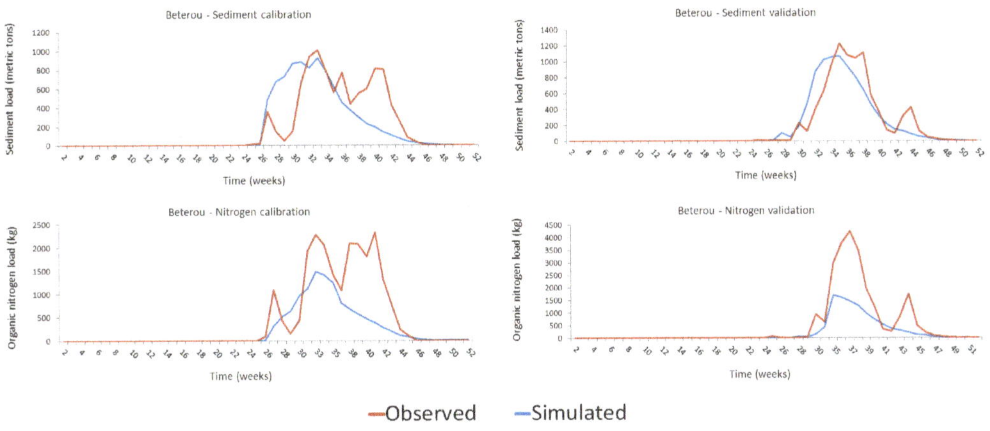

Figure 3. Comparing simulated and observed sediment loads and organic nitrogen loads during calibration and validation at Beterou monitoring station for the period 2008–2009.

4 Results

4.1 SWAT Landscape model calibration and validation results

Table 4 shows the statistical results of the model calibration and validation and Figs. 2 and 3 show the graphical results. There are no established absolute criteria for judging model performance. For this study, we used the criteria recommended by Moriasi et al. (2007). A watershed model is said to be performing satisfactorily if NSE > 0.50 and RSR < 0.70, PBIAS within the range -25 to 25 for streamflow, -55 to 55 for sediment, and -70 to 70 for nutrients. At different spatial scales (e.g. Affont-Pont, 1172 km^2; Igbomakoro, 2309 km^2; Beterou, 10 046 km^2), the model simulated hydrological processes satisfactorily as shown in Fig. 2. Seven out of 11 stations recorded NSE values greater than 0.5 during model validation of streamflow. Even though the NSE values for some monitoring stations were less than 0.5, all but one were greater than 0.0, indicating that the simulated streamflow was still a better predictor than the mean of

Table 4. Calibration and validation results for streamflow, sediment and organic nitrogen loads (Prefix H__ indicates results for streamflow calibration and validation; prefix S__ indicates results for sediment load calibration; N__ indicates results for organic nitrogen load calibration). NSE is Nash–Sutcliffe efficiency, PBIAS is percent bias, and RSR is the ratio of the root mean square error to the standard deviation of measured data.

Monitoring stations	Drainage area (km^2)	Calibration			Validation		
		NSE	PBIAS	RSR	NSE	PBIAS	RSR
Upstream stations							
H__Affon-Pont	1172	0.69	27.0	0.56	0.62	15.9	0.62
H__Aval-Sani	760	0.70	12.0	0.55	0.64	7.8	0.60
H__Bori	1608	0.65	−24.7	0.59	−0.49	−121.4	1.22
H__Tebou	522	0.47	43.5	0.72	0.58	20.3	0.65
Downstream stations							
H__Beterou	10 046	0.85	5.7	0.39	0.78	−17.8	0.47
H__Barerou	2128	0.71	20.8	0.54	0.72	−22.7	0.53
H__Cote-238	3040	0.69	3.5	0.56	0.68	−18.4	0.56
H__Igbomakoro	2309	0.76	11.3	0.49	0.71	−4.0	0.54
H__Sarmanga	1334	0.48	23.2	0.72	0.44	17.2	0.75
H__Aguimo	394	0.25	−20.9	0.87	0.12	−60.1	0.94
H__Wewe	297	0.42	21.6	0.76	0.42	−6.5	0.76
S__Beterou	10 046	0.45	6.9	0.74	0.83	2.55	0.42
N__Beterou	10 046	0.50	47.4	0.71	0.55	56.3	0.67

the observed values. Monitoring stations with larger drainage areas recorded higher NSE values than stations with smaller drainage areas. The PBIAS values in Table 4 show the level of bias in simulated streamflow. A negative PBIAS value indicates model overestimation, whereas a positive PBIAS value indicates model underestimation. The validation results show that the model largely underestimated streamflow at upstream stations and overestimated it downstream. The RSR results show varying levels of residual variation indicating the level of errors in simulated streamflow as compared to observed streamflow. The closer the RSR value is to zero, the lower the level of residual variation in simulated streamflow. During model validation, five stations recorded RSR values lower than 0.7. For sediment and nitrogen transport processes, the model performed satisfactorily. The statistical and graphical results of sediment load and organic nitrogen load during calibration and validation are shown in Fig. 3 and Table 4.

4.2 Spatial patterns of hydrological ecosystem services

Water supply by soil moisture is essential to reduce crop water stress in rainfed agricultural systems. If all other factors for crop growth (such as nutrients and temperature) remain constant, then a higher service capacity and higher service flow result in a higher crop yield. Computations of crop water supply were spatially restricted to upland agricultural fields. High service flow indicates the suitability of a spatial unit under assumed maize cultivation, whereas high service capacity indicates the potential suitability for crop cultivation irrespective of the crop type and not maize alone. The results of service capacity are indicative of the least number of days during a year crops would not experience water stress. Figure 4 reveals high spatial variability in service capacity and service flow in upland agricultural fields. Mean annual values of service capacity in upland agricultural fields ranged from 51 to 146 days yr^{-1} with a watershed-wide mean of 93 days yr^{-1} and standard deviation of 24 days yr^{-1}. The spatial distribution of mean annual values of service capacity and service flow in inland valley rice fields are not shown because of their significantly low total area (less than 1 % of the total cropland area). Mean annual values of service capacity in inland valleys ranged from 92 to 136 days yr^{-1} with a watershed-wide mean of 124 days yr^{-1} and standard deviation of 9 days yr^{-1}. Mean seasonal values of service flow in inland valleys ranged from 67 to 123 days GP^{-1} with a watershed-wide mean of 117 days GP^{-1} and a standard deviation of 12 days GP^{-1}. Overall, more than 95 % (approximately 1050 ha) of inland valley rice fields recorded mean seasonal values of service flow of at least 90 days, whereas less than 25 % (approximately 36 000 ha) of upland agricultural areas recorded mean seasonal values of service flow of at least 90 days.

The spatial distribution of mean annual values of service capacity and service flow of groundwater supply and surface water supply are shown in Figs. 5 and 6 respectively. Groundwater is the major source of water for household consumption (drinking and non-drinking purposes), with the service

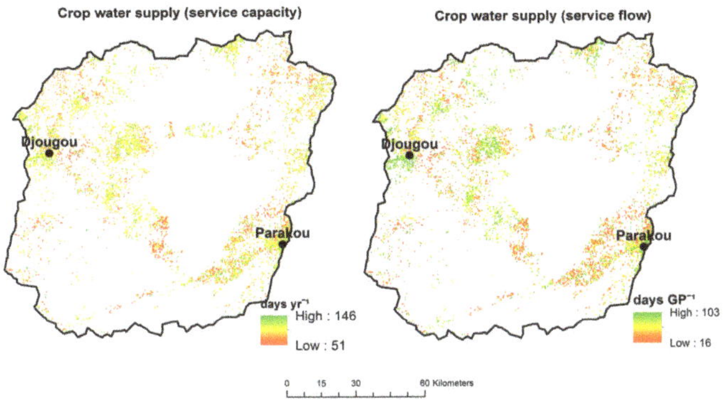

Figure 4. Spatial distribution of mean annual values of service capacity and mean seasonal values of service flow of crop water supply in upland agricultural areas in the Upper Ouémé watershed from 2001 to 2012 (GP indicates growing period).

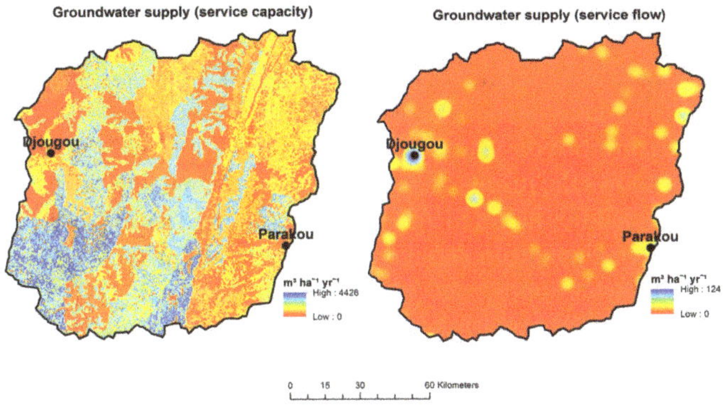

Figure 5. Spatial distribution of mean annual values of service capacity and service flow of groundwater supply in the Upper Ouémé watershed from 2001 to 2012.

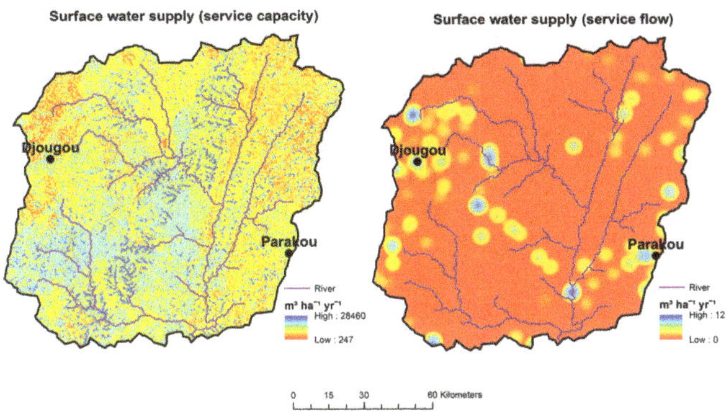

Figure 6. Spatial distribution of mean annual values of service capacity and service flow of surface water supply in the Upper Ouémé watershed from 2001 to 2012.

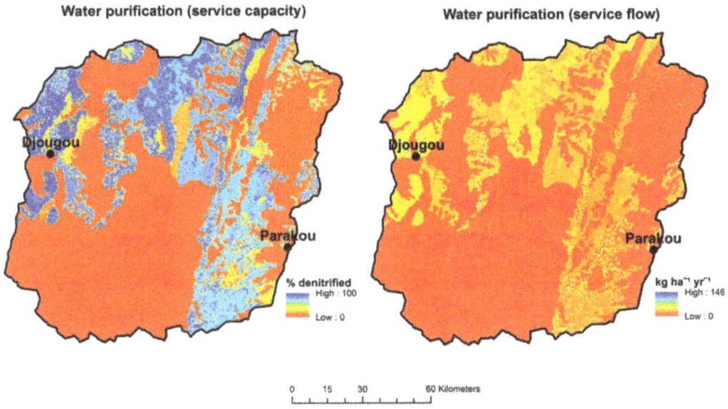

Figure 7. Spatial distribution of mean annual values of service capacity and service flow of water purification in the Upper Ouémé watershed from 2001 to 2012.

flow (groundwater extraction) significantly higher than the service flow of surface water supply (surface water extraction). High service flows of groundwater supply are concentrated in the most populous towns in the watershed. However, service flows in Parakou, which is the most populous city in the watershed, are relatively lower than in other areas such as Djougou. This is because the population in Parakou depends mainly on tap water sources. Service capacity of groundwater supply exhibited high spatial variability. High values of service capacity were concentrated in the south-western part of the watershed. For service capacity of surface water supply, Fig. 6 shows areas with a high propensity for generating water yield. These areas, referred to as hydrologically sensitive areas (HSAs) (Agnew et al., 2006), were not peculiar to a particular land cover type. They occurred in almost all land cover types. They occurred more frequently in savannah woodland and shrubland because approximately 80 % of the total watershed area is either one of this land cover type.

Water purification modelled as denitrification is essential to control the quantities of nitrate available for leaching and contaminating groundwater resources (Jarvis, 2000; Jahangir et al., 2012). Service capacity was measured as the percentage of nitrate that is denitrified and service flow was the rate of denitrification. The spatial distribution of mean annual values of service capacity and service flow of water purification are distinctly concentrated in the northern and eastern parts of the watershed, with the south-western parts recording zero values (Fig. 7). All barren land cover types also recorded zero values of service capacity and service flow. The zero values recorded are a result of the lack of soil saturation conditions and not the lack of nitrate availability. Soil saturation induces soil anaerobic conditions required for denitrification to take place. In areas where denitrification was recorded, the highest mean annual values of service flow were recorded in inland valley rice fields (12 $\mathrm{kg\,ha^{-1}\,yr^{-1}}$) and grasslands (7 $\mathrm{kg\,ha^{-1}\,yr^{-1}}$). The highest mean annual values of service capacity were also recorded in grasslands (55 % $\mathrm{yr^{-1}}$ and inland valley rice fields (35 % $\mathrm{yr^{-1}}$).

The spatial distributions of mean annual values of service capacity and service flow of soil erosion control are shown in Fig. 8. High service capacity indicates high potential for reduction in soil loss produced by the vegetation cover. The service flow, however, is a measure of the actual

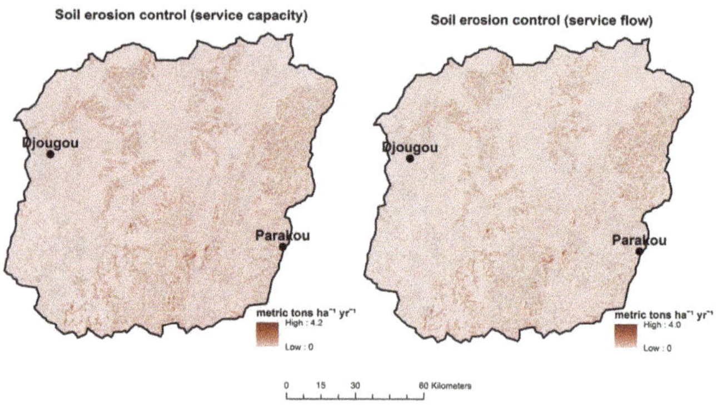

Figure 8. Spatial distribution of mean annual values of service capacity and service flow of soil erosion control in the Upper Ouémé watershed from 2001 to 2012.

reduction in soil loss under existing vegetation cover. Under existing vegetation cover, the mean annual rate of soil loss in the watershed was recorded at 0.01 metric t ha^{-1} yr^{-1} (standard deviation of 0.02 metric t ha^{-1} yr^{-1}). The mean annual rate of soil loss in the watershed will increase significantly to 0.05 metric t ha^{-1} yr^{-1} (standard deviation of 0.07 metric t ha^{-1} yr^{-1}) should there be complete loss of the existing vegetation cover. This value, 0.05 metric t ha^{-1} yr^{-1} (standard deviation of 0.07 metric t ha^{-1} yr^{-1}), can also be interpreted as the maximum potential reduction in soil loss (service capacity) that can be produced by the existing vegetation cover. Under existing vegetation cover and management conditions, however, the actual reduction in soil loss (service flow) was recorded at a watershed-wide mean annual value of 0.04 metric t ha^{-1} yr^{-1} (standard deviation of 0.07 metric t ha^{-1} yr^{-1}). For both service capacity and service flow, only about 0.04 % of the total area of the watershed recorded mean annual values greater than 1 metric t ha^{-1} yr^{-1}. These areas had the steepest slopes, indicating the importance of vegetation cover in soil erosion control in these areas. In forested areas, service flow was equal to service capacity, indicating that overall there was no net soil loss from forested areas.

4.3 Biophysical ecosystem accounts

The service capacity (Table 5) and service flow (Table 6) ecosystem accounting tables show the distribution of hydrological ecosystem services across the 11 SEAUs for the most current year of simulation, 2012. The total annual values of service capacity correlated with the spatial extent of an SEAU. Larger SEAUs recorded higher values than smaller SEAUs. However, the mean values for service capacity varied depending on the biophysical environment of an SEAU. For example, whereas the Beterou-Ouest SEAU is the largest, the highest mean service capacity of groundwater supply was recorded in the Sarmanga and Terou-Igbomakoro SEAUs. This signifies that the rate of groundwater recharge is highest in the Sarmanga and Terou-Igbomakoro SEAUs. The service flow table reveals that the ecohydrological conditions required for denitrification (water purification) do not occur in the Aguimo, Terou-Igbomakoro, Terou-Wanou, and Wewe SEAUs. However, a total of 77 000 m^3 of groundwater was extracted in the Terou-Igbomakoro and Wewe SEAUs in 2012. In the Aguimo and Terou-Wanou SEAUs, there is currently no groundwater extraction. For crop water supply, the tables also show the total area of land currently under crop cultivation in each SEAU. Upland agricultural areas provide over 99 % of total cropland area. The SEAUs with the largest upland agricultural areas did not necessarily record the highest service flow. For example, the highest service flow was recorded in Sarmanga and Terou-Igbomakoro. This signifies that maize cultivation in these SEAUs is less prone to water stress than in any other SEAU.

Temporal analysis of ecosystem accounts makes it possible to track ecosystem changes and measure the degree of sustainability, degradation or resilience. Decreasing the capacity of ecosystems to sustain human welfare over time is a measure of ecosystem degradation (EC et al., 2013). Figure 9 shows the results of trend analysis statistical tests of service capacities at the SEAU level. Increasing trends were observed in changes in service capacities of water purification, groundwater supply and surface water supply. For groundwater supply, increasing trends were observed in all SEAUs. The results in Fig. 9a are of the five SEAUs with the highest Mann–Kendall statistic. An increasing trend in changes in surface water supply was observed in four SEAUs, whereas an increasing trend in changes in water purification was observed only in the Aval-Sani SEAU. No trend was observed in changes in the service capacity of crop water supply in both upland agricultural fields and inland valleys in any of the SEAUs. No trend was observed either in changes in the service capacity of soil erosion control in any of the SEAUs.

5 Discussion

5.1 Model uncertainties and limitations

Model results to support decision-making are always associated with a certain degree of uncertainty. Uncertainty in ecohydrological modelling with SWAT may be from input data, model algorithms, model calibration and validation (parameter non-uniqueness) (Abbaspour et al., 2008). The major input uncertainty in our study was a result of resampling of spatial data from fine spatial resolutions to relatively coarse spatial resolutions in order to increase operational feasibility and computational efficiency of the grid-based SWAT Landscape model. We resampled land use/land cover data, a DEM and a soil map to a spatial resolution of 500 m × 500 m. Even though the spatial rigour of ecosystem accounting requires that modelling approaches that maintain adequate landscape spatial heterogeneity are more suitable, decisions on the

Table 5. Biophysical ecosystem account for service capacity at the SEAU (subwatershed ecosystem accounting unit) level in the Upper Ouémé watershed in 2012 (SD is standard deviation).

SEAU	Hydrological ecosystem service											
	Crop water supply				Household water supply				Water purification		Soil erosion control	
	Upland agricultural areas		Inland valley rice fields		Groundwater supply		Surface water supply					
	Area (10^3 ha)	Mean (SD) (days yr^{-1})	Area (ha)	Mean (SD) (days yr^{-1})	Total (10^6 m^3 yr^{-1} recharge)	Mean (SD) (10^3 m^3 ha^{-1} yr^{-1} recharge)	Total (10^6 m^3 yr^{-1} water yield)	Mean (SD) (10^3 m^3 ha^{-1} yr^{-1} water yield)	Total N added (10^3 kg)	(% N denitrified)	Total (10^3 metric t yr^{-1})	Mean (SD) (kg ha^{-1} yr^{-1})
Affon-Pont	20.2	103 (28)	200	149 (4)	121	26 (26)	624	133 (90)	2719	30	5.2	44 (74)
Aguimo	0.3	95 (34)	0	–	58	37 (28)	255	161 (112)	589	0	2.6	65 (102)
Aval-Sani	4.0	100 (32)	0	–	114	38 (20)	458	151 (89)	1370	36	3.5	45 (85)
Barerou	33.4	104 (29)	100	118 (18)	244	29 (20)	1328	156 (101)	4707	11	18.5	87 (101)
Beterou-Ouest	54.3	103 (29)	425	135 (16)	615	38 (30)	2526	155 (99)	8550	19	22.9	56 (102)
Bori	12.0	109 (30)	0	–	185	29 (20)	1082	168 (97)	3138	26	6.4	40 (79)
HVO	7.0	104 (32)	50	112 (33)	206	43 (40)	638	133 (80)	1953	15	8.3	69 (93)
Sarmanga	9.7	106 (24)	175	139 (12)	304	57 (37)	809	152 (82)	2382	13	4.5	34 (42)
Terou-Igbomakoro	4.0	108 (25)	50	138 (2)	222	57 (36)	591	151 (93)	1561	0	4.9	51 (91)
Terou-Wanou	0.8	91 (30)	25	96 (0)	73	54 (22)	170	126 (58)	514	0	2.5	74 (90)
Wewe	4.1	96 (30)	75	131 (20)	48	40 (33)	213	177 (117)	638	0	1.8	61 (182)
Total	149.8		1100		2190	–	8694	–	28 121	–	81.1	–

Table 6. Biophysical ecosystem account for service flow at the SEAU (subwatershed ecosystem accounting unit) level in the Upper Ouémé watershed in 2012 (GP is the length of the growing period between crop establishment and harvest; upland agricultural areas had a GP of 103 days; inland valley rice fields had a GP of 123 days; SD is standard deviation).

SEAU	Hydrological ecosystem service									
	Crop water supply				Household water supply		Water purification		Soil erosion control	
	Upland agricultural areas		Inland valley rice fields		Groundwater	Surface water				
	Area (10^3 ha)	Mean (SD) (days GP^{-1})	Area (ha)	Mean (SD) (days GP^{-1})	Total (10^3 $m^3\,yr^{-1}$ water extracted)	Total (10^3 $m^3\,yr^{-1}$ water extracted)	Total (10^3 kg N yr^{-1} denitrified)	Mean (SD) (kg ha^{-1} yr^{-1} denitrified)	Total (10^3 metric t yr^{-1})	Mean (SD) (kg ha^{-1} t yr^{-1})
Affon-Pont	20.2	59 (30)	200	123 (0)	123	65	810	6.9 (10)	4.4	38 (67)
Aguimo	0.3	52 (35)	0	–	0	0	0	0.0 (0)	2.3	58 (92)
Aval-Sani	4.0	64 (29)	0	–	8	0.2	498	6.5 (7)	3.2	42 (81)
Barerou	33.4	63 (31)	100	107 (17)	510	64	503	2.4 (6)	15.9	75 (92)
Beterou-Ouest	54.3	59 (31)	425	115 (22)	1124	219	1613	4.0 (9)	18.9	46 (90)
Bori	12.0	65 (32)	0	–	196	30	815	5.1 (7)	5.4	33 (67)
HVO	7.0	56 (32)	50	88 (35)	71	37	297	2.5 (5)	7.0	59 (79)
Sarmanga	9.7	69 (34)	175	119 (8)	532	66	317	2.3 (5)	4.0	30 (39)
Terou-Igbomakoro	4.0	69 (34)	50	123 (0)	95	36	0	0.0 (0)	4.4	45 (85)
Terou-Wanou	0.8	45 (35)	25	92 (0)	0	0	0	0.0 (0)	2.2	65 (83)
Wewe	4.1	63 (31)	75	107 (23)	41	41	0	0.0 (0)	1.5	51 (178)
Total	149.8	–	1100	–	2700	558.2	4853	–	69.2	–

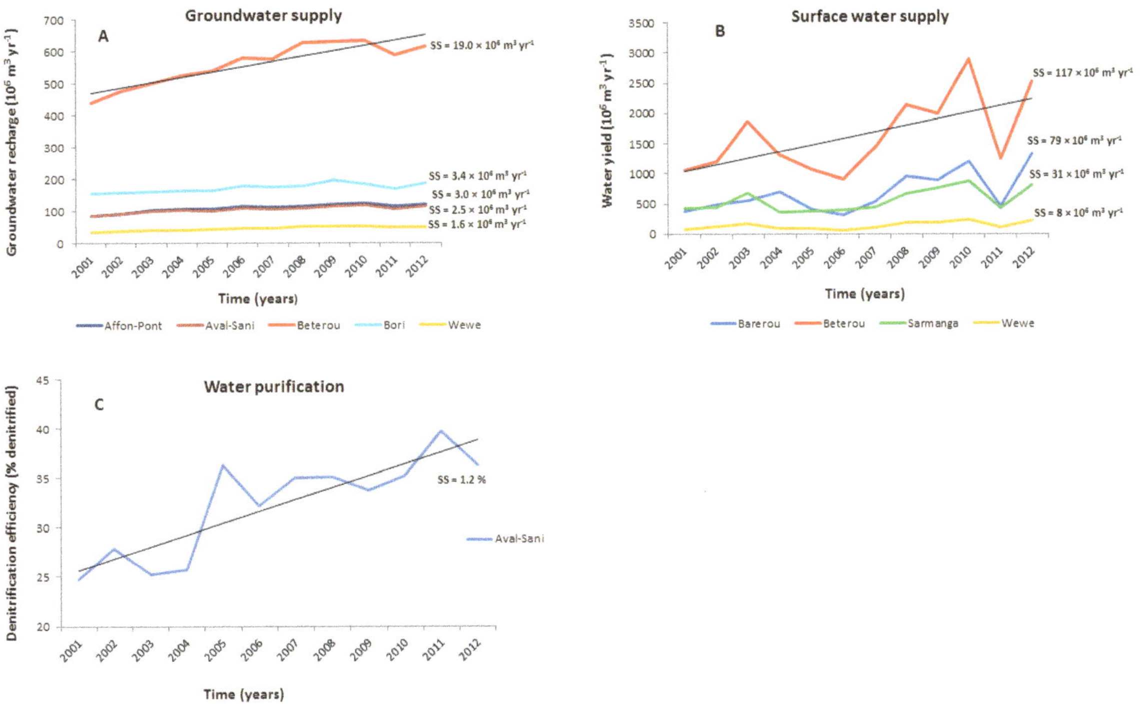

Figure 9. Trends in the service capacity of hydrological ecosystem services at the SEAU level in the Upper Ouémé watershed (SS is Sen's slope estimator, which is a measure of the magnitude of change of a trend). For each graph, a single trend line is drawn solely to illustrate the direction of a trend.

choice of spatial resolution should be made with model computational efficiency and operational feasibility in mind. For the SWAT model (and SWAT Landscape model), increasing spatial detail results in a considerable increase in computing time irrespective of the spatial discretization scheme employed (e.g. Arnold et al., 2010; Notter et al., 2012). In our case-study area, over 1 400 000 grid cells are generated at 1 ha resolution requiring over 2 days for each simulated year on 2.6 Ghz and 8 GB RAM. Computer storage capacity for the huge data outputs generated may also not be readily available. We acknowledge that in many regions of the world, high-resolution spatial input data may not be available at large spatial scales. However, for the grid-based set-up of the SWAT Landscape model, where such high-resolution spatial data are available, it may be necessary to compromise spatial explicitness to achieve operational feasibility. This introduces a certain amount of uncertainty with regards to spatial variation in ecohydrological processes; therefore, such decisions should be made taking into consideration the degree of spatial heterogeneity of landscape features. The need to compromise spatial detail for operational feasibility may limit the applicability of this model configuration for larger watersheds.

For larger watersheds, it is also extremely difficult to obtain spatially and temporally correct representations of the underlying ecohydrological processes and interactions. To achieve this, there is a need for multi-site calibration at different spatial scales with a sufficient length of time series of data to capture high and low flow years, annual, seasonal and monthly variations (Santhi et al., 2008). In our study, the model underestimated streamflow (especially peak flow) at some monitoring stations, whereas at other stations it overestimated streamflow. These biases in streamflow estimation lead to error propagation in the other components of the water balance such as soil moisture and actual evapotranspiration. Whereas the use of 11 years of daily streamflow data from 11 monitoring stations in the Upper Ouémé watershed reduces the uncertainties of modelled results, data for calibration and validation of sediment and nitrogen loads may not have been sufficient to enable the model to more accurately represent sediment and nitrogen transport processes. In evaluating model performance of sediment and nitrogen transport processes, we used only 2 years of data from a single monitoring station. Without multi-site calibration and validation, there remain large uncertainties in modelled results of sediment and transport processes at different spatial scales. In addition, without long-term temporal validation, there remain large uncertainties in the ability of the model to capture annual variability in these transport processes. Even with sufficient length of time series of multi-site data for calibration and validation, the problem of parameter non-uniqueness inherent in complex watershed models such as the SWAT model also introduces a degree of uncertainty in modelled results. Parameter non-uniqueness refers to the reproduction of similar observed ecohydrological signals by different input parameter sets. Therefore, even for so-called calibrated and validated SWAT models, there is always a degree of uncertainty introduced by parameter non-uniqueness. To reduce error propagation, non-uniqueness and consequently reduce parameter uncertainty requires the use of comprehensive data on different fluxes, loads and ecohydrological processes such as crop yield, soil moisture, groundwater level and evapotranspiration (Abbaspour et al., 2008) that are most of the time not readily available.

For this study, we used soil denitrification as an indicator of water purification service. Quantifying denitrification at watershed and subwatershed scales requires the use of models such as SWAT. It involves the simulation of a complex set of processes controlling denitrification that can broadly be classified as the prerequisite environmental/ecohydrological conditions, and microbial processes and dynamics. The SWAT model, however, provides only simplified representations of the complex set of processes controlling denitrification, and modelled estimates of denitrification rates remain highly uncertain (Boyer et al., 2006). The model only simulates the environmental/ecohydrological conditions and does not explicitly simulate microbial processes and dynamics. There is, therefore, an inherent assumption of spatial homogeneity with regards to denitrifier community species composition, quantities and activities across all land use types. Kramer et al. (2006) reported that specific land use and management types (such as organic, integrated and conventional agriculture) enhance or inhibit soil denitrifier activities affecting the rate of denitrification. In the SWAT model, however, spatial variability in denitrification is determined mainly by spatial variability in ecohydrological and abiotic controlling factors.

5.2 Lessons for ecosystem accounting

In ecosystem accounting, detailed and accurate land cover and land use data are important. Apart from their use as inputs in modelling ecosystem services, land cover classes are also used as ecosystem accounting units based on which ecosystem services are aggregated (Remme et al., 2014; Schröter et al., 2014). A single lumped land cover class for agricultural areas or croplands (be it as model input data or ecosystem accounting units) may be suitable when modelling and accounting for other ecosystem services (e.g. Remme et al., 2014; Schröter et al., 2014). However, when modelling and accounting for crop water supply, land cover and land use data with detailed and spatially disaggregated information on the types of crops grown in agricultural areas are needed. This is because different crops have different water requirements (Allen et al., 1998). In rainfed agricultural systems, crop water supply is the major limitation to crop production and is the main factor responsible for low yields in the seasonally dry and semiarid tropics and subtropics (Shaxson and Barber, 2003). However, in many of these regions, land cover and land use data with this level of detail are currently not available. Obtaining such information is complicated by the small plot sizes and cropping patterns varying from year to year. Our study area was no exception. Despite these constraints, the lack of detailed data reduces the accuracy and reliability of modelled results of service flow of crop water supply. In our study area, this limitation resulted in the simulation of only a single crop type in upland agricultural areas. Therefore, the results for service flow of crop water supply should be interpreted in the context of the crop simulated. However, because methodologies such as Allen et al. (1998) have been used extensively to compute the water requirements of various crops, our approach serves as a reference or baseline from which the service flow of crop water supply of a spatial unit could be estimated if a crop other than maize is grown.

A key feature of ecosystem accounting is the distinction between service capacity and service flow. The empirical distinction and separate spatial characterization of service capacity and service flow is essential in understanding the dynamics of service provision and in planning and devising sustainable management options. The distinction is also important for subsequent monetary valuation. Service capacity and service flow should be based on measurable indicators that have policy and management relevance. Indicators must also be able to represent cause–effect relations. For hydrological ecosystem services, selecting single indicators of service capacity that meet the above requirements and that sufficiently reflect ecosystem condition and their potential to provide service flows is difficult. This is because of the non-linear complex interactions among several ecohydrological processes that each relies on a suite of ecosystem components (van Oudenhoven et al., 2012; Villamagna et al., 2013). In this study, the service capacity indicators of crop water supply and household water supply meet the above requirements. For example, Ennaanay (2006) and Yan et al. (2013) reported that changes in land use and other ecosystem components alter the hydrological cycle, affecting patterns of evapotranspiration, infiltration, water retention, groundwater recharge and water yield. However, for services such as water purification and soil erosion control, the capacity indicators presented in this study are derived indicators and not actual physical processes. Such indicators do not convey information regarding key physical processes and therefore may not have management relevance. In such cases, a key question that arises is whether the underlying ecosystem components and processes should be weighted and aggregated to produce one composite indicator for service capacity (Edens and Hein, 2013). For example, soil erosion control is a function of surface runoff, slope, soil erodibility, cover and management factors, and support practice factors. Weighing and aggregation of ecosystem components and processes to establish a composite indicator for service capacity, however, is not straightforward and is challenging (Weber, 2007; Stoneham et al., 2012).

5.3 Implications for watershed and ecosystem management

Three of the key issues critical for watershed management and land use planning in an agricultural watershed such as the Upper Ouémé are nitrate leaching, non-point source pollution and alteration in streamflow regime. Nitrate leaching contaminates groundwater resources (Jarvis, 2000; Jahangir et al., 2012). Agricultural non-point source pollution leads to pollution of river networks (Agnew et al., 2006). Alteration of streamflow regime affects riverine ecological integrity and downstream water availability (Carlisle et al., 2011). Ecosystem accounting and spatial characterization of hydrological ecosystem service capacity and flow provide relevant information to help address these issues in policy-making. Such analyses can reveal high-risk areas (i.e. areas that would be affected by changes or continued trends in watershed ecohydrology) or high service production areas (i.e. areas that are crucial for maintaining water flow downstream). For example, our analyses reveal areas where the ecohydrological conditions required for denitrification do not occur but where there is currently groundwater extraction. These areas are high-risk areas of groundwater contamination from nitrate leaching. More crucially, there is currently crop cultivation in some of these areas. Agricultural intensification in these areas, therefore, will result in higher nitrate leaching and contamination of groundwater resources.

Furthermore, the grid-based set-up of the SWAT Landscape model enabled us to identify HSAs at a finer spatial resolution. Characterization of the spatiotemporal dynamics of HSAs is essential in controlling non-point source pollution and in maintaining streamflow regime. Hydrologically sensitive areas have a significant impact on key ecohydrological processes affecting interaction and transport of water, sediment, nutrients and pollutants. They also provide key landscape controls on riverine ecosystem integrity including aquatic flora and fauna and downstream water availability and quality. Agricultural intensification in HSAs has a higher potential for generating agricultural non-point source pollution (Agnew et al., 2006). Land use change in these areas can have a more significant impact on the streamflow regime. Such analyses can form the basis for establishing payment for ecosystem services (PES) schemes (Pagiola and Platais, 2007; Turpie et al., 2008). Watershed PES provides financial support to ecosystem management in high service production areas that are of particular relevance downstream (Lopa et al., 2012; Lu and He, 2014). We acknowledge that detailed ecohydrological modelling is only one of the considerations in establishing a watershed PES. Other considerations include transaction costs and the ability to pay of downstream water users. However, ecohydrological modelling can be used to support watershed PES schemes by providing a tool for upstream water managers to monitor the provision of hydrological ecosystem services or by identifying high service production areas that are potentially relevant for a new PES.

6 Conclusion

There are various components involved in ecosystem service delivery that need to be measured in order to better understand the full dynamics of service provision and to devise sustainable management options. Key amongst these components are service capacity and service flow. Empirical distinction of the service capacity and service flow of ecosystem services is a distinguishing feature of ecosystem accounting and is the basis for monetary accounting. In the case-study area, we have shown that despite the non-linear complex interactions among several ecohydrological processes, it is empirically feasible to distinguish between service capacity and service flow of hydrological ecosystem services. This requires appropriate decisions regarding physical and mathematical representation of ecohydrological processes, spatial heterogeneity of ecosystems, temporal resolution, and required model accuracy. The service flows we modelled are the contributions in time and space of ecosystems to productive and consumptive human activities leading to human benefits, whereas the service capacities we modelled reflect ecosystem condition and extent at a point in time, and the resulting potential to provide service flows. We demonstrated our approach by using a SWAT model, which has been configured with a grid-based landscape discretization and further enhanced to simulate water flow across the discretized landscape units, to map and quantify four hydrological ecosystem services vital to food and water security in the Upper Ouémé watershed in Benin. We set up ecosystem accounting tables for both service capacity and service flow and analysed trends in service capacities. For each hydrological ecosystem service, we were able to identify subwatershed ecosystem accounting units (SEAUs) where either service capacity or service flow is concentrated. We were also able to identify trends in changes in service capacity of hydrological ecosystem services for some SEAUs. Our approach can be extended and applied to other watersheds because it is based on the SWAT model, which has been tested extensively in different watersheds and landscapes. Our analyses show that integrating hydrological ecosystem services into an ecosystem accounting framework provides relevant information on watershed ecosystems and hydrological ecosystem services at appropriate scales suitable for decision-making.

Author contributions. C. Duku, L. Hein and S. J. Zwart conceived and designed the study; H. Rathjens developed the grid-based model code; C. Duku performed the simulations and analyses; C. Duku and L. Hein prepared the manuscript with contributions from S. J. Zwart and H. Rathjens.

Acknowledgements. This research was conducted at Wageningen University as part of the "Realizing the potential of inland valley lowlands in sub-Saharan Africa while maintaining their environmental services" project (RAP-IV). The project is implemented by the Africa Rice Center and its national partners and is funded by the European Commission through the International Fund for Agricultural Development (IFAD). We thank the IMPETUS project in Benin for making data available for this research through their public geoportal. We are grateful to Christophe Peugeot and the AMMA-CATCH regional observing system in Benin for providing us precipitation and streamflow data. Finally, we thank Dr. Aymar Bossa and Professor Bernd Diekkrüger for providing us sediment and nitrogen data for model calibration.

Edited by: E. Zehe

References

Abbaspour, K., Yang, J., Reichert, P., Vejdani, M., Haghighat, S., and Srinivasan, R.: SWAT-CUP, SWAT Calibration and Uncertainty Programs, Swiss Federal Institute of Aquatic Science and Technology (EAWAG), Zurich, Switzerland, 2008.

Agnew, L. J., Lyon, S. W., Gerard-Marchant, P., Collins, V. B., Lembo, A. J., Steenhuis, T. S., and Walter, M. T.: Identifying hydrologically sensitive areas: bridging the gap between science and application, J. Environ. Manage., 78, 63–76, doi:10.1016/j.jenvman.2005.04.021, 2006.

Allen, R. G., Pereira, S. L., Raes, D., and Smith, M.: Irrigation and Drainage Paper 56, FAO, Rome, Italy, 1998.

AMMA-CATCH Database, available at: http://bd.amma-catch.org/amma-catch2/main.jsf (last access: 28 May 2014), 2014.

Arnold, J. G., Srinivasan, R., Muttiah, R. S., and Williams, J. R.: Large area hydrologic modeling and assessment – Part I: Model development, J. Am. Water Resour. As., 34, 73–89, 1998.

Arnold, J. G., Allen, P. M., Volk, M., Williams, J. R., and Bosch, D. D.: Assessment of different representations of spatial variability on swat model performance, T. ASABE, 53, 1433–1443, 2010.

Arnold, J. G., Kiniry, J. R., Srinivasan, R., Williams, J. R., Haney, E. B., and Neitsch, S. L.: Soil Water and Assessment Tool Input/Output Documentation Version 2012, Texas Water Resources Institute, College Station, Texas, USA, 2013.

Bosch, D. D., Arnold, J. G., Volk, M., and Allen, P. M.: Simulation of a low-gradient coastal plain watershed using the swat landscape model, T. ASABE, 53, 1445–1456, 2010.

Bossa, A. Y., Diekkrüger, B., Giertz, S., Steup, G., Sintondji, L. O., Agbossou, E. K., and Hiepe, C.: Modeling the effects of crop patterns and management scenarios on N and P loads to surface water and groundwater in a semi-humid catchment (West Africa), Agr. Water Manage., 115, 20–37, doi:10.1016/j.agwat.2012.08.011, 2012.

Boyd, J. and Banzhaf, S.: What are ecosystem services? The need for standardized environmental accounting units, Ecol. Econ., 63, 616–626, doi:10.1016/j.ecolecon.2007.01.002, 2007.

Boyer, E. W., Alexander, R. B., Parton, W. J., Li, C. S., Butterbach-Bahl, K., Donner, S. D., Skaggs, R. W., and Del Gross, S. J.: Modeling denitrification in terrestrial and aquatic ecosystems at regional scales, Ecol. Appl., 16, 2123–2142, doi:10.1890/1051-0761(2006)016[2123:Mditaa]2.0.Co;2, 2006.

Brauman, K. A., Daily, G. C., Duarte, T. K., and Mooney, H. A.: The nature and value of ecosystem services: An overview highlighting hydrologic services, Ann. Rev. Env. Resour., 32, 67–98, doi:10.1146/annurev.energy.32.031306.102758, 2007.

Carlisle, D. M., Wolock, D. M., and Meador, M. R.: Alteration of streamflow magnitudes and potential ecological consequences: a multiregional assessment, Front. Ecol. Environ., 9, 264–270, doi:10.1890/100053, 2011.

EC, OECD (Organisation for Economic Co-operation and Development), UN (United Nations), and World Bank: System of Environmental-Economic Accounting 2012, Experimental Ecosystem Accounting, New York, USA, 2013.

Edens, B. and Hein, L.: Towards a consistent approach for ecosystem accounting, Ecol. Econ., 90, 41–52, doi:10.1016/j.ecolecon.2013.03.003, 2013.

Ennaanay, D.: Impacts of Land Use Changes on the Hydrologic Regime in the Minnesota River Basin, Major: Water Resources Science, 2006, University of Minnesota, USA, 2006.

ESRI: ArcGIS version 10.1, Redlands, California, USA, 2012.

FAO: Methodology and results for Africa, in: Report on the agro-ecological zones project, vol. 1., Food and Agriculture Organization, Rome, Italy, 1978.

FAO: Guidelines: land evaluation of rainfed agriculture, Soils Bulletin, 52, Food and Agriculture Organization, Rome, Italy, 1983.

Fisher, B., Turner, R. K., and Morling, P.: Defining and classifying ecosystem services for decision making, Ecol. Econ., 68, 643–653, doi:10.1016/j.ecolecon.2008.09.014, 2009.

Garbrecht, J. and Martz, L.: Topaz User Manual: Version 3.1, Technical Report. Grazinglands Research Laboratory, USDA, Agricultural Research Service, El Reno, Oklahoma, 2000.

Gassman, P. W., Reyes, M. R., Green, C. H., and Arnold, J. G.: The soil and water assessment tool: historical development, applications, and future research directions, T. ASABE, 50, 1211–1250, 2007.

Giertz, S., Steup, G., and Schonbrodt, S.: Use and constraints on the use of inland valley ecosystems in central Benin: results from an inland valley survey, Erdkunde, 66, 239–253, doi:10.3112/erdkunde.2012.03.04, 2012.

Gupta, H. V., Sorooshian, S., and Yapo, P. O.: Status of automatic calibration for hydrologic models: Comparison with multilevel expert calibration, J. Hydrol. Eng., 4, 135–143, 1999.

Guswa, A. J., Brauman, K. A., Brown, C., Hamel, P., Keeler, B. L., and Sayre, S. S.: Ecosystem services: Challenges and opportunities for hydrologic modeling to support decision making, Water Resour. Res., 50, 4535–4544, doi:10.1002/2014wr015497, 2014.

Hadjer, K., Klein, T., and Schopp, M.: Water consumption embedded in its social context, north-western Benin, Phys. Chem. Earth, 30, 357–364, doi:10.1016/j.pce.2005.06.014, 2005.

Hargreaves, G. L., Hargreaves, G. H., and Riley, J. P.: Agricultural benefits for Senegal river basin, J. Irrig. Drain E-ASCE, 111, 113–124, 1985.

INSAE: Troisième Recensement Général de la Population et de l'Habitation, Cotonou, Benin, 2003.

IWMI: Water for Food Water for Life: a Comprehensive Assessmesnt of Water Management in Agriculture, Earthscan, London, UK, 2007.

Jahangir, M. M. R., Khalil, M. I., Johnston, P., Cardenas, L. M., Hatch, D. J., Butler, M., Barrett, M., O'flaherty, V., and Richards, K. G.: Denitrification potential in subsoils: a mechanism to reduce nitrate leaching to groundwater, Agr. Ecosyst. Environ., 147, 13–23, doi:10.1016/j.agee.2011.04.015, 2012.

Jarvis, S. C.: Progress in studies of nitrate leaching from grassland soils, Soil Use Manage., 16, 152–156, 2000.

Jones, P. and Harris, I.: CRU TS3.21: Climatic Research Unit (CRU) Time-Series (TS) Version 3.21 of High Resolution Gridded Data of Month-by-month Variation in Climate (January 1901–December 2012), University of East Anglia Climatic Research Unit (CRU), NCAS British Atmospheric Data Centre, doi:10.5285/D0E1585D-3417-485F-87AE-4FCECF10A992, 2013.

Judex, M. and Thamm, H. P.: IMPETUS Atlas Benin. Research Results 2000–2007, 3rd Edn., Department of Geography, University of Bonn, Germany, 2008.

Kramer, S. B., Reganold, J. P., Glover, J. D., Bohannan, B. J. M., and Mooney, H. A.: Reduced nitrate leaching and enhanced denitrifier activity and efficiency in organically fertilized soils, P. Natl. Acad. Sci. USA, 103, 4522–4527, 2006.

Le Maitre, D. C., Milton, S. J., Jarmain, C., Colvin, C. A., Saayman, I., and Vlok, J. H. J.: Linking ecosystem services and water resources: landscape-scale hydrology of the Little Karoo, Front. Ecol. Environ., 5, 261–270, 2007.

Leh, M. D. K., Matlock, M. D., Cummings, E. C., and Nalley, L. L.: Quantifying and mapping multiple ecosystem services change in West Africa, Agr. Ecosyst. Environ., 165, 6–18, doi:10.1016/j.agee.2012.12.001, 2013.

Liquete, C., Maes, J., La Notte, A., and Bidoglio, G.: Securing water as a resource for society: an ecosystem services perspective, Ecohydrology and Hydrobiology, 11, 247–259, doi:10.2478/v10104-011-0044-1, 2011.

Liu, T., Merrill, N. H., Gold, A. J., Kellogg, D. Q., and Uchida, E.: Modeling the production of multiple ecosystem services from agricultural and forest landscapes in Rhode Island, Agricultural and Resource Economics Review, 42, 251–274, 2013.

Lopa, D., Mwanyoka, I., Jambiya, G., Massoud, T., Harrison, P., Ellis-Jones, M., Blomley, T., Leimona, B., van Noordwijk, M., and Burgess, N. D.: Towards operational payments for water ecosystem services in Tanzania: a case study from the Uluguru Mountains, Oryx, 46, 34–44, doi:10.1017/S0030605311001335, 2012.

Lu, Y. and He, T.: Assessing the effects of regional payment for watershed services program on water quality using an intervention analysis model, Sci. Total Environ., 493, 1056–1064, doi:10.1016/j.scitotenv.2014.06.096, 2014.

Maes, J., Egoh, B., Willemen, L., Liquete, C., Vihervaara, P., Schägner, J. P., Grizzetti, B., Drakou, E. G., Notte, A. L., Zulian, G., Bouraoui, F., Luisa Paracchini, M., Braat, L., and Bidoglio, G.: Mapping ecosystem services for policy support and decision making in the European Union, Ecosystem Services, 1, 31–39, doi:10.1016/j.ecoser.2012.06.004, 2012.

Maler, K. G., Aniyar, S., and Jansson, A.: Accounting for ecosystem services as a way to understand the requirements for sustainable development, P. Natl. Acad. Sci. USA, 105, 9501–9506, doi:10.1073/pnas.0708856105, 2008.

Martínez-Harms, M. J. and Balvanera, P.: Methods for mapping ecosystem service supply: a review, International Journal of Biodiversity Science, Ecosystem Services and Management, 8, 17–25, doi:10.1080/21513732.2012.663792, 2012.

Moriasi, D. N., Arnold, J. G., Van Liew, M. W., Bingner, R. L., Harmel, R. D., and Veith, T. L.: Model evaluation guidelines for systematic quantification of accuracy in watershed simulations, T. ASABE, 50, 885–900, 2007.

Naidoo, R., Balmford, A., Costanza, R., Fisher, B., Green, R. E., Lehner, B., Malcolm, T. R., and Ricketts, T. H.: Global mapping of ecosystem services and conservation priorities, P. Natl. Acad. Sci. USA, 105, 9495–9500, doi:10.1073/pnas.0707823105, 2008.

Nash, J. E. and Sutcliffe, J. V.: River flow forecasting through conceptual models: Part 1. A discussion of principles, J. Hydrol., 10, 282–290, 1970.

Neitsch, S. L., Arnold, J. G., Kiniry, J. R., and Williams, J. R.: Soil and Water Assessment Tool, Theoretical Documentation, Grassland, Soil and Water Resources Laboratory, Temple, TX, USA, 2009.

Notter, B., Hurni, H., Wiesmann, U., and Abbaspour, K. C.: Modelling water provision as an ecosystem service in a large East African river basin, Hydrol. Earth Syst. Sci., 16, 69–86, doi:10.5194/hess-16-69-2012, 2012.

Obst, C., Edens, B., and Hein, L.: Ecosystem services: accounting standards, Science, 342, p. 420, doi:10.1126/science.342.6157.420-a, 2013.

Orekan, V. O. A.: Implementation of the local land-use and land-cover change model CLUE-s for central Benin by using socio-economic and remote sensing data, PhD thesis, Mathematisch-Naturwissenschaftlichen Fakultät, Rheinischen Friedrich-Wilhelms-Universität Bonn, Bonn, Germany, 2007.

Pagiola, S. and Platais, G.: Payments for Environmental Services: from Theory to Practice, World Bank, Washington, 2007.

Pattanayak, S. K. and Kramer, R. A.: Worth of watersheds: a producer surplus approach for valuing drought mitigation in Eastern Indonesia, Environ. Dev. Econ., 6, 123–146, doi:10.1017/S1355770x01000079, 2001.

Rathjens, H. and Oppelt, N.: SWATgrid: an interface for setting up SWAT in a grid-based discretization scheme, Comput. Geosci.-UK, 45, 161–167, doi:10.1016/j.cageo.2011.11.004, 2012.

Rathjens, H., Oppelt, N., Bosch, D. D., Arnold, J. G., and Volk, M.: Development of a grid-based version of the SWAT landscape model, Hydrol. Process., 29, 900–914, doi:10.1002/hyp.10197, 2014.

Remme, R. P., Schröter, M., and Hein, L.: Developing spatial biophysical accounting for multiple ecosystem services, Ecosystem Services, 10, 6–18, 2014.

Rodenburg, J., Zwart, S. J., Kiepe, P., Narteh, L. T., Dogbe, W., and Wopereis, M. C. S.: Sustainable rice production in African inland valleys: seizing regional potentials through local approaches, Agr. Syst., 123, 1–11, doi:10.1016/j.agsy.2013.09.004, 2014.

Santhi, C., Kannan, N., Arnold, J. G., and Di Luzio, M.: Spatial calibration and temporal validation of flow for regional scale hydrologic modeling, J. Am. Water Resour. As., 44, 829–846, doi:10.1111/j.1752-1688.2008.00207.x, 2008.

Schröter, M., Barton, D. N., Remme, R. P., and Hein, L.: Accounting for capacity and flow of ecosystem services: a conceptual model and a case study for Telemark, Norway, Ecol. Indic., 36, 539–551, doi:10.1016/j.ecolind.2013.09.018, 2014.

Seppelt, R., Dormann, C. F., Eppink, F. V., Lautenbach, S., and Schmidt, S.: A quantitative review of ecosystem service studies: approaches, shortcomings and the road ahead, J. Appl. Ecol., 48, 630–636, doi:10.1111/j.1365-2664.2010.01952.x, 2011.

Shaxson, F. and Barber, R.: Optimizing soil moisture for plant production. The significance of soil porosity, FAO Soils Bulletin, FAO Soils Bulletin, 79, Rome, Italy, 2003.

Stoneham, G., O'Keefe, A., Eigenraam, M., and Bain, D.: Creating physical environmental asset accounts from markets for ecosystem conservation, Ecol. Econ., 82, 114–122, doi:10.1016/j.ecolecon.2012.06.017, 2012.

TEEB: The economics of ecosystems and biodiversity. Mainstreaming the economics of nature. A synthesis of the approach, conclusions and recommendations of TEEB, available at: www.teebweb.org, last access: 4 September 2014, 2010.

Terrado, M., Acuna, V., Ennaanay, D., Tallis, H., and Sabater, S.: Impact of climate extremes on hydrological ecosystem services in a heavily humanized Mediterranean basin, Ecol. Indic., 37, 199–209, doi:10.1016/j.ecolind.2013.01.016, 2014.

Tilman, D., Cassman, K. G., Matson, P. A., Naylor, R., and Polasky, S.: Agricultural sustainability and intensive production practices, Nature, 418, 671–677, doi:10.1038/nature01014, 2002.

Turpie, J. K., Marais, C., and Blignaut, J. N.: The working for water programme: evolution of a payments for ecosystem services mechanism that addresses both poverty and ecosystem service delivery in South Africa, Ecol. Econ., 65, 788–798, doi:10.1016/j.ecolecon.2007.12.024, 2008.

UN, European Commission, International Monetary Fund, Organisation for Economic Co-operation and Development, and World Bank: System of National Accounts 2008, New York, 2009.

van Oudenhoven, A. P. E., Petz, K., Alkemade, R., Hein, L., and de Groot, R. S.: Framework for systematic indicator selection to assess effects of land management on ecosystem services, Ecol. Indic., 21, 110–122, doi:10.1016/j.ecolind.2012.01.012, 2012.

Vigerstol, K. L. and Aukema, J. E.: A comparison of tools for modeling freshwater ecosystem services, J. Environ. Manage., 92, 2403–2409, doi:10.1016/j.jenvman.2011.06.040, 2011.

Villamagna, A. M., Angermeier, P. L., and Bennett, E. M.: Capacity, pressure, demand, and flow: a conceptual framework for analyzing ecosystem service provision and delivery, Ecol. Complex., 15, 114–121, doi:10.1016/j.ecocom.2013.07.004, 2013.

Volk, M., Arnold, J. G., Bosch, D. D., Allen, P. M., and Green, C. H.: Watershed configuration and simulation of landscape processes with the SWAT model, in: MODSIM 2007 International Congress on Modelling and Simulation, Christchurch, New Zealand, 10 December 2007 through 13 December 2007, edited by: Oxley, L. and Kulasiri, D., Modeling and Simulation Society of Australia and New Zealand, 1383–2389, 2007.

Weber, J. L.: Land and Ecosystem Accounts in the SEEA Revision, paper presented to the 13th meeting of the London Group, available at: http://unstats.un.org/unsd/envaccounting/londongroup/meeting13/LG13_25a.pdf, last access: 4 September 2014, 2007.

Willaarts, B. A., Volk, M., and Aguilera, P. A.: Assessing the ecosystem services supplied by freshwater flows in Mediterranean agroecosystems, Agr. Water Manage., 105, 21–31, doi:10.1016/j.agwat.2011.12.019, 2012.

Williams, J. R.: Sediment-yield prediction with universal equation using runoff energy factor, in: Present and prospective technology for predicting sediment yield and sources, Proceedings of the Sediment Yield Workshop, USDA Sedimentation Laboratory, Oxford, Mississippi, 28–30 November 1972, U.S. Department of Agriculture, Washington, D.C., 244–252, 1975.

Wolfe, A. H. and Patz, J. A.: Reactive nitrogen and human health: acute and long-term implications, Ambio, 31, 120–125, 2002.

Yan, B., Fang, N. F., Zhang, P. C., and Shi, Z. H.: Impacts of land use change on watershed streamflow and sediment yield: an assessment using hydrologic modelling and partial least squares regression, J. Hydrol., 484, 26–37, doi:10.1016/j.jhydrol.2013.01.008, 2013.

Zang, C. F., Liu, J., van der Velde, M., and Kraxner, F.: Assessment of spatial and temporal patterns of green and blue water flows under natural conditions in inland river basins in Northwest China, Hydrol. Earth Syst. Sci., 16, 2859–2870, doi:10.5194/hess-16-2859-2012, 2012.

7

Spatio-temporal assessment of annual water balance models for upper Ganga Basin

Anoop Kumar Shukla[1], **Shray Pathak**[1], **Lalit Pal**[1], **Chandra Shekhar Prasad Ojha**[1], **Ana Mijic**[2], **and Rahul Dev Garg**[1]

[1]Department of Civil Engineering, Indian Institute of Technology Roorkee, Uttarakhand, India
[2]Department of Civil and Environmental Engineering, Imperial College London, London, UK

Correspondence: Shray Pathak (shraypathak@gmail.com)

Abstract. The upper Ganga Basin in Uttarakhand, India, has high hydropower potential and plays an important role in the development of the state economy. Thus, an accurate knowledge of annual water yield is of paramount importance to this region. This paper deals with use of contemporary water yield estimation models such as the distributed Integrated Valuation of Ecosystem Services and Tradeoffs (InVEST) model and the Lumped Zhang model and their validation to identify the most suited one for water yield estimation in the upper Ganga Basin. In previous studies utilizing these models, water yield was estimated by considering a single value of some important model parameters for the entire basin, which in fact show distributed variation at a finer (pixel) scale. Therefore, in this study, pixel-level computations are performed to assess and ascertain the need for incorporating the spatial variation of such parameters in model applications. To validate the findings, the observed sub-basin discharge data are analyzed with the computed water yield for 4 decades, i.e., 1980, 1990, 2001 and 2015. The results obtained are in good agreement with the water yield obtained at the pixel scale.

1 Introduction

An accurate assessment of key ecosystem services (ES) such as water yield has gained focus in recent years in ES modeling, as fresh-water availability in a region is essential for agriculture, industry, human consumption, hydropower, etc. (Redhead et al., 2016). Hydrological ecosystem services generally include drinking water supply, power production, industrial use, irrigation and many more. The accurate estimation of water yield further facilitates in the identification of hotspots for storm-water harvesting in order to fulfill fresh-water demand in the region (Pathak et al., 2017). The hydrological ES are dependent on different factors, such as watershed characteristics (e.g., topography, land use and land cover – LULC, soil type and climatic condition. To incorporate these parameters into assessment and decision-making, there has been a proliferation of ecosystem-modeling tools and methods. Models for ES evaluation often focus on using globally available data, accepting large number of spatially explicit inputs producing spatially explicit output, and limiting the model structure to key biophysical processes involved in land use change (Guswa et al., 2014). The precise estimation of ES using these models is a complicated task owing to spatial variability and dependence of ES on various topographical and climatic factors. Further, the validation and uncertainty assessments in model outputs have proven to be key obstacles to the application of ES models. In the literature, studies focusing on comparison of different ES models have projected some light over the model output validation issues; however, a lack of studies highlighting the validation of these models for Indian river basins still exists. The benefits that can be derived from ES should be analyzed and quantified in a spatially explicit manner (Sánchez-Canales et al., 2012). The uncertainties involved in the determination of spatial and temporal distribution of the climatic variables, especially precipitation, constitute a major obstacle to the understanding of hydrological behavior at the catchment scales (Milly and Dunne, 2002).

The Integrated Valuation of ES and Tradeoffs (InVEST) model, developed by Natural Capital Project (Tallis et al., 2010), is a tool that provides a framework for planners and decision makers to assess trade-offs among ES and enables their comparison in various climate and land use change scenarios. The model includes a biophysical component, which facilitates the provision of fresh water or water yield from different parts of the landscape, and a valuation component, representing the benefits of water provisioning to people. The model works on simplified Budyko theory, which has a long history and still continues to receive attention in the hydrological literature (Budyko and Ronov, 1979; Zhang et al., 2001; Zhang et al., 2004; Ojha et al., 2008; Zhou et al., 2012; Donohue et al., 2012; Xu et al., 2013; Wang and Tang, 2014). The InVEST model applies a one-parameter formulation of the Budyko theory in a semi-distributed manner (Zhang et al., 2004). The model is capable of quantifying the water yield of a catchment under the influence of change in different drivers, viz. climate variables and catchment characteristics (e.g., land use change). Various studies have been carried out in the past demonstrating the application of the InVEST model to different river basins around the world. Sánchez-Canales et al. (2012) carried out a sensitivity analysis of three parameters, i.e., z (seasonal precipitation coefficient), precipitation (annual) and ET_0 (annual reference evapotranspiration), and using the InVEST model for a Mediterranean basin, they found precipitation to be the most sensitive parameter for the study region. Later, Terrado et al. (2014) applied the InVEST model for the heavily inhabited defined as Llobregat river basin. The model is applied for both extreme wet and dry conditions, and the role of climatic parameters is emphasized. Hoyer and Chang (2014) applied this model in the Tualatin and Yamhill basins of northwestern Oregon under a series of urbanization and climate-change scenarios. The results show that the climatic parameters have more sensitivity than other inputs for a water yield model. Hamel and Guswa (2015) applied the same water yield model for the Cape Fear catchment, North Carolina, and concluded that the precipitation is the most influencing parameter. Goyal and Khan (2017) employed the InVEST water yield model for the hilly catchment by considering two catchments, i.e., the Sutlej River catchment and Tungabhadra River catchment. The climate parameters, i.e., precipitation and ET_0, are observed to be the most influencing parameters for water yield in both the river basins. With the aforementioned studies, certain factors exist that limit the application of InVEST model such as the absence or inadequate comparison with observed data, the calibration of the model without prior identification of sensitive parameters and a lack of validation of the predictive capabilities in the context of land use and land cover change (Bai et al., 2012; Nelson et al., 2010; Su and Fu, 2013; Terrado et al., 2014).

The InVEST model operates on the principle of the Budyko theory (Budyko, 1958, 1974). Based on works of Schreiber (1904) and Ol'Dekop (1911), Budyko proposed formulations explaining the relationship between precipitation and potential evapotranspiration (PET) in order to couple water and energy balances, defined as the Budyko hypothesis. Several attempts have later been made to obtain an analytical solution of the Budyko hypothesis (Schreiber, 1904; Ol'Dekop, 1911; Turc, 1954; Mezentsev, 1955; Pike, 1964; Fu, 1981; Choudhury, 1999; Zhang et al., 2001, 2004; Porporato et al., 2004; Yang et al., 2008; Donohue et al., 2012; Wang and Tang, 2014; G. Zhou et al., 2015; S. Zhou et al., 2015). Among these studies, solutions provided by Fu (1981) called Fu's equation, gained significant attention as the work represented the effect of catchment properties on water balance components by incorporating an addition parameter "w". Fu's equation can provide a full picture of the evaporation mechanism at the annual timescale. Therefore, Fu's equation can be used through a top-down analysis for providing insight into the dynamic interactions among climate, soils, vegetation, and their controls on the annual water balance at the regional scale (Yang et al., 2007).

Considering the lack of studies on analysis and validation of ES on the Indian subcontinent, especially for Himalayan catchments, and to assess the applicability of various water-balance models to Himalayan catchments, the present work attempts to compute and analyze water yield in the upper Ganga Basin using a semi-distributed InVEST model and a Lumped Zhang model. The work primarily considers, in detail, the spatial variation of InVEST model parameters and uses different strategies to compute water yield. Accordingly, water yield is estimated for 4 years, i.e., 1980, 1990, 2001 and 2015 and the most appropriate strategy is identified. The parameters that are adopted as lumped at the basin scale in previous studies are estimated at the pixel scale in order to avoid the dependence of the model parameters on size of the catchment. In addition, pixel-level estimations of water yield are expected to be more accurate than output obtained using the conventional approach with basin-lumped output. The term "finer scale" in the paper represents the incorporation of spatial variations through the pixel-level estimation of parameters involved in InVEST model, which are otherwise taken as lumped. The work also compares the outcomes of spatially distributed water yield models and the conventionally used Lumped Zhang model.

2 Background theory

2.1 Water yield models

In this section, two water yield models, i.e., the InVEST water yield model, which is a distributed model, and the Lumped Zhang model, are described.

2.1.1 InVEST model

The InVEST water yield model (Tallis et al., 2010) is designed to provide information regarding the changes in the ecosystem that are likely to alter the flow. It is based on the Budyko theory, which is an empirical function that yields the ratio of actual to potential evapotranspiration (PET)

(Budyko, 1979). To describe the degree to which long-term catchment water balance deviates from the theoretical limits, a number of scholars have proposed one-parameter functions that can replicate the Budyko curve (Fu, 1981; Choudhury, 1999; Zhang et al., 2004; Wang and Tang, 2014). To observe and represent pixel-level changes to the landscape, InVEST model incorporates, explicitly, the spatial variability in precipitation, PET, soil depth and vegetation. The model operates at the grid scale and acquires the inputs in the raster format into a GIS environment such as ArcGIS.

The InVEST water yield model is based on an empirical function known as the Budyko curve (Budyko, 1974). Annual water yield, $Y(x)$, is determined at each pixel of a landscape as follows;

$$Y(x) = \left(1 - \frac{\mathrm{AET}(x)}{P(x)}\right) \times P(x), \tag{1}$$

where $\mathrm{AET}(x)$ is the actual annual evapotranspiration per pixel x and $P(x)$ is the annual precipitation per pixel x. Actual evapotranspiration (AET) is essentially determined by climatic factors (precipitation, temperature, etc.) and is mediated by catchment characteristics (vegetation cover, soil characteristics, topography, etc.). On the other hand, potential evapotranspiration (PET) represents the evaporating potential of the climate system at a specific location and time of year without the consideration of catchment characteristics and soil properties (Allen et al., 1998). Several attempts have been made in the past to establish a relationship between AET and PET, among which the solution provided by Fu (1981) has been adopted worldwide. Fu (1981) provided an analytical solution to the Budyko hypothesis and related AET with PET by incorporating a dimensionless parameter "w", which denotes the effect of catchment characteristics.

The InVEST model uses the expression of the Budyko curve proposed by Fu (1981) and Zhang et al. (2004). The ratio of mean annual PET to annual precipitation, known as index of dryness, is expressed as

$$\frac{\mathrm{AET}(x)}{P(x)} = 1 + \frac{\mathrm{PET}(x)}{P(x)} - \left[1 + \frac{\mathrm{PET}(x)}{P(x)}\right]^{\left(\frac{1}{w}\right)}, \tag{2}$$

where $\mathrm{PET}(x)$ is the annual potential evapotranspiration per pixel x (mm), and $w(x)$ is a non-physical parameter that influences the natural soil properties. The $\mathrm{PET}(x)$ is calculated using the following expression;

$$\mathrm{PET}(x) = K_c(x) \times \mathrm{ET}_0(x), \tag{3}$$

where $\mathrm{ET}_0(x)$ is the annual reference evapotranspiration per pixel x, which is computed based on evapotranspiration from alfalfa grass grown at that location using Eq. (6). $K_c(x)$ is the vegetation evapotranspiration coefficient that is influenced by the change in characteristics of land use and land cover at every pixel (Allen et al., 1998). The values of $\mathrm{ET}_0(x)$ are adjusted by $K_c(x)$ for each pixel over the map of land use and land cover. $w(x)$ is an empirical parameter, and the expression given by Donohue et al. (2012) for the InVEST model has been applied to define $w(x)$, which is expressed as follows:

$$w(x) = z \times \frac{\mathrm{AWC}(x)}{P(x)} + 1.25. \tag{4}$$

Thus, the minimum value of the parameter $w(x)$ is 1.25, corresponding to bare soil where the root depth is zero (Donohue et al., 2012). The Donohue model was originally developed for Australia, however, the online documentation on InVEST model states its application globally. The parameter z is known as the seasonality factor whose value varies from 1 to 30. It represents the nature of local precipitation and other hydrogeological parameters. The parameter $\mathrm{AWC}(x)$ depicts volumetric plant available water content expressed in depth (mm), which can be expressed by following formula for each pixel x:

$$\mathrm{AWC}(x) = \mathrm{Min.(Restricting\ layer\ depth,\ root\ depth)} \times \mathrm{PAWC}. \tag{5}$$

The root-restricting layer depth is defined as the depth of the soil up to which the soil can allow the penetration of the roots, and root depth is defined as the depth where 95 % of the root biomass occurs. Plant available water content (PAWC) is generally taken as the difference between the field capacity and the wilting point. It depends upon the soil properties and can be computed by the Soil-Plant-Air-Water (SPAW) software. In the study, PAWC is calculated using the method described by McKenzie et al. (2003). The modified Hargreaves method and Hargreaves method were employed for computing reference evapotranspiration for the study area at pixel scale.

The modified Hargreaves method is expressed as

$$\mathrm{ET}_0 = 0.0013 \times 0.408 \times \mathrm{RA} \times \left(T_{\mathrm{avg}} + 17.0\right) \times (\mathrm{TD} - 0.0123 \times P)^{0.76}, \tag{6}$$

where ET_0 is reference evapotranspiration, T_{avg} is the average daily temperature (°C) defined as the average of mean daily maximum and mean daily minimum temperature, TD (°C) is the temperature range computed as the difference between mean daily maximum and mean daily minimum temperature, and RA is extraterrestrial radiation ($\mathrm{MJ\,m^{-2}\,day^{-1}}$).

According to the Hargreaves method,

$$\mathrm{ET}_0 = 0.0023 \times 0.408 \times \mathrm{RA} \times \left(T_{\mathrm{avg}} + 17.8\right) \times \mathrm{TD}^{0.5}, \tag{7}$$

where terms involved in the equation means same as those in the modified Hargreaves method.

For computing the extraterrestrial radiation (RA), the following equation is used;

$$\mathrm{RA} = \frac{24(60)}{\pi} \times G_{\mathrm{sc}} \times d_{\mathrm{r}} \times [w_{\mathrm{s}} \sin(\varphi)\sin(\delta) + \cos(\varphi)\cos(\delta)\sin(w_{\mathrm{s}})], \quad (8)$$

where RA is extraterrestrial radiation (MJ m^{-2} day^{-1}), d_r is the inverse Earth–Sun relative distance, G_{sc} is the solar constant equal to 0.0820 MJ m^{-2} min^{-1}, w_s is sunset hour angle (rad), δ is the solar declination (rad) and φ is latitude (rad).

Determination of the parameter "w"

The dimensionless parameter w depends upon the local climatic variables such as the hydrological characteristics of the area, its rainfall intensity and topography. In the InVEST water yield model (Tallis et al., 2010), parameter w can be computed in three different ways. The first method is suggested by Donohue et al. (2012), in which parameter w is computed using Eq. (4) and where sensitivity parameter z is adopted as one fifth of the number of rain events per year. The second method is suggested by Xu et al. (2013), which compares w with latitude, the NDVI (normalized difference vegetation index), area, etc. The third method experiments with various selections of w (one value of w for the entire study region) until there is a good match between observed and computed water yield. Unfortunately, this method is not suited for a pixel-based analysis, as the number of pixels will be extremely large, making the method computationally intensive.

2.1.2 Lumped Zhang model

In this model, the mean value of different parameters is used as an input to compute the average value of the water yield for the whole watershed. The average actual evapotranspiration, potential evapotranspiration, w, precipitation, etc., are described by Zhang et al. (2004).

3 Study area and data

3.1 Study area

The Ganga river in India is ranked amongst the world's top 20 rivers in regards to the water discharge. The Ganga river is segregated into three zones, viz. the upper Ganga Basin, middle Ganga Basin and lower Ganga Basin. The area chosen for the present study, i.e., the upper Ganga Basin, is situated in the northern part of India within the geographical coordinates 29°48′–31°24′ N and 77°49′–80°22′ E, covering an area of 22 292.1 km^2 and reaching up to Haridwar. The altitude of the study area varies from 275 m in the plains to 7512 m in the Himalayan terrains. A region of approximately 433 km^2 of the basin is located under glacier landscape, and 288 km^2 of the region is located under a fluvial landscape.

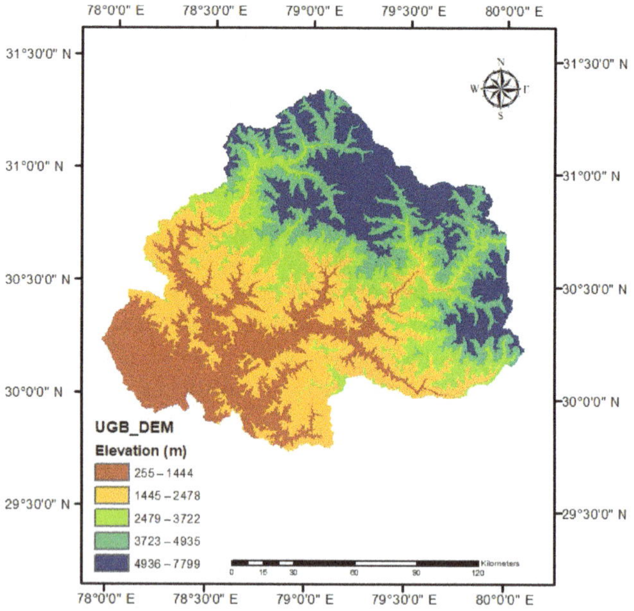

Figure 1. Graphical representation of the study area, the upper Ganga Basin.

About 60 % of the basin is utilized for agricultural practices, and 20 % of the basin is in the forest area, especially in the upper mountainous region. Nearly 2 % of the basin is permanently covered with snow in the mountain peaks. The most predominant soil groups found in the region are sand, clay, loam and their compositions. In the upper Ganga Basin, the average annual rainfall varies from 550 to 2500 mm (Bharati et al., 2011), where a major fraction of total annual rainfall is received during monsoon months (June–September). The geographical location and other information of the upper Ganga Basin are represented in Fig. 1.

3.2 Data

3.2.1 Precipitation and temperature

The daily time series of precipitation and temperature for the study area are acquired from India Meteorological Department (IMD) at a grid size of 0.25° and 1°, respectively. The upper Ganga Basin comes within the dataset latitude, which ranges from 29.5° N to 31.5° N, and its longitude, ranging from 77.75 to 80.25° E . The daily time series of precipitation was aggregated to obtain the annual time series at each grid point. Various analyses in the study are carried out for 4 years, i.e., 1980, 1990, 2001 and 2015.

3.2.2 Soil map

Spatial maps of soil were collected from the National Bureau of Soil Survey and Land Use Planning (NBSSLUP) at 1 : 250 000. Digital maps of soil available at a resolution of 1200 m × 1200 m were resampled to the resolution of land use data, i.e., 30 m × 30 m, using "resample" tool in ArcGIS in order to maintain the scale homogeneity. The attribute table of the raster layer contains fields like soil depth, soil texture, carbon content percentage, drainage, slope, erosion, soil temperature and mineralogy. The relevant features, i.e., soil depth and soil texture are converted into the raster image for the upper Ganga Basin.

3.2.3 Map of land use and land cover

Satellite images were acquired from different sensors of Landsat, viz. Landsat 3/4 Multispectral Scanner and Thematic Mapper (MSS/TM), Landsat 4 Thematic Mapper (TM), Landsat 7 Enhanced Thematic Mapper (ETM) and Landsat 8 Operational Land Imager (OLI) sensors for the years 1980, 1990, 2001 and 2015, respectively. The images are available at different resolutions and in several wavelength bands, from which green (G), red (R) and near-infrared (NIR) band images are combined to create a false color composite (FCC) for the study area in ERDAS Imagine. FCCs are then classified using supervised classification in ERDAS in six different classes, i.e., forest, water, agricultural, wasteland, snow and glacier, and built-up land. The classification of the area is based on their similar response under different bands. Each class is then recognized with the help of ground-truth and high-resolution satellite images.

4 Methodology

In the present work, five different strategies are employed to compute water yield. For the ease of presentation, these strategies are referred to as A–E. In strategy A, an average value of precipitation, temperature, extraterrestrial radiation and parameter w is used for the entire basin. This strategy is essentially based on Lumped Zhang model. Strategies B–E are designated, corresponding to a particular variation of the InVEST model where water yield is computed using different approach for estimating parameter w. For computing parameter w, relationships for large basins and for the global model from Xu et al. (2013) are given by Eqs. (9) and (10), respectively.
For large basins,

$$w = 0.69387 - 0.01042 \times \text{lat} + 2.81063 \times \text{NDVI} + 0.146186 \times \text{CTI}. \tag{9}$$

For the global model,

$$w = 3.50412 - 0.09311 \times \text{slp} - 0.03288 \times \text{lat} + 1.12312 \times \text{NDVI} - 0.00205 \times \text{long} - 0.00026 \times \text{elev}, \tag{10}$$

where, "slp" is the slope gradient, "lat" is the absolute latitude of basin center, "CTI" is the compound topographic index, "NDVI" is the normalized difference vegetation index, "lat" is latitude, "long" is longitude and "elev" is elevation.

In strategy B, the entire basin is considered for computing the parameter w for large basins, using Eq. (9), which is given by Xu et al. (2013). In strategy C, the parameter w is computed for entire basin using Eq. (10), which is given by Xu et al. (2013). In strategy D, parameter w is computed at each pixel in order to incorporate the spatial distribution of the hydrologic variables involved in the computations. In Strategy E, parameter z is computed according to the number of rain events in a year; subsequently, Eq. (4) is used to compute the parameter w.

For all the strategies, the extraterrestrial radiation (RA) parameter is computed for each month using Eq. (8), and a raster layer is generated. Precipitation data are obtained from Indian Meteorological Department (IMD) at a grid size of 0.25° for the study area. It has been interpreted and converted to the raster format by using the inverse distance weighted (IDW) interpolation technique in the ArcGIS environment for obtaining the values for all pixels at a resolution equal to the resolution of the Landsat satellite images. The temperature dataset is obtained from the IMD at a grid size of 1° × 1° for the study area and has also been converted to a raster format by using the IDW interpolation technique for obtaining the values for all pixels. Subsequently, the mean monthly value of average temperature (T_{avg}) and the difference between the mean daily maximum and mean daily minimum (TD) are obtained. The climate datasets used in the present study are of the finest resolution available so far for the study region. Gridded datasets of temperature and precipitation used in the present study have been developed using quality-controlled stations and well-proven interpolation techniques. Further details about the datasets of precipitation and temperature are given in Srivastava et al. (2009) and Pai et al. (2014), respectively.

The modified Hargreaves method is applied for obtaining the value of reference evapotranspiration at each pixel for each month (Droogers et al., 2002). To compute potential evapotranspiration, the yearly values obtained for the reference evapotranspiration are multiplied by the vegetation evapotranspiration coefficient (K_c), which depends on the LULC characteristics, as expressed in Eq. (3). The value of K_c is taken from Allen et al. (1998), as shown in Table 1. In this study, K_c is taken in the same was for all 4 years, as shown in Table 1, and is used to obtain potential evapotranspiration, which is subsequently used to obtain annual water yield at each pixel of the study area.

5 Results

5.1 Reference evapotranspiration, $ET_0(x)$

Reference Evapotranspiration (ET_0) is computed for the upper Ganga Basin using a high-resolution monthly climate dataset. The modified Hargreaves method is applied for obtaining the values of reference evapotranspiration at each

Table 1. Value of K_c corresponding to the classes of land use and land cover.

S. no.	Land use and land cover	Percentage cover (1980)	Percentage cover (1990)	Percentage cover (2001)	Percentage cover (2015)	K_c
1	Forest	17.84	16.32	15.78	15.19	1
2	Water	21.87	21.27	19.47	17.65	1
3	Wastelands	51.1	52.36	54.18	55.46	0.2
4	Built-up area	2.07	2.14	2.27	2.49	0.4
5	Agricultural	3.67	4.04	3.76	4.22	0.75
6	Snow and glacier	3.45	3.87	4.54	4.99	2

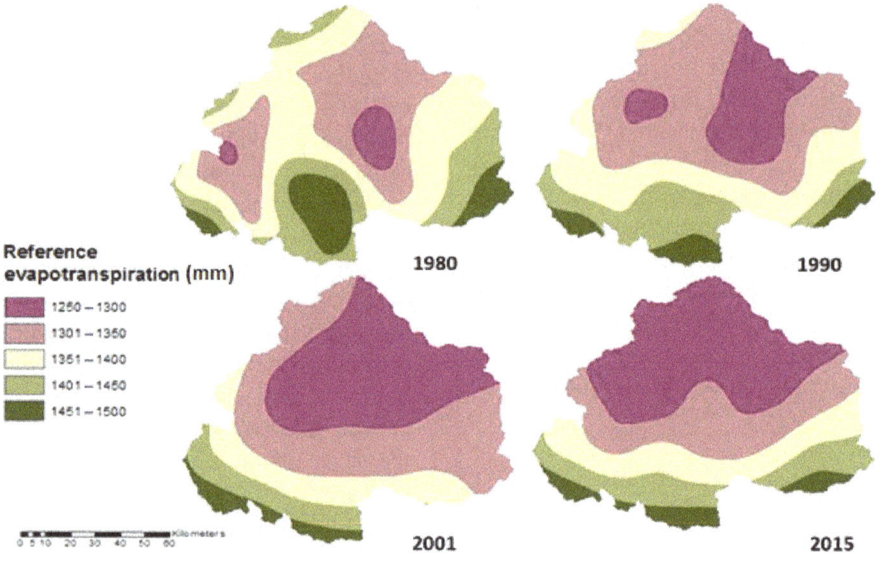

Figure 2. Reference evapotranspiration (mm) of the upper Ganga Basin for the years 1980, 1990, 2001 and 2015.

pixel for each month (Droogers and Allen, 2002). ET_0 is a function of RA, precipitation, T_{avg} and TD, which are computed pixel-wise for each month of the years 1980, 1990, 2001 and 2015. Some of the months, i.e., July, July and August 1990; June, July and August 2001; and June, July and August 2015, showed negative values of reference evapotranspiration from applying the modified Hargreaves method. For these months, Hargreaves method is applied for obtaining the positive values. Subsequently, all mean monthly values are added up to get the mean annual values of evapotranspiration for the years 1980, 1990, 2001 and 2015, as represented in Fig. 2.

5.2 Potential evapotranspiration, PET(x)

The annual values obtained for the ET_0 are multiplied by the vegetation evapotranspiration coefficient (K_c), which varies with the characteristics of land use and land cover, as expressed in Eq. (3). The value of the K_c is taken from Allen et al. (1998). The values of the vegetation evapotranspiration coefficient are taken from Table 1. Thus, the potential evapotranspiration is computed for upper Ganga Basin for the years 1980, 1990, 2001 and 2015, as represented in Fig. 3.

5.3 Water yield, $Y(x)$

As described in the methodology, five different strategies, viz. A–E, are used to estimate water yield for the upper Ganga Basin.

Strategy A: water yield computed using the Lumped Zhang model

Here, the basin average values of all the input parameters are considered, and water yield is computed for the upper Ganga Basin for the years 1980, 1990, 2001 and 2015, which are obtained as 658.52, 925.68, 603.71 and 1194.25 mm, respectively.

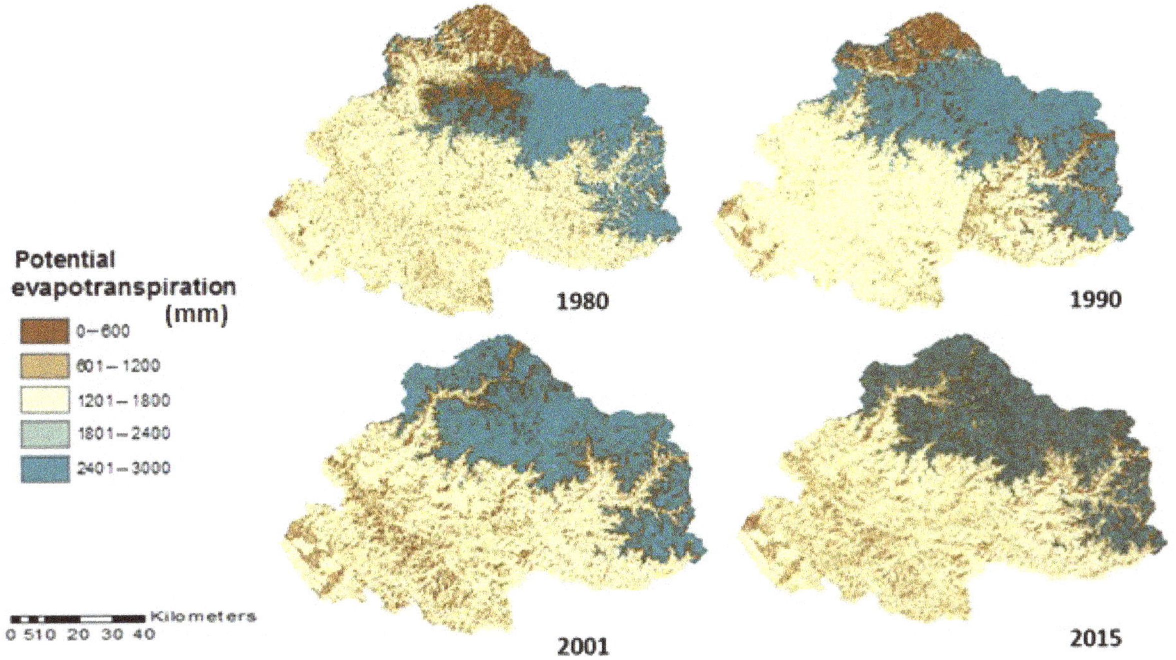

Figure 3. Potential evapotranspiration (mm) of the upper Ganga Basin for the years 1980, 1990, 2001 and 2015.

Strategy B: water yield obtained by taking the single weighted mean value of parameter "w" from Xu et al. (2013) for large basins

In this strategy, water yield is computed by considering a single value of the parameter w for the whole basin using Eq. (9). The weighted mean value for parameter w for the years 1980, 1990, 2001 and 2015 are obtained as 1.507, 1.541, 1.403 and 1.507, respectively. The spatial distribution of the water yield for the upper Ganga Basin computed using strategy B is represented in Fig. 4. The mean values of water yield as obtained using this method for the years 1980, 1990, 2001 and 2015 are 755.65, 959.48, 742.39 and 1131.42 mm, respectively.

Strategy C: water yield obtained by taking a single weighted mean value of parameter "w" from Xu et al. (2013) for the global model

In this strategy, water yield is computed by considering a single value of parameter w for the entire upper Ganga Basin using Eq. (10). The weighted mean value of parameter w for the years 1980, 1990, 2001 and 2015 are obtained as −0.967, −0.955, −1.010 and −0.968, respectively. The spatial distribution of the water yield for the upper Ganga Basin as computed using strategy C is shown in Fig. 5. The mean values of water yield for the years 1980, 1990, 2001 and 2015 are 1239.92, 1549.46, 1149.93 and 1754.59 mm, respectively.

Strategy D: water yield obtained using the pixel-level estimation of parameter "w" from Xu et al. (2013)

In this strategy, the values of parameter w are estimated at the pixel level. The water yield computed for the years 1980, 1990, 2001 and 2015 for upper Ganga Basin is shown in Fig. 6. The mean values of water yield as computed using strategy D for the years 1980, 1990, 2001 and 2015 are 1240.02, 1549.44, 1149.89 and 1754.62 mm, respectively.

Strategy E: water yield obtained using the pixel-level estimation of parameter "w" from Donohue et al. (2012)

Equation (4) represents the parameter w as a function of parameter "z", AWC and precipitation. The parameter w in the equation used in strategy E has been proposed by Donohue et al. (2012), which is also cited in online documentation of InVEST model; however, the final equation used for estimating water yield is obtained from the InVEST model. Considering this fact, Donohue et al. (2012) has been cited in strategy E. The water yield as computed using strategy E for the upper Ganga Basin for different years is shown in Fig. 7. The mean values of water yield for the years 1980, 1990, 2001 and 2015 are 1241.09, 1552.38, 1153.95 and 1753.53 mm, respectively.

Table 2. Comparison of model-estimated PET and AET with a global dataset from different sources.

Parameter (mm)	Year	Source 1 (GLDAS)	Source 2 (CRU)	Strategy A (Lumped Zhang model)	InVEST model			
					Strategy B (Large model)	Strategy C (Global model)	Strategy D (Xu et al. 2013)	Strategy E (Donohue et al., 2012)
AET	1980	555.0355		696.84	486.07	679.52	679.68	680.01
	1990	646.168		815.02	592.3	735.23	735.27	736.25
	2001	588.084		680.76	408.86	548.28	548.39	550.38
	2015	716.8316		900.11	625.41	743.48	743.52	744.34
PET	1980		1175.964	1376.64	1382.12	1382.12	1382.12	1382.12
	1990		1156.497	1456.16	1461.86	1461.86	1461.86	1461.86
	2001		1184.847	1457.08	1462.96	1462.96	1462.96	1462.96
	2015		1156.686	1544.20	1550.42	1550.42	1550.42	1550.42

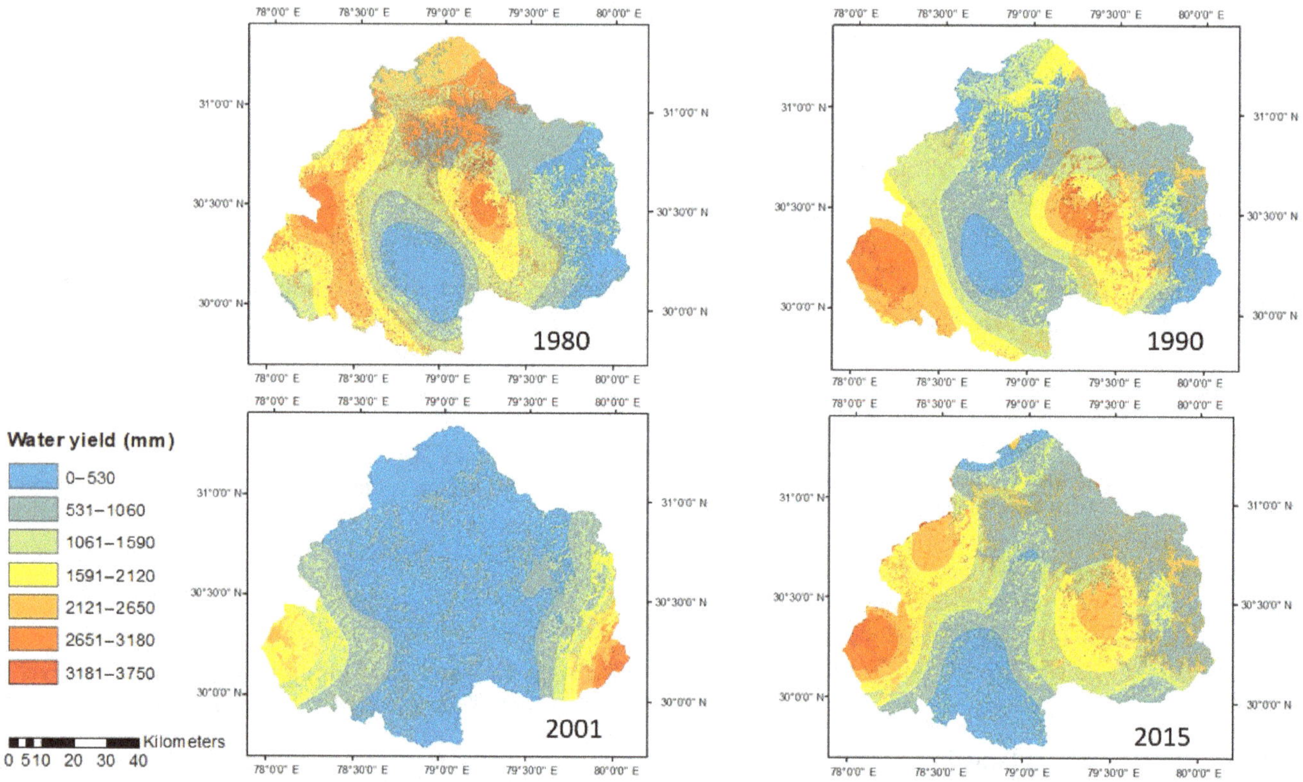

Figure 4. Water yield obtained by taking the single weighted mean value of parameter w from Xu et al. (2013) for large basins.

5.4 Validation of ET and water yield estimates

For validation of model outputs, the basin's average annual values of PET and AET estimated using various strategies are compared with the corresponding basin average values obtained from available global datasets (Table 2). Model-simulated AET values are obtained from the Global Land Data Assimilation System (GLDAS) ET dataset from Noah model outputs. Basin average values of PET are obtained from the Climate Research Unit's (CRU's) PET datasets (CRU TS v. 4.01) available at resolution of 0.5°. From the comparison, both AET (GLDAS) and PET (CRU TS) values are found to be in fair agreement with the globally estimated values (Table 2). Spatial maps of global datasets of AET and PET are shown in Figs. 8 and 9, respectively.

The validation of water yield obtained from various strategies is performed at the Rishikesh gauging site of the upper Ganga Basin (Fig. 10). The discharge data of the basin are

Table 3. Observed vs. computed water yield for various proposed strategies for Rishikesh sub-basin.

Strategies	1980	1990	2001	2015
Observed discharge (mm)	1831.31	2422.43	2187.22	2835.81
Observed discharge (mm) (after reducing approx. 32 % melting snow contribution)	1245.29	1647.25	1487.31	1928.35
Water yield strategy A (mm)	652.47	914.35	598.25	1189.72
Water yield strategy B (mm)	745.38	917.77	697.75	1092.17
Water yield strategy C (mm)	1229.90	1506.82	1102.62	1718.17
Water yield strategy D (mm)	1229.99	1506.74	1102.61	1718.18
Water yield strategy E (mm)	1230.77	1508.88	1106.86	1720.16

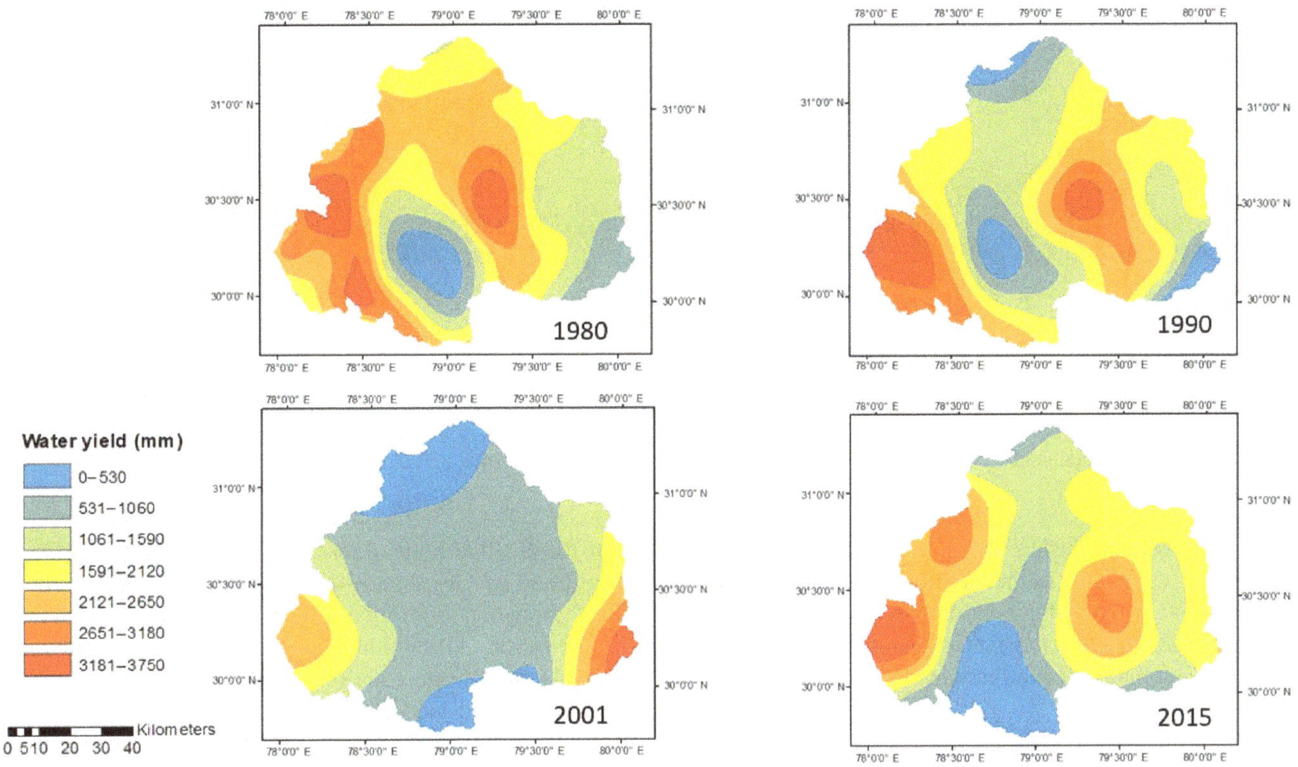

Figure 5. Water yield obtained by taking the single weighted mean value of parameter "w" from Xu et al. (2013) for the global model.

obtained from Irrigation Department of the state of Uttarakhand. The discharge observed in the basin is generated from precipitation as well as snowfall in the region, where 32 % of the discharge has been removed, because it is contributed to by glacier ice melt, as explained by Maurya et al. (2011) for our study area. The aforementioned fraction of discharge had been quantified using an isotope study that separates the contribution of glacier melt in quantifying discharge (Maurya et al., 2011). A comparison of the water yield computed and observed for the study region for different years by various proposed strategies is shown in Table 3.

As can be seen in Table 3, values of water yield estimated using strategies A to E are systematically increasing but are not steady in nature, as water yield estimated using strategy A and B lies in the range 650–750 mm, whereas water yield from strategies C–E lie in range of 1229–1231 mm for the years 1980 (see Table 3). Similar results are also evident for other years, too. Also, water yield estimated using strategies C–E are more or less the same for a given year, because these strategies involve pixel-based estimations of water yield considering spatial variation in the Budyko parameters. The parameters involved in the Budyko model, such as w, are dependent on various factors, such as catchment characteristics, vegetation cover, etc., as well as climate seasonality (Li et al., 2013). Ahn and Merwade (2017) have analyzed the relationship between basin characteristics and parameter w for 175 stations spread across the USA. Considering their study, no precise conclusion can be drawn regarding

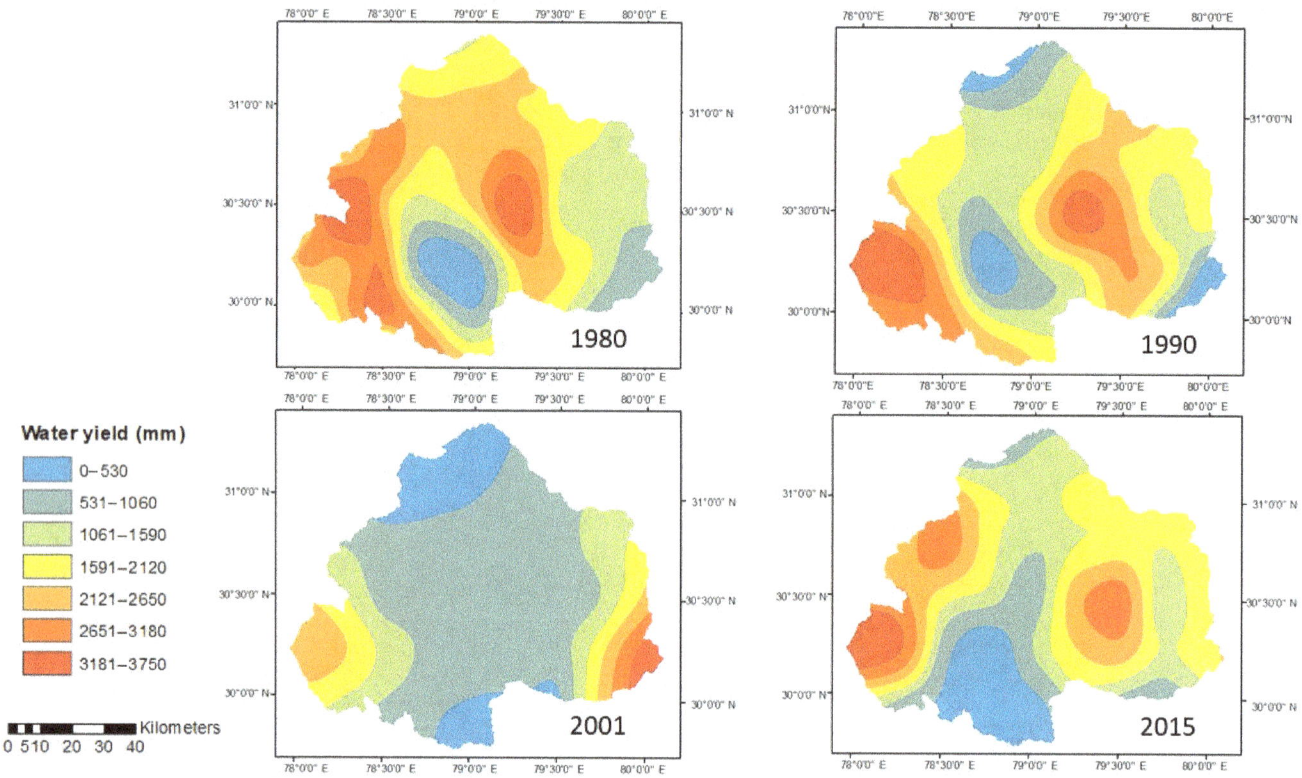

Figure 6. Water yield obtained by computing pixel-wise value of parameter w from Xu et al. (2013).

relationship between basin characteristics and the value of parameter w, especially in the case of basin-area characteristics. Moreover, no definite relationship has been yet identified between basin characteristics and model parameters, and this is a subject matter for further study.

6 Discussion

The study aimed to apply the InVEST water yield model to compute the water yield for upper Ganga Basin having highly variable topography consisting of hilly, plain and snow-covered areas. The InVEST model is based on the Budyko theory, which requires low amounts of data and low levels of expertise, thus making it acceptable worldwide. The mean monthly precipitation, temperature, monthly value of difference of the mean daily maximum and mean daily minimum, and extraterrestrial radiation parameters for the upper Ganga Basin of all 4 years, i.e., 1980, 1990, 2001 and 2015, are converted into the raster format for various analyses. The monthly reference evapotranspiration is thus computed using input parameters in GIS environment by applying the modified Hargreaves equation for all the months, except for a few months in which the modified Hargreaves equation gives negative results for the reference evapotranspiration. For those months, the Hargreaves method is applied to obtain the positive value of reference evapotranspiration, as also suggested by Goyal and Khan (2017). Reference evapotranspiration when multiplied with K_c gives the potential evapotranspiration. All monthly values are added up to obtain the annual value of reference evapotranspiration. K_c is a function of land use and land cover; thus, supervised classification is done to prepare the raster map of land use and land cover for the upper Ganga Basin. Subsequently, the annual value of potential evapotranspiration is obtained for the study area for the years 1980, 1990, 2001 and 2015.

The paper employs various methodologies for water yield estimation, as discussed in the methodology section for the upper Ganga Basin. Thus, water yield is computed both from the InVEST model as well as the Lumped Zhang model. The value of the parameter w is computed using four different approaches, i.e., the mean single value obtained from Xu et al. (2013) for large basins, mean single value obtained from Xu et al. (2013) for the global model, pixel-level estimated value of parameter w from Xu et al. (2013) and pixel-wise value of parameter w from Donohue et al. (2012). Although the upper Ganga Basin lies in large basin category as per the definition from Xu et al. (2013), the yield computed using global model is in good agreement with the observed data for the region. In the study, the pixel-level estimation of parameter w is made in order to incorporate the spatial variability of the parameter involved in water yield estimation. Thus, two

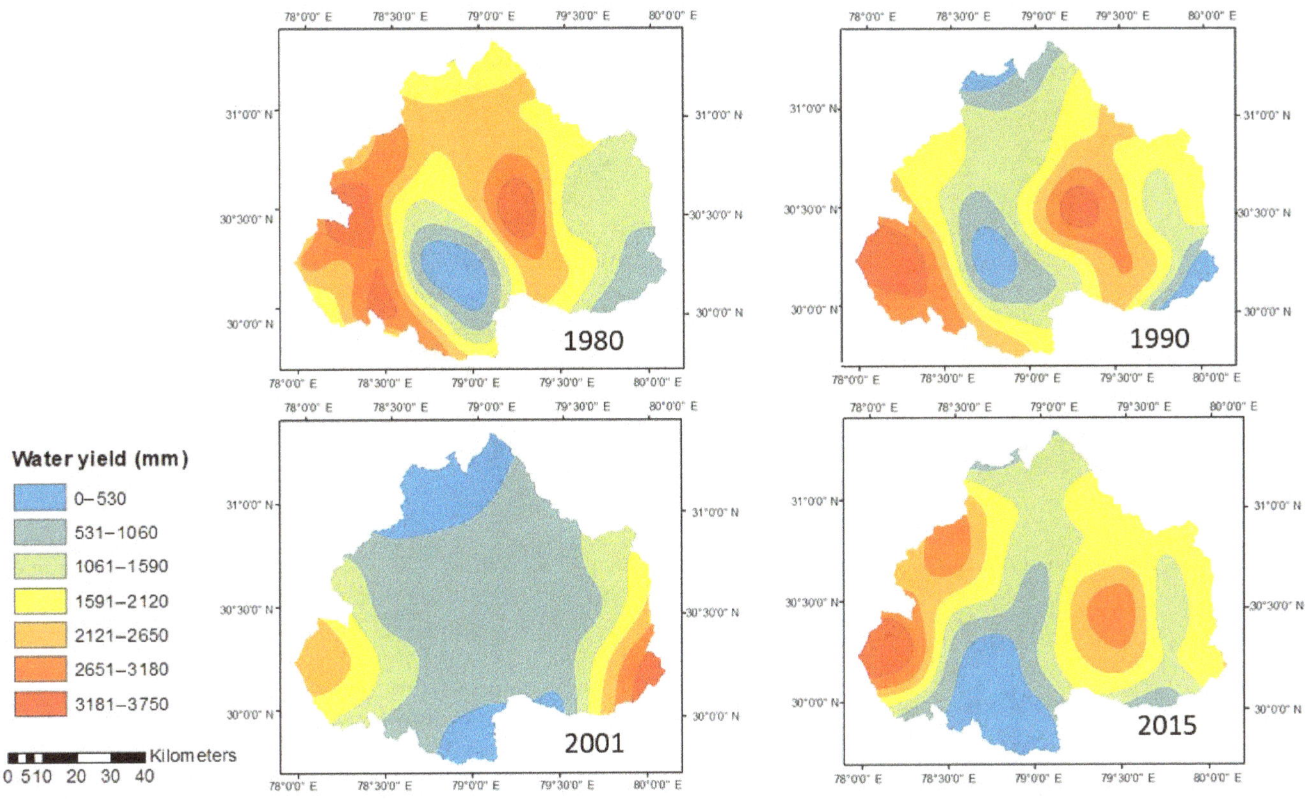

Figure 7. Water yield obtained by computing pixel-wise value of parameter "w" from Donohue et al. (2012).

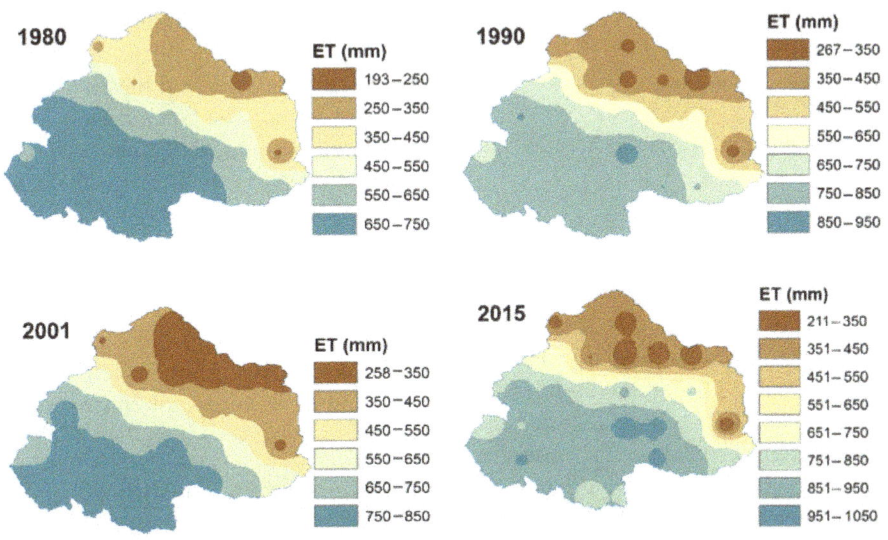

Figure 8. Spatial distribution of AET obtained from GLDAS Noah output datasets.

pixel-wise values of parameter w are computed for the upper Ganga Basin for years 1980, 1990, 2001 and 2015 by considering two approaches given by Xu et al. (2013) and the approach given by Donohue et al. (2012). Also, the basin-lumped water yield is computed using Lumped Zhang model, which considers the single mean values for entire basin of all the parameters involved in the computation of water yield.

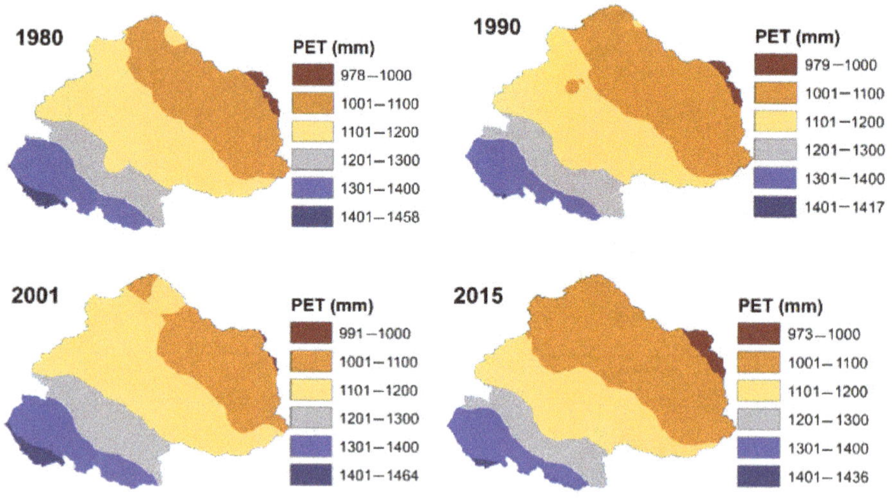

Figure 9. Spatial distribution of PET obtained from CRU datasets.

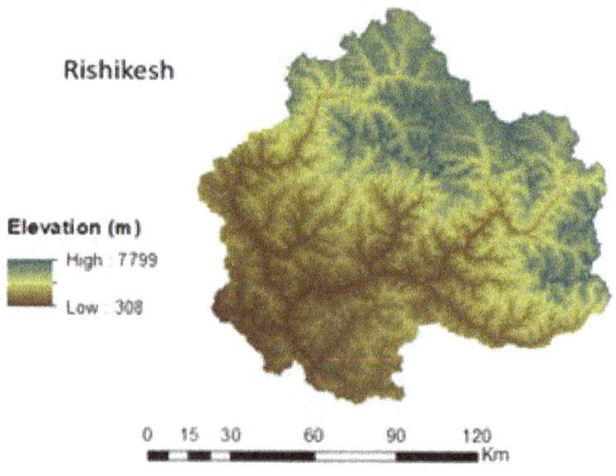

Figure 10. Graphical representation of sub-basin Rishikesh.

The water yield is computed in five different ways for the upper Ganga Basin for the years 1980, 1990, 2001 and 2015.

At the Rishikesh gauging site, surface runoff data are obtained by extracting the snowmelt from the discharge data, as the melting snow contributes about 32 % of total runoff in the Himalayan basins (Maurya et al., 2011). For validating the water yield obtained from different strategies, the observed yield is compared with the computed water yield based on different proposed strategies for the years 1980, 1990, 2001 and 2015, as represented in Table 3. The results obtained from Donohue et al. (2012) and Xu et al. (2013) are computed at pixel level (Strategy C–E); thus, they exhibit better performance than other approaches and are in good agreement with the observed data. These results exhibit the superiority of pixel-level computation to hydrological analyses for a watershed. The parameters involved in the Budyko model are dependent on various factors, such as basin characteristics (size, topography, stream length, slope, etc.), climate seasonality, etc. (Li et al., 2013). Again, the factors affecting model parameters vary both spatially and temporally. Moreover, the relationship between these factors and model parameters are not yet well defined (Ahn and Merwade, 2017). In such scenarios, adopting a hypothesis by assuming either of these controlling factors (such as "w") to be spatially or temporally constant is inappropriate. Considering these facts, the present study attempts to incorporate the spatial variability of model parameter for estimation of water yield at the pixel level. As the computations are made at pixel level (on a grid of size $30\,\mathrm{m} \times 30\,\mathrm{m}$), the assumption of dependence of model parameters on the size of the catchment may also be disregarded. The computations made in the present work are based on empirical equations; however, the application of these equations has been well documented worldwide for estimations of various water balance components at various basin scales (Zhang et al., 2008; Ma et al., 2008; Ning et al., 2017; Rouholahnejad Freund and Kirchner, 2017; Wang and Zhou, 2016). Hence, it is recommended that for such a large basin, it is required to compute all the parameters involved in the computations of water yield at the pixel scale rather than adopting mean values for entire watershed.

7 Summary and conclusions

The present study aimed to apply the InVEST annual water yield model, a tool that is gaining interest in the ecosystem services community, in the upper Ganga Basin. While such simple models have low requirements for data and level of expertise, practical applications of such a model with single representative values of the model parameter for the entire basin do not provide accurate estimates of water yield. Performing pixel scale computation of water yield in the study indicates a better performance, and the results obtained show better agreement with the observed water yield. As far as parameter w is concerned, the global model works better than other representations of parameter w available in literature, especially in the upper Ganga Basin. In the study, the water yield is computed using five different strategies, and results are validated with the observed data at the outlet of the upper Ganga Basin. The present study attempts to quantify annual water yield at the pixel level, making the computations independent of the size of catchment. Therefore, the proposed methodology is expected to perform well for a catchment of any given size. Changes in catchment water storage over time are required to be quantified in order to validate the applicability of Budyko's model to long-term data for the studied catchment. Earlier, some of the important parameters defining water yield used to be computed at a basin-level scale, which caused errors in the results.

The study attempts to incorporate the spatial variability of parameters involved in the model through the pixel-level estimation of parameters that are otherwise taken as lumped in the previous studies. Study results show that the estimated water yield, considering spatial variability in model parameters, is in better agreement with the observed water yield compared to the water yield estimated when considering the parameters to be lumped over the study region. Further, the computations of various parameters are made at the pixel level; therefore, the estimates of water balance components using this approach are expected to be independent of the assumption of dependence of parameters on catchment size. As the relationship between Budyko's model parameters and their controlling factors has not been well defined (Ahn and Merwade, 2017), the study emphasizes water yield estimation using pixel-based computations. The study outcomes can be summarized as follows: (i) between two approaches used in the study, i.e., considering the entire basin and pixel-level approach, the pixel-level approach is found to provide better results; and (ii) in pixel-based computations, results are further improved with the use of a parameter w based on a global model rather than regional models of parameter w, especially for large basins in the Himalayan region.

Data availability. The meteorological data products are provided by the Indian Meteorological Department on the basis of payment. It can be purchased from the following URL: http://www.imdpune.gov.in/ndc_new/Request.html (India Meteorological Department, 2018). The hydrological data in upper Ganga basin is provided by the Uttarakhand Irrigation Department, which is available for research purposes only. Satellite datasets are acquired from the USGS web portal (https://earthexplorer.usgs.gov/, Earth Explorer – USGS, 2018). The soil maps are provided by the National Bureau of Soil Survey and Land Use Planning, India on the basis of payment from the following URL: https://www.nbsslup.in/publications.html (ICAR, 2018).

Author contributions. AKP assisted with data collection, data processing and data analysis; SP with data analysis and writing, the analysis of results, and the review, revision and proofreading of the paper; LP with data analysis and writing, the analysis of results, and the review, revision and proofreading the paper; CSPO with the analysis of results and the review, revision, and supervision of the whole work and proofread the paper; AM supervised the whole work; and RDG assisted with the review, revision and supervision of the whole work.

Competing interests. The authors declare that they have no conflict of interest.

Special issue statement. This article is part of the special issue "The changing water cycle of the Indo–Gangetic Plain". It does not belong to a conference.

Acknowledgements. Authors are thankful to the executive engineer of the Irrigation Department of Uttarakhand for providing the discharge data for the Rishikesh sub-basin of upper Ganga Basin.

Edited by: Pradeep P. Mujumdar

References

Ahn, K. H. and Merwade, V.: The Integrated Impact of Basin Characteristics on Changes in Hydrological Variables, in: chap. 12 in "Sustainable Water Resources Management", American Society of Civil Engineers (ASCE), Virginia, USA, 317–336, 2017.

Allen, R. G., Pereira, L. S., Raes, D., and Smith, M.: Crop evapotranspiration – Guidelines for computing crop water requirements, FAO Irrigation and drainage paper 56, FAO, Rome, 300, D05109, 1998.

Bai, Y., Ouyang, Y., and Pang, J. S.: Biofuel supply chain design under competitive agricultural land use and feedstock market equilibrium, Energy Econ., 34, 1623–1633, 2012.

Bharati, L., Lacombe, G., Gurung, P., Jayakody, P., Hoanh, C. T., and Smakhtin, V.: The impacts of water infrastructure and climate change on the hydrology of the Upper Ganges River Basin, in: Vol. 142, IWMI, Colombo, Sri Lanka, 2011.

Budyko, M. I.: The Heat Balance of the Earth, Leningrad, 1956, Translation by N. A. Stepanova, US Weather Bureau, Washington, 1–254, 1958.

Budyko, M. I.: Climate and Life, Academic Press, New York, USA, 1–507, 1974.

Budyko, M. I. and Ronov, A. B.: Evolution of chemical composition of the atmosphere during the Phanerozoic, Geokhimiya, 5, 643–653, 1979.

Choudhury, B.: Evaluation of an empirical equation for annual evaporation using field observations and results from a biophysical model, J. Hydrol., 216, 99–110, 1999.

Donohue, R. J., Roderick, M. L., and McVicar, T. R.: Roots, storms and soil pores: Incorporating key ecohydrological processes into Budyko's hydrological model, J. Hydrol., 436, 35–50, 2012.

Droogers, P. and Allen, R. G.: Estimating reference evapotranspiration under inaccurate data conditions, Irrig. Drain. Syst., 16, 33–45, 2002.

Earth Explorer – USGS: available at: https://earthexplorer.usgs.gov/, last access: 15 September 2018.

Fu, B. P.: On the calculation of the evaporation from land surface, Sci. Atmos. Sin., 5, 23–31, 1981.

Goyal, M. K. and Khan, M.: Assessment of spatially explicit annual water-balance model for Sutlej River Basin in eastern Himalayas and Tungabhadra River Basin in peninsular India, Hydrol. Res., 48, 542–558, 2017.

Guswa, A. J., Brauman, K. A., Brown, C., Hamel, P., Keeler, B. L., and Sayre, S. S.: Ecosystem services: Challenges and opportunities for hydrologic modeling to support decision making, Water Resour. Res., 50, 4535–4544, 2014.

Hamel, P. and Guswa, A. J.: Uncertainty analysis of a spatially explicit annual water-balance model: case study of the Cape Fear basin, North Carolina, Hydrol. Earth Syst. Sci., 19, 839–853, https://doi.org/10.5194/hess-19-839-2015, 2015.

Hoyer, R. and Chang, H.: Assessment of freshwater ecosystem services in the Tualatin and Yamhill basins under climate change and urbanization, Appl. Geogr., 53, 402–416, 2014.

India Meteorological Department: available at: http://www.imdpune.gov.in/ndc_new/Request.html, last access: 10 October 2018.

ICAR: ICAR-NBSS&LUP Publications 2012–17, available at: https://www.nbsslup.in/publications.html, last access: October 2018.

Khatami, S. and Khazaei, B.: Benefits of GIS Application in Hydrological Modeling: A Brief summary, VATTEN – J. Water Manage. Res., 70, 41–49, 2014.

Li, D., Pan, M., Cong, Z., Zhang, L., and Wood, E.: Vegetation control on water and energy balance within the Budyko framework, Water Resour. Res., 49, 969–976, 2013.

Ma, Z. M., Kang, S. Z., Zhang, L., Tong, L., and Su, X. L.: Analysis of impacts of climate variability and human activity on streamflow for a river basin in arid region of northwest China, J. Hydrol., 352, 239–249, 2008.

Maurya, A. S., Shah, M., Deshpande, R. D., Bhardwaj, R. M., Prasad, A., and Gupta, S. K.: Hydrograph separation and precipitation source identification using stable water isotopes and conductivity: River Ganga at Himalayan foothills, Hydrol. Process., 25, 1521–1530, 2011.

McKenzie, N. J., Gallant, J., and Gregory, L.: Estimating water storage capacities in soil at catchment scales, CRC for Catchment Hydrology, CRC Publications, Monash University, Monash, Australia, 2003.

Mezentsev, V.: More on the computation of total evaporation (Yechio raz o rastchetie srednevo summarnovo ispareniia), Meteorog. i Gridrolog., 5, 24–26, 1955.

Milly, P. C. D. and Dunne, K. A.: Macroscale water fluxes 2. Water and energy supply control of their interannual variability, Water Resour. Res., 38, 241–249, 2002.

Nelson, E., Sander, H., Hawthorne, P., Conte, M., Ennaanay, D., Wolny, S., Manson, S., and Polasky, S.: Projecting global land-use change and its effect on ecosystem service provision and biodiversity with simple models, PloS One, 5, E14327, https://doi.org/10.1371/journal.pone.0014327, 2010.

Ning, T., Li, Z., and Liu, W.: Vegetation dynamics and climate seasonality jointly control the interannual catchment water balance in the Loess Plateau under the Budyko framework, Hydrol. Earth Syst. Sci., 21, 1515–1526, https://doi.org/10.5194/hess-21-1515-2017, 2017.

Ojha, C. S. P., Bhunya, P., and Berndtsson, R.: Engineering Hydrology, 1st Edn., Oxford University Press, Oxford, UK, 1–459, 2008.

Ol'Dekop, E. M.: On Evaporation from the Surface of River Basins, Univ. of Tartu, Tartu, Estonia, 1–208, 1911.

Pai, D. S., Sridhar, L., Rajeevan, M., Sreejith, O. P., Satbhai, N. S., and Mukhopadhyay, B.: Development of a new high spatial resolution (0.25 × 0.25) long period (1901–2010) daily gridded rainfall data set over India and its comparison with existing data sets over the region, Mausam, 65, 1–18, 2014.

Pathak, S., Ojha, C. S. P., Zevenbergen, C., and Garg, R. D.: Ranking of Storm Water Harvesting Sites Using Heuristic and Non-Heuristic Weighing Approaches, Water, 9, 710, 2017.

Pike, J. G.: The estimation of annual run-off from meteorological data in a tropical climate, J. Hydrol., 2, 116–123, 1964.

Porporato, A., Daly, E., and Rodriguez-Iturbe, I.: Soil water balance and ecosystem response to climate change, Am. Nat., 164, 625–632, 2004.

Redhead, J. W., Stratford, C., Sharps, K., Jones, L., Ziv, G., Clarke, D., Oliver, T. H., and Bullock, J. M.: Empirical validation of the InVEST water yield ecosystem service model at a national scale, Sci. Total Environ., 569, 1418–1426, 2016.

Rouholahnejad Freund, E. and Kirchner, J. W.: A Budyko framework for estimating how spatial heterogeneity and lateral moisture redistribution affect average evapotranspiration rates as seen from the atmosphere, Hydrol. Earth Syst. Sci., 21, 217–233, https://doi.org/10.5194/hess-21-217-2017, 2017.

Sánchez-Canales, M., Benito, A. L., Passuello, A., Terrado, M., Ziv, G., Acuña, V., Schuhmacher, M., and Elorza, F. J.: Sensitivity analysis of ecosystem service valuation in a Mediterranean watershed, Sci. Total Environ., 440, 140–153, 2012.

Schreiber, P.: Über die Beziehungen zwischen dem Niederschlag und der Wasserführung der Flüsse in Mitteleuropa, Z. Meteorol., 21, 441–452, 1904.

Srivastava, A. K., Rajeevan, M., and Kshirsagar, S. R.: Development of a high resolution daily gridded temperature data set (1969–2005) for the Indian region, Atmos. Sci. Lett., 10, 249–254, 2009.

Su, C. and Fu, B.: Evolution of ecosystem services in the Chinese Loess Plateau under climatic and land use changes, Global Planet. Change, 101, 119–128, 2013.

Tallis, H. T., Ricketts, T., Nelson, E., Ennaanay, D., Wolny, S., Olwero, N., Vigerstol, K., Pennington, D., Mendoza, G., Aukema, J., and Foster, J.: InVEST 1.004 beta User's Guide, The Natural Capital Project, Stanford University, USA, available at: https://naturalcapitalproject.stanford.edu/invest/ (last access: October 2018), 2010.

Terrado, M., Acuña, V., Ennaanay, D., Tallis, H., and Sabater, S.: Impact of climate extremes on hydrological ecosystem services in a heavily humanized Mediterranean basin, Ecol. Indic., 37, 199–209, 2014.

Turc, L.: Le bilan d'eau des sols: relations entre les précipitations, l'évaporation et l'écoulement, Annales Agronomiques A, 20, 491–595, 1954.

Wang, D. and Tang, Y.: A one-parameter Budyko model for water balance captures emergent behavior in darwinian hydrologic models, Geophys. Res. Lett., 41, 4569–4577, 2014.

Wang, X.-S. and Zhou, Y.: Shift of annual water balance in the Budyko space for catchments with groundwater-dependent evapotranspiration, Hydrol. Earth Syst. Sci., 20, 3673–3690, https://doi.org/10.5194/hess-20-3673-2016, 2016.

Xu, X., Liu, W., Scanlon, B. R., Zhang, L., and Pan, M.: Local and global factors controlling water-energy balances within the Budyko framework, Geophys. Res. Lett., 40, 6123–6129, 2013.

Yang, D., Sun, F., Liu, Z., Cong, Z., Ni, G., and Lei, Z.: Analyzing spatial and temporal variability of annual water-energy balance in nonhumid regions of China using the Budyko hypothesis, Water Resour. Res., 43, W04426, https://doi.org/10.1029/2006WR005224, 2007.

Yang, H., Yang, D., Lei, Z., and Sun, F.: New analytical derivation of the mean annual water-energy balance equation, Water Resour. Res., 44, W03410, https://doi.org/10.1029/2007WR006135, 2008.

Zhang, L., Dawes, W. R., and Walker, G. R.: Response of mean annual evapotranspiration to vegetation changes at catchment scale, Water Resour. Res., 37, 701–708, 2001.

Zhang, L., Hickel, K., Dawes, W. R., Chiew, F. H., Western, A. W., and Briggs, P. R.: A rational function approach for estimating mean annual evapotranspiration, Water Resour. Res., 40, W02502, https://doi.org/10.1029/2003WR002710, 2004.

Zhang, L., Potter, N., Hickel, K., Zhang, Y. Q., and Shao, Q. X.: Water balance modeling over variable time scales based on the Budyko framework – Model development and testing, J. Hydrol., 360, 117–131, 2008.

Zhou, G., Wei, X., Chen, X., Zhou, P., Liu, X., Xiao, Y., Sun, G., Scott, D. F., Zhou, S., Han, L., and Su, Y.: Global pattern for the effect of climate and land cover on water yield, Nat. Commun., 6, 5918, https://doi.org/10.1038/ncomms6918, 2015.

Zhou, S., Yu, B., Huang, Y., and Wang, G.: The complementary relationship and generation of the Budyko functions, Geophys. Res. Lett., 42, 1781–1790, 2015.

Zhou, X., Zhang, Y., Wang, Y., Zhang, H., Vaze, J., Zhang, L., Yang, Y., and Zhou, Y.: Benchmarking global land surface models against the observed mean annual runoff from 150 large basins, J. Hydrol., 470, 269–279, 2012.

8

A pore-size classification for peat bogs derived from unsaturated hydraulic properties

Tobias Karl David Weber[1,a], **Sascha Christian Iden**[1], **and Wolfgang Durner**[1]

[1]Soil Science and Soil Physics Division, Institute of Geoecology, Technische Universität Braunschweig, Langer Kamp 19c, 38106 Braunschweig, Germany
[a]now at: Institute of Soil Science and Land Evaluation, University of Hohenheim, 70593 Stuttgart, Germany

Correspondence to: Tobias Karl David Weber (tobias.weber@uni-hohenheim.de.de)

Abstract. In ombrotrophic peatlands, the moisture content of the acrotelm (vadoze zone) controls oxygen diffusion rates, redox state, and the turnover of organic matter. Thus, variably saturated flow processes determine whether peatlands act as sinks or sources of atmospheric carbon, and modelling these processes is crucial to assess effects of changed environmental conditions on the future development of these ecosystems. We show that the Richards equation can be used to accurately describe the moisture dynamics under evaporative conditions in variably saturated peat soil, encompassing the transition from the topmost living moss layer to the decomposed peat as part of the vadose zone. Soil hydraulic properties (SHP) were identified by inverse simulation of evaporation experiments on samples from the entire acrotelm. To obtain consistent descriptions of the observations, the traditional van Genuchten–Mualem model was extended to account for non-capillary water storage and flow. We found that the SHP of the uppermost moss layer reflect a pore-size distribution (PSD) that combines three distinct pore systems of the *Sphagnum* moss. For deeper samples, acrotelm pedogenesis changes the shape of the SHP due to the collapse of inter-plant pores and an infill with smaller particles. This leads to gradually more homogeneous and bi-modal PSDs with increasing depth, which in turn can serve as a proxy for increasing state of pedogenesis in peatlands. From this, we derive a nomenclature and size classification for the pore spaces of *Sphagnum* mosses and define inter-, intra-, and inner-plant pore spaces, with effective pore diameters of > 300, 300–30, and 30–10 μm, respectively.

1 Introduction

Sphagnum moss is a dominant and keystone species in temperate and boreal peat bog development (Kuhry and Vitt, 1996). As an ecosystem engineer, *Sphagnum* moss (Jones et al., 1994) induces vertical growth of peat bogs and may lead to a manifestation of the ombrotrophication process (Balyea, 2009; Rydin and Jeglum, 2016), and has led to considerable terrestrial carbon accumulation during the Holocene (Frolking and Roulet, 2007). The ability of *Sphagnum* spp. to photosynthesize critically depends on the water regime and is limited to the topmost centimetres of the bog profile where growth occurs (Clymo, 1973). Under field conditions, soil hydraulic properties (SHP) control the water regime in the topmost, growing part of the peat profile and are thus a critical factor for moss growth and survival (Hájek and Beckett, 2008).

The importance of capillary, film, and vapour flow processes for upward water fluxes in moss and peat has been emphasized by Hayward and Clymo (1982) and Price et al. (2009), while SHP accounting for these processes have only recently been identified by Weber et al. (2017a) for a limited number of samples.

In natural and undisturbed peat bog ecosystems there is a transition from living moss near the surface to heavily decomposed moss peat at greater depths (Clymo and Hayward, 1982; Limpens et al., 2008; Morris et al., 2011), with a concomitant change in SHP (e.g. Quinton et al., 2008; Price et al., 2008; McCarter and Price, 2012, 2014; Morris et al., 2015; Weber et al., 2017a). As *Sphagnum* decomposes, the fibrous material is biochemically broken up into smaller solid fragments (Rezanezhad et al., 2016). This change in physical structure reduces the pore sizes by a collapse of large pores and fine material filling the voids of the organic soil matrix. An increase in overburden leads to vertical shrinkage due to compression and a resulting increase in dry bulk density with depth (Clymo, 1978; Johnson et al., 1990; Price, 2003). The resulting strong decrease in saturated hydraulic conductivity, K_s ($\mathrm{L\,T^{-1}}$) (Boelter, 1969; Ingram, 1978; Quinton et al., 2008), is positively correlated with the state of decomposition as shown by Ingram (1978), Clymo (1970, 1984), Clymo and Hayward (1982), Hayward and Clymo (1982), and Morris et al. (2015). Exceptions to this rule have been observed in cases where pipe flow (Holden, 2005), fire disturbances (Sherwood et al., 2013), and rapid climate change resulting in changes in vegetation and subsequent peat deposition history (Rydin and Jeglum; 2016; Hedwall et al., 2017) occur. In summary, the afore-mentioned processes constitute the entire continuous pedogenesis shaping the soil profile as an ongoing process (Blume et al., 2016). It should be noted that this includes the actively growing *Sphagnum* mosses, as Weber et al. (2017a) define the actively growing and living part of the *Sphagnum* mosses as part of the vadose zone, i.e. as part of the soil profile.

In soil hydrology, the Richards equation is widely used as a process model for simulating temporally and spatially variable water contents and fluxes in soils. Recently, Weber et al. (2017a) demonstrated that the Richards equation is an adequate process model for unsaturated water flow under evaporative conditions even for the uppermost layer of a bog, where the living *Sphagnum* moss can be treated as part of the soil. The solution of the Richards equation requires the parameterization of SHP. For systems with depth-dependent SHP like peat bogs, one can assume either a homogeneous system and identify effective SHP, or parameterize the SHP as depth-dependent functions (Durner et al., 2008; Vereecken et al., 2007). In peatland hydrology, the saturated hydraulic conductivity K_s is the most frequently measured hydraulic parameter for different depths. However, the saturated conductivity is not necessarily a good predictor for the unsaturated hydraulic conductivity curve, since the contribution of macro-porosity to preferential flow can be substantial in the saturated case. While the correlation of the saturated hydraulic properties, i.e. K_s and porosity, with depth has been well established, corresponding changes in the SHP, namely the water retention characteristic (WRC) and particularly the unsaturated hydraulic conductivity curve (HCC) with depth, have neither been extensively investigated nor been treated in numerical simulations.

In ecohydrological modelling of peatland mires containing *Sphagnum*, unimodal expressions of the WRC still dominate (e.g. Dimitrov et al., 2010; Sulman et al., 2012; Wu and Blodau, 2013; Mezbahuddin et al., 2016). However, the investigation of the WRC of organic soils has revealed bimodal pore-size distributions which have been found for different but limited pressure head ranges (Rezanezhad et al., 2009, 2010, 2016; Quinton et al., 2008, 2009; Kettridge and Binley, 2011), and have so far rarely been expressed in terms of effective SHP over the full pressure head range.

Weber et al. (2017a) applied the highly sensitive method of inverse modelling evaporation experiments performed on *Sphagnum* moss and peat over a wide moisture range to parameterize SHP. The WRCs over the pressure head range from −1500 to 0 cm revealed a trimodal pore-size distribution (PSD) and clearly showed the necessity to account for adsorbed water in drying soil. Samples from two depths of the acrotelm were analysed and the identified SHP differed systematically and in a complex manner for the two depths. The HCC could not be described by simply scaling the saturated hydraulic conductivity, and showed the relevance of capillary and film components for water flow in organic soils. The trimodal nature of the PSD of the sample with living *Sphagnum* moss was linked to three conceptual pore spaces: pores between plant individuals (inter-plant pores), pores between branches and leaves of individual plants (intra-plant pores), and plant-internal water stored in hyaline cells (inner-plant pores). The deeper sample with decomposed moss showed a bimodal PSD with pores associated with the organic soil matrix and pores associated with the hyaline cells and its skeletal remnants (Fig. 1).

From this we state three hypotheses for this study. First, we expect a gradual shift in the characteristic pore sizes of the inter-plant and inner-plant pores with depth, due to increasing compaction and decomposition of the living moss material. Second, we expect a gradual disappearance of the distinct modalities (particularly the one that indicates the inter-plant pores) and an associated narrowing of the effective pore-size distribution with depth. Finally, we hypothesize that the variability of replicate samples from the same depth decreases with increasing depth, which is associated with the disappearance of the inter-plant pores. The last hypothesis is based on the assumption that samples become more homogeneous due to pedogenic processes such as biochemical degradation and compaction, and that the typical representative elementary volume for the different pore classes differs, being largest for the inter-plant pores.

Correspondingly, the research aims of this study are (i) to investigate in detail the gradual change in SHP with depth, encompassing the transition from living moss to peat, (ii) to analyse and quantify the spatial variability in SHP with depth, and (iii) to provide a unifying framework to distinguish between different pore systems in *Sphagnum* moss and peat which is based on the PSD and distinct pressure head ranges. To achieve these goals, a series of 28 laboratory evaporation experiments was conducted on undisturbed samples from the topmost 40 cm of a bog profile. The measured data were evaluated by inverse modelling using the Richards equation as a process model with the objective of identifying SHP for living moss and *Sphagnum*-derived peat. The results are presented in terms of average SHP for each depth, PSDs, and air-entry pressure heads of the individual pore systems.

Figure 1. *Sphagnum* moss structures and soil pore sizes. **(a)** *Sphagnum* lawn with visible bleaching due to desiccation of the capitula (in the German language *Sphagnum* is also referred to as "Bleichmoos", which translates as "bleaching moss"), **(a, b)** images of *Sphagnum cuspidatum H. Klinggr.* to exemplify the **(b)** sampled and slightly spread out individuals with visible inter-connectedness of branches, **(c)** close-up of the capitulum with pending branches and leaves, and **(d)** scanning electroscope microscopy image of *Sphagnum* leaves on a branch with visible dark circles as the opening to the hyaline cells (courtesy of Reuven Simhayov). Right panel: drawing of *Sphagnum magellanicum* (courtesy of Bent Johnsen. ©Department of Biology, University of Copenhagen) with (e) individual plant, (f) branch surface with skeletal structure in green, (g) branch with leaves, (h) cross section of a leaf with large hyaline cells, and (i) individual leaf.

2 Materials and methods

2.1 Sampling site and sample preparation

The samples were collected at an ombrotrophic peat bog, the Odersprungmoor, Harz Mountains, central Germany (UTM 32U 608000 mE 5737000 mN; 800 to 821 m a.s.l.). The Odersprungmoor formed on a saddle with an average downslope of 3 % in the SE–NW direction. In the SW–NE direction it is located in a gentle trough position (Fig. 2; Jensen, 1990). The long-term average temperature is 6.8 °C and annual precipitation is 1270 mm with a high inter-annual variance. The open bog is almost elliptical in shape and plan, with the longer axes oriented in the SE–NW direction, and has an area of 16.9 ha which is surrounded by spruce forest. The predominant bog vegetation consists of *Sphagnum magellanicum Brid.* and *Sphagnum rubellum Wilson* with interdispersed *Eriophorum angustifolium Honck.* Further extensive vegetation mapping was carried out by Baumann (2009). Figure 2 provides a detailed spatial reference and an overview of the peat types in the bog. The Odersprungmoor shows features of a poor fen in some small areas where it is slightly influenced by minerotrophic water, which only occurs on a small strip on the north-western flank (indicated by the arrows, Fig. 1, bottom panels). Most of the incoming water from the shallow soils formed on granite in the north-east is diverted past the bog along the northern rim of the bog towards the north-west (Border and Biester, 2015); thus, our sampling location is situated in the ombrotrophic part of the bog. Broder and Biester (2015) provided information on the geochemical composition of the substrate and pore waters which supports this. Weber et al. (2017b) give a detailed account of the carbon, water, and energy fluxes of the site during the vegetation period of 2013, a year with exceptionally dry summertime conditions in comparison to the 136-year long term record measured by the meteorological station of the German Weather Service (DWD) at nearby Braunlage.

In the acrotelm, a profile characterization with depth is possible as follows: the first 15 cm are composed of living moss, in 15–30 cm dead plant remnants dominate with some parts of the plants still visible, and the 30–40 cm samples contain the dark coloured decomposed moss peat. After the first 15 cm of living moss, a continuous increase in the state of decomposition is measurable (Broder and Biester, 2015). From the profile, 28 cylindrical samples of 250 cm^3 volume and 5 cm height were obtained from frozen large sampling blocks using a drill rig (Quinton et al., 2009). We will refer to the samples by the respective mid sampling depths, i.e. 2.5 ($n=5$), 7.5 ($n=6$), 12.5 ($n=3$), 17.5 ($n=3$), 22.5 ($n=3$), 32.5 ($n=4$), and 37.5 ($n=4$) cm (number of replicates in brackets). Freezing is considered to be a negligible source of error (Branham and Strack, 2014). A comprehensive description of sampling protocol, sample treatment, and methodology is provided by Weber et al. (2017a).

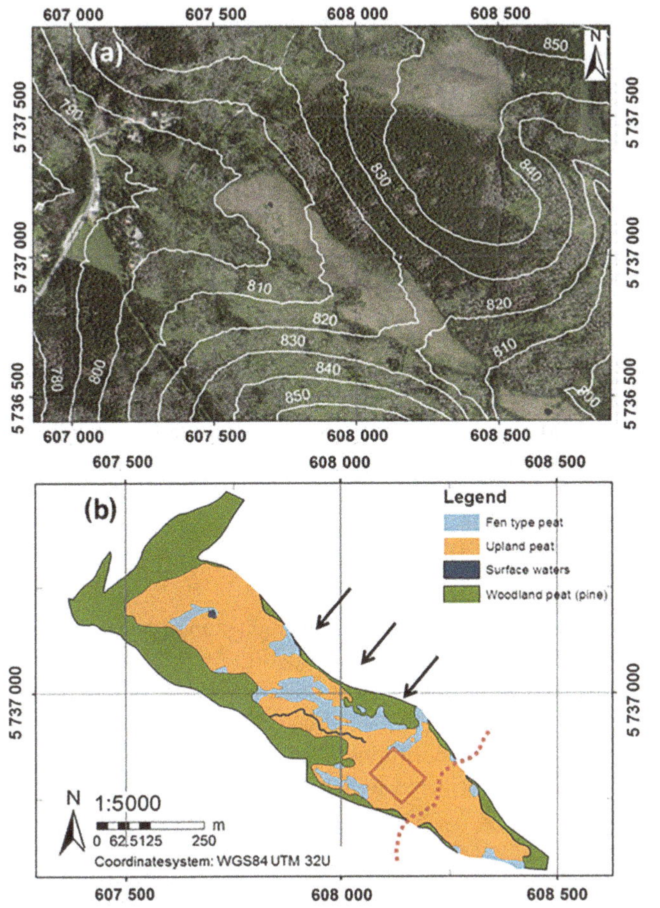

Figure 2. (a) Satellite image and elevation contours (m a.s.l.) of the Odersprungmoor; **(b)** vegetation mapping and characterization of the peatland. Black arrows indicate minerotrophic water influx into the peatland resulting in a higher proportion of fen type vegetation, the red dashed line is the approximate watershed divide, and the red box marks the sampling area.

2.2 Identification of soil hydraulic properties

Initially, the samples were saturated using de-aired and deionized water over a period of 48 h. The saturated hydraulic conductivity K_s (cm day^{-1}) of all samples was determined by the falling head method (KSAT device version 1.4; UMS GmbH, Munich, Germany). Subsequently, transient evaporation experiments were carried out (Wendroth et al., 1993; Schwärzel et al., 2006) on samples 5 cm i.h. and 8 cm i.d., starting with fully saturated samples that were exposed to free evaporation in the lab. Matric potentials were measured at two depths, and water fluxes were derived from weight changes with time. The measured data were evaluated by inverse modelling using the Hydrus-1D code to solve the Richards equation numerically (Šimůnek et al., 2016). Model parameters were identified by minimizing a weighted least-squares objective function (OF), which contained the time series of pressure heads at two depths (1.25 and 3.75 cm) and the sample's average water content. Details are given by Weber et al. (2017a). Six different parameterizations of the SHP which can be categorized into two main groups were tested. The first model group is based on the widely used van Genuchten–Mualem model (van Genuchten, 1980), which conceptualizes all pore water to be contained and conducted in capillaries. We used unimodal (VGM1), bimodal (VGM2), and trimodal (VGM3) van Genuchten models to parameterize the SHP. The uni- and multi-modal van Genuchten models are used widely in soil hydrology and the equations are therefore not repeated here in detail, but can be found e.g. in Priesack and Durner (2006). The second model group consists of the Peters–Durner–Iden (PDI) model of the SHP. The model is physically more comprehensive in that it ensures zero water content at oven dryness, and accounts for water flow in completely filled capillaries, partly filled capillaries, adsorbed water films, and isothermal vapour diffusion. We used the parameterization derived by Peters (2013), in the modified form published by Iden and Durner (2014), and Peters (2014). The PDI retention curve model has an identical number of free parameters as the VGM. For the conductivity curve, two additional parameters are required which express (i) the relative contribution of a non-capillary conductivity to the total unsaturated conductivity, and (ii) the steepness of the decrease in the non-capillary conductivity curve with increasing suction. Similar to the VGM models, the PDI models were applied in uni-, bi- and tri-modal form (PDI equations are given in the Appendix).

The data from the evaporation experiments on the 28 samples were evaluated by inverse modelling using all six models of the SHP (VGM1, VGM2, VGM3, PDI1, PDI2, PDI3). In the parameter optimization, the saturated conductivity parameter, K_s, was fixed to the measured value for each individual sample. This resulted in six sets of estimated model parameters, corresponding SHP, and model performance criteria per sample. Model selection was based on model ranking using the Akaike information criterion corrected for small sample size, AICc, as defined in Ye et al. (2008).

2.3 Calculation of mean soil hydraulic properties

After identifying the best parameterization of the SHP for each sample depth based on the AICc, the SHP of the replicates were averaged to obtain representative, mean SHP for each depth. Due to the non-linear dependence of the SHP on the model parameters, average SHP were not calculated by averaging parameters. The averaging was done by first generating 197 equidistant support points on the pF interval [−3, 6.8] (since the pressure head ranges across orders of magnitude in drying soil, we will use, for convenience, the pF unit in the remainder of this article. It is defined as the common logarithm of the negative pressure head h (cm), i.e. pF $= \log_{10}(-h)$ (–)). Subsequently, point values of the WRC and HCC were calculated at the support points for each replicate based on its respective best fitted model. The resulting

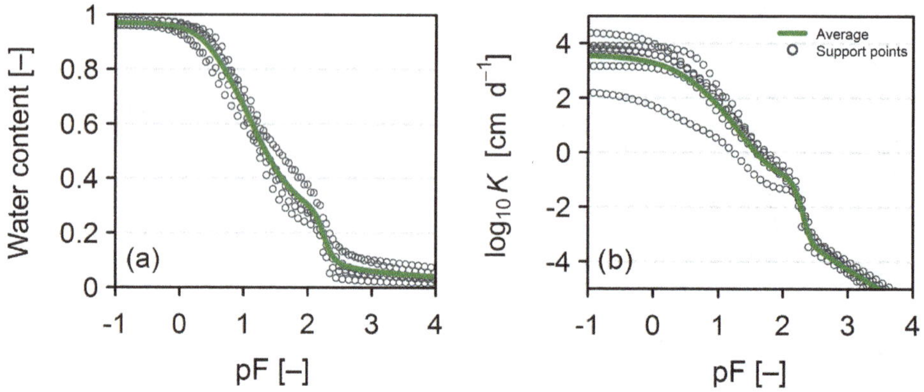

Figure 3. Example of the averaging process for **(a)** water retention curves and **(b)** hydraulic conductivity curves, shown for depth 7.5 cm (five replicates). Despite the variability of the individual curves, the key structural features are preserved.

datasets of the replicates were binned into one dataset and a new set of average PDI3 parameters was estimated by non-linear least-squares fitting. Fitting was done sequentially: first the average WRC model parameters were estimated and, in a second step, the retention parameters were kept constant and the parameters of the HCC were estimated using $\log_{10}$ conductivity data in the objective function. The sequential parameter estimation circumvents the need to weight the data groups of $\theta(h)$ and $\log_{10}K(h)$, whereas if measured WRC and HCC data are used to estimate SHP model parameters simultaneously, it involves a weighted multi-objective problem. An example of the averaging method is illustrated in Fig. 3. The iterative least-squares minimization was done with the differential evolution algorithm (Price et al., 2006) as implemented in the DEoptim R package (Mullen et al., 2011).

2.4 Pore-size distributions and classification of pore sizes

According to the Young–Laplace equation (Jury and Horton, 2004), the pressure head at which a capillary tube drains is inversely proportional to the equivalent effective diameter, d_{eff} (L), of that capillary which is calculated by

$$d_{\mathrm{eff}} = \frac{4\gamma \cos\beta}{|h|\rho_{\mathrm{w}} g} \tag{1}$$

where γ is the surface tension (M T^{-2}) of water at 20 °C, h is the pressure head (L), ρ_{w} is the density of water (M L^{-3}), g is the gravitational acceleration (L T^{-2}), and β (–) is the contact angle between water and the solids which we here set to 0° (Valat et al., 1991), i.e. complete wettability is assumed during monotonic drying (Goetz and Price, 2015).

The PSD function, $f(h)$, was defined as the change in capillary saturation with log pressure head (Durner, 1994):

$$f(h) = -\frac{\mathrm{d}S_{\mathrm{c}}}{\mathrm{d[pF]}} = -\log(10)|h|\frac{\mathrm{d}S_{\mathrm{c}}}{\mathrm{d}h}, \tag{2}$$

where S_{c} is the capillary saturation function defined by Eqs. (A2) and (A3) in the Appendix. In multimodal pore systems, this can be applied to the capillary saturation functions of each pore system $S_{\mathrm{c},i}$ (–), which define the individual pore systems that are superimposed in the effective PSD.

The analysis of the systematic changes in SHP with depth is based on Eqs. (1) and (2). We characterize the depth evolution of d_{eff} by the evolution of the median pore diameter, d_{50}, and the ratio of the 75th to 25th quantiles, d_{75}/d_{25}, of the pore-size distributions (Eq. 2). This is in analogy to characterizations of particle-size distributions.

A size classification of the individual pore domains was based on pressure heads at which air entry occurs into the different pore systems defining the multimodal PSD. In a first step, air-entry pressure heads, $h_{\mathrm{ae},i}$, were determined for each superimposed effective PSD of each depth. From the seven determined values each for the first, second, and third PSDs of the depth-averaged SHP, as described in Sect. 2.3, a profile average for each modality was computed to determine the air entry.

In the case of the van Genuchten saturation model used in the PDI models, the air-entry pressure is usually approximated as the reciprocal of parameter α (Assouline et al., 1998). However, the steepness of the WRC is controlled by the shape parameter n, and only for relatively large values of n is the air-entry pressure indeed approximately equal to α^{-1}. Therefore, we define the air-entry pressure of a pore system as the pressure head where $S_{\mathrm{c},i} = 0.95$. For the weighted saturation function of the van Genuchten model, the equation to calculate the approximate air-entry value pressure of the ith pore system is given by the equation

$$h_{\mathrm{ae},i} = \alpha_i^{-1} \cdot \left(0.95^{-1/m_i} - 1\right)^{1/n_i}. \tag{3}$$

This equation can equally be used for the PDI model. Equation (3) shows that the air-entry pressure depends on not only shape parameter α, but additionally on parameter n, which characterizes the width of the PSD.

As mentioned above, three pressure heads are calculated delimiting the three pore domains, as averages over all depths.

3 Results and discussion

3.1 Model performance

The statistical metrics for all 168 inverse simulations (28 samples times 6 models) are listed in Table S1 in the Supplement, characterizing model performance by the AICc enabling model comparisons. The values for the best performing model for each sample are summarized in Table 1. According to the AICc, the PDI3 model performed best for 24 out of the 28 samples. For the remaining four samples, the VGM3 model performed best in three cases and the PDI2 in one. Thus, the PDI3 leads to the best model prediction in the majority of replicates regardless of sampling depths. We recall that this is achieved by estimating one additional conductivity curve parameter in the PDI compared to the VGM models with the same modality. This finding challenges the de facto standard usage of the VGM1 and VGM2 models for parameterizing the SHP of *Sphagnum* moss and peat and extends the theoretical considerations, results, and conclusions in Weber et al. (2017a) to multiple depths in the bog profile. It furthermore corroborates the fact that even model parameters of trimodal functions can be uniquely identified. Note that the use of an incorrect model, despite a lower number of parameters, often leads to very high parameter uncertainties like in Dettman and Bechtold (2016). As a consequence, the averaging of the SHP among the replicates was based on the PDI3 model for all depths.

3.2 Soil hydraulic properties and pore-size distributions of all samples

Figure 4 presents the WRCs (left panels), HCCs (middle panels), and PSDs (right panels) of all measured samples, grouped for the seven sampling depths (denoted A to G, from the top to bottom panels). A comparison of the SHP at the different depths reveals three main trends. First, the variability between the replicates decreases with increasing depth, i.e. the SHP of the replicates become more similar. Second, for all depths a comparable modality of relatively small pores is discernible, which starts to drain at about pF = 2 (−100 cm). Superimposed onto this pore system, there is a second and third modality indicating larger pore systems for the near-surface samples, with the third modality reflecting a pore domain close to water saturation. With increasing depth the variability in the two larger pore domains decreases considerably. Third, the HCCs for all depths appear very similar in the more unsaturated moisture range (pF > 2), but diverge towards saturation. One possible reason for the scatter for pF < 2 is the insensitivity of evaporation experiments to changes in the $K(h)$ function in wet soil. This insensitivity occurs if the evaporation rate is of the order of $K(h)$ or smaller, leading to small gradients in pressure head which cannot be resolved by tensiometers (Peters and Durner, 2008). A more reliable determination of the HCC close to water saturation could be obtained by the tension disk method (Klute and Dirckson, 1986; McCarter et al., 2017), multi-step outflow (Durner and Iden, 2011), or multi-step flux measurements (Weller et al., 2011). However, these methods have their own challenges when applied to organic soils, and the latter two have, to the best of our knowledge, not been applied to obtain SHP for organic media so far, although Qi et al. (2011) used multi-step outflow experiments to determine the WRC.

3.3 Change in pore-system characteristics with depth

To better identify the pore-system changes with depth, we compare mean SHP for all depths in Fig. 5. The trends with depth which have been observed for the individual samples now become much clearer. The seven mean WRCs (Fig. 5, left panel) show an increased water holding capacity with increasing depth across a wide pressure head range, from saturation up to pF = 3.2 (a pressure head of −1500 cm). This finding is supported by literature data for conditions near water saturation (Price et al., 2008; Price and Whittington, 2010; McCarter and Price, 2012; Goetz and Price; 2015). Concurrently, the HCCs show a pronounced decrease in the saturated conductivity and systematically higher unsaturated conductivities with increasing depth (Fig. 5, right panel), with a crossing point of the conductivity functions at around pF = 1.

From an ecophysiological point of view, it is interesting to analyse the relationship between SHP and the resilience against desiccation. In fact, all samples have a high water capacity around pF = 2. During meteorological conditions favouring prolonged periods of drying, the decrease in pressure head is slowed down by the release of this water, and low pressure heads leading to desiccation and cessation of photosynthesis (Hájek and Beckett, 2008) are reached only slowly. Commensurate with the WRC, the slope of the HCC around pF = 2 decreases as water drains from hyaline cells and their skeleton material, and hydraulic conductivity remains at values of approximately 10^{-2} cm day^{-1} until approx. pF = 2.5, where it drops at all depths rapidly below 10^{-4} cm day^{-1}. The relatively high hydraulic conductivity in the pressure head range until pF = 2.5 ensures an upward flow of water which contributes to the effective desiccation tolerance of the vegetation under field conditions.

We further observe that the WRCs shift with depth from a trimodal pore-size distribution to a bimodal one with a distinct expression of the two remaining pore systems. This is nicely recognizable from the plots of the PSDs in Fig. 6. To calculate meaningful pore-size parameters, we separated the

Table 1. Statistical evaluation results of the inverse parameter estimation for 31 samples of eight mid sampling depths.

Sample name	Mid sampling depth (−)	OF value[1] (cm)	$RMSE_w^2$ 0.75 ($\times 10^4$)	$RMSE_w^3$ 0.25 (cm)	AIC_c^4 (−) (cm)	ΔAIC_c^5 (−)	Best model[6] (−)
A.1	2.5	0.82	5.1	2.4	2348	417	PDI3
A.2	2.5	1.60	7.1	2.7	2734	795	PDI3
A.3	2.5	0.58	6.5	2.1	2688	394	PDI3
A.4	2.5	1.03	8.2	5.1	3518	321	PDI3
A.5	2.5	0.78	6.8	4.5	3051	432	PDI3
B.1	7.5	2.03	3.8	2.9	1766	806	PDI3
B.2	7.5	1.51	4.9	2.3	2156	954	PDI3
B.3	7.5	1.05	6.6	3.0	2600	413	PDI3
B.4	7.5	0.91	7.4	4.2	2595	782	PDI3
B.5	7.5	0.61	8.4	3.9	3692	350	PDI3
B.6	7.5	1.41	4.3	3.7	3408	1566	VGM3
C.1	12.5	1.17	4.6	2.3	1856	642	PDI3
C.2	12.5	1.29	8.6	8.0	5562	27	PDI2
C.3	12.5	1.12	5.0	4.6	3594	797	PDI3
D.1	17.5	0.95	2.6	4.9	2362	185	PDI3
D.2	17.5	2.31	3.5	3.8	3209	1727	PDI3
D.3	17.5	2.06	1.9	2.4	1727	2788	PDI3
E.1	22.5	1.81	2.2	2.3	1724	1182	PDI3
E.2	22.5	2.45	4.1	2.1	2418	957	PDI3
E.3	22.5	3.99	2.6	3.7	2684	1654	PDI3
F.1	32.5	3.69	3.0	2.9	1966	958	VGM3
F.2	32.5	2.52	2.1	3.9	2347	186	PDI3
F.3	32.5	1.26	4.3	1.5	2102	447	PDI3
F.4	32.5	0.95	4.8	3.5	2166	94	VGM3
G.1	37.5	2.79	3.8	4.3	2640	653	PDI3
G.2	37.5	1.90	3.0	3.2	2162	470	PDI3
G.3	37.5	0.54	2.6	2.7	1724	1424	PDI3
G.4	37.5	0.38	4.6	2.1	2445	156	PDI3

[1] Objective function value (Eq. 20 in Weber et al., 2017a). [2] Root mean square weighted error of the matric potential measurements at a height of 3.75 cm. The median $RMSE_w$ value over all depths is 3.0 cm. [3] Root mean square weighted error of the matric potential measurements at a height of 1.25 cm. The median $RMSE_w$ value over all depths is 4.3 cm. [4] Corrected Akaike information criterion. [5] Difference in AICc between the best and second best models. [6] Trimodal van Genuchten–Mualem model (VGM3) and trimodal PDI variant (PDI3).

fine pore system, characterized by the capillary saturation function with the parameters w_3, a_3, and n_3, from the system of larger pores, defined by the superposition of the two saturation functions with parameters w_2, a_2, and n_2, and w_1, a_1, and n_1, and calculated the median diameters of the two pore systems for each depth as the characteristic pore size. These characteristic pore sizes are indicated in Fig. 6 by the dashed vertical lines. From visual inspection, and confirmed by the values of the fitted parameters of the third pore subsystem (Table 2), it becomes obvious that the fine pore system remains virtually unchanged with depth, encompassing about 15 to 20 % of the total pore space, $w_3(\theta_s - \theta_r)$. This pore system is characterized by a well-defined maximum of its pore-size distribution at pF = 2.3, which does not change with depth (Fig. 6). Also, the width of this fine pore system shows no change with depth. In contrast, the median equivalent diameter, d_{50} (L), from the superimposed first and second modalities drifts gradually from larger to finer pores (Fig. 6).

To further carve out the depth dependence of key characteristics of the pore systems, we show the depth dependence of saturated conductivity and two characteristic pore-size parameters in Fig. 7. The near log-linear decrease in measured saturated hydraulic conductivity K_s (Fig. 7a) is in agreement with established trends in these ecosystems (Morris et al., 2015). The shift in the median pore size of the larger

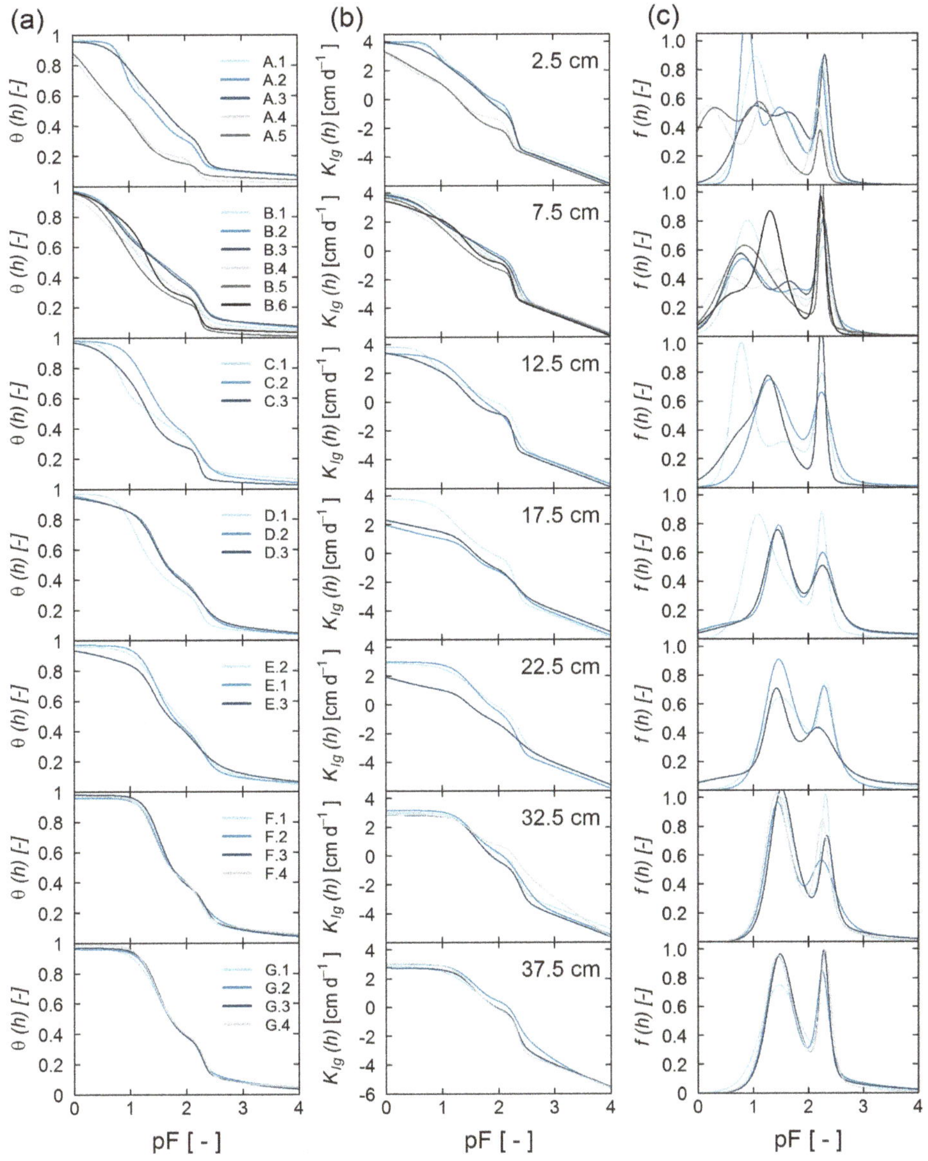

Figure 4. Left panels: water retention curves; centre panels: hydraulic conductivity curves; and right panels: pore-size distributions (of all 28 samples, K_{lg} refers to the common logarithm of the conductivity value).

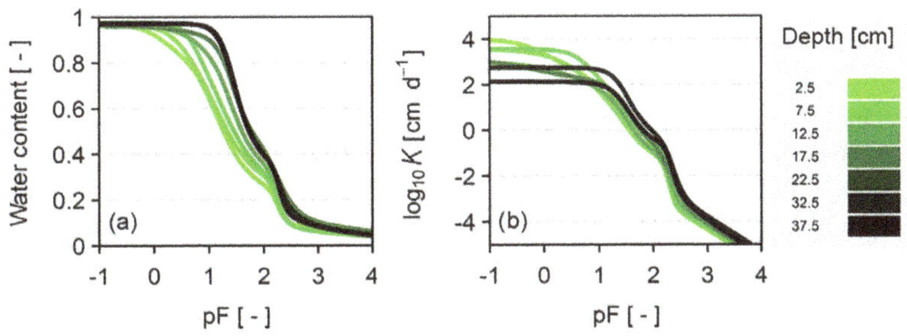

Figure 5. Mean **(a)** retention and **(b)** conductivity curves for the depths 2.5 cm to 37.5 cm in 5 cm increments.

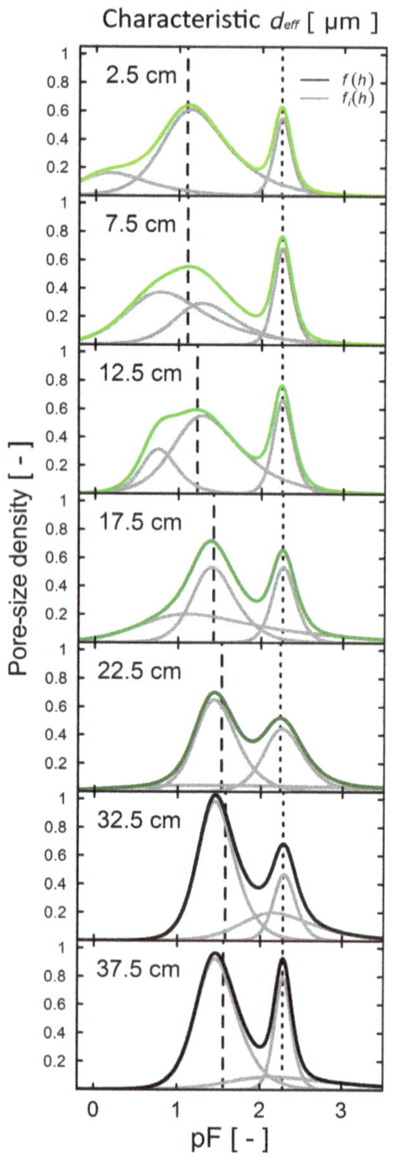

Figure 6. Continuous pore-size distributions calculated from Eq. (4) per depth showing the overlap of the three pore spaces. The dashed lines represent the median (d_{50}) of the first and second pore domains, the increase of which positively correlates with increasing depth. The dotted lines mark the mode of the third modality which remains constant regardless of depth.

pore system indicates the strongest transition between 10 and 25 cm (Fig. 7b). The width of the associated pore system, expressed by the ratio of the first and third quartiles of the PSD, d_{75}/d_{25} (–), shows a distinct and continuous narrowing with depth (Fig. 7c).

As a side note, we would like to comment on an interesting aspect of using the different parameterizations of the HCC. We found that the tortuosity–connectivity parameter τ in this study is positive for each individual sample parameterization with the PDI3 (Table S1). The same is true for the calculated mean functions at all depths (Table 2). A positive value of τ is in agreement with its conceptual meaning (Mualem, 1976; Peters et al., 2011). This contrasts with reports from the literature (Price et al., 2008; Price and Whittington, 2010) and our findings with van Genuchten type functions, where τ is often negative (Table S1). Obviously, the use of a structurally more correct model of the SHP that accounts for partially filled capillaries prevents parameter τ from becoming negative. Conversely, if an inadequate model of the SHP is applied, τ has a tendency to become increasingly small and even negative because it induces the HCC to be less steep in the medium to dry range. Although this effect may be welcome to describe measurements, it is obtained by a physically implausible value of the shape parameter (Hoffman-Riem et al., 1999).

Another important feature of the identified values of τ is their relatively small variation with depth.

The gradual change in PSDs with depth summarized by Fig. 6 can be explained by the origin of the samples and the pedogenic processes they have undergone, i.e. biochemical degradation breaking up fibres into smaller pieces (e.g. Rezanezhad et al., 2016), compaction (Price and Schlotzhauer, 1999), the shrinkage and swelling of bog soils (Schlotzhauer and Price, 1999; Price et al., 2003; Price, 2003; Glaser et al., 2004), and freeze–thaw processes (Meiers et al., 2013). These findings are corroborated by a positive relationship between depth and advance in decomposition at this site (Broder and Biester, 2015). Obviously, these processes lead to a homogenization of the organic material, a disappearance of the largest pores, and thus to a smaller spatial variability of the SHP at greater depths. While in the upper 15 cm the living moss reveals a large natural phenotypic variability over orders of centimetres, the dead plant remnants at 15–30 cm have already been subject to some compaction and the inter-plant pores are filled with smaller particles. The most decomposed samples from 30 to 40 cm, which lie beneath the water table most of the time, are the most homogeneous and have completely lost the distinct trimodality in the PSD, which was evident for the samples from near the surface. The deepest samples are also the most homogeneous in colour and solid particle size. It is worthwhile noting that the PSDs are basically bimodal, but this does not contradict the superiority of using a trimodal parameterization (see Sect. 3.1 and Table 1) as indicated by the AICc. The superposition of PSDs can improve the overall description, even if they do not represent two distinct pore systems with different characteristic pore sizes. This was also discussed for the bimodal van Genuchten model by Zurmühl and Durner (1998) and for organic soils by Weber et al. (2017a). Without contradiction of the presented model, we point out that preferential flow may be an important driver for saturated flow on the very large scale.

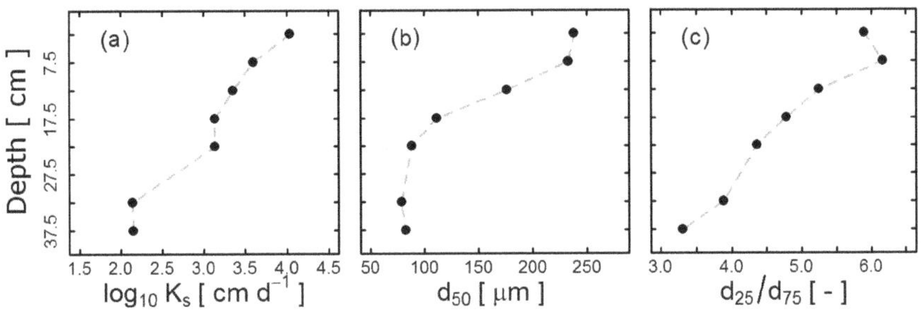

Figure 7. **(a)** Trends of key pore-system characteristics with depth: **(a)** near-exponential decrease in K_s with depth, **(b)** decrease in characteristic pore diameter with depth, expressed as the median pore diameter d_{50} of the second modality, and **(c)** reduction of the width of the pore-size distribution with depth, expressed by d_{75}/d_{25}.

Table 2. PDI3-model parameter* values of the depth-averaged effective soil hydraulic properties.

Depth (cm)	θ_r (–)	θ_s (–)	α_1 (cm^{-1})	n_1 (–)	w_1 (–)	α_2 (cm-^{1})	n_2 (–)	w_2 (–)	α_3 (cm^{-1})	n_3 (–)	lgK_s (cm day^{-1})	τ (–)	lgω (–)
2.5	0.12	0.97	0.96	2.0	0.18	0.1	2.1	0.66	0.006	6.4	4.0	0.6	−5.9
7.5	0.09	0.97	0.26	1.8	0.50	0.07	2.1	0.30	0.006	6.2	3.6	0.6	−4.2
12.5	0.10	0.98	0.19	3.3	0.20	0.07	2.0	0.60	0.006	7.5	3.4	1	−4.2
17.5	0.06	0.97	0.18	1.4	0.42	0.05	2.9	0.37	0.006	4.9	3.1	1.0	−5.9
22.5	0.04	0.96	0.44	1.1	0.24	0.04	2.9	0.46	0.006	3.1	2.8	1.7	−6.0
32.5	0.09	0.97	0.04	3.4	0.55	0.02	1.7	0.25	0.005	5.3	2.1	0.2	−5.6
37.5	0.06	0.97	0.04	3.0	0.61	0.02	1.5	0.17	0.005	6.6	2.1	0.3	−5.7

* θ_r: residual water content, θ_s: saturated water content, α_i and n_i: shape parameters of the three superimposed capillary bundles, respectively, w_i: weighting coefficient of the capillary bundles, K_s: saturated hydraulic conductivity, ω: non-capillary conductivity, τ: tortuosity parameter. The common logarithm to the base 10 is denoted by lg. The parameters of all individual samples can be found in the Supplement.

4 Proposal of a pore-size classification for *Sphagnum* moss and peat

The results of our study suggest a pore-size classification that is based on the three different pore domains characterized by the air-entry pressures (Eq. 3, Table 3). This classification relates the PSD to different physical components of moss and peat and hereby also directly links the SHP to pedogenesis. The resulting averaged air-entry pressures for the first, second, and third pore domains are given as pF 0, 1, and 2, respectively, and in the following we give a detailed description of the pore water domains and the residual water (Fig. 8, Table 4). I–III are based on capillary water storage and IV is based on non-capillary (adsorptive) water storage.

The *first domain* (*I*) extends from saturation to a pressure head of approx. −10 cm (pF = 1), which is marked by the mean $\overline{h}_{ae,2}$, corresponding to effective pore diameters of > 300 µm. At this point the HCCs of different depths converge in the sense that for pF > 1 the scatter is significantly reduced and the individual curves run in parallel with a range of approximately 1 order of magnitude (Fig. 5). This pressure head corresponds to a state where the highly heterogeneous *inter-plant pores* (living and slightly decayed *Sphagnum*) have drained (Fig. 1).

Table 3. Air-entry pressures of the three pore systems of the effective SHP for all sampling depths (left column) and arithmetic means to delimit the individual pore domains.

Depth (cm)	$h_{ae,1}$ (cm)	$h_{ae,2}$ (cm)	$h_{ae,3}$ (cm)
2.5	−0.3	−3.4	−108
7.5	−1.2	−4.9	−107
12.5	−2.4	−4.7	−104
17.5	−1.7	−9.3	−96
22.5	−1.8	−10.5	−72
32.5	−11.8*	−31.8	−117
37.5	−10.8*	−20.2	−132
Mean	−1.5	−8.0	−105
SD	−0.7	−3.2	−17.1

* As previously established, the depths 32.5 and 37.5 cm show no macro porosity and are excluded from calculating the mean and standard deviation for the average $\overline{h}_{ae,1}$ value. The mean $\overline{h}_{ae,2}$ value is calculated from the $h_{ae,2}$ values of the sampling depths 2.5–27.5 cm, and the two $h_{ae,1}$ values at depths 32.5 and 37.5 cm. The mean $\overline{h}_{ae,3}$ value is calculated based on all depths.

Table 4. Overview of the proposed pore domain nomenclature with corresponding pressure head delimitations and effective pore diameters.

Name of pore domain	Pore domain number	Effective pore diameter (μm)	Pressure head (cm)	pF (−)
Inter-plant pores	I	> 300	> −10	1
Intra-plant and outer-plant matrix[1] pores	II	300 to 30	−10 to −100	1–2
Inner-plant and plant matrix[1] pores	III	30 to 10	−100 to −300	2–2.5
non-capillary, cytoplasmic and apoplastic water	IV	n/a[2]	< −1000	> pF 3

[1] The plant matrix refers to the skeleton of the *Sphagnum*, which in its living form contains intact hyaline cells and the decomposition of cell wall results in the remaining, more persistent, plant skeleton (Fig. 1). [2] It is not appropriate to calculate an effective radius for water which is not contained in capillaries.

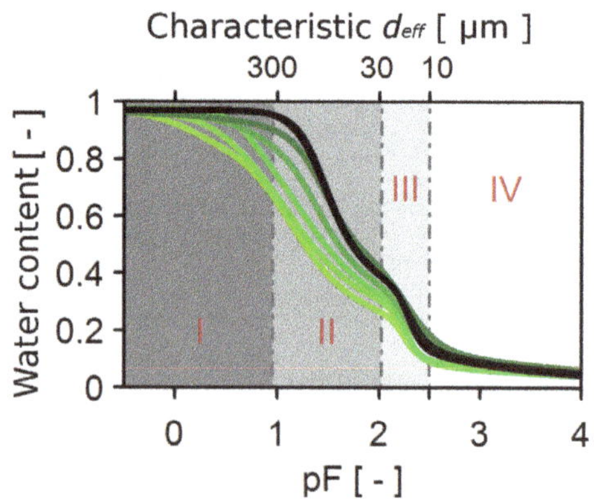

Figure 8. Four pore domains represented in the soil water retention curve. The Roman letters I–III denote different capillary water domains for which a detailed explanation is given in the text and values are presented in Table 4. The region denoted by IV is non-capillary (residual) water.

The *second domain* (*II*) extends from pF = 1 to pF = 2, corresponding to effective pore diameters between 300 and 30 μm. It comprises the *intra-plant pore space* in the sampling depths 2.5 to 27.5 cm where macroscopic living and undecomposed moss fragments exist. At the remaining sampling depths decomposition is advanced, but the skeletal structures containing the hyaline cells remain (Fig. 1). Correspondingly the water in the second domain is stored here in the outer skeletal spaces of the peat (Rezanezhad et al., 2016).

Finally, the *third pore domain* (*II*) is represented by intact and decayed hyaline cells with effective pore radii of 30 to 10 μm. As shown in Fig. 1, the hyaline cells can still have intact cell membranes, but appear to decay more rapidly than the structures bearing them. In fact, the distances between two structural elements (skeleton) making up the hyaline cells is, effectively, marginally larger than the hyaline cell opening.

Thus, in the absence of macropores, two size classes can be differentiated in decomposed *Sphagnum* peat; the *outer-plant matrix* pores > 30 μm, and the *inner-plant matrix* pores with effective pore radii 30 to 10 μm (Table 4). The above-mentioned hyaline cells and their skeletal structures are drained by pF = 2.5, which is the pressure head where the *fourth domain* (*IV*) starts. At these pressures of pF > 2.5, the water conductance can no longer be described by capillary flow theory.

In our analysis we assume that shrinkage does not affect our key findings. Shrinkage was observed to be around 5–8 % on samples from the same depths and same bog, as referenced to the initial volume. Since the delimitations of our pore-size classification span orders of magnitude in pressure head and related effective pore radii, we believe that shrinkage will not have a considerable influence on the derived soil hydraulic properties for the small scale.

In line with these definitions, a pressure head delimitation of pore water into an active porosity (*inter- and intra-plant and inter-plant matrix pore space*) and an inactive porosity (*inner-plant and inner-plant matrix*) at a pressure head of $h = -100$ cm is suggested. Finally, a word of caution with respect to generalizing our findings: Michaelis (2011) describes 286 species of *Sphagnum*, occurring globally; thus, our results might not be applicable to all species, since the phenology of decaying *Sphagnum* might be different between species.

5 Conclusions

Based on measured data from transient evaporation experiments and an evaluation by inverse modelling using the Richards equation as a process model, we identified SHP for 28 samples from an acrotelm of an ombrotrophic, *Sphagnum* moss dominated bog. To parameterize the SHP, we compared six different models. The successful description of the observed data was possible by (a) considering three modalities representing the underlying PSD and (b) including the contribution of non-capillary water in the SHP models. Combined, this results in an encompassing parameterization across the full moisture range. The continuous information from fully parameterized models, available from the presented work, greatly extends the level of detail in comparison to previous works.

We averaged the SHP for each of the seven sampled depths and found, compared to insights from biological studies, that these mean hydraulic properties reflect the physiological nature of the pore-size domains in the living moss carpets of peat bogs. Larger heterogeneities in identified SHP between depth replicates could be shown for samples from the living moss; however, this effect reduced considerably with increasing depth and a concomitant reduction of PSDs from tri- to bi-modal. Both effects, the homogenization and reduction in modality, can be explained by pedogenic processes.

A unifying nomenclature to describe and report results from research on the vadose zone of ombrotrophic peatlands is proposed. With this, the pore spaces in *Sphagnum* moss and peat can be classified according to size ranges of the effective pore diameters. These size ranges are derived from an analysis of the shape of SHP and underlying PSDs from the uppermost 40 cm of the acrotelm. In living *Sphagnum* moss, the three pore spaces are thus proposed to be referred to as *inter-*, *intra-*, and *inner-plant pore spaces*, with effective pore diameters of > 300, 300 to 30, and 30 to 10 µm, respectively. For deeper samples, the pedogenesis of the acrotelm has had a considerable impact on the shape of the SHP. First, the collapse of the inter-plant pores and their filling with smaller particles led to more homogeneous and gradually bimodal PSDs with increasing depth. Much of the hyaline cell structure and skeleton remains less affected by compaction and is resilient to decomposition. These remaining larger pores with effective pore diameters of > 300 µm and fine pores of 30 to 10 µm correspond more to the plant matrix than the fully visible plant and were thus described as *outer-plant matrix* pores and the *inner-plant matrix*, respectively.

Following these delimitations and with depth-continuous differentiation between the larger pores and inner-plant pores, we conclude that, based on this soil hydrological perspective, the water storage domain suggests a partitioning into active and inactive porosities when considering solute transport and is delimited at a pressure head of $h = -100$ cm. The proposed domain classification was inferred from a soil hydrological modelling approach and, with the discussion of the physiological meaning, we associate them with measurements and reflections in the seminal biological studies by Hayward and Clymo (1982), Lewis (1988), and van Breemen (1995).

Future research should include testing the Richards equation under various boundary conditions to verify that the identified SHP can be used for predictive purposes. Additionally, the research should be conducted to establish whether proxies such as bulk density, C / N ratios, and/or vegetation structure can be used to predict SHP accurately. Lastly, scaling approaches are expected to help to describe in numerical simulations the gradual and continuous change in SHP with depth.

Author contributions. TW designed the study and experiments and carried them out. SI developed the model code and jointly performed the simulations with TW. TW prepared the manuscript with contributions from all co-authors. WD oversaw the formal analysis, acquired the funds, and provided all the necessary resources for this research.

Competing interests. The authors declare that they have no conflict of interest.

Acknowledgements. We wish to express our gratitude to the Niedersächsische Technische Hochschule for grant TD 2.1.4 Top-Down-Project GeoFluxes and for the research scholarship awarded to Tobias Karl David Weber by the German Academic Exchange Service. Thanks to Benedikt Scharnagl and Efstathios Diamantopolous for fruitful discussions, to H.-U. Kison of the Harz National Park for granting site access, as well as to our esteemed fellow geoecologists Daniela Reineke, Lennart Rolfes, and Matthias Spieckerman for their diligent work in the field and lab. For the images we wish to thank Reuven Simhayov for providing and Christina Smeaton for creating the scanning electron microscope image, and to Bent Johnsen for the drawing of *Sphagnum magellanicum* in Fig. 1 (©Department of Biology, University of Copenhagen). We additionally greatly thank Andy Baird and Donald Siegel for their constructive reviews and discussion which helped us to improve the manuscript, and Brian Berkowitz for the editorial work.

Edited by: Brian Berkowitz

References

Assouline, S., Tessier, D., and Bruand, A.: A conceptual model of the soil water retention curve, Water Resour. Res., 34, 223–231, 1998.

Balyea, L. R.: Nonlinear Dynamics of Peatlands and Potential Feedbacks on the Climate System, in: Carbon Cycling in Northern Peatlands, Geophysical Monograph 184, Geophysical Monograph Series, edited by: Baird, A. J., Belyea, L. R., Comas, X., Reeve, A. S., and Slater, L. D., American Geophysical Union, Washington, D.C., 2009.

Baumann, K.: Entwicklung der Moorvegetation im Nationalpark Harz, in: Schriftenreihe aus dem Nationalpark Harz, Band 4, Nationalpark Harz, Wernigerode, 2009.

Blume, H.-P., Brümmer, G. W., Fleige, H., Horn, R., Kandeler, E., Kögel-Knabner, I., Kretzschmar, R., Stahr, K., and Wilke, B.-M.: Scheffer/Schachtschabel Soil Science, Springer, Berlin, Heidelberg, 618 pp., https://doi.org/10.1007/978-3-642-30942-7, 2016.

Boelter, D. H.: Physical Properties of Peat as related to degree of decomposition, Soil Sci Soc Am Proc, 33, 606-609, 1969.

Branham, J. E. and Strack, M.: Saturated hydraulic conductivity in Sphagnum-dominated peatlands: do microforms matter?, Hydrol. Process., 28, 4161–4387, https://doi.org/10.1002/hyp.10228 2014.

Broder, T. and Biester, H.: Hydrologic controls on DOC, As and Pb export from a polluted peatland – the importance of heavy rain events, antecedent moisture conditions and hydrological connectivity, Biogeosciences, 12, 4651–4664, https://doi.org/10.5194/bg-12-4651-2015, 2015.

Clymo, R. S.: The growth of Sphagnum: methods of measurement, J. Ecol., 58, 13–40, 1970.

Clymo, R. S.: The growth of Sphagnum: some effects of environment, J. Ecol., 61, 849–869, 1973.

Clymo, R. S.: A model of peat-bog growth. Production Ecology of some British Moors and Montane Grasslands, edited by: Heal, O. W., Perkins, D. F., and Brown, W. M., Springer, Berlin, 187–223, 1978.

Clymo, R. S.: The Limits to Peat Bog Growth, Philos. T. Roy. Soc. B, 303, 605–654, 1984.

Clymo, R. S. and Hayward, P. M.: Ecology of Sphagnum, in: Bryophyte Ecology, Chapman and Hall, London, UK, 229–289, 1982.

Dettmann, U. and Bechtold, M.: Deriving effective soil water retention characteristics from shallow water table fluctuations in peatlands, Vadose Zone J., 15, 1–13, https://doi.org/10.2136/vzj2016.04.0029, 2016.

Dimitrov, D. D., Grant, R. F., Lafleur, P. M., and Humphreys, E. R.: Modeling the effects of hydrology on ecosystem respiration at Mer Bleue bog, J. Geophys. Res., 115, G04043, https://doi.org/10.1029/2010JG001312, 2010.

Durner, W.: Hydraulic conductivity estimation for soils with heterogeneous pore structure, Water Resour. Res., 30, 211–223, 1994.

Durner, W. and Iden, S. C.: Extended multistep outflow method for the accurate determination of soil hydraulic properties near water saturation, Water Resour. Res., 47, W08526, https://doi.org/10.1029/2011WR010632, 2011.

Durner, W., Jansen, U., and Iden, S. C.: Effective hydraulic properties of layered soils at the lysimeter scale determined by inverse modelling, Eur. J. Soil Sci., 59, 114–124, https://doi.org/10.1111/j.1365-2389.2007.00972.x, 2008.

Frolking, S. and Roulet, N. T.: Holocene radiative forcing impact of northern peatland carbon accumulation and methane emissions, Global Change Biol., 13, 1079–1088, 2007.

Glaser, P. H., Chasar, L. I., Chanton, J. P., Morin, P., Rosenberry, D. O., Siegel, D. I., Ruud, O., and Reeve, A. S.: Surface deformations as indicators of deep ebullition fluxes in a large northern peatland, Global Biogeochem. Cy., 18, GB1003, https://doi.org/10.1029/2003GB002069, 2004.

Goetz, J. and Price, J. S.: Role of morphological structure and layering of Sphagnum and Tomenthypnum mosses on moss productivity and evaporation rates, Can. J. Soil Sci., 95, 109–124, https://doi.org/10.4141/CJSS-2014-092, 2015.

Hájek, T. and Beckett, R. P.: Effect of Water Content Components on Desiccation and Recovery in Sphagnum Mosses, Ann. Bot., 101, 165–173, https://doi.org/10.1093/aob/mcm287, 2008.

Hayward, P. M. and Clymo, R. S.: Profiles of Water Content and Pore Size in Sphagnum and Peat, and their Relation to Peat Bog Ecology, P. Roy. Soc. B, 215, 299–325, https://doi.org/10.1098/rspb.1982.0044, 1982.

Hedwall, P.-O., Brunet, J., and Rydin, H.: Peatland plant communities under global change: negative feedback loops counteract shifts in species composition, Ecology, 98, 150–161, https://doi.org/10.1002/ecy.1627, 2017.

Hoffmann-Riem, H., van Genuchten, M. T., and Flühler, H.: A general model of the hydraulic conductivity of unsaturated soils, in: Proceedings of International Workshop, Characterization and Measurements of the Hydraulic Properties of Unsaturated Porous Media, edited by: van Genuchten, M. T., Leij, F. J., and Wu, L., University of California, Riverside, CA, 31–42, 1999.

Holden, J.: Piping and woody plants in peatlands: Cause or effect?, Water Resour. Res., 41, W06009, https://doi.org/10.1029/2004WR003909, 2005.

Iden, S. C. and Durner, W.: Comment on "Simple consistent models for water retention and hydraulic conductivity in the complete moisture range" by A. Peters, Water Resour. Res., 50, 1–5, https://doi.org/10.1002/2014WR015937, 2014.

Ingram, H. A. P.: Soil layers in mires: function and terminology, J. Soil Sci., 29, 224–227, 1978.

Jensen, U.: Die Moore des Hochharzes. Spezieller Teil, Niedersächsisches Landesverwaltungsamt, Hannover, Germany, 1990.

Johnson, L. C., Damman, A. W. H., and Malmer, N.: Sphagnum macrostructure as an indicator of decay and compaction in peat cores from an ombrotrophic south Swedish peat-bog, J. Ecol., 78, 633–647, 1990.

Jones, C. G., Lawton, J. H., and Shachak, M.: Organisms as ecosystem engineers, Oikos, 69, 373–386, 1994.

Jury, W. and Horton, R.: Soil Physics, 6th Edition John Wiley & Sons, New York, p. 390, 2004.

Kettridge, N. and Binley, A.: Characterization of peat structure using X-ray computed tomography and its control on ebullition of biogenic gas bubbles, J. Geophys. Res., 116, G01024, https://doi.org/10.1029/2010JG001478, 2011.

Klute, A. and Dirksen, C.: Hydraulic conductivity and diffusivity: laboratory methods, in: Methods of Soil Analysis, Part 1. Physical and Mineralogical Methods, edited by: Klute, A., American Society of Agronomy, Inc., Madison, WI, 687–734, 1986.

Kuhry, P. and Vitt, D.: Fossil carbon/nitrogen ratios as a measure of peat decomposition, Ecology, 77, 271–275, 1996.

Lewis, A. M.: A test of the air-seeding hypothesis using Sphagnum hyalocysts, Plant Physiol., 87, 577–582, https://doi.org/10.1104/87.3.577, 1988.

Limpens, J., Berendse, F., Blodau, C., Canadell, J. G., Freeman, C., Holden, J., Roulet, N., Rydin, H., and Schaepman-Strub, G.: Peatlands and the carbon cycle: from local processes to global implications – a synthesis, Biogeosciences, 5, 1475–1491, https://doi.org/10.5194/bg-5-1475-2008, 2008.

McCarter, C. P. R. and Price, J. S.: Ecohydrology of Sphagnum moss hummocks: mechanisms of capitula water supply and simulated effects of evaporation, Ecohydrology, 7, 33–44, https://doi.org/10.1002/eco.1313, 2012.

McCarter, C. P. R. and Price, J. S.: The hydrology of the Bois-des-Bel peatland restoration: hydrophysical properties limiting connectivity between regenerated Sphagnum and remnant vacuum harvested peat deposit, Ecohydrology, 8, 173–187, https://doi.org/10.1002/eco.1498, 2014.

McCarter, C. P. R., Ketcheson, S., Weber, T. K. D., Whittington, P., Scarlett, S., and Price, J. S.: A modified technique for measuring unsaturated hydraulic conductivity in Sphagnum moss and peat, Soil Sci. Soc. Am. J., 81, 747–757, https://doi.org/10.2136/sssaj2017.01.0006, 2017.

Meiers, G. P., Barbour, S. L., Qualizza, C. V., and Dobchuk, B. S.: Evolution of the Hydraulic Conductivity of Reclamation Covers over Sodic/Saline Mining Overburden, J. Geotech. Geoenviron. Eng., 137, 968–976, https://doi.org/10.1061/(ASCE)GT.1943-5606.0000523, 2013.

Mezbahuddin, M., Grant, R. F., and Flanagan, L. B.: Modeling hydrological controls on variations in peat water content, water table depth, and surface energy exchange of a boreal western Canadian fen peatland, J. Geophys. Res.-Biogeo., 121, 2216–2242, https://doi.org/10.1002/2016JG003501, 2016.

Michaelis, D.: Die Sphagnum-Arten der Welt, Bibliotheca Botanica, 160, p. 408, 2011.

Morris, P. J., Baird, A. J., and Belyea, L. R.: Ecohydrological feedbacks in peatland development: a theoretical modelling study, J. Ecol., 99, 1190–1201, https://doi.org/10.1111/j.1365-2745.2011.01842.x, 2011.

Morris, P. J., Baird, A. J., and Belyea, L. R.: Bridging the gap between models and measurements of peat hydraulic conductivity, Water Resour. Res., 51, 5353–5364, https://doi.org/10.1002/2015WR017264, 2015.

Mualem, Y.: New model for redicting hydraulic conductivity of unsaturated porous, Water Resour. Res., 12, 513–522, 1976.

Mullen, K., Ardia, D., Gil, D., Windover, D., and Cline, J.: 'DEoptim': An R Package for Global Optimization by Differential Evolution, J. Stat. Softw., 40, 1–26, 2011.

Peters, A.: Simple consistent models for water retention and hydraulic conductivity in the complete moisture range, Water Resour. Res., 49, 6765–6780, https://doi.org/10.1002/wrcr.20548, 2013.

Peters, A.: Reply to comment by S. Iden and W. Durner on "Simple consistent models for water retention and hydraulic conductivity in the complete moisture range", Water Resour. Res., 50, 7535–7539, https://doi.org/10.1002/2014WR016107, 2014.

Peters, A. and Durner, W.: Simplified evaporation method for determining soil hydraulic properties, J. Hydrol., 356, 147–162, https://doi.org/10.1016/j.jhydrol.2008.04.016, 2008.

Peters, A., Durner, W., and Wessolek, G.: Consistent parameter constraints for soil hydraulic functions, Adv. Water Resour., 34, 1352–1365, https://doi.org/10.1016/j.advwatres.2011.07.006, 2011.

Price, J. S.: Role and character of seasonal peat soil deformation on the hydrology of undisturbed and cutover peatlands, Water Resour. Res., 39, 1241, https://doi.org/10.1029/2002WR001302, 2003.

Price, J. S. and Schlotzhauer, S. M.: Importance of shrinkage and compression in determining water storage changes in peat: the case of a mined peatland, Hydrol. Process., 13, 2591–2601, 1999.

Price, J. S. and Whittington, P. N.: Water flow in Sphagnum hummocks: Mesocosm 836 measurements and modelling, J. Hydrol., 381, 333–340, 2010.

Price, J. S., Heathwaite, A. L., and Baird, A. J.: Hydrological processes in abandoned and restored peatlands: An overview of management approaches, Wetlands Ecol. Manage., 11, 65–83, 2003.

Price, J. S., Whittington, P. N., Elrick, D. E., Strack, M., Brunet, N., and Faux, E.: A method to determine unsaturated undecomposed Sphagnum moss, Soil Sci. Soc. Am. J., 72, 487–491, https://doi.org/10.2136/sssaj2007.0111N, 2008.

Price, J. S., Edwards, T. W. D., Yi, Y., and Whittington, P. N.: Physical and isotopic characterization of evaporation from Sphagnum moss, J. Hydrol., 369, 175–182, 2009.

Price, K. V., Storn, R. M., and Lampinen, J. A.: Differential Evolution – A Practical Approach to Global Optimization, Springer-Verlag, Berlin, Heidelberg, 2006.

Priesack, E. and Durner, W.: Closed-form expression for the multimodal unsaturated conductivity function, Vadose Zone J., 5, 121–124, https://doi.org/10.2136/vzj2005.0066, 2006.

Qi, G., Michel, J.-C., Boivin, P., and Charpentier, S.: A Laboratory Device for Continual Measurement of Water Retention and Shrink/Swell Properties during Drying/Wetting Cycles, Hortscience, 46, 1298–1302, 2011.

Quinton, W. L., Hayashi, M., and Carey, S. K.: Peat hydraulic conductivity in cold regions and its relation to pore size and geometry, 22, 2829–2837, https://doi.org/10.1002/hyp.7027, 2008.

Quinton, W. L., Elliot, T. R., Price, J. S., Rezanezhad, F., and Heck, R. J.: Measuring physical and hydraulic properties of peat from X-ray tomography, Geoderma, 153, 269–277, 2009.

Rezanezhad, F., Quinton, W. L., Price, J. S., Elrick, D., Elliot, T. R., and Heck, R. J.: Examining the effect of pore size distribution and shape on flow through unsaturated peat using computed tomography, Hydrol. Earth Syst. Sci., 13, 1993–2002, https://doi.org/10.5194/hess-13-1993-2009, 2009.

Rezanezhad, F., Quinton, W. L., Price, J. S., Elliot, T. R., Elrick, D., and Shook, K. R.: Influence of pore size and geometry on peat unsaturated hydraulic conductivity computed from 3D computed tomography image analysis, Hydrol. Process., 24, 2983–2994, 2010.

Rezanezhad, F., Price, J. S., Quinton, W. L., Lennartz, B., Milojevic, T., and Van Cappellen, P.: Structure of peat soils and implications for water storage, flow and solute transport: A review update for geochemists, Chem. Geol., 429, 75–84, 2016.

Rydin, H. and Jeglum, J. K.: The Biology of Peatlands, 2nd Edn., OUP, Oxford, p. 400, 2016.

Saito, H., Šimůnek, J., and Mohanty, B. P.: Numerical analysis of coupled water, vapor, and heat transport in the Vadose Zone, Vadose Zone J., 5, 784–800, https://doi.org/10.2136/vzj2006.0007, 2006.

Schlotzhauer, S. M. and Price, J. S.: Soil water flow dynamics in a managed cutover peat field, Quebec: Field and laboratory investigations, Water Resour. Res., 35, 3675–3683, 1999.

Schwärzel, K., Šimůnek, J., Stoffregen, H., Wessolek, G., and van Genuchten, M. T.: Estimation of the unsaturated hydraulic conductivity of peat soils: Laboratory versus field data, Vadose Zone J., 5, 628–640, 2006.

Sherwood, J. H., Kettridge, N., Thompson, D. K., Morris, P. J., Silins, U., and Waddington, J. M.: Effect of drainage and wildfire on peat hydrophysical properties, Hydrol. Process., 27, 1866–1874, https://doi.org/10.1002/hyp.9820, 2013.

Šimůnek, J., van Genuchten, M. T., and Miroslav, Š.: Recent Developments and Applications of the HYDRUS Computer Software Packages, Vadose Zone J., 15, https://doi.org/10.2136/vzj2016.04.0033, 2016.

Sulman, B. N., Desai, A. R., Schroeder, N. M., Ricciuto, D., Barr, A., Richardson, A. D., Flanagan, L. B., Lafleur, P. M., Tian, H., Chen, G., Grant, R. F., Poulter, B., Verbeeck, H., Ciais, P., Ringeval, B., Baker, I. T., Schaefer, K., Luo, Y., and Weng, E.: Impact of hydrological variations on modeling of peatland CO_2 fluxes: Results from the North American Carbon Program site synthesis, J. Geophys. Res., 117, G01031, https://doi.org/10.1029/2011JG001862, 2012.

Tokunaga, T. K.: Hydraulic properties of adsorbed water films in unsaturated porous media, Water Resour. Res., 45, W06415, https://doi.org/10.1029/2009WR007734, 2009.

Valat, B., Jouany, C., and Rivere, L. M.: Characterization of the wetting and properties of air-dried peats and composts, Soil Sci., 152, 100–107, 1991.

van Breemen, N.: How Sphagnum bogs down other plants, Tree, 10, 270–275, 1995.

van Genuchten, M. T.: Closed-form equation for predicting the hydraulic conductivity of unsaturated soils, Soil Sci. Soc. Am. J., 44, 892–898, 1980.

Vereecken, H., Kasteel, R., Vanderborght, J., and Harter, J.: Upscaling Hydraulic Properties and Soil Water Flow Processes in Heterogeneous Soils: A Review, Vadose Zone J., 6, 1–28, https://doi.org/10.2136/vzj2006.0055, 2007.

Weber, T. K. D., Iden, S. C., and Durner, W.: Unsaturated hydraulic properties of Sphagnum moss and peat reveal trimodal pore-size distributions, Water Resour. Res., 53, 415–434, https://doi.org/10.1002/2016WR019707, 2017a.

Weber, T. K. D., Gerling, L., Reineke, S., Weber, S., Durner, W., and Iden, S. C.: Robust inverse modeling of net ecosystem exchange in a mountainous peatland: influence of distributional assumptions on estimated parameters and total carbon fluxes, J. Adv. Model. Earth Syst., in review, 2017b.

Weller, U., Ippisch, O., Köhne, M., and Vogel, H.-J.: Direct measurement of unsaturated conductivity including hydraulic nonequilibrium and hysteresis, Vadose Zone J., 10, 654–661, https://doi.org/10.2136/vzj2010.0074, 2011.

Wendroth, O., Ehlers, W., Hopmans, J. W., Kage, H., Halbertsma, J., and Wösten, J. H. M.: Reevaluation of the Evaporation Method for Determining Hydraulic Functions in Unsaturated Soils, Soil Sci. Soc. Am. J., 57, 1436–1443, https://doi.org/10.2136/sssaj1993.03615995005700060007x, 1993.

Wu, Y. and Blodau, C.: PEATBOG: a biogeochemical model for analyzing coupled carbon and nitrogen dynamics in northern peatlands, Geosci. Model Dev., 6, 1173–1207, https://doi.org/10.5194/gmd-6-1173-2013, 2013.

Ye, M., Meyer, P. D., and Neuman, S. P.: On model selection criteria in multimodel analysis, Water Resour. Res., 44, W03428, https://doi.org/10.1029/2008WR006803, 2008.

Zurmühl, T. and Durner, W.: Determination of Parameters for Bimodal Hydraulic Functions by Inverse Modelling, Soil Sci. Soc. Am. J., 62, 874–880, https://doi.org/10.2136/sssaj1998.03615995006200040004x, 1998.

Appendix A: PDI model equations

The WRC of the PDI model family (Peters, 2013, 2014; Iden and Durner, 2014) is given by the equation

$$\theta(h) = (\theta_s - \theta_r)\, S_c(h) + \theta_r S_{nc}(h) \tag{A1}$$

where θ (–) is volumetric water content, h (cm) is pressure head, and $S_c(h)$ (–) and $S_{nc}(h)$ (–) are the saturation functions for capillary and non-capillary water, respectively. $S_c(h)$ (–) is defined as

$$S_c(h) = \frac{\Gamma(h) - \Gamma_0}{1 - \Gamma_0} \tag{A2}$$

where $\Gamma_0 = \Gamma(h_0)$, h_0 (cm) is the pressure head at oven dryness, and $\Gamma(h)$ is the multimodal van Genuchten saturation function (Durner, 1994)

$$\Gamma(h) = \sum_{i=1}^{k} \Gamma_i(h) = \sum_{i=1}^{k} w_i \left[1 + (-\alpha_i h)^{n_i}\right]^{-m_i} \tag{A3}$$

with shape parameters α_i (cm^{-1}) and n_i, the constraint $m_i = 1 - 1/n_i$, the number of modes or pore systems, k (–), and weights w_i which sum to unity. The non-capillary saturation function is given as

$$S_{nc}(h) = 1 + \frac{1}{x_a + x_0}\left\{x - x_a + \ln\left[1 + \exp\left(\frac{x_a - x}{b}\right)\right]\right\} \tag{A4}$$

where $x = \log_{10}(-h)$, $x_a = \log_{10}(-h_a)$, $x_0 = \log_{10}(-h_0)$, and $h_a = 1/\alpha_1$. Parameter b is a smoothing parameter for which empirical equations are given in Iden and Durner (2014). The HCC is defined as

$$K(h) = (1-\omega) K_s K_{rc}(S_c) + \omega K_s K_{rnc}(S_{nc}) + K_v(h) \tag{A5}$$

where ω (–) is the relative proportion of water flow in films and corners, $K_{\mathrm{rc}}(h)$ (–) is relative hydraulic conductivity caused by flow in completely filled capillaries, $K_{\mathrm{rnc}}(h)$ (–) is relative hydraulic conductivity caused by flow in films and corners (incompletely filled capillaries), and $K_{\mathrm{v}}(h)$ is the effective hydraulic conductivity for isothermal vapour diffusion. The equation for $K_{\mathrm{v}}(h)$ can be found in Saito et al. (2006) and the closed-form equations for the two liquid conductivities are (Peters, 2013, 2014)

$$K_{\mathrm{rc}}\left(S_{\mathrm{c}}\right)=S_{\mathrm{c}}^{\tau}\left[1-\left(\frac{\sum_{i=1}^{k} w_i\alpha_i\left(1-\Gamma_i^{1/m_i}\right)^{m_i}}{\sum_{i=1}^{k} w_i\alpha_i\left(1-\Gamma_{i,0}^{1/m_i}\right)^{m_i}}\right)^{m}\right]^2 \quad \text{(A6)}$$

and

$$K_{\mathrm{rnc}}\left(S_{\mathrm{nc}}\right)=\left(\frac{h_0}{h_a}\right)^{a\left(1-S_{\mathrm{nc}}\right)} \quad \text{(A7)}$$

where $\Gamma_{i,0}=\Gamma_i(h_0)$ and $a=1.5$ is an empirical parameter for film flow (Tokunaga, 2009).

9

Modelling the water balance of Lake Victoria (East Africa)

Inne Vanderkelen[1], **Nicole P. M. van Lipzig**[2], **and Wim Thiery**[1,3]

[1]Department of Hydrology and Hydraulic Engineering, Vrije Universiteit Brussel, Brussels, Belgium
[2]Department of Earth and Environmental Sciences, KU Leuven, Leuven, Belgium
[3]Institute for Atmospheric and Climate Science, ETH Zurich, Zurich, Switzerland

Correspondence: Inne Vanderkelen (inne.vanderkelen@vub.be)

Abstract. Lake Victoria, the second largest freshwater lake in the world, is one of the major sources of the Nile river. The outlet to the Nile is controlled by two hydropower dams of which the allowed discharge is dictated by the Agreed Curve, an equation relating outflow to lake level. Some regional climate models project a decrease in precipitation and an increase in evaporation over Lake Victoria, with potential important implications for its water balance and resulting level. Yet, little is known about the potential consequences of climate change for the water balance of Lake Victoria. In this second part of a two-paper series, we feed a new water balance model for Lake Victoria presented in the first part with climate simulations available through the COordinated Regional Climate Downscaling Experiment (CORDEX) Africa framework. Our results reveal that most regional climate models are not capable of giving a realistic representation of the water balance of Lake Victoria and therefore require bias correction. For two emission scenarios (RCPs 4.5 and 8.5), the decrease in precipitation over the lake and an increase in evaporation are compensated by an increase in basin precipitation leading to more inflow. The future lake level projections show that the dam management scenario and not the emission scenario is the main controlling factor of the future water level evolution. Moreover, inter-model uncertainties are larger than emission scenario uncertainties. The comparison of four idealized future management scenarios pursuing certain policy objectives (electricity generation, navigation reliability and environmental conservation) uncovers that the only sustainable management scenario is mimicking natural lake level fluctuations by regulating outflow according to the Agreed Curve. The associated outflow encompasses, however, ranges from 14 m^3 day^{-1} (−85 %) to 200 m^3 day^{-1} (+100 %) within this ensemble, highlighting that future hydropower generation and downstream water availability may strongly change in the next decades even if dam management adheres to he Agreed Curve. Our results overall underline that managing the dam according to the Agreed Curve is a key prerequisite for sustainable future lake levels, but that under this management scenario, climate change might potentially induce profound changes in lake level and outflow volume.

1 Introduction

Lake Victoria is directly sustaining 30 million people living in its basin and 200 000 fishermen operating from its shores. Therefore, the water level fluctuations of the lake are of major importance. Declining water levels may affect the local communities by their ability to access water, to fish and to transport goods (Semazzi, 2011). Further downstream, the livelihood of about 300 million people in the Nile Basin is supported by the natural resources of Lake Victoria, as it is one of the two major sources of the Nile (Semazzi, 2011). Originating from Lake Victoria, the White Nile provides a more constant flow during the year, providing 70 % to 90 % of the total Nile discharge during the dry season in the Ethiopian highlands, where the second major source is located (Di Baldassarre et al., 2011). A drop in the water level could imply a decreasing outflow, which may have major implications downstream in the Nile Basin. The countries of the Nile Basin require sufficient water resources for their future development and wellbeing, considering the population growth and economic development (Deconinck, 2009; Taye et al., 2011). Consequently, there are a lot of tensions between countries in the Nile Basin. The outflow of the lake is controlled by the Nalubaale and Kiira dams for hydropower generation, located in Jinja (Fig. 1). Human strategies towards regulating the outflow might therefore play a crucial role in the downstream Nile Basin water resources and associated political tensions. This will be even more relevant in the light of climate change, where water will become an even more important resource, both around the lake and downstream in the Nile Basin (Taye et al., 2011). While the released outflow affects the lake level, climate-driven lake level fluctuations also influence the outflow volume released by the dam. The hydropower potential, and thus energy availability in the region, strongly depends on the interplay between outflow and lake levels. Information on the future evolution of the levels and outflow volumes of Lake Victoria is thus vital for future generations living along its shores.

The water levels of Lake Victoria are determined by the water balance (WB) of the lake, consisting of precipitation on the lake, lake evaporation, inflow by tributary rivers and dam outflow. Since the construction of the dam complex in 1954, a rating curve called the "Agreed Curve" was established relating the lake level and outflow in natural conditions (Sene, 2000):

$$Q_{\text{out}} = 66.3(L - 7.96)^{2.01}. \quad (1)$$

In this equation, the dam outflow Q_{out} ($\text{m}^3\,\text{day}^{-1}$) is calculated based on the lake level, L (m), as directly measured at the dam. The Agreed Curve can be used to calculate outflow volumes based on lake levels which lie in the historically observed range going from 10 to 13.5 m measured at the dam (Sutcliffe and Parks, 1999).

The climate in East Africa experiences large interannual variability in precipitation (Nicholson, 2017). The region is a hotspot for climate change, as it is very likely that climate change will have a major influence on precipitation (Nicholson, 2017; Kent et al., 2015; Otieno and Anyah, 2013; Souverijns et al., 2016). Precipitation over the Lake Victoria Basin (LVB) experiences a seasonal cycle with two main rainfall seasons: the long rains in March, April and May and the short rains in September, October and November (Williams et al., 2015). In the last decades, the long rain seasons in East Africa have experienced a series of dry anomalies (Lyon and Dewitt, 2012; Rowell et al., 2015; Souverijns et al., 2016; Nicholson, 2016, 2017; Thiery et al., 2016), while there was no trend observed for the short rains due to a large year-to-year variability (Rowell et al., 2015). This drying trend of the long rains is in contrast with climate model projections for the upcoming decades, projecting an increase in precipitation over East Africa (Otieno and Anyah, 2013; Kent et al., 2015). This apparent contradiction has been called the East African climate paradox of which the causes remain unclear (Rowell et al., 2015). To find explanations, (Rowell et al., 2015) stated that more research is needed on the reliability of climate projections over East Africa, for the attribution of changing anthropogenic aerosol emissions and for the role of natural variability in recent droughts. However, (Philip et al., 2017) found that the severe drought in northern and central Ethiopia in 2015 is attributable to natural variability and therefore conclude that there is no paradox for this type of event.

Future climate simulations with regional climate models (RCMs) project a decreasing mean precipitation and an increasing evaporation over Lake Victoria (Thiery et al., 2016; Souverijns et al., 2016). Compared to global climate models (GCMs), RCMs have a high spatial resolution and are therefore able to represent regional- and local-scale forcings (Kim et al., 2014; Giorgi et al., 2009). In East Africa, accounting for local variations in topography, vegetation, lakes, soils and coastlines is of major importance, as these variations have a significant effect on the regional climate. Over the LVB, models with sufficiently high resolution are needed to reproduce key circulation features such as the lake–land breeze system (Williams et al., 2015). High-resolution ($\sim$7 km grid spacing) coupled lake–land–atmosphere climate simulations for the African Great Lakes region with the Consortium for Small-Scale Modelling in climate mode (COSMO-CLM^2) RCM were therefore performed by (Thiery et al., 2015). These simulations outperform state-of-the-art regional climate simulations for Africa conducted at $\sim$ 50 km grid spacing, because of the coupling to land surface and lake models and enhanced model resolution, which allows for a better representation of the fine-scale circulation and precipitation patterns (Thiery et al., 2015, 2016).

At the moment, almost no research is dedicated to the potential consequences of climate change for the WB of Lake Victoria. This is remarkable, since the evolution of the future levels of Lake Victoria is vital information for the future generations living on its coasts. Tate et al. (2004) used simulations with one fully coupled GCM to model future fluctuations of the lake level. Model disparities were, however, very high over East Africa and the GCM was not capable to sufficiently model precipitation over the African Great Lakes (Tate et al., 2004). Therefore, the results from Tate et al. (2004) serve as an illustration of the sensitivity of lake levels and outflows to climate change scenarios, rather than as an actual lake level projection. Recently, high-resolution ensemble climate projections became available for Africa through the COordinated Regional Climate Downscaling Experiment (CORDEX; Nikulin et al., 2012). Operating at 0.44° ($\sim$ 50 km) horizontal resolution, these models attempt to resolve the lake and its mesoscale circulation. When these simulations are used as input for a water balance model (WBM) for Lake Victoria, future lake level projections can be generated. The ensemble approach ensures that the model spread incorporates uncertainties associated

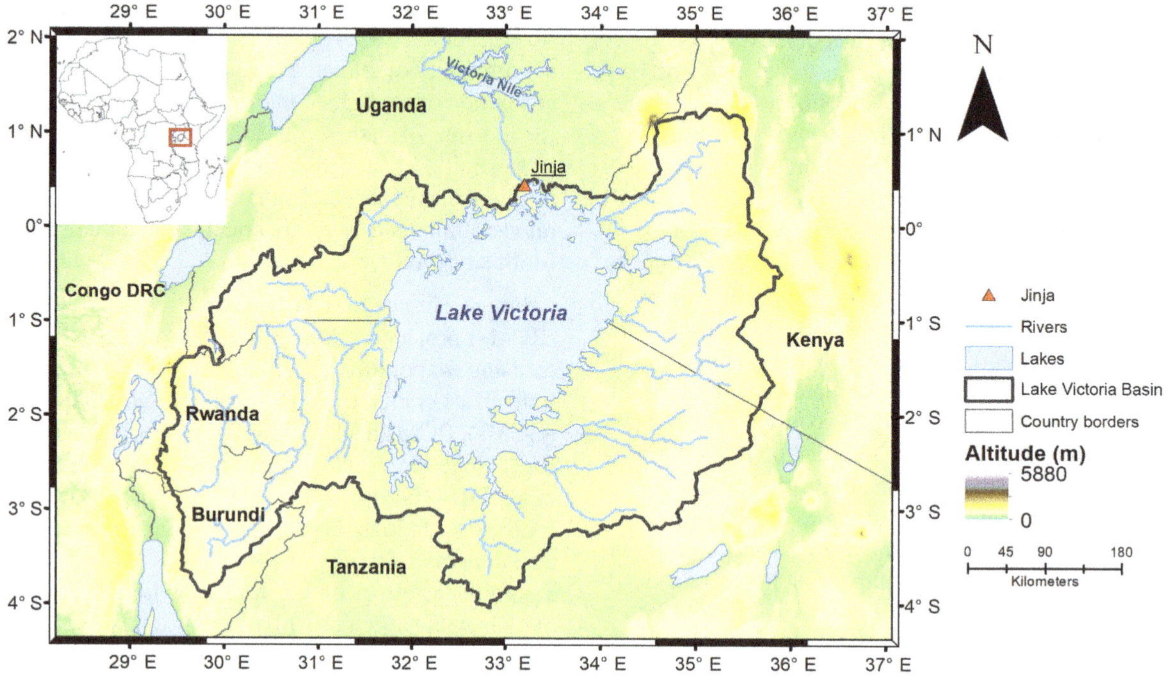

Figure 1. Map of Lake Victoria and its basin with surface heights from the Shuttle Radar Topography Mission (SRTM). Taken from Vanderkelen et al. (2018).

with individual model deficiencies, concentration pathways and natural variability.

Here, we use the WBM constructed in the first part of this two-paper series (Vanderkelen et al., 2018) and drive it with climate simulations from the CORDEX over the African domain from 1950 to 2100. First, we assess the ability of RCMs from the CORDEX ensemble to reproduce the historical lake level of Lake Victoria. Based on this analysis it appears that it is not possible to use a subset of models for which the WB closes. Therefore, a bias correction based on observations is applied on these simulations. Last, the future evolution of the water level of Lake Victoria under various climate change scenarios is investigated, together with the role of different human management strategies at the outflow dams based on three policy objectives (electricity generation, navigation reliability and environmental conservation).

2 Data and methodology

2.1 CORDEX ensemble

In recent years, RCM downscaling of the Coupled Model Intercomparison Project Phase 5 (CMIP5) GCMs have become available through the CORDEX framework. The CORDEX-Africa project provides simulations over the African domain, which includes the whole African continent with a spatial resolution of 0.44° by 0.44° and a daily output frequency (Nikulin et al., 2012). In CORDEX-Africa, there are currently simulations with six RCMs available (*CCLM4-8-17, CRCM5, HIRHAM5, RACMO22T, RCA4* and *REMO2009*) for the historical (1950–2005) and future period (2006–2100) under representative concentration pathways (RCPs) 2.6, 4.5 and 8.5. In the *CRCM5* and *RCA4* models, lakes are represented by a one-dimensional lake model FLake (Samuelsson et al., 2013; Hernández-Díaz et al., 2012; Martynov et al., 2012), while the other RCMs have no lake model embedded. By using different GCMs as initial and lateral boundary conditions, there are in total 25 historical model simulations, 11 simulations for RCP 2.6, 21 simulations for RCP 4.5 and 20 simulations for RCP 8.5 (see Table A1 in the Appendix). The use of model ensembles is essential, because separate simulations show larger biases than ensemble means when they are compared to the observed climate (Nikulin et al., 2012; Endris et al., 2013; Kim et al., 2014; Davin et al., 2016). Next to the historical and RCP simulations, the CORDEX framework also provides an evaluation simulation for each RCM, driven by the European Centre for Medium-Range Weather Forecasts (ECMWF) ERA-Interim reanalysis as initial and lateral boundary conditions for the period 1990–2008. Here we use these reanalysis downscaling simulations to evaluate the skill of the RCMs by comparing them with observations.

2.2 Water balance model

For a detailed description of the WBM used in this paper, we refer to Vanderkelen et al. (2018). In summary, the WB is modelled following Eq. (2) in which the change in lake level per day $\mathrm{d}L/\mathrm{d}t$ (m day^{-1}) is calculated based on the precipitation on the lake P (m day^{-1}), the evaporation from the lake E (m day^{-1}), the inflow from tributary rivers Q_{in} (m^3 day^{-1}) and the dam outflow Q_{out} (m^3 day^{-1}) divided by the lake surface area A (6.83×10^{10} m^2).

$$\frac{\mathrm{d}L}{\mathrm{d}t} = P - E + \frac{Q_{\text{in}} - Q_{\text{out}}}{A} \tag{2}$$

First, relevant spatial variables provided by the CORDEX simulations are regridded using a nearest neighbour remapping to the WBM grid with a resolution of 0.065° by 0.065° (about 7 by 7 km) containing Lake Victoria and its basin. P is computed by taking the daily mean over the lake cells of the regridded CORDEX precipitation. Basin and lake cells are defined by using masks based on lake and basin shape files. E is calculated from the latent heat flux (W m^{-2}) simulated by the CORDEX models, using the latent heat of vaporization, which is assumed to be constant at 2.5×10^6 J kg^{-1}. This term is aggregated in the same way as the precipitation term. The inflow term is calculated from the daily gridded basin precipitation with the "curve number" method as described in Vanderkelen et al. (2018) using daily basin precipitation retrieved from the CORDEX precipitation simulations. The same land cover classes based on the Global Land Cover 2000 dataset (GLC 2000; Mayaux et al., 2003) and the hydrological soil groups are applied to all CORDEX simulations. Note that this approach does not account for potential influences of future land use changes on runoff. However, the curve number does change based on the antecedent moisture condition for a particular day, derived from the preceding 5-day precipitation.

First, a set of WBM integrations is conducted by driving the WBM with the six CORDEX evaluation simulations. As these simulations (hereafter referred to as evaluation simulations) are driven by "ideal" boundary conditions, this WBM simulation allows us to examine the ability of the RCMs to represent precipitation, lake evaporation, inflow and the resulting lake levels during the period when observations for these terms are available (1990–2008). Therefore, the outflow is given by observations too. Second, the WBM is driven by the 25 historical CORDEX simulations for the period 1951–2005. Finally, future WBM runs are conducted following the RCP 2.6, 4.5 and 8.5 scenarios. One GCM-RCM combination is excluded from the analysis because of inconsistencies between the historical and future simulations (see Appendix A). To be comparable, the input terms to the transient WBM need to adhere to the same calendar. Therefore, the number of days are adjusted in a number of historical and future CORDEX simulations, as described in Appendix B. Finally, the WBM requires an initial lake level. The evaluation simulations start with the observed lake level in 1990 (1135 m a.s.l.) and the historical simulations with the observed lake level in 1950 (1133.7 m a.s.l.). For the future simulations, the last lake level calculated by the corresponding historical simulation is used.

2.3 Dam management scenarios

The evaluation of WBM simulations use recorded outflow values. In the historical WBM simulations, the outflow is calculated based on the Agreed Curve. While observed outflow volumes are available for the historical period, these cannot be used in the WBM as RCMs driven by GCMs represent the general climatology and do not account for the actual observed weather, reflected in the recorded outflow volumes. Considering the known deviations of water release from the Agreed Curve during the period 2000–2006 (Kull, 2006; Vanderkelen et al., 2018), future outflow is subject to uncertainty. Therefore, we start from three main policy objectives concerning the environmental conservation, navigation reliability and constant electricity generation to determine future dam management scenarios. These policy objectives lead to four idealized dam management scenarios: (i) managing outflow following the Agreed Curve, reflecting natural conditions by mimicking natural outflow, (ii) managing outflow so that the lake level remains constant, to keep the lake accessible for fishing boats from the harbours in shallow bays, and (iii) managing outflow to provide a constant supply of hydropower from the dams: one scenario prescribing the historical mean production of hydropower and the other an elevated hydropower production, reflecting the supply needed to meet the rising power demand in Uganda (Adeyemi and Asere, 2014). These scenarios are highly simplified and reflect very different management; they were chosen to investigate the effect of extreme dam management scenarios on the lake level fluctuations of Lake Victoria. Each scenario will thereby be applied with lake levels constrained to their physical boundaries.

A first assumption is that future outflow starts following the Agreed Curve again. In this *Agreed Curve scenario*, daily outflow is calculated following Eq. (1) based on the lake level of the previous day. Lake level fluctuations are restricted to fluctuate within the historically observed range (10 to 13.5 m measured at the dam, corresponding to 1130 to 1136.5 m a.s.l.), as this is the range for which the Agreed Curve is known. If the lake level of the previous day drops below the lower limit, the outflow is set to 0 m^3 day^{-1} and if the lake level rises above the upper limit, all additional water is discharged.

Another possibility is to manage future outflow in such a way that the lake level remains constant and the WB is closed. In this *constant lake level scenario*, daily outflow is calculated as residual of the WB, with the lake level kept constant at the last known lake level, ranging between 1134.5 and 1135.2 m for the different simulations. If the WB is negative, there is no outflow, but the lake level is allowed to decrease. When the WB is positive again, the lake level is first restored to its predefined constant height. The possible remainder from the positive WB results then in outflow for that day. By consequence, the lake level in this scenario is constant apart from short negative excursions.

In the last defined scenarios, future outflow is regulated in order to provide a constant hydropower production without interruptions while lake levels fluctuate in their physical range. In the study of Koch et al. (2013), hydropower production of a reservoir is quantified as

$$P_{\mathrm{el}} = Q_{\mathrm{out}} \times h \times k \tag{3}$$

with Q_{out} the outflow of the reservoir in $\mathrm{m^3\,day^{-1}}$, P_{el} the electricity produced (kW), h the water head (m) and k the efficiency factor ($\mathrm{kN\,m^{-3}}$). After rearranging this equation and adding a constraint to maximum outflow, the outflow needed to produce a firm amount of electricity is given by

$$Q_{\mathrm{out}} = \mathrm{MIN}\left(\frac{P_{\mathrm{el}}}{h \times k}, \mathrm{Cap_{hpp}}\right) \tag{4}$$

with $\mathrm{Cap_{hpp}}$ the maximum turbine flow capacity ($\mathrm{m^3\,s^{-1}}$). Lake Victoria is controlled by both the Nalubaale and Kiira dams. As these dams operate parallel to each other, we simplified the analysis by assuming only one dam regulating the outflow, with the combined hydropower capacity of both real dams. This results in a $\mathrm{Cap_{hpp}}$ of $1150\,\mathrm{m^3\,s^{-1}}$ (the average of $1200\,\mathrm{m^3\,s^{-1}}$ for Nalubaale and $1100\,\mathrm{m^3\,s^{-1}}$ for Kiira; Kizza and Mugume, 2006). h is assumed to be equal to the relative lake level, as measured at the dam. For the efficiency factor k a value of $13.77\,\mathrm{kN\,m^{-3}}$ is used, calculated from Eq. (2) using the values for maximum turbine flow $\mathrm{Cap_{hpp}}$, maximum water head ($h_{\mathrm{max}} = 24\,\mathrm{m}$) and the sum of maximum electricity production, 380 MW (180 MW for Nalubaale and 200 MW for the Kiira dam; Kull, 2006). Based on this, two management scenarios providing a constant hydropower production (HPP) are defined. The *historical HPP management scenario* prescribes a daily HPP equal to the mean historical HPP (P_{el} = 168 MW), which is calculated using Eq. (3) with the historically observed mean outflow ($88 \times 10^6\,\mathrm{m^3\,day^{-1}}$) and the mean relative lake level (11.9 m). Second, in the *high HPP management scenario*, HPP is set equal to the electricity produced in the year in which the outflow was maximum (P_{el} = 247 MW, in 1964 with an outflow of $138 \times 10^6\,\mathrm{m^3\,day^{-1}}$).

In both the constant lake level scenario and the HPP management scenarios, we impose physical constraints to the lake level fluctuations: the lake level should fluctuate between 0 and 26 m as measured at the dam, 1122.9 and 1146.9 m a.s.l. (the height of the dam is 31 m with a safety level of 7 m). Similar to the limits of the Agreed Curve, the outflow is set to $0\,\mathrm{m^3\,day^{-1}}$ whenever the lake level drops below 1122.9 m a.s.l. and all additional water is discharged by the sluice gates if the lake level rises above 1146.9 m a.s.l. These are merely theoretical limits. In extreme cases, lake levels could drop under the lower limit if there is more water evaporating than precipitating and flowing in the lake.

2.4 Bias-correction method

In this study, we applied a bias correction on the three WB terms derived from the CORDEX simulations (daily mean lake precipitation, lake evaporation and inflow). For every historical simulation from the CORDEX ensemble, a transformation function is calculated based on the WB terms from the observational WBM presented in Vanderkelen et al. (2018). This is done for the overlapping period of 22 years ranging from 1983 until 2005. Next, the transformation functions specific to each simulation are applied on the full historical simulation and on the available corresponding future simulations for all three RCPs. Finally, the resulting lake levels are calculated by forcing the WBM with the bias-corrected WB terms. The main assumption of bias correction is that the bias in the RCM simulations is stationary for all scenarios (historical, RCP 2.6, 4.5 and 8.5).

WB closure is adhered with two of the seven tested bias-correction methods of the *qmap* package, provided in the R language by Gudmundsson et al. (2012). The first method uses a linear parametric transformation to model the quantile–quantile relation between the observed and modelled data according to

$$\hat{P}_{\mathrm{o}} = a + bP_{\mathrm{m}} \tag{5}$$

with $\hat{P}_{\mathrm{o}}$ the best estimate of P_{o}, the distribution of the observed values, P_{m} the modelled values, and a and b calibration parameters. An overview of the a and b parameters generated per WB term for the different simulations can be found in Table D1 in the Appendix. The second method is the non-parametric quantile mapping method and is described in Sect. E in the Appendix.

3 Results

3.1 Evaluation water balance simulations

Results of the evaluation WBM runs are compared directly to the terms used in the observational WBM (Vanderkelen et al., 2018). Precipitation observations are retrieved from the Precipitation Estimation Remotely Sensed Information using Artificial Neural Networks – Climate Data Record (PERSIANN-CDR; Ashouri et al., 2015). Evaporation is estimated from the latent heat flux output of the high-resolution reanalysis downscaling of the COSMO-CLM^2 RCM (Thiery et al., 2015).

The modelled annual precipitation over the study area during the evaluation period (1990–2008) shows different spatial distributions (Fig. 2). Compared to the observed precipitation, the majority of the models (CCLM4-8-17,

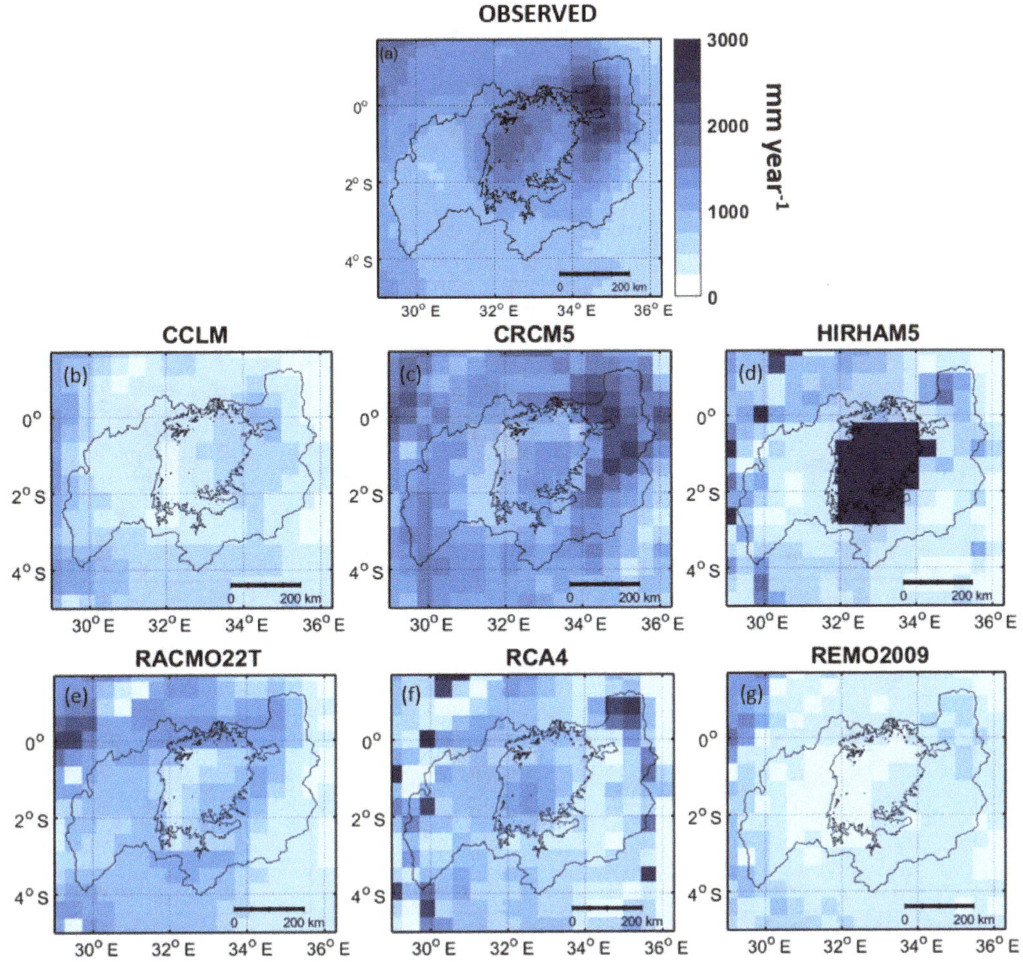

Figure 2. Annual accumulated precipitation during the period 1993–2008 as derived from PERSIANN-CDR **(a)** the CORDEX evaluation simulations **(b–g)**.

RACMO22T, RCA4 and REMO2009) underestimates the amount of lake precipitation up to −79 % compared to the reference. Only HIRHAM5 gives a large overestimation of precipitation over the lake of +78 % compared to the reference. The mean annual lake evaporation varies a lot among the models as well (Fig. 3). The evaporation amount is underestimated by CCLM4-8-17, RACMO22T and REMO2009 up to −71 % compared to the reference and overestimated by HIRHAM5 and RCA4 up to +39 %. Similar conclusions can be drawn from the comparison of the histograms of observed and simulated lake precipitation (Fig. 4). The difference between the distributions is also found for inflow and the resulting lake levels. The seasonal cycles of the WB terms – lake precipitation, evaporation and inflow (Fig. 5b, c and d) – show the same over- and underestimations compared to the observed values. This indicates that even RCM downscaling reanalysis data still entail important precipitation and evaporation biases. In most cases, the biases in precipitation and evaporation result in drifting lake levels (Fig. 5a). The overestimation of HIRHAM5 in the precipitation term is too large to be compensated by its overestimated evaporation term. Therefore, the modelled HIRHAM5 lake levels shows an unrealistic increase. The lake levels modelled with CCLM4-8-17, RACMO22T, RCA4 and REMO2009 show a large, unrealistic drop up to 13.3 m, which is mainly due to the underestimation of lake precipitation and inflow. Overall, only CRCM5 is able to represent the lake level within an acceptable range.

3.2 Simulations with bias-corrected water balance terms

As only CRCM5 is able to close the WB and there are only two simulations following RCP 4.5 with this RCM available, it is not possible to base the analysis of lake level projections on an observationally constrained RCM ensemble. Therefore, we applied two bias-correction methods to be able to use all CORDEX simulations. After applying the linear parametric transformation on the CORDEX evaluation simulations, the seasonal cycle of the lake precipitation, lake

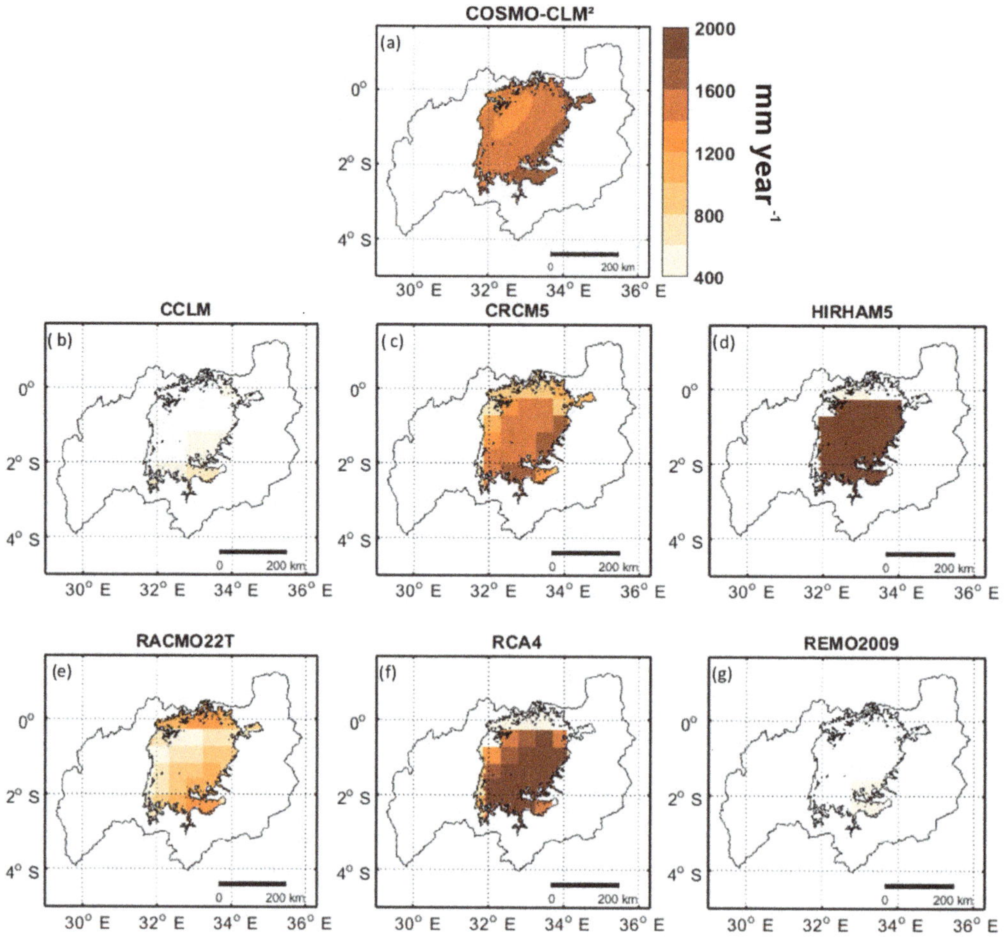

Figure 3. Annual accumulated evaporation during the period 1993–2008 as derived from COSMO-CLM^2 **(a)** the CORDEX evaluation simulations **(b–g)**.

evaporation and inflow term approximates the observations (Fig. 6). Consequently, the resulting lake levels lie within the range of observed lake levels. As the linear parametric bias-correction method provides a closed WB for all six RCMs, it is applied on the WB terms of the historical and future CORDEX simulations. Using the second bias-correction method, with empirical quantiles, also leads to WB closure. Applying this bias-correction method on the historical and future CORDEX simulations yields similar results as the linear parametric transformation. Therefore, only results from the linear parametric transformation are shown hereafter. Results with the empirical quantile bias-correction method are presented in the Appendix.

After applying the linear parametric bias correction, we quantified the future change of the three WB terms according to the three RCP scenarios for all simulations. This is achieved by computing the difference between the future (mean of the period 2071–2100) and the historical (mean of the period 1971–2000) simulations (Fig. 7). The climate change signal for lake precipitation differs between the simulations in every RCP scenario (Fig. 7a, b, c). For some simulations, lake precipitation demonstrates a strong decrease (e.g. CCLM4-8-17 driven by EC-EARTH, CRCM5 driven by CanESM2 and REMO2009 driven by EC-EARTH), while other simulations show an increase (e.g. RACMO22T driven by HadGEM2-ES and REMO2009 driven by CM5A-LR). The model simulations with a smaller increase or decrease also vary in sign. In contrast to lake precipitation, the lake evaporation signal is more consistent for the different simulations, with generally an increasing trend (Fig. 7d, e and f). The future changes in the inflow term are generally consistent as well and show an increase in inflow under all three RCP scenarios. As lake inflow is directly related to precipitation, the increase in inflow can be attributed to the increase in precipitation over the LVB. For all three WB terms, the width of the 95 % confidence intervals is larger for strong climate change signals.

The signs of the signals are broadly consistent with the signs of the original, non-bias-corrected WB terms (see Fig. E1 in the Appendix). The amplitude of the signals, how-

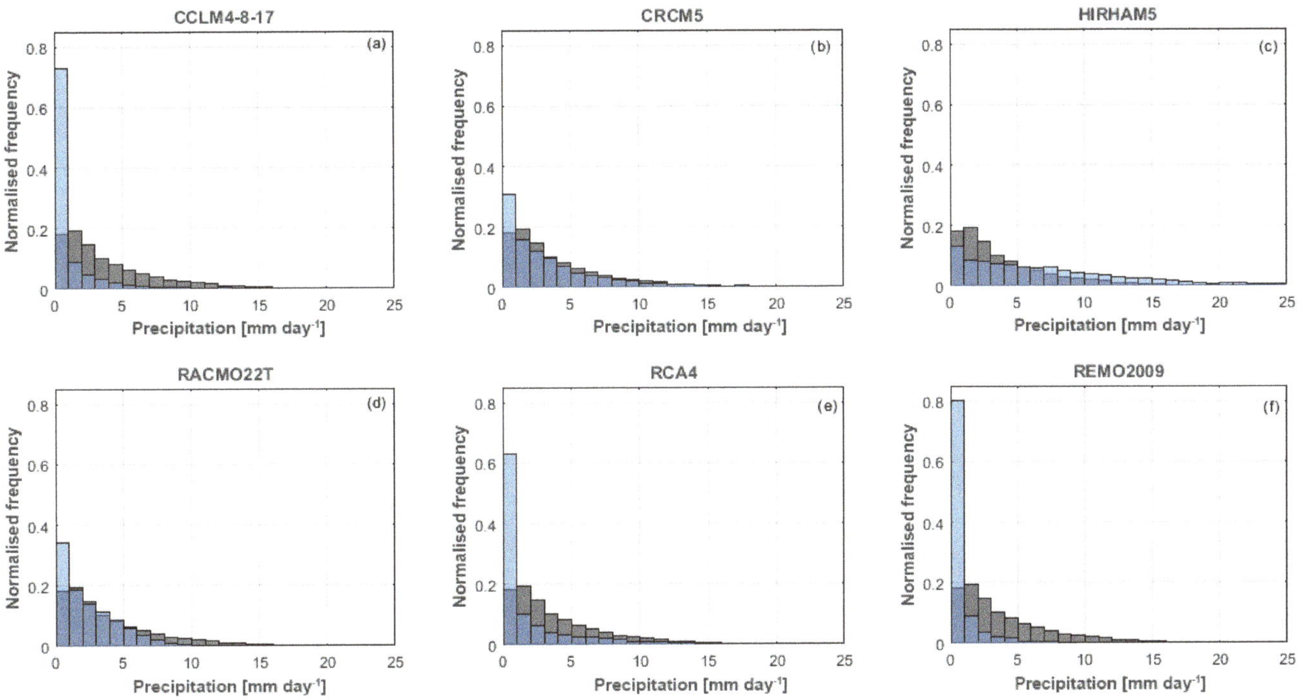

Figure 4. Histograms of lake precipitation derived from the CORDEX evaluation simulations for the period 1993–2008 (observed distributions, derived from PERSIANN-CDR are indicated in grey).

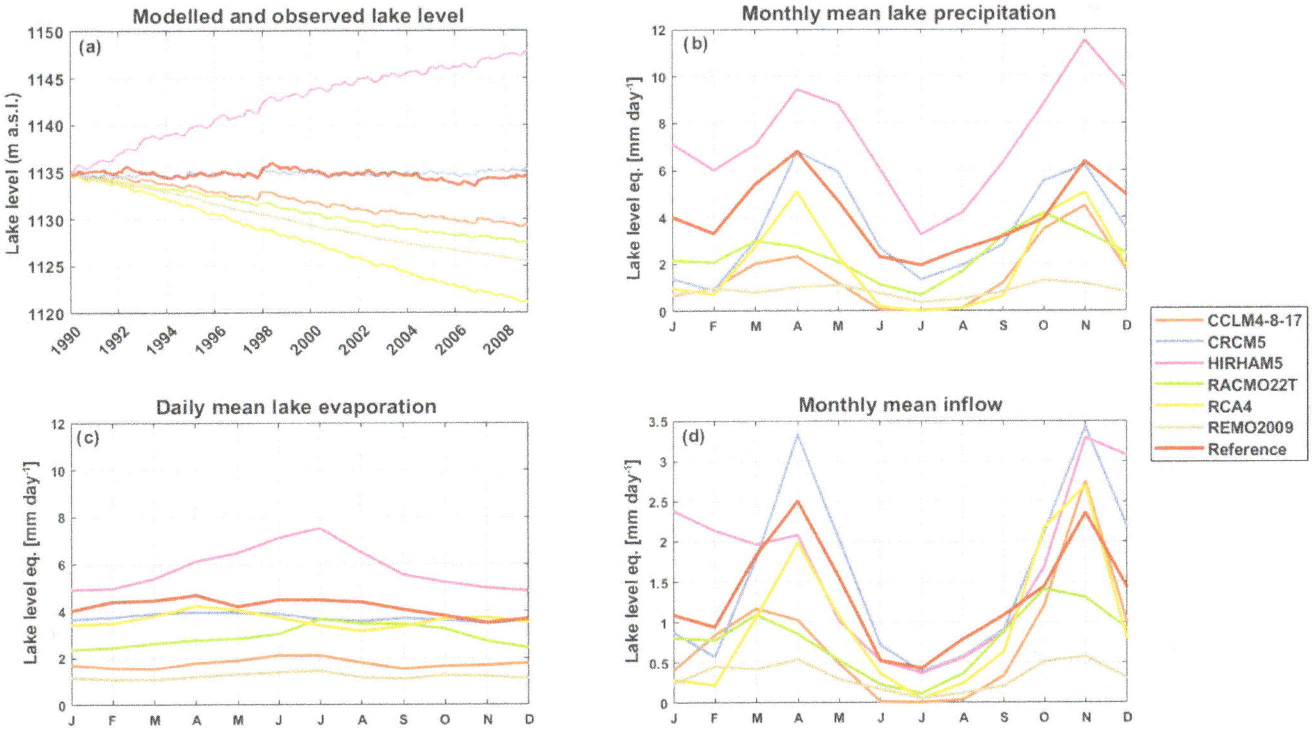

Figure 5. Modelled lake levels **(a)**, seasonal precipitation **(b)**, evaporation **(c)** and inflow **(d)** according to the CORDEX evaluation simulations without bias correction. Note the different y axis scales.

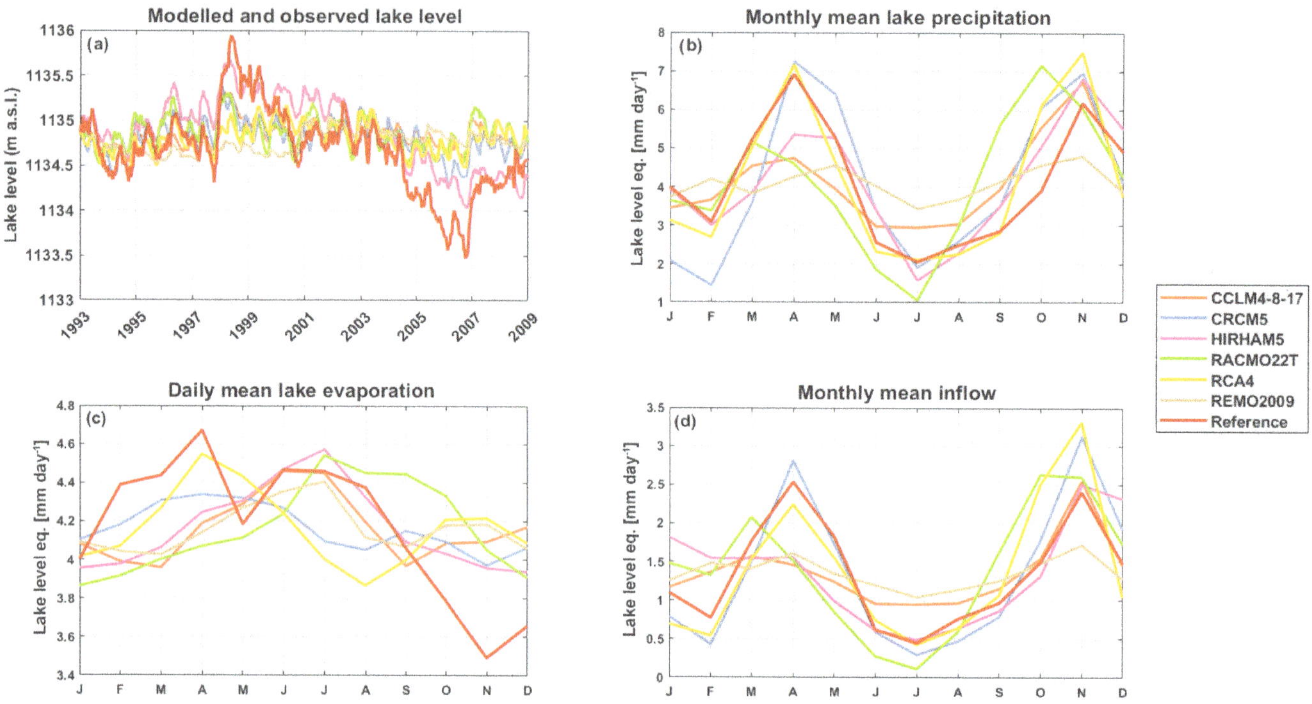

Figure 6. Modelled lake levels **(a)**, seasonal precipitation **(b)**, evaporation **(c)** and inflow **(d)** according to the CORDEX evaluation simulations, bias corrected using parametric linear transformations. Note the different *y* axis scales.

ever, generally decreases after applying the bias correction. This decrease is larger for the simulations with the more extreme signals, with important effects on the multi-model means.

3.3 Future lake level and outflow projections

Lake level projections following different dam management scenarios are computed from the CORDEX simulations, bias corrected with a linear parametric transformation (Fig. 8). Future lake levels according to the constant lake level scenario are not represented in this figure, as they are constant at their 2006 level per definition. Under the historical HPP scenario, the ensemble mean projects a lake level increase of 1.03 m for RCP 2.6, 2.2 m for RCP 8.5 and a decline of −0.36 m for RCP 4.5 in 2100, compared to the 2006 level (Fig. 8a). The ensemble mean lake level projections thereby all stay within the observed range. The model uncertainty increases over time for three RCPs with an enveloped width ranging up to 24 m. The uncertainty range encompasses both rising and decreasing lake levels (e.g. up to +12 and −12 m in 2100 under RCP 8.5). The different simulations tend to diverge with a more or less constant ensemble mean as a result, as is shown by their interquartile range (Fig. C1 in the Appendix). Overall, lake level projections thus encompass large uncertainties within this management scenario. In the high HPP scenario the ensemble mean projections of the three RCPs project a decrease in lake level (−3.9 m for RCP 2.6, −3.2 m for RCP 4.5 and −1.5 m for RCP 8.5; Fig. 8b). The model uncertainty again increases with time for the three RCPs, consistent with the rising spread in the historical HPP management scenario.

In the Agreed Curve scenario, the outflow is adjusted every day based on the lake level of the previous day. In this scenario, the modelled lake levels stay within the range of the historical fluctuations and show no clear trend (Fig. 8c). The seasonal cycles in lake level are clearly visible in the ensemble mean. In 2100, the uncertainty has increased to 1.1 m for RCP 2.6, 3.1 m for RCP 4.5 and 3.9 m for RCP 8.5. It is not surprising that the lake level modelled with this scenario stays within the historical range, as the approximated WB equilibrium is maintained by adjusting the outflow based on the lake level on a daily basis. Moreover, the Agreed Curve relation between lake level and outflow is originally made to mimic natural outflow, accounting for the natural climate variability (Sene, 2000).

The outflow projections following the Agreed Curve scenario fluctuate around the historically observed outflow volume for RCPs 2.6 and 4.5. For RCP 8.5 the model projects a slight increase towards the end of the century, resulting in an averaged outflow volume being $6.7 \times 10^6\ m^3\ day^{-1}$ (+7.6 %) higher than the historically observed outflow. This increase in outflow results from outflow volumes for some simulations that are larger than prescribed by the Agreed Curve, as the lake level for these simulations reaches the prescribed max-

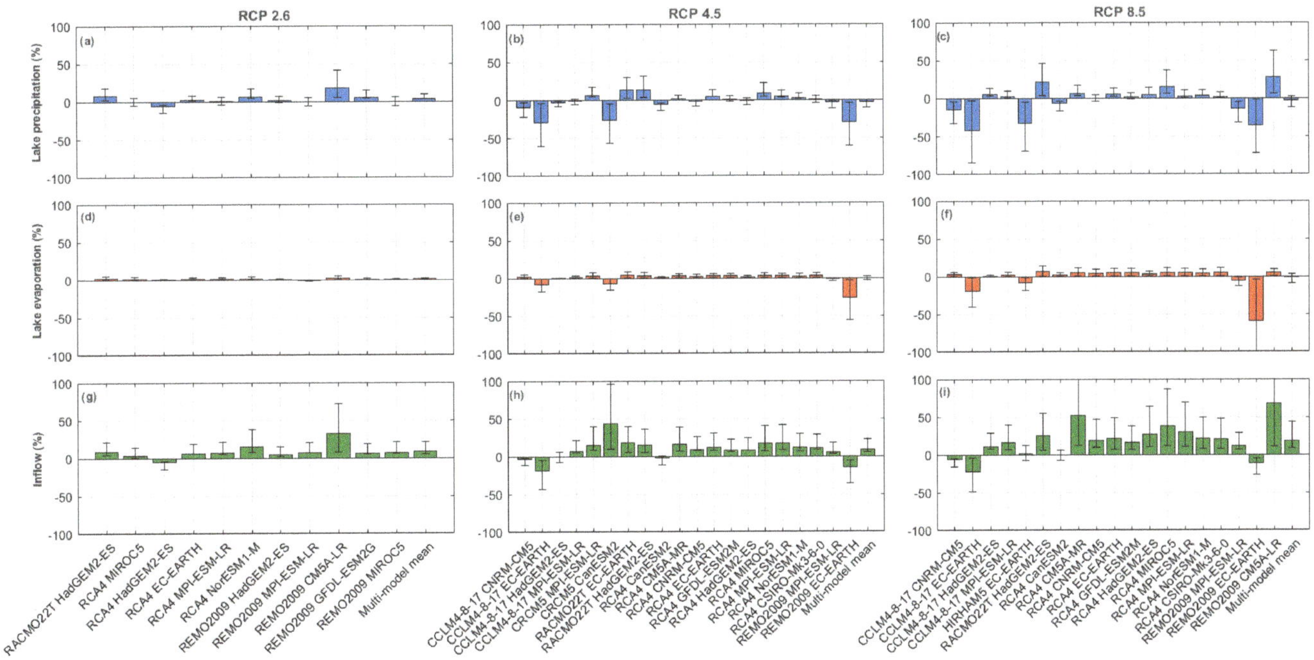

Figure 7. Bar plots showing the relative projected climate chance following RCPs 2.6, 4.5 and 8.5 for lake precipitation **(a–c)**, lake evaporation **(d–f)** and inflow **(g–i)** for the CORDEX simulations bias corrected with the linear parametric transformation. The climate change signal is defined as the mean difference between the future (2071–2100) and the historic (1971–2000) simulations. The whiskers indicate the 95 % confidence interval of the change based on the 30-year annual difference.

imum lake level and all additional water is discharged. The uncertainty in the projections is therefore very large, ranging from 14×10^6 to $209 \times 10^6\ m^3\ day^{-1}$ for RCP 8.5, whereby the latter constitutes more than double the historically observed outflow ($88 \times 10^6\ m^3\ day^{-1}$).

In the constant lake level scenario, the outflow volume varies following a spiky pattern (Fig. 9b). Since the outflow is altered each day to maintain a constant lake level given the precipitation, inflow and evaporation terms of that day, the outflow volume greatly varies on daily timescales with an average standard deviation of $172 \times 10^6\ m^3\ day^{-1}$. Fig. 9b therefore shows annually averaged daily outflow values. The three RCPs show no clear trend, but uncertainties range up to $253 \times 10^6\ m^3\ day^{-1}$. Fluctuating around a multi-model mean of $88.9 \times 10^6\ m^3$ for RCP 2.6, $90.3 \times 10^6\ m^3$ for RCP 4.5 and $95.5 \times 10^6\ m^3$ for RCP 8.5, annual average outflow volumes are higher than historical outflow, with an average outflow volume of $88 \times 10^6\ m^3\ day^{-1}$, measured from 1950 until 2006.

Finally, outflow volumes following the historical HPP scenario decrease slightly until roughly 2055, whereas outflow volumes following the high HPP scenario remain constant until 2055 (Fig. 9c and d). Afterwards, the ensemble mean outflow projections start to diverge and uncertainties strongly increases. From this moment, most lake level projections reach the imposed minimum or maximum levels (Sect. 2.3). No outflow is released for projections which drop to the minimum level, while all additional water is discharged for lake levels which reach the maximum level. Consequently, with outflow volumes reaching these extremes, it is not possible to maintain a constant hydropower supply, despite the management scenario being designed for this aim.

4 Discussion

4.1 Model quality and projected changes

None of the RCMs, except for CRCM5, are able to provide reliable estimations of the WB terms in the LVB (Figs. 4 and 5). Endris et al. (2013) found that most RCMs simulate the main precipitation features reasonably well in East Africa. Over the whole CORDEX-Africa domain, all RCMs capture the main elements of the seasonal mean precipitation distribution and its cycle, but also significant biases are present in individual models depending on season and region (Nikulin et al., 2012; Kim et al., 2014). Here, a specific region is investigated wherein lakes act as main driving features of the regional climate (Thiery et al., 2015, 2016, 2017; Docquier et al., 2016). Therefore, model performance is primarily determined by how lakes are resolved in the models. A correct representation of lake surface temperatures in the models is crucial to account for the lake–climate interactions and associated mesoscale circulations (Stepanenko et al., 2013; Thiery et al., 2014a, b). The good performance of CRCM5

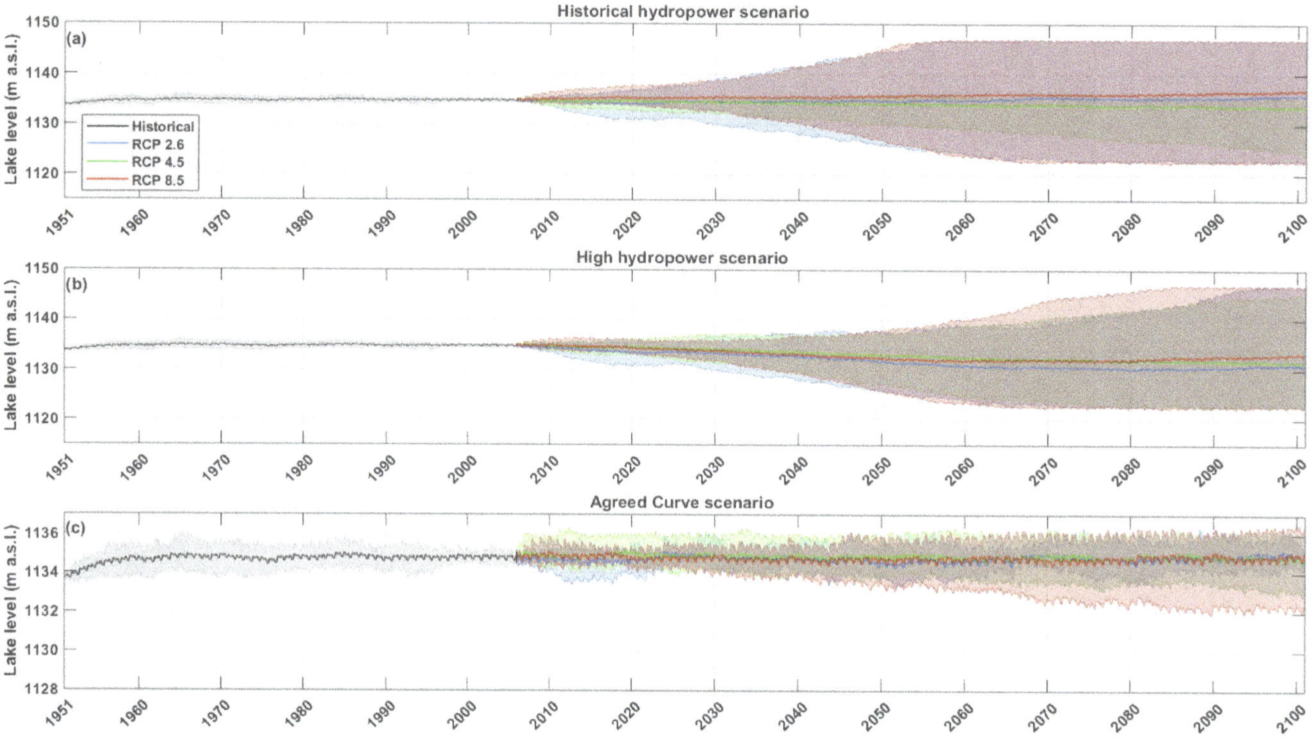

Figure 8. Lake level projections for the *historical HPP management scenario* (P_{el} is 168 MW) **(a)**, the *high HPP management scenario* (P_{el} is 247 MW) **(b)** and the *Agreed Curve scenario* **(c)**. The solid line shows the ensemble means and the envelope the 5th–95th percentiles of the CORDEX ensemble simulations, bias corrected using the parametric linear transformation method. Note the different y axis scale for panel **(c)**.

can be attributed to the presence of the lake model FLake, ensuring a realistic representation of lake surface temperatures, while the other models have no lake model embedded. The fact that RCA4, which also uses FLake as lake model, gives no accurate representation of the WB terms in the LVB is most likely due to other model biases apart from the lake model.

Thiery et al. (2016) performed high-resolution simulations ($\sim$ 7 km grid spacing) with the coupled land–lake–atmosphere model COSMO-CLM^2 under RCP 8.5 over the LVB. In these simulations, the precipitation shows a decrease of −7.5 % towards the end of the century over the lake surfaces of the African Great Lake region, which is a higher decrease than found in this study (−2.3 %). The increase in lake evaporation following RCP 8.5 according to (Thiery et al., 2016) (+142 mm year^{-1}) confirms the sign of the evaporation signal of most CORDEX simulations of this study (Fig. E1f). However, the effect is not visible in the bias-corrected multimodel mean (−0.07 mm year^{-1}), which results from the negative signal of three CORDEX simulations (Fig. E1f). While COSMO-CLM^2 generally outperforms CORDEX-Africa simulations, these simulations are only available in time slices and could therefore not be used to drive the WBM.

The decrease in lake precipitation for RCPs 4.5 and 8.5 (respectively −2.5 % and −2.3 %) is not visible in the lake level projections following the Agreed Curve or the HPP scenarios. This is due to the fact that the decrease in lake precipitation is largely compensated in the total WB by the increase in lake inflow, which is determined by the increase in precipitation over the LVB. In the bias-corrected simulations using the linear parametric transformation, the deficit in lake precipitation is compensated for 184 % (RCP 4.5) and 450 % (RCP 8.5) by an increase in inflow. In RCP 2.6, this effect is not present.

4.2 Water management and climate change

The analysis reveals that the management scenario has an important influence on the future lake levels and outflow volumes. The physical lake level constraint means that the effect of unsustainable management scenarios is reflected in the outflow volumes, which become highly variable once the lake level limit is reached. In both constant HPP scenarios, various simulations reach these limits, leading to very high or almost no outflow anymore, and a hydropower production which does not meet the goal for which it was designed. The multi-model mean lake level projections following these management scenarios appear sustainable, but are

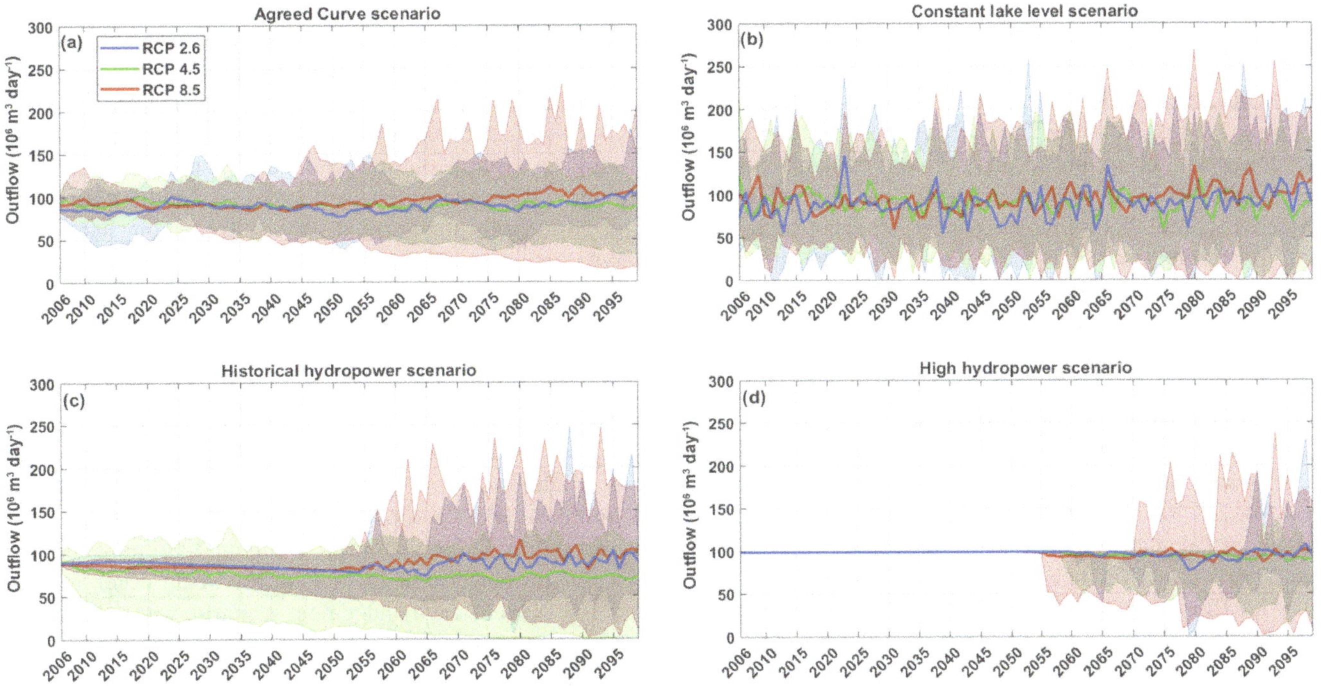

Figure 9. Outflow projections (annual averaged) for the *Agreed Curve scenario* **(a)**, the *constant lake level scenario* **(b)**, the *historical HPP management scenario* **(c)** and the *high HPP management scenario* **(d)**. The solid line shows the ensemble means and the envelope the 5th–95th percentiles of the CORDEX ensemble simulations, bias corrected using the parametric linear transformation method.

the result of averaging two branches of drifting lake levels, as illustrated by the interquartile range being large compared to the Agreed Curve (see Fig. C1 in the Appendix). Lake levels diverge towards their limits, leading to extreme situations where the lake extent is altered substantially, causing various impacts along the shoreline, such as reduced accessibility to fishing grounds and harbours located in shallow bays. Therefore, the dam management scenarios aiming at constant hydropower production are not sustainable. Furthermore, to meet the goal of a constant lake level, the outflow volumes have to be highly variable, which is not realistic if hydropower generation is pursued. Therefore the constant lake level scenario can also be qualified as unsustainable.

If the released dam outflow follows the Agreed Curve, the lake level will reflect the climatic conditions and it will fluctuate within its natural range. Moreover, the corresponding outflow following the multi-model mean also stays within the observed range. However, the uncertainties in the outflow simulations reach up to 229 % of the projected multi-model mean, highlighting that future lake level trajectories may strongly differ even under a single climate change realization. This has important implications for the potential hydropower generation and water availability downstream: while the lowest projected outflow volumes close to zero inhibit hydropower potential and exacerbate hydrological drought in the White Nile, increased outflow volumes could lead to more flooding downstream. Hence, strong changes in downstream water availability may occur in the next decades even if dam managers adhere to the Agreed Curve. Nevertheless, considering the multi-model mean, the Agreed Curve scenario can be denoted as a sustainable management scenario. However, violations against the prescribed outflow can have important consequences for the lake levels, as shown by the observed drop in lake level in 2004–2005, which was 48 % attributable to an enhanced dam outflow (Vanderkelen et al., 2018). In Uganda, hydropower provides up to 90 % of the electricity generated (Adeyemi and Asere, 2014). There is a rapidly growing gap between electricity supply and a rising demand, as the electricity consumption per capita in Uganda is among the lowest in the world. The Kiira and Nalubaale hydropower stations, managing Lake Victoria's outflow, are the largest power generators in the country (Adeyemi and Asere, 2014). The third largest capacity is provided by the new Bujagali hydropower dam located 8 km downstream of Lake Victoria. Therefore, operations at those dams will become even more important in the future. If there are again violations against the Agreed Curve because of the increasing hydropower demand, this may have substantial consequences for the future evolution of the lake level. A relatively stable lake level is, however, necessary for local water availability providing resources to the 30 million people living in its basin and to the 200 000 fisherman operating from its shores (Semazzi, 2011).

Within each management scenarios, the climate model uncertainty appears to be larger than the uncertainty related to the emission scenario. This could be seen by the large spread around the multi-model mean and the coinciding RCP curves and spread (Figs. 8 and 9). The spread according to the RCP 2.6 scenario is the smallest. This scenario contains only 11 simulations, while there are 19 simulations following RCP 4.5 and 17 simulations following RCP 8.5. The future projections provide no clear differentiation between the three RCP scenarios, indicating that uncertainties associated with the model deficiencies and initial conditions play a more important role. Therefore, to further refine lake level projections presented in this study, it is of vital importance to account for model deficiencies and natural variability.

Apart from the large climate model uncertainties, this approach using the WB model has some other shortcomings. First, we do not account for future land cover changes in the inflow calculations, as a static land cover map for the year 2000 is used (Vanderkelen et al., 2018). Changes in land cover are, however, very important in future simulations, as they affect the curve number and therefore the amount of runoff (Ryken et al., 2014). Moreover, future changes in land use could induce changes in precipitation in tropical regions (Akkermans et al., 2014; Lejeune et al., 2015). Second, the employed management scenarios are based on three simple assumptions. The management scenario exerts a major influence on future lake level fluctuations and future lake levels appeared to be sustainable only if the Agreed Curve is followed. As the importance of dam management in response to rising hydropower demand increases, more sophisticated management scenarios accounting for the rising hydropower demand could be developed and examined for their ability to preserve historical lake level fluctuations.

5 Conclusions

In this second part of a two-paper series, a water balance model (WBM) developed for Lake Victoria is forced with climate projections from the COordinated Regional Climate Downscaling Experiment (CORDEX) ensemble following three representative concentration pathways (RCPs 2.6, 4.5 and 8.5). Lake level fluctuations are projected up to 2100 using four different dam management scenarios, which emerge from three policy objectives.

This study identified the following key messages. (i) RCMs incorporated in the CORDEX ensemble are typically not able to reproduce Lake Victoria's WB, and therefore require bias correction. Applying the bias correction closes the WB and results in realistically simulated lake levels. (ii) The projected decrease in lake precipitation under RCPs 4.5 and 8.5 is compensated by an increase in lake inflow, which is directly determined by precipitation over the Lake Victoria Basin. (iii) The choice of management strategy will determine whether the lake level evolution remains sustainable or not. Idealized dam management pursuing a constant hydropower production (electricity policy objective) leads to unsustainable lake levels and outflow fluctuations. The management scenario in which the lake level is kept constant, targeting a reliable navigation policy, leads to highly variable outflow volumes, which is not realistic in terms of hydropower production. (iv) When outflow is managed following the Agreed Curve, mimicking natural outflow pursuing environmental policy targets, the evolution of lake level and outflow remains sustainable for most realizations. (v) The outflow projections following the Agreed Curve, however, encompass large uncertainties, ranging from 14×10^6 to $209 \times 10^6\ \mathrm{m^3\ day^{-1}}$ (up to 229 % of the historically observed outflow). Although the multi-model mean projected lake levels demonstrate no clear trend, these large uncertainties show that even if the Agreed Curve is followed, future lake levels and outflow volumes could potentially rise or drop drastically, with profound potential implications for local hydropower potential and downstream water availability. (vi) Next to the management scenario, we found that climate model uncertainty (RCM-GCM combination) is larger than the uncertainty related to the emission scenario.

In this study, we provide the first indications of potential consequences of climate change for the water level of Lake Victoria. The large biases and uncertainties present in the projections stress the need for an adequate representation of lakes in RCMs to be able to make reliable climate impact studies in the African Great Lakes region. Finally, the evolution of future lake levels of Lake Victoria are primarily determined by the decisions made at the dam. Therefore, the dam management of Lake Victoria is of major concern to ensure the future of the people living in the basin, the future hydropower generation and water availability downstream.

Author contributions. IV, NPMvL and WT designed the study. IV performed the analysis and wrote the manuscript. All authors commented on the manuscript.

Competing interests. The authors declare that they have no conflict of interest.

Special issue statement. This article is part of the special issue "Modelling lakes in the climate system (GMD/HESS inter-journal SI)". It is a result of the 5th workshop on "Parameterization of Lakes in Numerical Weather Prediction and Climate Modelling", Berlin, Germany, 16–19 October 2017.

Acknowledgements. We acknowledge the CLM-community (https://www.clm-community.eu/, last access: 16 October 2018) for developing COSMO CLM^2 and making the model code available. Computational resources and services used for the COSMO-CLM^2

simulations were provided by the VSC (Flemish Supercomputer Centre), funded by the Hercules Foundation and the Flemish Government department EWI. In addition, we are grateful to the World Climate Research Programme (WRCP) for initiating and coordinating the CORDEX-Africa initiative and to the modelling centres for making their downscaling results publicly available through ESGF. Wim Thiery was supported by an ETH Zurich postdoctoral fellowship (Fel-45 15-1). The Uniscientia Foundation and the ETH Zurich Foundation are thanked for their support to this research.

Edited by: Miguel Potes

References

Adeyemi, K. O. and Asere, A. A.: A Review of the Energy Situation in Uganda, International Journal of Scientific and Research Publications, 4, 2250–3153, https://doi.org/10.1260/014459806779398811, 2014.

Akkermans, T., Thiery, W., and Van Lipzig, N. P. M.: The Regional Climate Impact of a Realistic Future Deforestation Scenario in the Congo Basin, J. Climate, 27, 2714–2734, https://doi.org/10.1175/JCLI-D-13-00361.1, 2014.

Ashouri, H., Hsu, K. L., Sorooshian, S., Braithwaite, D. K., Knapp, K. R., Cecil, L. D., Nelson, B. R., and Prat, O. P.: PERSIANN-CDR: Daily precipitation climate data record from multisatellite observations for hydrological and climate studies, B. Am. Meteorol. Soc., 96, 69–83, https://doi.org/10.1175/BAMS-D-13-00068.1, 2015.

Boé, J., Terray, L., Habets, F., and Martin, E.: Statistical and dynamical downscaling of the Seine basin climate for hydro-meteorological studies, Int. J. Climatol., https://doi.org/10.1002/joc.1602, 27, 1643–1655, 2007.

Cannon, A. J.: Multivariate quantile mapping bias correction: an N-dimensional probability density function transform for climate model simulations of multiple variables, Clim. Dynam., 50, 31–49, https://doi.org/10.1007/s00382-017-3580-6, 2017.

Christensen, J. H., Boberg, F., Christensen, O. B., and Lucas-Picher, P.: On the need for bias correction of regional climate change projections of temperature and precipitation, Geophys. Res. Lett., 35, 20709, https://doi.org/10.1029/2008GL035694, 2008.

CORDEX-Africa simulation data: available at http://www.cordex.org/data-access/esgf/, last access: 16 October 2018.

Davin, E. L., Maisonnave, E., and Seneviratne, S. I.: Is land surface processes representation a possible weak link in current Regional Climate Models?, Environ. Res. Lett, 11, 074027, https://doi.org/10.1088/1748-9326/11/7/074027, 2016.

Deconinck, S.: Security as a threat to development: the geopolitics of water scarcity in the Nile River basin, Focus Paper, Royal High Insitute for Defence, Brussels, Belgium, 2009.

Di Baldassarre, G., Elshamy, M., Van Griensven, A., Soliman, E., Kigobe, M., Ndomba, P., Mutemi, J., Mutua, F., Moges, S., Xuan, Y., Solomatine, D., and Uhlenbrook, S.: Future hydrology and climate in the River Nile basin: a review, Hydrolog. Sci. J., 56, 199–211, https://doi.org/10.1080/02626667.2011.557378, 2011.

Docquier, D., Thiery, W., Lhermitte, S., and van Lipzig, N. P. M.: Multi-year wind dynamics around Lake Tanganyika, 47, 3191–3202, https://doi.org/10.1007/s00382-016-3020-z, 2016.

Endris, H. S., Omondi, P., Jain, S., Lennard, C., Hewitson, B., Chang'a, L., Awange, J. L., Dosio, A., Ketiem, P., Nikulin, G., Panitz, H. J., Büchner, M., Stordal, F., and Tazalika, L.: Assessment of the performance of CORDEX regional climate models in simulating East African rainfall, J. Climate, 26, 8453–8475, https://doi.org/10.1175/JCLI-D-12-00708.1, 2013.

Giorgi, F., Jones, C., and Asrar, G. R.: Addressing climate information needs at the regional level: The CORDEX framework, World Meteorological Organization Bulletin, 58, 175–183, 2009.

Gudmundsson, L., Bremnes, J. B., Haugen, J. E., and Engen-Skaugen, T.: Technical Note: Downscaling RCM precipitation to the station scale using statistical transformations – a comparison of methods, Hydrol. Earth Syst. Sci., 16, 3383–3390, https://doi.org/10.5194/hess-16-3383-2012, 2012.

Gudmundsson, L.: qmap R package, available at: https://cran.r-project.org/web/packages/qmap/index.html, last access: 16 October 2018.

Hernández-Díaz, L., Laprise, R., Laxmi, S., Martynov, A., Winger, K., and Dugas, B.: Climate simulation over CORDEX Africa domain using the fifth-generation Canadian Regional Climate Model (CRCM5), Clim. Dynam., 40, 1415–1433, https://doi.org/10.1007/s00382-012-1387-z, 2012.

Kent, C., Chadwick, R., and Rowell, D. P.: Understanding uncertainties in future projections of seasonal tropical precipitation, J. Climate, 28, 4390–4413, https://doi.org/10.1175/JCLI-D-14-00613.1, 2015.

Kim, J., Waliser, D. E., Mattmann, C. A., Goodale, C. E., Hart, A. F., Zimdars, P. A., Crichton, D. J., Jones, C., Nikulin, G., Hewitson, B., Jack, C., Lennard, C., and Favre, A.: Evaluation of the CORDEX-Africa multi-RCM hindcast: Systematic model errors, Clim. Dynam., 42, 1189–1202, https://doi.org/10.1007/s00382-013-1751-7, 2014.

Kizza, M. and Mugume, S.: The Impact of a Potential Dam Break on the Hydro Electric Power Generation: Case of: Owen Falls Dam Break Simulation, Uganda, in: International Conference on Advances in Engineering and Technology, edited by: Mwakali, J. and Taban-Wani, G., 711–722, Entebbe, Uganda, 2006.

Koch, H., Liersch, S., and Hattermann, F. F.: Integrating water resources management in eco-hydrological modelling, Water Sci. Technol., 67, 1525–1533, https://doi.org/10.2166/wst.2013.022, 2013.

Kull, D.: Connections Between Recent Water Level Drops in Lake Victoria, Dam Operations and Drought, Daniel Kull, Nairobi, available at: https://www.internationalrivers.org/sites/default/files/attached-files/full_report_pdf.pdf (last access: 19 October 2018), 2006.

Lange, S.: Bias correction of surface downwelling longwave and shortwave radiation for the EWEMBI dataset, Earth Syst. Dynam., 9, 627–645, https://doi.org/10.5194/esd-9-627-2018, 2018.

Lejeune, Q., Davin, E. L., Guillod, B. P., and Seneviratne, S. I.: Influence of Amazonian deforestation on the future evolution of regional surface fluxes, circulation, surface temperature and precipitation, Clim. Dynam., 44, 2769–2786, https://doi.org/10.1007/s00382-014-2203-8, 2015.

Lyon, B. and Dewitt, D. G.: A recent and abrupt decline in the East African long rains, Geophys. Res. Lett., 39, 1–5, https://doi.org/10.1029/2011GL050337, 2012.

Maraun, D., Wetterhall, F., Ireson, A. M., Chandler, R. E., Kendon, E. J., Widmann, M., Brienen, S., Rust, H. W., Sauter, T., Themel, M., Venema, V. K., Chun, K. P., Goodess, C. M., Jones, R. G., Onof, C., Vrac, M., and Thiele-Eich, I.: Precipitation downscaling under climate change: Recent developments to bridge the gap between dynamical models and the end user, Rev. Geophys., 48, 2009RG000314, https://doi.org/10.1029/2009RG000314, 2010.

Maraun, D., Shepherd, T. G., Widmann, M., Zappa, G., Walton, D., Gutiérrez, J. M., Hagemann, S., Richter, I., Soares, P. M., Hall, A., and Mearns, L. O.: Towards process-informed bias correction of climate change simulations, Nat. Clim. Change, 7, 764–773, https://doi.org/10.1038/nclimate3418, 2017.

Martynov, A., Laprise, R., Sushama, L., Winger, K., Šeparović, L., and Dugas, B.: Reanalysis-driven climate simulation over CORDEX North America domain using the Canadian Regional Climate Model, version 5: Model performance evaluation, Clim. Dynam., 41, 2973–3005, https://doi.org/10.1007/s00382-013-1778-9, 2012.

Mayaux, P., Massart, M., Cutsem, C. V., Cabral, A., Nonguierma, A., Diallo, O., Pretorius, C., Thompson, M., Cherlet, M., Defourny, P., Vasconcelos, M., Gregorio, A. D., Grandi, G. D., and Belward, A.: A land cover map of Africa, EUR 20665 EN, European Commission, Joint Research Centre, Luxembourg, 2003.

Nicholson, S. E.: An analysis of recent rainfall conditions in eastern Africa, Int. J. Climatol., 36, 526–532, https://doi.org/10.1002/joc.4358, 2016.

Nicholson, S. E.: Climate and climatic variability of rainfall over eastern Africa, Rev. Geophys., 55, 590–635, https://doi.org/10.1002/2016RG000544, 2017.

Nikulin, G., Jones, C., Giorgi, F., Asrar, G., Büchner, M., Cerezo-Mota, R., Christensen, O. B., Déqué, M., Fernandez, J., Hänsler, A., van Meijgaard, E., Samuelsson, P., Sylla, M. B., and Sushama, L.: Precipitation climatology in an ensemble of CORDEX-Africa regional climate simulations, J. Climate, 25, 6057–6078, https://doi.org/10.1175/JCLI-D-11-00375.1, 2012.

Otieno, V. O. and Anyah, R. O.: CMIP5 simulated climate conditions of the Greater Horn of Africa (GHA). Part II: Projected climate, Clim. Dynam., 41, 2099–2113, https://doi.org/10.1007/s00382-013-1694-z, 2013.

Panofsky, H. and Brier, G. W.: Some Apllications of Statistics to Meteorology, The Pennsylvania State University Press, Philadelphia, 1968.

Philip, S., Kew, S. F., van Oldenborgh, G. J., Otto, F., O'Keefe, S., Haustein, K., King, A., Zegeye, A., Eshetu, Z., Hailemariam, K., Singh, R., Jjemba, E., Funk, C., Cullen, H., Philip, S., Kew, S. F., van Oldenborgh, G. J., Otto, F., O'Keefe, S., Haustein, K., King, A., Zegeye, A., Eshetu, Z., Hailemariam, K., Singh, R., Jjemba, E., Funk, C., and Cullen, H.: Attribution analysis of the Ethiopian drought of 2015, J. Climate, 31, 2465–2486, https://doi.org/10.1175/JCLI-D-17-0274.1, 2017.

Rowell, D. P., Booth, B. B. B., and Nicholson, S. E.: Reconciling Past and Future Rainfall Trends over East Africa, J. Climate, 28, 9768–9788, https://doi.org/10.1175/JCLI-D-15-0140.1, 2015.

Ryken, N., Vanmaercke, M., Wanyama, J., Isabirye, M., Vanonckelen, S., Deckers, J., and Poesen, J.: Impact of papyrus wetland encroachment on spatial and temporal variabilities of stream flow and sediment export from wet tropical catchments, Sci. Total Environ., 511, 756–766, https://doi.org/10.1016/j.scitotenv.2014.12.048, 2014.

Samuelsson, P., Gollvik, S., Jansson, C., Kupiainen, M., Kourzeneva, E., and Berg, W. J. V. D.: The surface processes of the Rossby Centre regional atmospheric climate model (RCA4), Meteorology, 157, available at: https://www.smhi.se/polopoly_fs/1.89803!/Menu/general/extGroup/attachmentColHold/mainCol1/file/meteorologi_157.pdf (last access: 19 October 2018), 2013.

Semazzi, F. H. M.: Enhancing Safety of Navigation and Efficient Exploitation of Natural Resources over Lake Victoria and Its Basin by Strengthening Meteorological Services on the Lake, p. 104, Technical Report, North Carolina State University Climate Modeling Laboratory, 2011.

Sene, K. J.: Theoretical estimates for the influence of Lake Victoria on flows in the upper White Nile, Hydrolog. Sci. J., 45, 125–145, https://doi.org/10.1080/02626660009492310, 2000.

Souverijns, N., Thiery, W., Demuzere, M., and Lipzig, N. P. M. V.: Drivers of future changes in East African precipitation Drivers of future changes in East African precipitation, Environ. Res. Lett, 11, 114011, https://doi.org/10.1088/1748-9326/11/11/114011, 2016.

Stepanenko, V. M., Martynov, A., Jöhnk, K. D., Subin, Z. M., Perroud, M., Fang, X., Beyrich, F., Mironov, D., and Goyette, S.: A one-dimensional model intercomparison study of thermal regime of a shallow, turbid midlatitude lake, Geosci. Model Dev., 6, 1337–1352, https://doi.org/10.5194/gmd-6-1337-2013, 2013.

Sutcliffe, J. and Parks, Y.: The Hydrology of the Nile, IAHS Special Publication, 5, 192, 1999.

Tate, E., Sutcliffe, J., Conway, D., and Farquharson, F.: Water balance of Lake Victoria: update to 2000 and climate change modelling to 2100, Hydrolog. Sci. J., 49, 563–574, https://doi.org/10.1623/hysj.49.4.563.54422, 2004.

Taye, M. T., Ntegeka, V., Ogiramoi, N. P., and Willems, P.: Assessment of climate change impact on hydrological extremes in two source regions of the Nile River Basin, Hydrol. Earth Syst. Sci., 15, 209–222, https://doi.org/10.5194/hess-15-209-2011, 2011.

Teutschbein, C. and Seibert, J.: Is bias correction of regional climate model (RCM) simulations possible for non-stationary conditions?, Hydrol. Earth Syst. Sci., 17, 5061–5077, https://doi.org/10.5194/hess-17-5061-2013, 2013.

Themeßl, M. J., Gobiet, A., and Leuprecht, A.: Empirical-statistical downscaling and error correction of daily precipitation from regional climate models, Int. J. Climatol., 31, 1530–1544, https://doi.org/10.1002/joc.2168, 2011.

Themeßl, M. J., Gobiet, A., and Heinrich, G.: Empirical-statistical downscaling and error correction of regional climate models and its impact on the climate change signal, Climatic Change, 112, 449–468, https://doi.org/10.1007/s10584-011-0224-4, 2012.

Thiery, W., Martynov, A., Darchambeau, F., Descy, J.-P., Plisnier, P.-D., Sushama, L., and van Lipzig, N. P. M.: Understanding the performance of the FLake model over two African Great Lakes, Geosci. Model Dev., 7, 317–337, https://doi.org/10.5194/gmd-7-317-2014, 2014a.

Thiery, W., Stepanenko, V. M., Fang, X., Jöhnk, K. D., Li, Z., Martynov, A., Perroud, M., Subin, Z. M., Darchambeau, F., Mironov, D., and van Lipzig, N. P. M.: LakeMIP Kivu: evaluating the representation of a large, deep tropical lake by a set of one-dimensional lake models, Tellus A, 66, 21390, https://doi.org/10.3402/tellusa.v66.21390, 2014b.

Thiery, W., Davin, E. L., Panitz, H.-J., Demuzere, M., Lhermitte, S., and van Lipzig, N. P. M.: The Impact of the African Great Lakes on the Regional Climate, J. Climate, 28, 4061–4085, https://doi.org/10.1175/JCLI-D-14-00565.1, 2015.

Thiery, W., Davin, E. L., Seneviratne, S. I., Bedka, K., Lhermitte, S., and van Lipzig, N. P. M.: Hazardous thunderstorm intensification over Lake Victoria, Nat. Commun., 7, https://doi.org/10.1038/ncomms12786, 12786, 2016.

Thiery, W., Gudmundsson, L., Bedka, K., Semazzi, F. H., Lhermitte, S., Willems, P., van Lipzig, N. P. M., and Seneviratne, S. I.: Early warnings of hazardous thunderstorms over Lake Victoria, Environ. Res. Lett., 12, 2–5, https://doi.org/10.1088/1748-9326/aa7521, 2017.

Vanderkelen, I.: Water Balance Model of Lake Victoria, https://doi.org/10.5281/zenodo.1464820, 2018.

Vanderkelen, I., van Lipzig, N. P. M., and Thiery, W.: Modelling the water balance of Lake Victoria (East Africa) – Part 1: Observational analysis, Hydrol. Earth Syst. Sci., 22, 5509–5525, https://doi.org/10.5194/hess-22-5509-2018, 2018.

Vrac, M. and Friederichs, P.: Multivariate-intervariable, spatial, and temporal-bias correction, J. Climate, 28, 218–237, https://doi.org/10.1175/JCLI-D-14-00059.1, 2015.

Williams, K., Chamberlain, J., Buontempo, C., and Bain, C.: Regional climate model performance in the Lake Victoria basin, Clim. Dynam., 44, 1699–1713, https://doi.org/10.1007/s00382-014-2201-x, 2015.

Wood, A. W., Leung, L. R., Sridhar, V., and Lettenmaier, D. P.: Hydrologic implications of dynamical and statistical approaches to downscaling climate model outputs, Climate Change, 62, 189–216, 2004.

Appendix A: Details on used CORDEX simulations

The CORDEX ensemble simulations used in this study are listed in Table A1. From all available simulations, the simulation of HIRHAM5 driven by EC-EARTH following RCP 4.5 is not used because it exhibits discrepancies between its historical and future simulation. These discrepancies are non-physical and inhibit the application of a bias correction.

Table A1. Overview of the different CORDEX simulations and their availability (Y: yes, N: no). (*data not used because of discrepancy between historical and future simulation).

RCM	Driving GCM	RCP 2.6	RCP 4.5	RCP 8.5
CCLM4-8-17	EC-EARTH	N	Y	Y
CCLM4-8-17	HasGEM2-ES	N	Y	Y
CCLM4-8-17	MPI-ESM-LR	N	Y	Y
CCLM4-8-17	CNRM-CM5	N	Y	Y
CRCM5	MPI-ESM-LR	N	Y	N
CRCM5	CanESM2	N	Y	N
HIRHAM5	EC-EARTH	N	Y*	Y
RACMO22T	EC-EARTH	N	Y	Y
RACMO22T	HadGEM2-ES	Y	Y	Y
RCA4	CanESM2	N	Y	Y
RCA4	EC-EARTH	Y	Y	Y
RCA4	MIROC5	Y	Y	Y
RCA4	HadGEM2-ES	Y	Y	Y
RCA4	NorESM1-M	Y	Y	Y
RCA4	GFDL-ESM2M	N	Y	Y
RCA4	CM5A-MR	N	Y	Y
RCA4	CNRM-CM5	N	Y	Y
RCA4	MPI-ESM-LR	Y	Y	Y
RCA4	CSIRO-Mk3-6-0	N	Y	Y
REMO2009	MIROC5	Y	N	N
REMO2009	GFDL-ESM2G	Y	N	N
REMO2009	CM5A-LR	Y	N	Y
REMO2009	HadGEM2-ES	Y	N	N
REMO2009	EC-EARTH	N	Y	Y
REMO2009	MPI-ESM-LR	Y	Y	Y

Appendix B: Correction of CORDEX ensemble members for number of days

Not all simulations from the CORDEX ensemble have the same number of days. As a fixed number of days is a necessary condition to compare the WBM simulations, a correction was applied on the daily WB terms of some simulations.

First, simulations driven by HadGEM2-ES (CCLM4-8-17, RACMO22T, RCA4, REMO2009), have 30-day months and only go until 2099. To account for the missing days, 5 extra days are added for every 72 days in the year, starting after the 36th day. The index of these 5 extra days within each year are given in Table B1. The added days are the average of the respective WB terms during the previous and next day. In addition, we accounted for the fact that these model simulations do not include the year 2100, by repeating the year 2099. The simulations with HadGEM2-ES for RCP 4.5 have no December month in the year 2099. This is also the case for the HadGEM2-ES CCLM4-8-17 simulation for RCP 8.5. In both cases, December 2099 is added by repeating the month November of the same year.

Finally, in all simulations that do not account for leap years (RCMs driven by CanESM2, NorESM1-M, MIROC5, GFDL-ESM2M, CM5A-MR and CSIRO-Mk3-6-0), an extra day in the leap years is added by taking the average WB term value of the days corresponding to the 28 February and the 1 March. Overall, compared to the total number of days of the future projections (34 698 days), we note that corrections on single days (up to 888 days depending on the simulations) have a little influence on the outcomes of this study.

Table B1. Indices where extra days are added per year

Original index	36	108	180	252	324
Index of added day	37	110	183	256	329

Appendix C: Interquartile ranges of lake level projections

The interquartile range of the lake level projections (Fig. C1) compared with their 5th and 95th percentile envelope shows a large decrease in uncertainty range for outflow following the Agreed Curve, while the uncertainty following the HPP management scenarios knows a smaller decrease.

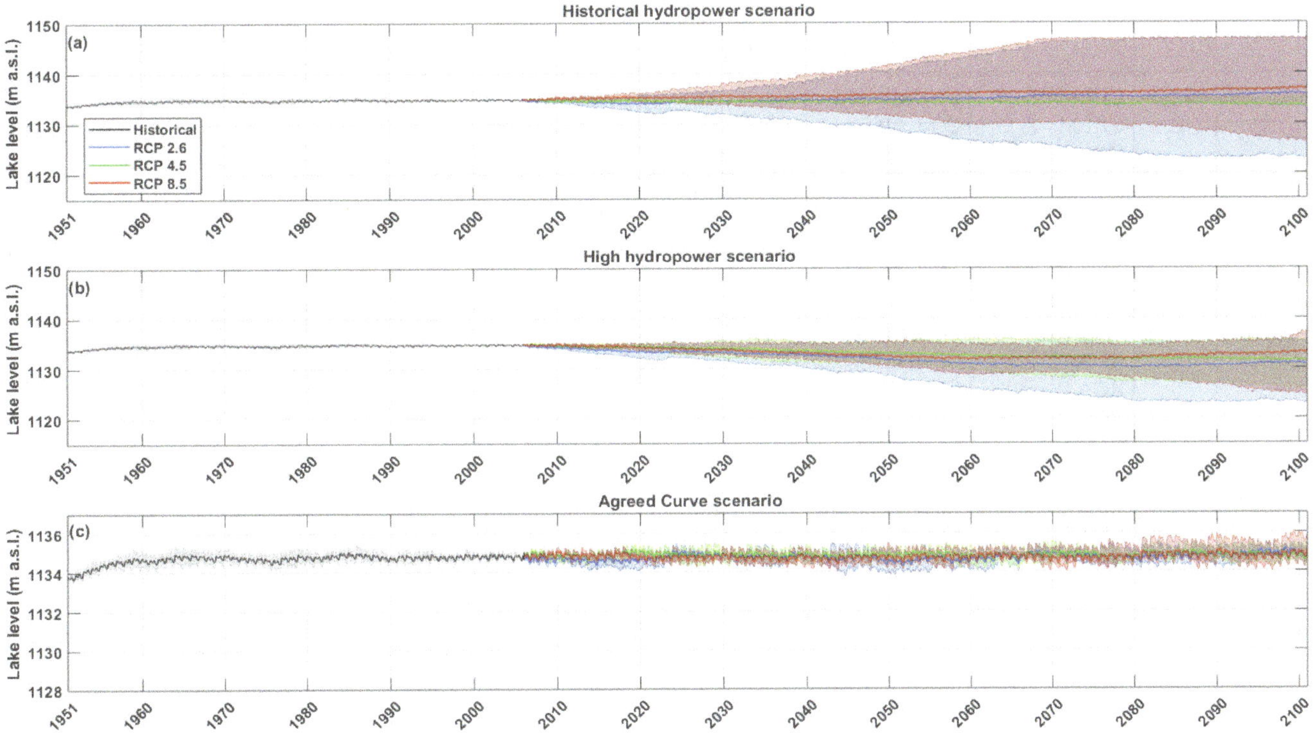

Figure C1. As in Fig. 8, but the envelope shows the interquartile range of the CORDEX simulations.

Appendix D: Overview of the parameters used in the linear parametric transformation

Table D1 shows the *a* and *b* calibration parameters for the different CORDEX simulations used in the linear parametric transformation to bias correct the lake precipitation, evaporation and inflow terms of the WB.

Table D1. Parameters *a* and *b* of the linear parametric transformation of the WB terms for the different CORDEX simulations (Eq. 5).

RCM	Driving GCM	Lake precipitation a (10^{-3})	Lake precipitation b	Lake evaporation a (10^{-3})	Lake evaporation b	Inflow a (10^{6})	Inflow b
CCLM4-8-17	CNRM-CM5	0.65	0.350	2.30	0.476	27.0	0.257
CCLM4-8-17	EC-EARTH	1.68	0.574	2.82	0.536	38.1	0.315
CCLM4-8-17	HadGEM2-ES	1.69	0.425	2.88	0.339	47.6	0.404
CCLM4-8-17	MPI-ESM-LR	1.20	0.342	2.05	0.556	36.6	0.261
CRCM5	MPI-ESM-LR	1.06	0.810	1.71	0.676	−16.5	0.705
CRCM5	CanESM2	−1.54	0.850	2.86	0.655	−15.0	0.647
RACMO22T	EC-EARTH	−0.01	0.528	1.95	0.206	−31.1	2.041
RACMO22T	HadGEM2-ES	0.43	1.823	2.98	0.944	−16.0	2.225
HIRHAM5	EC-EARTH	1.04	1.630	1.59	0.424	0.90	0.730
RCA4	CanESM2	1.78	0.926	1.78	0.658	16.0	0.759
RCA4	CM5A-MR	1.81	0.860	1.86	0.629	24.4	0.691
RCA4	CNRM-CM5	1.95	0.982	1.78	0.643	27.6	0.916
RCA4	EC-EARTH	1.64	0.794	1.66	0.706	19.0	0.657
RCA4	GFDL-ESM2M	2.04	0.878	1.73	0.699	36.8	0.652
RCA4	HadGEM2-ES	2.40	1.125	2.70	0.433	36.5	1.177
RCA4	MIROC5	1.73	0.885	2.00	0.593	23.9	0.767
RCA4	MPI-ESM-LR	1.79	0.889	1.76	0.657	21.8	0.812
RCA4	NorESM1-M	2.04	0.994	1.81	0.657	31.3	0.946
RCA4	CSIRO-Mk3-6-0	1.74	1.048	1.89	0.648	22.9	0.971
REMO2009	HadGEM2-ES	0.16	0.425	2.90	0.241	33.5	0.525
REMO2009	MPI-ESM-LR	1.23	0.639	2.50	0.519	42.2	0.814
REMO2009	EC-EARTH	1.76	0.923	2.59	0.703	40.7	1.013
REMO2009	CM5A-LR	0.73	0.414	2.81	0.271	19.7	0.453
REMO2009	GFDL-ESM2G	0.51	0.415	2.58	0.327	13.3	0.470
REMO2009	MIROC5	0.57	0.481	2.88	0.279	23.1	0.524

Appendix E: Simulations with empirical quantile bias correction

Next to the bias-correction method using a linear parametric transformation (see Sect. 2.4), WB closure was adhered with a second method which is the non-parametric quantile mapping method, a common approach for statistical transformation (e.g. Panofsky and Brier, 1968; Wood et al., 2004; Boé et al., 2007; Themeßl et al., 2011, 2012). Following Gudmundsson et al. (2012) and Boé et al. (2007), this method uses the cumulative density function (CDF) based on the empirical quantiles from the observed variable to transform the modelled variable. First, the CDFs of the three WB terms following each historical simulation in the overlapping period (the reference simulations) are matched with the cumulative density function of the WB terms from the observational WBM (observations). This generates a correction function, relating the quantiles of both distributions. Next, this correction function is used to unbias the WB term simulations for the whole simulation period quantile by quantile (Boé et al., 2007).

When a bias correction based on empirical quantiles is used, very similar results are found (compare Figs. E2, E3, E4 and E6). Based on this, we conclude that the bias-correction methods has very little effect on the results presented in this study.

In this study, applying a bias correction on the WB terms of the CORDEX simulations was necessary to be able to make lake level and outflow projections, as subsetting is not possible. RCMs are often bias corrected, as their simulations inhibit errors (Christensen et al., 2008; Teutschbein and Seibert, 2013; Maraun et al., 2010; Themeßl et al., 2012; Lange, 2018). Both linear parametric transformation and the quantile mapping bias-correction methods are used. The advantage of the first is the simplicity and transparency of the method (Teutschbein and Seibert, 2013). The quantile mapping method, on the other hand, is a non-parametric method and is able to correct for errors in variability as well (Themeßl et al., 2011). Yet, no substantial differences could be noted between the resulting projections of both methods, which supports that there is no single optimal way to correct for RCM biases (Themeßl et al., 2011). It is, however, important to consider the limitations concerning the bias-correction methods. In both methods, each WB term is corrected independently, whereas biases may not be independent among the terms, which may be important in the context of climate change (Boé et al., 2007). The consistency between the variables could be preserved by using a more sophisticated method using a multivariate bias correction (Cannon, 2017; Vrac and Friederichs, 2015). However, Maraun et al. (2017) showed that bias correction could lead to improbable climate change signals and cannot overcome large model errors.

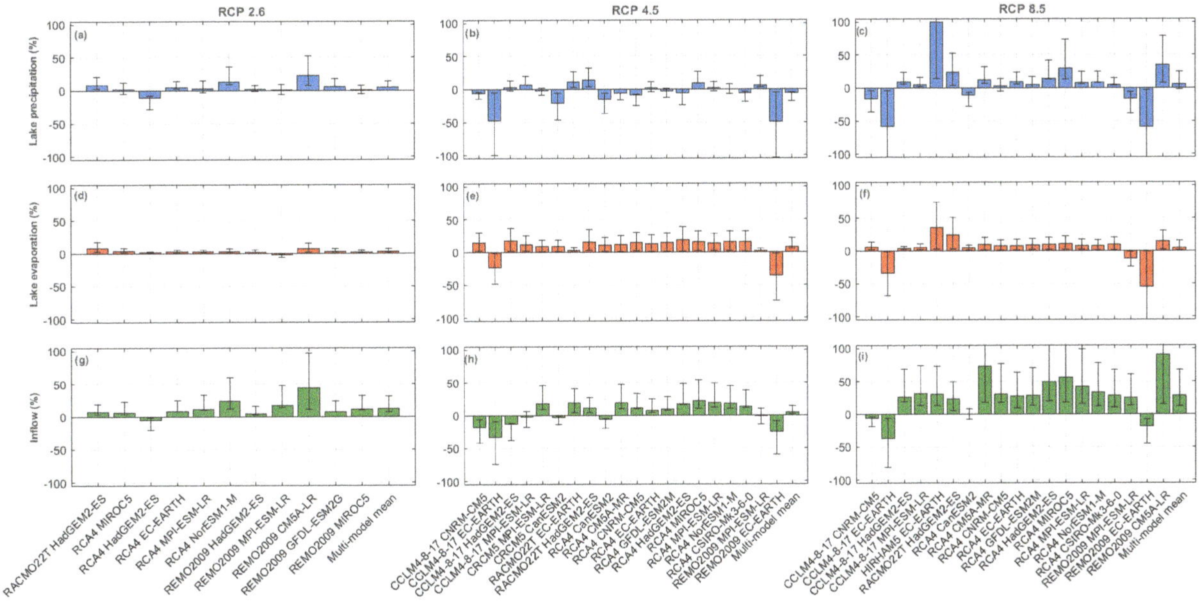

Figure E1. Bar plots showing the relative projected climate chance following RCPs 2.6, 4.5 and 8.5 for lake precipitation **(a–c)**, lake evaporation **(d–f)** and inflow **(g–i)** for the CORDEX simulations without bias correction. The climate change signal is defined as the difference between the future (2071–2100) and the historical (1971–2000) simulations. The whiskers indicate the 95 % confidence interval of the change based on the 30-year annual difference.

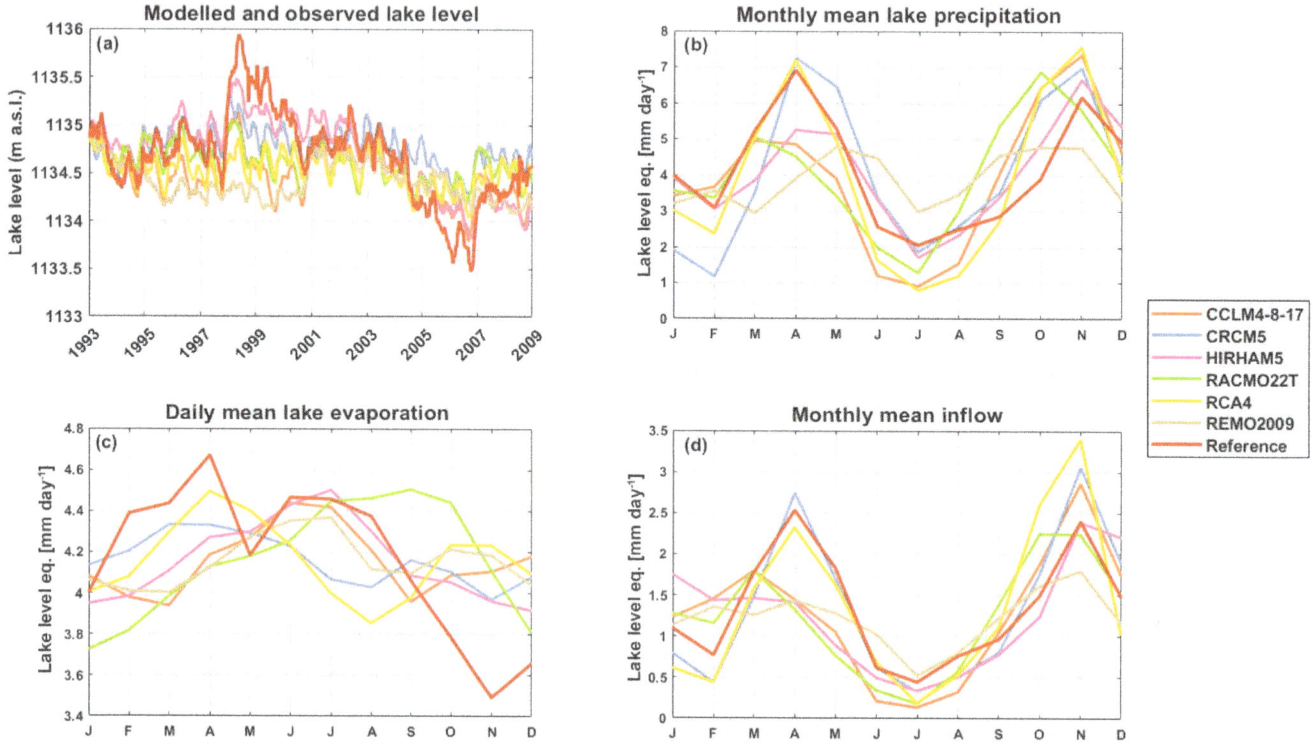

Figure E2. As in Fig. 6, but bias corrected using empirical quantiles.

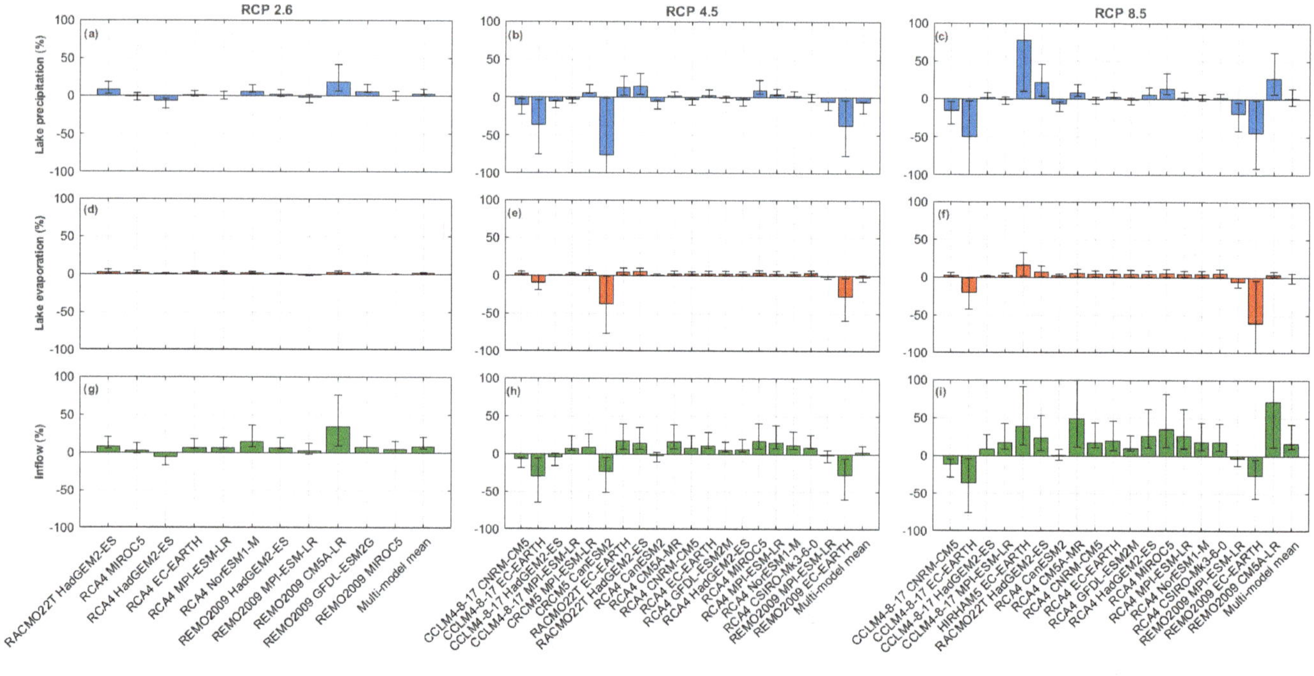

Figure E3. As in Fig. 7, but bias corrected using empirical quantiles.

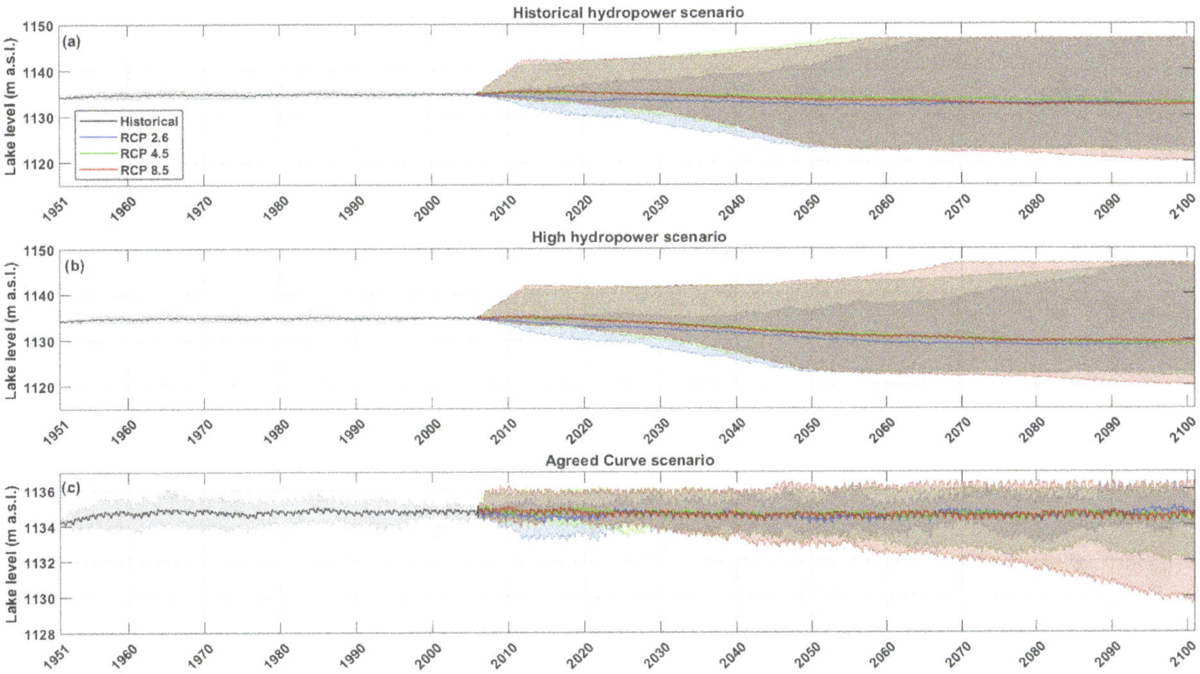

Figure E4. As in Fig. 8, but bias corrected using empirical quantiles.

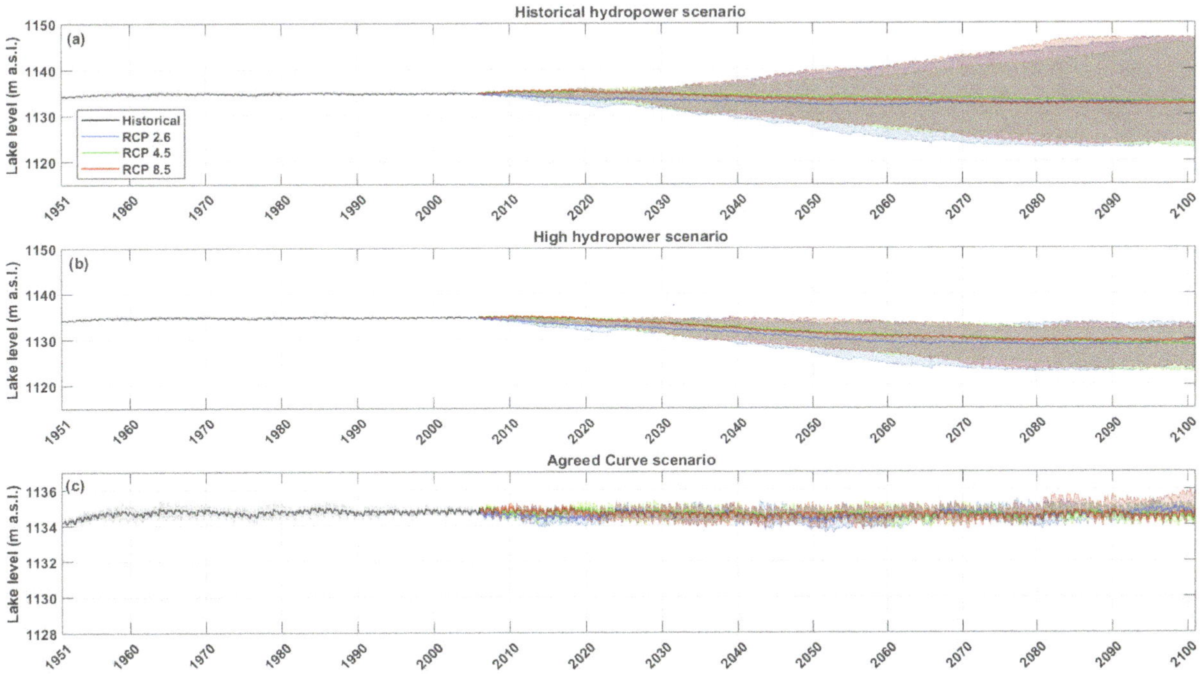

Figure E5. As in Fig. C1, but bias corrected using empirical quantiles.

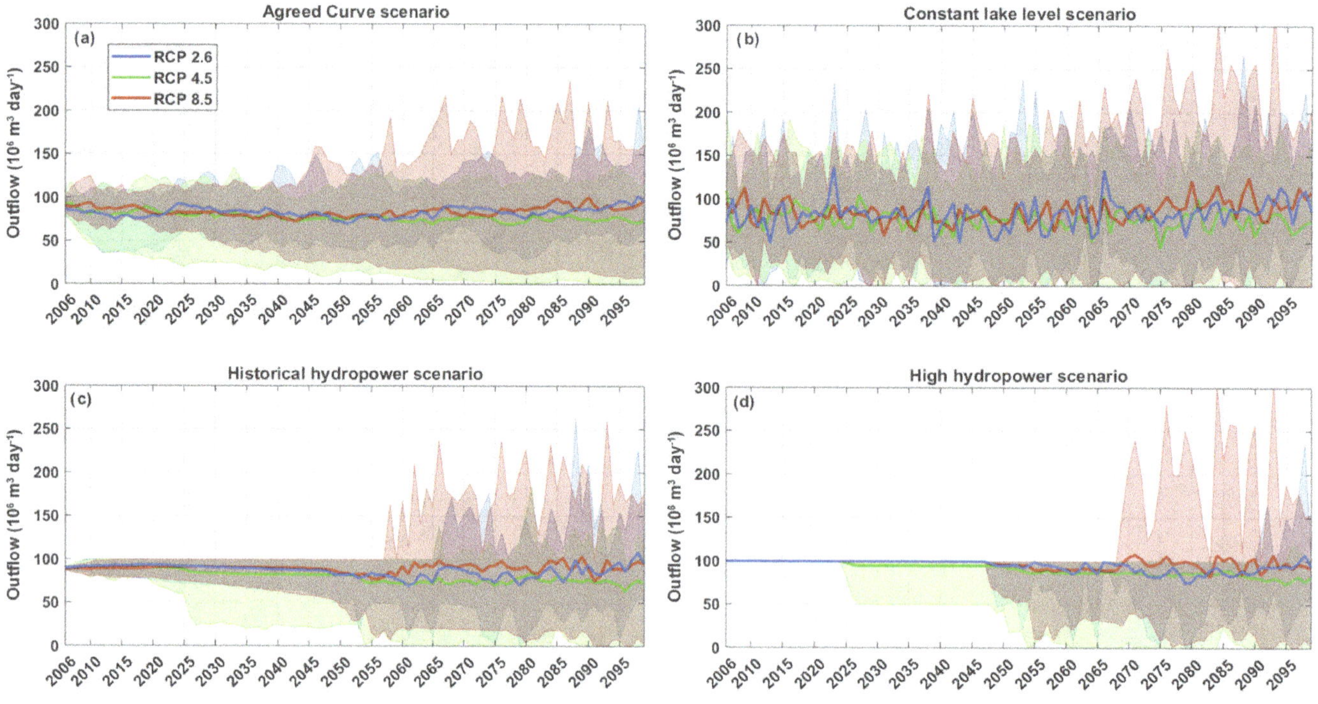

Figure E6. As in Fig. 9, but bias corrected using empirical quantiles.

10

Water scarcity under various socio-economic pathways and its potential effects on food production in the Yellow River basin

Yuanyuan Yin[1], Qiuhong Tang[1,2], Xingcai Liu[1], and Xuejun Zhang[1]

[1]Key Laboratory of Water Cycle and Related Land Surface Processes, Institute of Geographic Sciences and Natural Resources Research, Chinese Academy of Sciences, Beijing, China

[2]University of Chinese Academy of Sciences, Beijing, China

Correspondence to: Qiuhong Tang (tangqh@igsnrr.ac.cn)

Abstract. Increasing population and socio-economic development have put great pressure on water resources of the Yellow River (YR) basin. The anticipated climate and socio-economic changes may further increase water stress. Many studies have investigated the changes in renewable water resources under various climate change scenarios, but few have considered the joint pressure from both climate change and socio-economic development. In this study, we assess water scarcity under various socio-economic pathways with emphasis on the impact of water scarcity on food production. The water demands in the 21st century are estimated based on the newly developed shared socio-economic pathways (SSPs) and renewable water supply is estimated using the climate projections under the Representative Concentration Pathway (RCP) 8.5 scenario. The assessment predicts that the renewable water resources would decrease slightly then increase. The domestic and industrial water withdrawals are projected to increase in the next a few decades and then remain at the high level or decrease slightly during the 21st century. The increase in water withdrawals will put the middle and lower reaches in a condition of severe water scarcity beginning in the next a few decades. If 40 % of the renewable water resources were used to sustain ecosystems, a portion of irrigated land would have to be converted to rain-fed agriculture, which would lead to a 2–11 % reduction in food production.

This study highlights the links between water, food and ecosystems in a changing environment and sug-gests that trade-offs should be considered when developing regional adaptation strategies.

1 Introduction

The Yellow River (YR) is the second-longest river in China and is regarded as the cradle of Chinese civilization. The YR plays an important role in the development of the regional economy as the major source of freshwater for a large amount of people living there. As of 2010, there were 113.7 million inhabitants and 12.6 million hectares of cultivated land in the basin (Yellow River Conservancy Commission, YRCC, 2013). In addition, the lower reaches of the river supply freshwater for 2.86 million hectares of irrigated area and a population of 54.73 million located outside the basin (Fu et al., 2004). Increasing population and socio-economic development have put great pressure on the water resources of the basin. Anticipated climate and socio-economic changes may further increase water scarcity. The water managers of the basin will face great challenges meeting the human and environmental requirements for water. This water crisis in the YR basin has received much attention for many years.

Climate change and human water use are two major reasons for water crisis in the YR basin (Fu et al., 2004; Tang et al., 2008a; Wang et al., 2012). Numerous studies have investigated the changes in water supply due to climate change. Since the 1950s, streamflow of the river has decreased partly because of the decrease in precipitation and increase in temperature (Tang et al., 2008b; Xu, 2011; Wang et al., 2012). Some recent studies showed that there has been a substantial recovery of natural runoff over the past decade as a response to changes in precipitation, radiation, and wind speed (Tang et al., 2013; Liu et al., 2014). Climate projections suggest that temperature will continue to rise, but renewable water resources might decrease over the next few decades (Leng et al., 2015). Renewable water resources of the YR are likely to decrease due to both precipitation decrease and temperature increase over the next few decades (Li et al., 2012; Davie et al., 2013; Haddeland et al., 2014). However, water resources might increase by the end of the 21st century due to an increase in precipitation (Liu et al., 2011; Leng et al., 2015). The change in water availability under climate change suggests the need for adaptation.

Along with rapid economic development and population growth, water withdrawals from the YR basin for industrial and household use have increased significantly. Water consumption for irrigation has induced a streamflow decrease by about half in the past half century (Tang et al., 2007; Shi et al., 2012). The lower reaches of the YR frequently ran dry (i.e., no streamflow in the low flow season) in the 1980s and 1990s (Tang et al., 2008b). Thereafter, the YRCC implemented a water flow regulation rule, which enforced an upper limit on water withdrawals for the eight provinces that rely on water supply from the river (Cai and Rosegrant, 2004). The expected increase in economic prosperity together with a growing population, both within and outside of the basin, will increase water demand from the river and thus water scarcity may impose further constraints on development and social well being (Schewe et al., 2014). As water becomes increasingly scarce, there will be more competitions and conflicts among different water use sectors and regions (provinces). The current water flow regulation rule, which has been enforced since the late 1980s, might not be applicable in the 21st century.

Many studies have investigated the changes in water supply under various climate change but few have considered the joint pressure from both climate change and socio-economic development (Tang and Oki, 2016). It becomes important to develop qualitative scenario storylines to assess future water scarcity in a changing environment at the regional scale. These storylines would facilitate assessment of the water use competitions among different sectors and regions and thus aid development of adaptation strategy for the river basin. A few studies have tried to describe the main characteristics of future climate change scenarios and development pathways at the global scale (Elliott et al., 2014; Schewe et al., 2014). These efforts, though important, are too coarse for vulnerability assessment at the regional scale in the following aspects. First, the global studies do not consider the water flow regulation rule implemented by the local river administration, which sets limits on water withdrawals for each subbasin. Second, the global studies usually set a strict criterion on discharge reduction for human water use such as 40 % (Schewe et al., 2014) while human water use in the YR basin has already exceeded the criterion (YRCC, 2013). Third, the global studies often exhibit considerable biases in water supply assessment as most global models are not validated using streamflow observations. Fourth, the YR river supplies water for irrigation districts not only inside the river basin but also those located outside of the basin. The water demands outside the basin are generally not considered in the global scale assessments. In this study, we use streamflow observations in the YR basin to bias-corrected global model outputs (Ho et al., 2012; Hawkins et al., 2013), and present a multi-model analysis of water supply and demand narratives under different climate change scenarios and socio-economic pathways at the sub-basin scale (Fig. 1 and Table 1). The objectives of the analysis are (i) to describe the water supply and demand changes in a changing environment, (ii) to identify the possible time horizon when current management practices may no longer be sustainable, (iii) to investigate the contributions of different water demand sectors to water scarcity, and (iv) to assess the potential impacts of water scarcity on agricultural production.

2 Study area and data

2.1 Study area

The YR originates from the northern foothills of the Tibetan Plateau, runs through nine provinces and autonomous regions, and discharges into the Bohai Gulf (Fig. 1). Total area of the basin is 75 200 km^2. The YR basin lies in a temperate continental climate zone, and most parts of the basin belong to arid or semi-arid regions. The mean annual temperature in 1981–2010 ranges spatially from −5. to 15. °C in the basin, and it increases from north to south as a consequence of the decrease in latitude to the south (Fig. 2a). Precipitation has large spatial variation within the whole river basin. The mean annual precipitation in 1981–2010 ranges spatially from 60 to 900 mm in the basin, and shows an increasing trend from northwest to southeast (Fig. 2b). The temperature and precipitation are projected to increase during the 21st century under the Representative Concentration Pathway (RCP) 8.5 emission scenario (see Fig. S1 in the Supplement).

There are six land cover types in the basin (Fig. 2c). The dominant land cover types are grasslands (47.6 %), croplands (26.1 %), and forest and shrublands (13.4 %). The urban and built-up lands are concentrated along the river. The croplands are mainly distributed in the lower reach of the YR. The land cover changes influence the hydrological cycle (Tang et al., 2008b, 2012), and may alter runoff (Sterling et al., 2013). However, interactions among land cover change, climate change, and hydrological cycle are complicated. To focus on runoff responses to climatic variations, the fixed land cover map was used in this study.

In 2010, the population within the basin boundary was more than 100 million, representing about 9 % of China's population. The basin's gross domestic product (GDP) represented 8 % of China's GDP in 2010. Both population and GDP are concentrated along the river (Fig. 2d and e). The projected population increases first and then decreases during the 21st century (see Fig. S2a in the Supplement). China's

Table 1. The eight sub-basins of the Yellow River (YR) basin.

Sub-basins		Area ($\times 10^3$ km^2)	Water use proportion (%)	Irrigated area (km^2)	Rain-fed area (km^2)	Note
Upper reaches	I	127	0.57	219	27	Above LYX station
	II	87	8.23	2680	1706	LYX to LZ
	III	157	37.45	23 692	2106	LZ to HKZ
Middle reaches	IV	107	3.5	1591	3940	HKZ to LM
	V	185	16.64	25 422	12 311	LM to SMX
	VI	40	6.16	5717	2956	SMX to HYK
Lower reaches	VII	50.6	27.2	42 824	2430	HYK to LJ, including irrigation districts outside the basin but receiving water from YR
Endorheic basin	VIII	42	0.25	446	225	Endorheic basin

Note: the sub-basins and names of hydrological stations are given in Fig. 1.

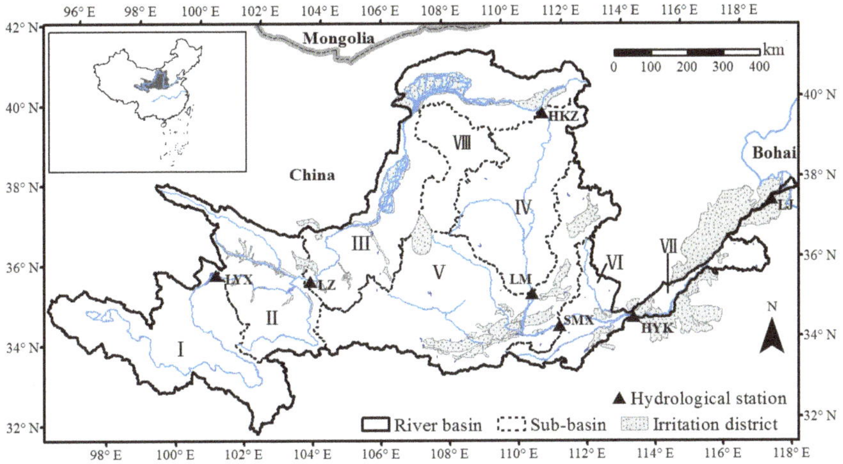

Figure 1. The Yellow River (YR) basin, the eight sub-basins, and location of Longyangxia (LYX), Lanzhou (LZ), Hekouzhen (HKZ), Longmen (LM), Sanmenxia (SMX), Huanyuankou (HYK), and Lijin (LJ) hydrological stations. The insert panel shows the location of the YR basin in China.

population has been greatly affected by its fertility policy (Peng, 2011). Under the current fertility policy, China's population would continue to grow, and then begin to decrease as the aging of the population accelerates (Chen and Liu, 2009; Peng, 2011). The range of projected population at the end of the 21st century varies from 50 million to more than 100 million, with shared socio-economic pathway (SSP) 5 at the bottom of the range and SSP3 at the top. The range of projected GDP at the end of the 21st century varies from 21 000 billion Yuan to more than 40 000 billion Yuan. SSP5, with its focus on development, has the highest GDP projections and SSP3, representing the scenario with lowest international cooperation, has the lowest income projection (O'Neill et al., 2017).

2.2 Data

The data used in this study are summarized in Table 2. The simulated runoff data for the period 1971–2099, and the simulated irrigation water use and crop yield data for the period 1981–2099 were obtained from the Inter-Sectoral Impact Model Intercomparision Project (ISI-MIP; Warszawski et al., 2014). These simulated data were provided at a spatial resolution of 0.5° × 0.5°. The runoff data were produced by 10 global gridded hydrological models (GGHMs), namely, DBH, H08, LPJmL, MacPDM, MATSIRO, MPI-HM, PCR-GLOBWB, VIC, WaterGAP, and WBM (see Table S1 in the Supplement). The irrigation water use and crop yield data were produced by six global gridded crop models (GGCMs), namely, EPIC, GEPIC, LPJmL, LPJ-GUESS, pDSSAT, and PEGASUS (see Table S2 in the Supplement). Forcing data, bias-corrected by the ISI-MIP team for the GGHMs and GGCMs, were derived from climate projections of five global climate models (GCMs), namely, HadGEM2-ES, IPSL-CM5A-LR, MIROC-ESM-CHEM, GFDL-ESM2M, and NorESM1 (see Table S3 in the Supplement) under the RCP 8.5 scenario (Warszawski

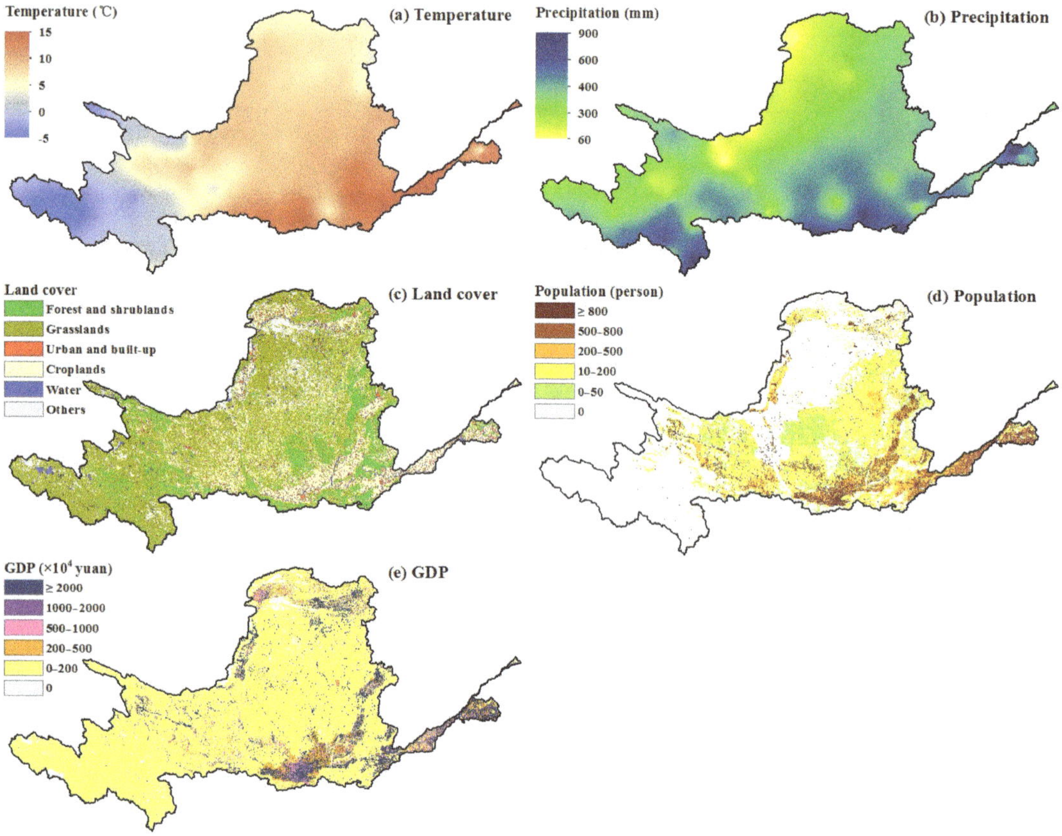

Figure 2. (a) Mean temperature (1981–2010), **(b)** annual mean precipitation (1981–2010), **(c)** land cover in 2010, **(d)** population in 2010, and **(e)** gross domestic product (GDP) in 2010 in the YR basin.

Table 2. Datasets used in this study.

Datasets		Spatial and temporal resolution	Source
Simulated runoff data		0.5° × 0.5°, 1971–2099	The Inter-Sectoral Impact Model Intercomparison Project (ISI-MIP)
Simulated yield data		0.5° × 0.5°, 1971–2099	
Simulated irrigation water data		0.5° × 0.5°, 1971–2099	
Rain-fed and irrigation area data		0.5° × 0.5°, 2000	
Population data	1 km grid population dataset of China	1 km × 1 km, 2005	Institute of Geographic Sciences and Natural Resources Research
	Historical population data of China	Country, 1981–2013	National Bureau of Statistics of China
	SSP population data*	0.5° × 0.5°, 2010–2099	ISI-MIP
GDP data	1 km grid GDP dataset of China	1 km × 1 km, 2005	Institute of Geographic Sciences and Natural Resources Research
	Historical GDP data of China	Country, 1981–2013	National Bureau of Statistics of China
	SSP GDP data*	Country, 2010–2099	Organization for Economic Co-operation and Development (OECD)
Official exchange rate data		Country, 2005	World Bank
Observed runoff data		1971–2000	Hydrological Bureau of the Ministry of Water Resources of China

* SSP is short for shared socio-economic pathways.

et al., 2014). The global irrigated and rain-fed crop area data (MIRCA2000), which consist of all major food crops, such as wheat, rice, maize, and soybean, were also obtained from ISI-MIP. The MIRCA2000 dataset refers to the crop area over the period of 1998–2002 (Portmann et al., 2010). According to the survey by the Ministry of Water Resources (MWR) of China, the cropland area of the YR basin in the 2000s is 16 million ha. The MIRCA2000 dataset shows that the cropland area is 16.27 million ha, a value quite close to the MWR estimate. Although the cropland area may change in the future due to local adaptation to environmental change, the projection of land use change is beyond the scope of this paper. The cropland map is fixed throughout the 21st century in this study.

The gridded population and GDP datasets over China were provided by the Institute of Geographic Sciences and Resources Research (IGSRR), Chinese Academy of Sciences (CAS). The population and GDP datasets refer to the conditions in 2005 (Fu et al., 2014; Huang et al., 2014). The datasets were developed based on remote-sensing-derived land use data and the statistical population and GDP data of each county in China. The population and GDP data were provided with a spatial resolution of 1 km and were resampled to 0.5° in this study with ArcGIS. The annual total population and GDP data of China during 1981–2013 were obtained from the National Bureau of Statistics of China (NBC). Using a simple linear downscaling method (Gaffin et al., 2004), we downscaled the annual total population and GDP data to the gridded maps. The future water demand should be closely related to the growth of GDP and population in the basin, and the SSPs offer the possibility for describing different conditions in terms of future sectoral water demand. Quantitative projections for population and GDP were developed for the 2010–2099 period based on the SSP Scenario Database, available at https://secure.iiasa.ac.at/web-apps/ene/SspDb. The population and GDP projections were provided at country level at 5-year intervals. The country-level population data were gridded to 0.5° according to the 2010 Gridded Population of the World (GPWv3) dataset provided by the Center for International Earth Science Information Network (CIESIN), Columbia University. The country-level GDP data were provided in US dollars at 5-year intervals. The GDP data were converted to Chinese Yuan using the official exchange rate provided by the World Bank. The GDP data from the SSP Scenario Database were regridded to the 0.5° GDP of China grid and were linearly interpolated in time to obtain annual values (Gaffin et al., 2004). The assumption underlying the downscaling method is that the annual growth rate of GDP at each grid, at any year, is equal to the growth rate of China. The observed runoff data over 1971–2000 of four major hydrologic stations at the main stream of the YR (Lanzhou, LZ; Longmen, LM; Sanmenxia, SMX; and Huanyuankou, HYK) were collected from the Hydrological Year Book of the Ministry of Water Resources of China.

3 Method

The river basin was divided into eight sub-basins in order to understand the regional patterns of water abundance and scarcity (Fig. 1). The area of sub-basins varies from 40 000 to 185 000 km^2 (Table 1). There are seven sub-basins along the main stream of the basin and one endorheic basin (sub-basin VIII) that does not flow to the main stream of the river. Because the irrigation districts located outside of the basin in lower reaches get their water supply from the river, the sub-basin in the lowest reaches (sub-basin VII) consists of one part in the river basin and these irrigation districts.

The mean annual runoff, including both the subsurface and surface runoff, is assumed to be the renewable water resources (Oki and Kanae, 2006). Because the bias of GGHMs is usually large (Hattermann et al., 2017), we used the streamflow observations at the YR basin to bias correct the model simulated runoff (see "Supplemental Methodology" in the Supplement). We compared the bias-corrected runoff of 10 GGHMs with the streamflow observations at the four selected hydrological stations (see Fig. S3 in the Supplement). The bias-corrected runoff can reproduce the streamflow well in the reference period (1971–2000). The bias-corrected runoff was aggregated for each sub-basin and the river basin as a whole (see Fig. S4 in the Supplement). In order to maintain the river in a desired environmental condition, we assumed that only a part of the renewable water resources could be appropriated by humans. The net water withdrawal, i.e., water withdrawal minus return flow, accounts for 53 % of renewable water resources in the 1980s (Zhang et al., 2004) and 72 % in the 2010s (YRCC, 2013). Annual water supply was estimated as annual runoff multiplied by the ratio of human water appropriation (RHWA); i.e., the proportion of renewable water resources that is allowed to be used by humans. Because water supply is largely determined by the RHWA, water supply was estimated with RHWA values of 50, 60, and 70 %. The water flow regulation rule currently implemented by YRCC (YRCC, 2013) sets the upper limit on water withdrawals for each sub-basin. According to the rule, the maximum water use proportion is prescribed for each sub-basin (Table 1). The annual water supply was calculated for each GCM-GGHM pair. There were 5 GCMs and 10 GGHMs, making 50 model pairs. The multi-model-ensemble median of water supply from all the available model pairs was calculated.

On the water demand side, the consumptive agricultural, domestic, and industrial water demands were considered. Agricultural water demand consists of the demands for irrigation and livestock. As the livestock demand is relatively small and the related statistical data were unavailable in the basin (YRCC, 2013), only irrigation demand was considered. The irrigation water demand was estimated by the GGCM. Table S2 shows an overview of the six GGCMs. Four crops, i.e., wheat, rice, maize, and soybean, were taken into account because these crops accounted for over 80 % of total cropland area in the YR basin. The irrigation water demands were aggregated for each sub-basin and the river basin as a whole to get irrigation water withdrawal (IrrWW). The multi-model-ensemble median of IrrWW from all the available GCM-GGCM pairs (five GCMs × six GGCMs) was calculated.

Net domestic and industrial water withdrawals were linked to the main driving forces of water in the domestic and industrial sectors, i.e., population, GDP, and electricity production (Alcamo et al., 2003). Following Hanasaki et al. (2013a), the annual net domestic water withdrawal was calculated using Eq. (1).

$$W_{\mathrm{dom}} = \mathrm{Pop} \times \left(i_{\mathrm{dom},t0} + s_{\mathrm{dom,cat}} \times (t - t0)\right) \times 0.365, \qquad (1)$$

where W_{dom} is the net domestic water withdrawal ($\mathrm{m^3\,yr^{-1}}$), Pop is the population, $i_{\mathrm{dom},t0}$ is the domestic water intensity for the base year ($\mathrm{L\,day^{-1}\,person^{-1}\,yr^{-1}}$), $s_{\mathrm{dom,cat}}$ is the domestic water intensity change rate ($\mathrm{L\,person^{-1}\,day^{-1}\,yr^{-1}}$), and the multiplier 0.365 is applied for unit conversion.

Industrial water use includes two main components: water withdrawal for the manufacturing sector and for cooling the thermoelectric plants in the electricity sector. The manufacturing water withdrawal is positively correlated with the economic metric manufacturing gross value added (Dziegielewski et al., 2002). Following Flörke et al. (2013) and Wada et al. (2016), the annual net industrial water withdrawal depends on the value added of manufacturing sectors and the water use intensities (Eq. 2).

$$W_{\mathrm{ind}} = \mathrm{GDP_{manu}} \times i_{\mathrm{ind},t0} \times \left(1 - s_{\mathrm{ind,cat}}\right)^{(t-t0)}, \qquad (2)$$

where W_{ind} is the net industrial water withdrawal ($\mathrm{m^3\,yr^{-1}}$), $\mathrm{GDP_{manu}}$ is the value added of manufacturing sectors (Chinese Yuan), $i_{\mathrm{ind},t0}$ is the industrial water intensity for the base year ($\mathrm{m^3}$ per 10 000 Yuan), and $s_{\mathrm{ind,cat}}$ is the industrial water intensity change rate (%). The value added of manufacturing sectors is calculated by multiplying the GDP projection by the share of manufacturing gross value added in total GDP for non-OECD countries from the UNEP GEO4 Driver Scenarios (Hughes, 2005). Change in water withdrawal for thermal power industry is not considered in this study for two reasons. First, water conservation technology, such as dry cooling has been widely adopted in northern China and thermal power industry, is not the major water user in the YR basin (Zhang et al., 2016). Second, the projection of future change is subject to large uncertainties as increase in demand is complicated with the advanced water conservation technologies (Zhang et al., 2016).

The change rates of domestic and industrial water intensity are dependent on the technology scenario of the SSPs. A high-technology scenario was set for SSP1 and SSP5, medium for SSP2, and low for SSP3 and SSP4 (O'Neill et al., 2017). For domestic water use, SSP1 and SSP5 would be more efficient, whereas SSP3 and SSP4 would be less efficient; SSP2 would be intermediate between the two groups (Hanasaki et al., 2013a). The domestic water intensity change rate was proposed in Table S4 (Hanasaki et al., 2013a). For industrial water use, the change rate is set at 1.1 % for SSP1 and SSP2, 0.6 % for SSP2 and SSP4, and 0.3 % for SSP3 (Wada et al., 2016). In this study, the base year is 2005. The domestic water intensity for the base year is 83.6 ($\mathrm{L\,day^{-1}\,person^{-1}\,yr^{-1}}$) while the industrial water intensity for the base year is 205.4 ($\mathrm{m^3}$ per 10 000 Yuan) in the YR basin.

Annual water demand was calculated as the sum of net water withdrawal requirement for agricultural, domestic, and industrial uses. In order to measure water supply stress, the water supply stress index (WaSSI) was used. WaSSI is defined as the ratio of annual water demand to annual water supply for a specific watershed. It measures whether water supplies are sufficient for all net withdrawal requirements within a watershed to be met concurrently. The WaSSI was calculated for each sub-basin and the whole basin to assess the water abundance/scarcity condition. To investigate the contributions of different water demand sectors to water scarcity, WaSSI was calculated for the major sectors, i.e., domestic, industrial, and agricultural (denoted as irrigation hereafter because only irrigation was considered) sectors, at the end of the 21st century. If the WaSSI is projected to be greater than 1, water resources cannot sustain the socio-economic development and water scarcity occurs. The greater the WaSSI value is, the greater water scarcity is expected. We assume that irrigated agriculture has the lowest priority of all water consumers under water scarcity. If water scarcity occurs in a given year for a specific watershed, irrigation will be constrained by reducing the irrigated fraction of the cropland (Elliott et al., 2014); the agricultural production of the watershed, calculated as calorie content of the major crop yields, will be the sum of production over the expanded rain-fed fraction of the cropland and the shrunken irrigated fraction. If water is sufficient in a given year for a given sub-basin, we will assume that no rain-fed areas will be converted for irrigation.

The water supply and demands were assessed for each year during 1981–2099. However, the 30-year moving averages were also computed and illustrated. The 30-year window ensures that year-to-year variability does not dominate the signal. The center year of the 30-year moving average was used to denote the 30-year period. For example, the average of the historical period of 1981–2010 was denoted as 1995.

4 Results

4.1 Changes of water supply

Figure 3 shows the estimated water supply in the YR basin and eight sub-basins in the 21st century. With an increase in the RHWA, the water supply is projected to increase during

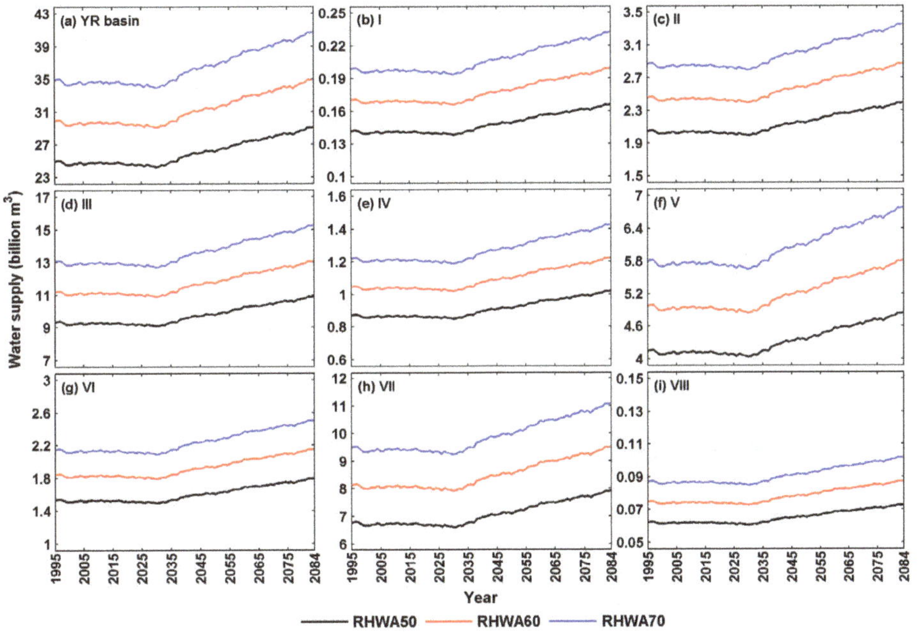

Figure 3. Annual water supplies in the YR basin and eight sub-basins during the 21st century. The water supply was estimated with RHWA values of 50, 60, and 70 %, respectively.

the 21st century in the YR basin. The average water supply is 24.7, 29.8, and 34.8 billion m^3 per year during the historical period under three RHWA – 50, 60 and 70 %. The water supply is projected to decrease slightly from 1995–2035 due to an increase in air temperature (see Fig. S1 in the Supplement), and is projected to increase from 2036 to 2084 due to an increase in precipitation under all RHWAs (see Fig. S1 in the Supplement). The result is consistent with the conclusions from Zhao et al. (2009). The water supply is projected to be 29.1, 34.9, and 40.6 billion m^3 per year at the end of the 21st century under RHWA50, RHWA60, and RHWA70, respectively, increasing by about 17.8 % from that during the historical period. The water supply is projected to first decrease and then increase in all the sub-basins during the 21st century. The average water supply of sub-basin III has the maximum value of 9.3 billion m^3 per year during the historical period and rises to 10.9 billion m^3 per year by the end of the 21st century under RHWA50. The average water supply of sub-basin VIII has the minimum value of 0.06 billion m^3 per year during the historical period and rises to 0.73 billion m^3 per year by the end of the 21st century under RHWA50.

4.2 Changes of total and sectoral water demand

Figures 4, S5, and S6 show the estimated total and sectoral (domestic, industrial, and irrigation) water demand in the YR basin and eight sub-basins under five SSPs in the 21st century. In the YR basin, the total water demand is projected to increase from 24 billion $m^3\,yr^{-1}$ in 1995 to close to 27.8, 33.1, 23.8, 30.0, and 30.3 billion $m^3\,yr^{-1}$ in 2084 under SSP1, SSP2, SSP3, SSP4, and SSP5, respectively. This increase is primarily driven by the growth in the industrial water withdrawal, accounting for at least 32 % of the total in 2084. Irrigation is the dominant water use sector during the period 1995–2084. Domestic water withdrawal is projected to increase before 2025, as both population and domestic water use intensities will increase, and then to decrease due to a decrease in population (see Fig. S2a in the Supplement). Industrial water withdrawal is projected to rapidly increase before 2050, as the value added of manufacturing sectors will increase (see Fig. S2b in the Supplement), and then is projected to decrease slightly due to a decrease in industrial water use intensities. The irrigation water withdrawal is projected to decrease from 20 billion $m^3\,yr^{-1}$ in 1995 to close to 17 billion $m^3\,yr^{-1}$ in 2084 under RCP 8.5. Irrigation water withdrawal is projected to increase slightly during 1995–2030, and is projected to decrease during 2031–2084. The total water demand, both domestic and industrial water withdrawals, are also projected to increase and then decrease in each sub-basin during the 21st century under all SSPs. In the sub-basins III, V, VII, and VIII, although the industrial water withdrawal would increase rapidly, irrigation is always the dominant water use sector during the 21st century.

In the sub-basins I, II, IV, and V, the irrigation and domestic water withdrawal are the dominant water use sectors at the begin-ning of the 21st century, while the industry would become the dominant water use sector after the 2030s.

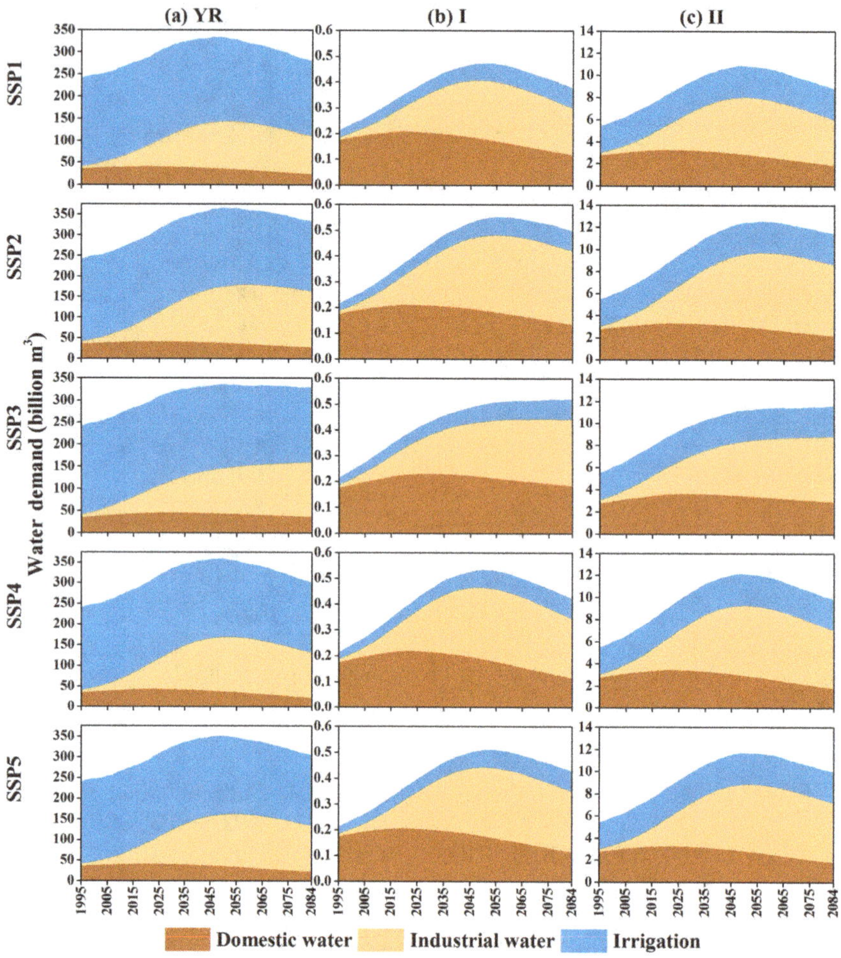

Figure 4. Estimated sectoral (domestic, industrial, and irrigation) and total annual water demand (million $m^3\ yr^{-1}$) in the YR basin, and sub-basin I and II during the 21st century.

4.3 Water abundance/scarcity and sectoral contributions to water scarcity

Figure 5 shows the average annual WaSSI for the YR basin and eight sub-basins throughout the 21st century under the five different SSPs. The WaSSI is projected to increase due to an increase in demand during the 21st century. Under RHWA50, the YR basin is projected to have a WaSSI greater than 1 after 2000s for all SSPs, which means that water demand outstrips supply. The WaSSI is projected to decrease with an increase in the RHWA. Under RHWA70, water scarcity would not occur in the 21st century for all SSPs. The upper reaches of the YR basin (sub-basins I, II, and III) are projected to have a WaSSI less than 1, meaning that water supply would be more than water demand during the 21st century for all SSPs under all RHWAs. The endorheic basin in the YR basin (sub-basin VIII) is the only region in which the WaSSI is always larger than 1, meaning that the water would be scarce during the 21st century for all SSPs under all RHWAs. In the middle and lower reaches of the YR basin (sub-basins IV, V, VI, and VII), the WaSSI would begin to be large than 1 at the beginning of the 21st century under RHWA50. With an increase in the RHWA, water scarcity would occur later. If the RHWA reaches 70 %, water supply will be more than water demand during 1995–2084 in sub-basins IV under all SSPs.

Figure 6 shows the WaSSI calculated as the ratio of annual water demand and sectoral (domestic, industrial, and irrigation) water withdrawals to annual water supply under RHWA50 for the YR basin and eight sub-basins at the end of the 21st century under the five different SSPs. In the YR basin, the WaSSI calculated as annual water demand to water supply is larger than 1 under all SSPs except SSP1, meaning that the water scarcity would occur at the end of the 21st century. Among the three different water demand sectors, irrigation is projected to contribute most (about half) to WaSSI for all SSPs, and the domestic sector is projected to have the smallest contribution to WaSSI (less than 0.1) for all SSPs except SSP3. Both WaSSI and contributions from the water demand sectors to water scarcity would decrease with an in-

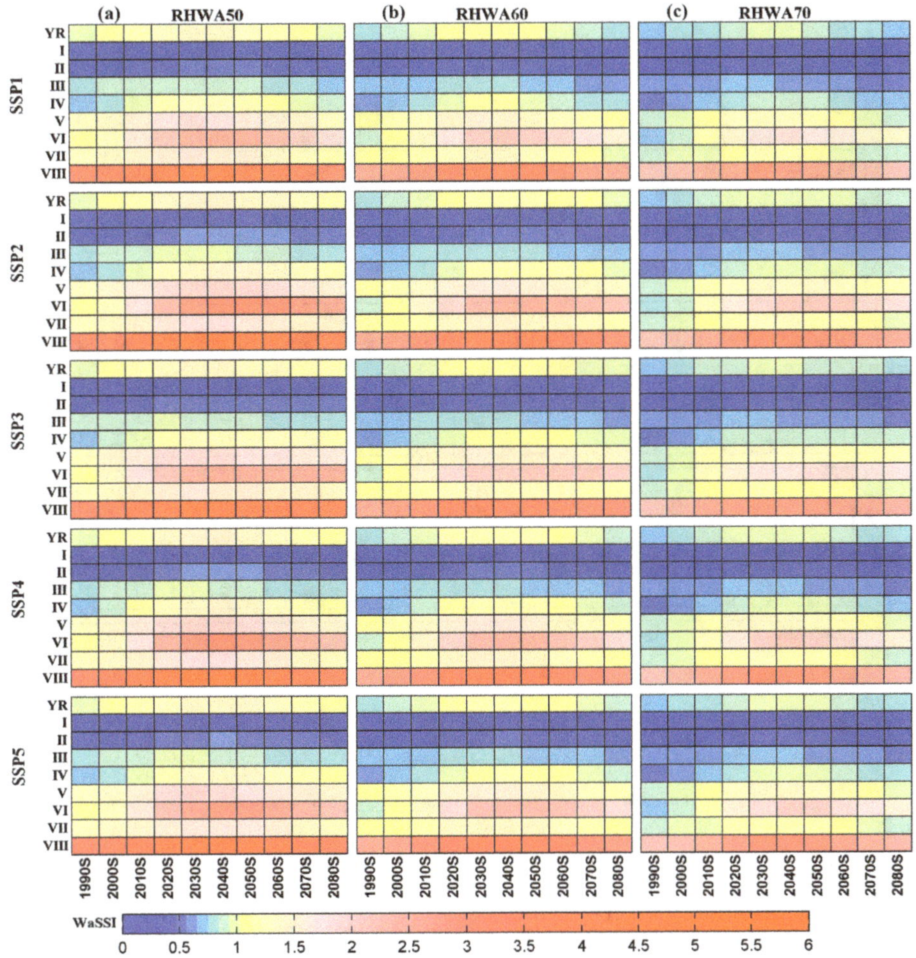

Figure 5. Average annual water supply stress index (WaSSI) for the YR basin and eight sub-basins throughout the 21st century under five different SSPs. Water supply was estimated with an RHWA value of 50, 60, and 70 % in the left, center, and right column, respectively. The WaSSI is calculated for each decade. Water scarcity occurs in a given basin if WaSSI is greater than 1.

crease in the RHWA. With an increase in the RHWA, more water can be used to meet water demands. Consequently, WaSSI becomes smaller with higher RHWA (see Figs. S7 and S8 in the Supplement). In sub-basins III, V, VII, and III, irrigation is the main contributing factor to WaSSI, and the industrial sector is another important contributing factor. An increase in GDP would make the industrial sector become the main contributing factor to WaSSI for sub-basins I, II, IV, and VI. Because both population and GDP are concentrated in the middle and lower reaches, the estimated WaSSI is larger than one in those sub-basins, but smaller than one for the sub-basins in the upper reaches.

4.4 Agricultural loss due to irrigation water scarcity

The climate change and irrigation water shortage in the YR basin would have significant implications for the food security of these regions. Considering the CO_2 fertilization effect, agricultural production will be enhanced by climate change, and is projected to increase by close to 15.1 % by the middle of the 21st century compared with production during the historical period in the YR basin (Fig. 7). Irrigation water scarcity could necessitate the reversion of cropland from irrigated to rain-fed management, and would lead to decreased agricultural production. Under RHWA50, irrigation water scarcity in the basin could necessitate the reversion of about half of the cropland from irrigated to rain-fed management by the middle of the 21st century. Considering the CO_2 fertilization effect, irrigation water scarcity would lead to 15.7, 25.4, 17.7, 22.7, and 21 % of present-day total production reduction under SSP1, SSP2, SSP3, SSP4, and SSP5, respectively, in 2050 (Fig. 7). The change rate of production is projected to decrease with an increase in the RHWA. Under RHWA60, the reduction of agriculture production in 2050 might be 1–11 % of present-day total production reduction. Under RHWA70, the reduction of agriculture production in 2050 would not occur under all SSPs. Considering the impacts of climate and water supply stress, the reduction

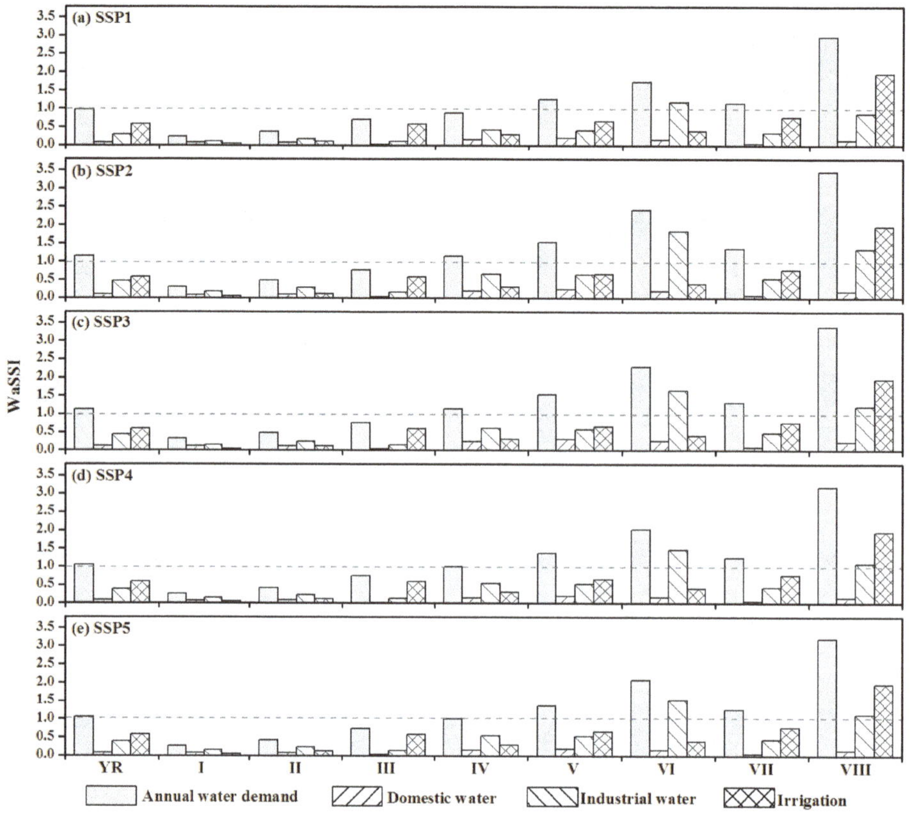

Figure 6. WaSSI calculated as the ratio of annual water demand and sectoral (domestic, industrial, and irrigation) water withdrawals to annual water supply for the YR basin and eight sub-basins at the end of the 21st century under the five different SSPs. The annual water supply was estimated with an RHWA value of 50 % (RHWA50).

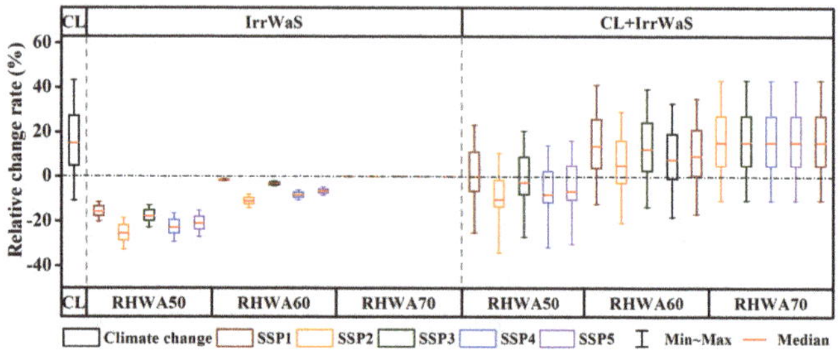

Figure 7. Comparison of relative change of agriculture production with only climate impact (CL), only irrigation water scarcity impact (IrrWaS), and combined climate and irrigation water scarcity impact (CL + IrrWaS) in the YR basin in the 2050s (%).

of agriculture production in 2050 occurs for all SSPs under RHWA50. Under RHWA60 and RHWA70, the agriculture production is projected to increase under each SSP by the middle of the 21st century.

5 Discussion

The renewable water resources will be affected by projected changes in precipitation and temperature (Schewe et al., 2014), and in the RHWA. The water supply in the YR basin would first decrease and then increase due to the impact of temperature and precipitation changes over the 21st century (see Fig. S1 in the Supplement). However, the true water shortage might be larger because the CMIP5 models may overestimate the magnitude of precipitation in the YR basin during the 21st century (Chen and Frauenfeld, 2014). The RHWA of the YR basin has increased to 75.6 % during the beginning of the 21st century (Shi et al., 2012) from about 50 % in 1980s (Zhang et al., 2004). An increase in RHWA tends to result in increase in water supply and reductions in irrigation water scarcity and loss of agriculture production (Figs. 1 and 7). Therefore, improvement of the RHWA could alleviate the water shortages in this region. However, because of the different geographical and economic conditions among the sub-basins, the impact of the RHWA should be considered when we analyze the water resources of the sub-basins.

To quantify the domestic and industrial water withdrawal is difficult because the future water withdrawals are influenced by a combination of social, economic, and political factors. However, a few hydrologic modeling frameworks have integrated methods to estimate the water withdrawals, e.g., H08 (Hanasaki et al., 2010, 2013a, b), PCR-GLOBWB (Wada et al., 2011, 2014, 2016), and WaterGAP (Flörke et al., 2013). The differences in these approaches result in significantly different projections even with the same set of scenario assumptions (Wada et al., 2016). Our study does not consider the change in water for the thermal power industry, which accounts for about 30 % of industrial water use in the YR basin (Zhang et al., 2016); it might lead to an underestimation of the contribution of industrial water withdrawal to water scarcity.

With the currently implemented water flow regulation rule, water is projected to be scarce in sub-basins located in the middle and lower reaches of the YR basin characterized by a generally large population and GDP, whereas water is projected to be abundant in sub-basins located in the upper reaches of the YR basin characterized by a small population and GDP during the 21st century. In order to alleviate the water shortages in the middle and lower reaches, a new water flow regulation rule should be adopted.

In order to solve the problem of water resources shortages in the more arid and industrialized northern China, the South-to-North Water Diversion Project has been undertaken. One aim of the project is to channel the fresh water from the Yangtze River in southern China to the YR basin (YRCC, 2013). By 2030, about 9.7 billion m^3 of fresh water from the Yangtze River will have been drawn to the YR basin (YRCC, 2013). This could alleviate the water shortage in the YR basin to some degree.

6 Conclusions

In this study, we assessed the change in renewable water resources of the YR basin under climate change and the changes in domestic and industrial water withdrawals in the basin under socio-economic change in the 21st century. The results show that the renewable water resources are projected to first slightly decrease and then increase in the YR basin and each sub-basin with an increase in temperature and precipitation in the 21st century under RCP 8.5. Irrigation is the dominant water use sector before the 2030s, but irrigation and industry sectors are the dominant water users thereafter. With social and economic development, domestic and water withdrawals are projected to increase first and then remain at a high level or decrease slightly during the 21st century.

Water is always scarce in the endorheic basin, whereas water is always abundant in the sub-basins located in the upper reaches of the YR basin in the 21st century under all RHWAs and SSPs. Due to a water withdrawal increase in industrial sectors, the available water resources cannot sustain all water use sectors in the next few decades in the YR basin and the sub-basins located in the middle and lower reaches of the basin. The water resources shortage is most serious under SSP2, and the use of 60 % renewable water resources cannot sustain all the water use sectors in the YR basin. With the three water demand sectors considered, the industrial water withdrawal is the main contributing factor to water scarcity in sub-basin I, II, IV, and V, whereas the irrigation water withdrawal is the main contributing factor to water scarcity in sub-basin III, V, VII, and VIII.

Although climate change may have a positive impact on agriculture through the CO_2 fertilization effect in most regions of the YR basin (Yin et al., 2015), irrigation water scarcity would lead to the net loss of agricultural production. With the CO_2 fertilization effect, if more than 40 % of the renewable water resources are used to sustain ecosystems, a portion of irrigated land would have to be converted to rainfed agriculture, which would lead to a 2–11 % reduction in food production. It should be noted that the change in water use for the thermal power industry was not considered in this study. This might cause underestimation of the water stress index. Nevertheless, this study highlights the linkage between water and food security in a changing environment in the YR basin, and suggests that a trade-off should be considered when developing regional adaptation strategies.

7 Data availability

The SSP population data and model outputs of GGCMs and GGHMs were provided by the Inter-Sectoral Impact Model Intercomparison Project (https://www.isimip.org, ISIMIP, 2016). The SSP GDP data were provided by Organization for Economic Co-operation and Development (http://www.oecd.org/, OECD, 2016). The official exchange rate data of China were provided by the World Bank (http://data.worldbank.org/, World Bank, 2015). The historical population and GDP data of China were obtained from the National Bureau of Statistics of China (http://www.stats.gov.cn/tjsj/, NBSC, 2015). The 1 km grid population dataset and 1 km grid GDP dataset of China were provided by Data Center for Resources and Environmental Sciences, Chinese Academy of Sciences (http://www.resdc.cn, RESDC, 2015).

Competing interests. The authors declare that they have no conflict of interest.

Acknowledgements. This research is supported by the National Natural Science Foundation of China (41425002), the Key Research Program of the Chinese Academy of Sciences (ZDRW-ZS-2016-6-4), and the National Youth Top-notch Talent Support Program in China. We would like to thank the anonymous reviewers and the editor for their valuable and constructive suggestions. Thanks are due to Huijuan Cui for her comments.

Edited by: L. Wang

References

Alcamo, J., Döll, P., Henrichs, T., Kaspar, F., Lehner, B., Rösch, T., and Siebert, S.: Development and testing of the WaterGAP2 global model of water use and availability, Hydrolog. Sci. J., 48, 317–337, doi:10.1623/hysj.48.3.317.45290, 2003.

Cai, X. and Rosegrant, M. W.: Optional water development strategies for the Yellow River basin: Balancing agricultural and ecological water demands, Water Resour. Res., 40, W08S04, doi:10.1029/2003WR002488, 2004.

Chen, L. and Frauenfeld, O. W.: A comprehensive evaluation of precipitation simulations over China based on CMIP5 multimodel ensemble projections. J. Geophys. Res.-Atmos., 119, 5767–5786, doi:10.1002/2013JD021190, 2014.

Chen, W. and Liu, J. J.: Future population trends in China: 2005–2050. Centre of Policy Studies (COPS), Monash University, Wellington Road, Australia, 2009.

Davie, J. C. S., Falloon, P. D., Kahana, R., Dankers, R., Betts, R., Portmann, F. T., Wisser, D., Clark, D. B., Ito, A., Masaki, Y., Nishina, K., Fekete, B., Tessler, Z., Wada, Y., Liu, X., Tang, Q., Hagemann, S., Stacke, T., Pavlick, R., Schaphoff, S., Gosling, S. N., Franssen, W., and Arnell, N.: Comparing projections of future changes in runoff from hydrological and biome models in ISI-MIP, Earth Syst. Dynam., 4, 359–374, doi:10.5194/esd-4-359-2013, 2013.

Dziegielewski, B., Sharma, S. C., Bik, T. J., Margono, H., and Yang, X.: Analysis of water use trends in the Unites States: 1950-1995. Special Report 28, Illinois Water Resources Center, University of Illinois, USA, 2002.

Elliott, J., Deryng, D., Müller, C., Frieler, K., Konzmann, M, Gerten, D., Glotter, M., Flörke, M., Wada, Y., Best, N., Eisner, S., Fekete, B. M., Folberth, C., Foster, I., Gosling, S. N., Haddeland, I., Khabarov, N., Ludwing, F., Masaki, Y., Olin, S., Rosenzweig, C., Ruane, A. C., Satoh, Y., Schmid, E., Stacke, T., Tang, Q., and Wisser, D.: Constraints and potentials of future irrigation water availability on agricultural production under climate change, P. Natl. Acad. Sci. USA, 111, 3239–3244, doi:10.1073/pnas.1222474110, 2014.

Flörke M., Kynast E., Bärlund I., Eisner S., Wimmer F., and Alcamo J.: Domestic and industrial water uses of the past 60 years as a mirror of socio-economic development: A global simulation study, Global Environ. Chang., 23, 144–156, doi:10.1016/j.gloenvcha.2012.10.018, 2013.

Fu, G. B., Chen, S. L., Liu, C. M., and Shepard, D.: Hydro-climatic trends of the Yellow River basin for the last 50 years, Climatic Change, 65, 149–178, doi:10.1023/B:CLIM.0000037491.95395.bb, 2004.

Fu, J. Y., Jiang, D., and Huang, Y. H.: 1 km grid population dataset of China (2005, 2010) (Supplement), Acta Geogr. Sin., 69, 136–139, doi:10.3974/geodb.2014.01.06.V1, 2014.

Gaffin, S. R., Rosenzweig, C., Xing, X. S., and Yetman, G.: Downscaling and geo-spatial gridding of socio-economic projections from the IPCC Special Report on Emissions Scenarios (SRES), Global Environ. Chang., 14, 105–123, doi:10.1016/j.gloenvcha.2004.02.004, 2004.

Haddeland, I., Heinke, J., Biemans, H., Eisner, S., Flörke, M., Hanasaki, N., Konzmann, M., Ludwig, F., Masaki, Y., Schewe, J., Stacke, T., Tessler, Z. D., Wada, Y., and Wisser, D.: Global water resources affected by human interventions and climate change, P. Natl. Acad. Sci. USA, 111, 3251–3256, doi:10.1073/pnas.1222475110, 2014.

Hanasaki, N., Inuzuka, T., Kanae, S., and Okim, T.: An estimation of global virtual water flow and sources of water withdrawal for major crops and livestock products using a global hydrological model, J. Hydrol., 382, 232–244, doi:10.1016/j.jhydrol.2009.09.028, 2010.

Hanasaki, N., Fujimori, S., Yamamoto, T., Yoshikawa, S., Masaki, Y., Hijioka, Y., Kainuma, M., Kanamori, Y., Masui, T., Takahashi, K., and Kanae, S.: A global water scarcity assessment under Shared Socio-economic Pathways – Part 1: Water use, Hydrol. Earth Syst. Sci., 17, 2375–2391, doi:10.5194/hess-17-2375-2013, 2013a.

Hanasaki, N., Fujimori, S., Yamamoto, T., Yoshikawa, S., Masaki, Y., Hijioka, Y., Kainuma, M., Kanamori, Y., Masui, T., Takahashi, K., and Kanae, S.: A global water scarcity assessment under Shared Socio-economic Pathways – Part 2: Water availability and scarcity, Hydrol. Earth Syst. Sci., 17, 2393–2413, doi:10.5194/hess-17-2393-2013, 2013b.

Hattermann, F. F., Krysanova, V., Gosling, S., Dankers, R., Daggupati, P., Donnelly, C., Flörke, M., Huang, S., Motovilov, Y., Buda, S., Yang, T., Müller, C.,Leng, G., Tang, Q., Portmann, F. T., Hagemann, S., Gerten, D., Wada, Y., Masaki, Y., Alemayehu, T., Satoh, Y., and Samaniego, L.: Cross-scale intercomparison of climate change impacts simulated by regional and global hydrological models in eleven large river basins, Climatic Change, doi:10.1007/s10584-016-1829-4, accepted, 2017.

Hawkins, E., Osborne, T. M., Ho, C. K., and Challinor, A. J.: Calibration and bias correction of climate projections for crop modeling: An idealized case study over Europe, Agr. Forest Meteorol., 170, 19–31, 2013.

Ho, C. K., Stephenson, D. B., Collins, M., Ferro, C., and Brown, S.: Calibration strategies: A source of additional uncertainty in climate change projections, B. Am. Meteorol. Soc., 93, 21–26, doi:10.1175/2011BAMS3110.1, 2012.

Huang, Y. H., Jiang, D., and Fu, J. Y.: 1 km grid GDP data of China (2005, 2010) (Supplement), Acta Geogr. Sin., 69, 140–143, doi:10.3974/geodb.2014.01.07.V1, 2014.

Hughes, B. B.: UNEP GEO4 diver scenarios (fifth draft), Josef Korbel School of International Studies, University of Denver, Colorado, USA, 2005.

ISI-MIP: The Inter-Sectoral Impact Model Inter-comparison Project, ISI-MIP Output Data, available at: https://www.isimip.org, last access: 12 November 2016.

Leng, G., Tang, Q., Huang, M., Hong, Y., and Ruby, L.: Projected changes in mean and interannual variability of surface water over continental China, Sci. China Ser. D, 58, 739–754, doi:10.1007/s11430-014-4987-0, 2015.

Li, L., Shen, H. Y., Dai, S., Xiao, J. S., and Shi, X. H.: Response of runoff to climate change and its future tendency in the source region of Yellow River, J. Geogr. Sci., 23, 431–440, doi:10.1007/s11442-012-0937-y, 2012.

Liu, L. L., Liu, Z. F., Ren, X. Y., Fischer, T., and Xu, Y.: Hydrological impacts of climate change in the Yellow River Basin for the 21st century using hydrological model and statistical downscaling model, Quatern. Int., 244, 211–220, doi:10.1016/j.quaint.2010.12.001, 2011.

Liu, X., Zhang, X.-J., Tang, Q., and Zhang, X.-Z.: Effects of surface wind speed decline on modeled hydrological conditions in China, Hydrol. Earth Syst. Sci., 18, 2803–2813, doi:10.5194/hess-18-2803-2014, 2014.

NBSC: National Bureau of Statistics of China, The historical population and GDP data of China, available at: http://www.stats.gov.cn/tjsj/, last access: 8 November 2015.

OECD: Organization for Economic Co-operation and Development, SSP Scenario Database-GDP, available at: http://www.oecd.org/, last access: 12 November 2016.

Oki, T. and Kanae, S.: Global hydrological cycles and world water resources, Science, 313, 1068–1072, doi:10.1126/science.1128845, 2006.

O'Neill, B. C., Kriegler, E., Ebi, K. L., Kemp-Benedict, E., Riahi, K., Rothman, D. S., van Ruijven, B. J., van Vuuren, D. P., Birkmann, J., Kok, K., Levy, M., and Solecki, W.: The roads ahead: Narratives for shared socioeconomic pathways describing world futures in the 21st century, Global Environ. Chang., 42, 169–180, doi:10.1016/j.gloenvcha.2015.01.004, 2017.

Peng, X.: China's demographic history and future challenges, Science, 333, 581–587, 2011.

Portmann, F. T., Siebert, S., and Döll, P.: MIRCA2000 – Global monthly irrigated and rain-fed crop areas around the year 2000: A new high-resolution data set for agricultural and hydro- logical modeling, Global Biogeochem. Cy., 24, 1–24, doi:10.1029/2008GB003435, 2010.

RESDC: Data Center for Resources and Environmental Sciences, Chinese Academy of Sciences, The 1 km grid population dataset and 1 km grid GDP dataset of China, available at: http://www.resdc.cn, last access: 8 November 2015.

Schewe, J., Heike, J., Gerten, D., Haddeland, I., Arnell, N. W., Clark, D. B., Dankers, R., Eisner, S., Fekete, B. M., Colón-González F. J., Gosling, S. N., Kim, H., Liu, X., Masaki, Y., Portmann, F. T., Satoh, Y., Stacke, T., Tang, Q., Wada, Y., Wisser, D., Albrecht, T., Frieler, K., Piontek, F., Warszawski, L., and Kabat, P.: Multimodel assessment of water scarcity under climate change, P. Natl. Acad. Sci. USA, 111, 3245–3250, doi:10.1073/pnas.1222460110, 2014.

Shi, C. X., Zhou, Y. Y., Fan, X. L., and Shao, W. W.: A study on the annual runoff change and its relationship with water and soil conservation practices and climate change in the middle Yellow River basin, Catena, 100, 31–41, doi:10.1016/j.catena.2012.08.007, 2012.

Sterling, S. M., Ducharne, A., and Polcher J.: The impact of global land-cover change on the terrestrial water cycle, Nat. Clim. Change, 3, 385–390, doi:10.1038/NCLIMATE1690, 2013.

Tang, Q. and Oki, T. (Eds.): Terrestrial Water Cycle and Climate Change: Natural and Human-Induced Impacts, American Geophysical Union (AGU) Geophysical Monograph Series, Vol. 221, John Wiley & Sons, New Jersey, USA, 2016.

Tang, Q., Oki, T., Kanae, S., and Hu, H.: The influence of precipitation variability and partial irrigation within grid cells on a hydrological simulation, J. Hydrometeorol., 8, 499–512, doi:10.1175/JHM589.1, 2007.

Tang, Q., Oki, T., Kanae, S., and Hu, H.: Hydrological cycles change in the Yellow River basin during the last half of the twentieth century, J. Climate, 21, 1790–1806, doi:10.1175/2007JCLI1854.1, 2008a.

Tang, Q., Oki, T., Kanae, S., and Hu, H.: A spatial analysis of hydro-climatic and vegetation condition trends in the Yellow River basin, Hydrol. Process., 22, 451–458, doi:10.1002/hyp.6624, 2008b.

Tang, Q., Vivoni, E. R., Muñoz-Arriola, F., and Lettenmaier, D. P.: Predictability of evapotranspiration patterns using remotely sensed vegetation dynamics during the North American monsoon, J. Hydrometeorol., 13, 103–121, doi:10.1175/JHM-D-11-032.1, 2012.

Tang, Y., Tang, Q., Tian, F., Zhang, Z., and Liu, G.: Responses of natural runoff to recent climatic variations in the Yellow River basin, China, Hydrol. Earth Syst. Sci., 17, 4471–4480, doi:10.5194/hess-17-4471-2013, 2013.

Wada, Y., van Beek, L. P. H., Viviroli, D., Dürr, H. H., Weingartner, R., and Bierkens, M. F. P.: Global monthly water stress 2: Water demand and severity of water stress, Water Resour. Res., 47, W07518, doi:10.1029/2010WR009792, 2011.

Wada, Y., Wisser, D., and Bierkens, M. F. P.: Global modeling of withdrawal, allocation and consumptive use of surface water and groundwater resources, Earth Syst. Dynam., 5, 15–40, doi:10.5194/esd-5-15-2014, 2014.

Wada, Y., Flörke, M., Hanasaki, N., Eisner, S., Fischer, G., Tramberend, S., Satoh, Y., van Vliet, M. T. H., Yillia, P., Ringler, C., Burek, P., and Wiberg, D.: Modeling global water use for the 21st century: the Water Futures and Solutions (WFaS) initiative and its approaches, Geosci. Model Dev., 9, 175–222, doi:10.5194/gmd-9-175-2016, 2016.

Wang, S. J., Yan, M., Yan, Y. X., Shi, C. X., and He, L.: Contributions of climate change and human activities to the changes in runoff increment in different sections of the Yellow River, Quatern. Int., 282, 66–77, doi:10.1016/j.quaint.2012.07.011, 2012.

Warszawski, L., Frieler, K., Huber, V., Piontek, F., Serdeczny, O., and Schewe, J.: The Inter-Sectoral Impact Model Intercomparison Projection (ISI-MIP): Project framework, P. Natl. Acad. Sci. USA, 111, 3228–3232, doi:10.1073/pnas.1312330110, 2014.

World Bank: The Official Exchange Rate Data of China, available at: http://data.worldbank.org/, last access: 1 December 2015.

Xu, J.: Variation in annual runoff of the Wudinghe River as influenced by climate change and human activity, Quatern. Int., 244, 230–237, doi:10.1016/j.quaint.2010.09.014, 2011.

Yin, Y., Tang, Q., and Liu, X.: A multi-model analysis of change in potential yield of major crops in China under climate change, Earth Syst. Dynam., 6, 45–59, doi:10.5194/esd-6-45-2015, 2015.

YRCC (Yellow River Conservancy Commission): Comprehensive planning of Yellow River Basin (2012–2030), The Yellow River Water Conservancy Press, Zhengzhou, China, 2013 (in Chinese).

Zhang, C., Zhong, L., Fu, X., Wang, J., and Wu, Z.: Revealing water stress by the thermal power industry in China based on a high spatial resolution water withdrawal and consumption inventory, Environ. Sci. Technol., 50, 1642–1652, doi:10.1021/acs.est.5b05374, 2016.

Zhang, H. M., Niu, Y. G., Wang, B. X., and Li, S. M.: The Yellow River water resources problems and countermeasures, Hydrology, 24, 26–31, 2004 (in Chinese).

Zhao, F. F., Xu, Z. X., Zhang, L., and Zou, D. P.: Streamflow response to climate variability and human activities in the upper catchment of the Yellow River Basin, Sci. China Ser. E, 52, 3249–3256, doi:10.1007/s11431-009-0354-3, 2009.

11

Hydrological dynamics of water sources in a Mediterranean lagoon

C. Stumpp[1], A. Ekdal[2], I. E. Gönenc[3], and P. Maloszewski[1]

[1]Helmholtz Zentrum München, German Research Center for Environmental Health, Institute of Groundwater Ecology, Neuherberg, Germany
[2]Istanbul Technical University, Environmental Engineering Department, Maslak, Istanbul, Turkey
[3]IGEM Research & Consulting Co., Istanbul, Turkey

Correspondence to: C. Stumpp (christine.stumpp@helmholtz-muenchen.de)

Abstract. Lagoons are important ecosystems occupying large coastal areas worldwide. Lagoons contain various mixtures of marine and freshwater sources which are highly dynamic in time. However, it often remains a challenge to identify and quantify dynamic changes of water sources, particularly in heterogeneous lagoon systems like the Köycegiz–Dalyan lagoon (KDL), which is located at the south-west of Turkey on the Mediterranean Sea coast. The objective of this study was to quantify different contributions of potential water sources i.e. surface water, groundwater and seawater in the lagoon and how these water sources changed over time and space. In the wet- and dry-season stable isotopes of water, chloride concentration (Cl^-) and salinity were measured in two depths in the lagoon and surrounding water bodies (sea, lake, groundwater). Different components of water sources were quantified with a three component endmember mixing analysis. Differences in Cl^- and stable isotopes over time indicated the dynamic behaviour of the system. Generally, none of the groundwater samples was impacted by water of the Mediterranean Sea. During the wet season, most of the lagoon water (> 95 %) was influenced by freshwater and vertically well mixed. During the dry season, high Cl^- in the deeper sampling locations indicated a high contribution of marine water throughout the entire lagoon system due to saltwater intrusion. However, a distinct layering in the lagoon was obvious from low Cl^- and depleted isotope contents close to the surface supporting freshwater inflow into the system even during the dry season. Besides temporal dynamics also spatial heterogeneities were identified. Changes in water sources were most evident in the main lagoon channel compared to more isolate lagoon lakes, which were influenced by marine water even in the wet season, and compared to side branches indicating slower turnover times. We found that environmental tracers helped to quantify highly dynamic and heterogeneous contributions of different water sources in the Köycegiz–Dalyan lagoon.

1 Introduction

Lagoons are important ecosystems occupying 13 % of the coastal areas worldwide (Barnes, 1980). Along the Mediterranean coastline, more than 100 lagoons are found but only little hydrological and biological data of most of these ecosystems are available (Pérez-Ruzafa et al., 2011a). Generally, lagoons are shallow, coastal water bodies with marine water influence. Mostly they have limited connectivity to the open sea through coastal barriers or connecting inlets. Further freshwater input can come from upstream rivers or groundwater. Lagoons are important ecosystems being a habitat for rare species like seagrass, fishes and turtles, and with a high productivity and diversity (Alongi, 1998; Pérez-Ruzafa et al., 2011b; Remane and Schlieper, 1971). Lagoons not only are valued for fauna and flora, but also due to recreational and industrial purposes by society. These societal values are, however, difficult to quantify (Anthony et al., 2009) also due to conflicts of interest in lagoon ecosystems (fishery, aquaculture, tourism) (Pérez-Ruzafa et al., 2011a). Particularly in the context of global change lagoon ecosystems require a proper management for a sustainable use and to protect the ecosystem (Kløve et al., 2014; Anthony et al., 2009). Here, management must not impact the quality and quantity of the lagoon water in terms of chemical and ecological status on the one hand. On the other hand, also groundwater management (drinking water/irrigation) must not impact lagoons depending on groundwater and vice versa. For example, pumping of groundwater can influence the quality of the withdrawn drinking/irrigation water due to increased marine water influence or due to the mobilization of groundwater from deeper layers. In addition, groundwater withdrawal can change the fraction of freshwater source in the lagoon water body which strongly would influence its functions as a habitat for species specifically adapted to the environment. This example highlights the vulnerability of lagoon systems. It shows the strong need to protect and manage these ecosystems and to identify seawater intrusions and groundwater dependencies in the lagoon catchment area.

Here, a detailed knowledge about the water sources and water dynamics in lagoon ecosystems is fundamental before studying further ecological and chemical processes. It has been shown that lagoon systems are heterogeneous and dynamic systems. The quality of the water and subsequent seawater quality or adjacent groundwater quality strongly depends on the water flow and origin of water and nutrients (Gattacceca et al., 2009; Niencheski et al., 2007; Santos et al., 2008a, b). Land use can impact the interaction of lagoon with surrounding groundwater resulting in lagoon infiltration processes under pine tree plantations compared to negligible interactions under natural dune vegetation (Schmidt et al., 2011). Such spatial impacts can be identified using environment tracer methods. Further, they can also be used to study temporal dynamics of water sources and hydrological processes like seasonal changes in evaporation and seawater contribution (Lecuyer et al., 2012; Schmidt et al., 2011). Salinity and stable isotopes of water were used to identify spatiotemporal changes of water in the Akyatan lagoon, Turkey (Lecuyer et al., 2012). Assuming two different end members, river and seawater, it was found that in the wet season the contribution of freshwater and seawater was 62 and 38 % on average. Throughout spring to autumn, progressive evaporation of lagoon water results in hypersaline conditions with strongly enriched isotope values suggesting limited input of freshwater in the system (Lecuyer et al., 2012). Still, it remains unknown in many lagoon systems what the contribution of different water sources is and how they change not only over time, i.e. wet and dry seasons, but also over space, i.e. both horizontal, spatial locations in the lagoon and vertical, depth locations in the lagoon; the latter is of particular interest in wetland type lagoon systems or lagoons with stratification expecting poorly mixed hydrological systems. Therefore, the objective of the current study was (i) to identify and quantify different water sources in a lagoon, (ii) how they change over time and space, and (iii) thus how heterogeneous and dynamic the hydrology of the lagoon and adjacent groundwater was. We achieved these objectives by applying environmental tracer methods and developing a three component endmember mixing approach. Different sources of water (seawater, groundwater, lake water) were identified at different locations in the lagoon, including top and bottom water column depths, for wet and dry season. Thus, the novelty of this study is to present an environmental tracer method identifying and quantifying both temporal dynamics (wet and dry season) and spatial heterogeneities (depth of the water column and distance to coastline) of water sources in a wetland type lagoon system. With improved, detailed understanding of heterogeneous and dynamic hydrological processes in groundwater influenced lagoon ecosystems, targeted strategies to better manage may be developed.

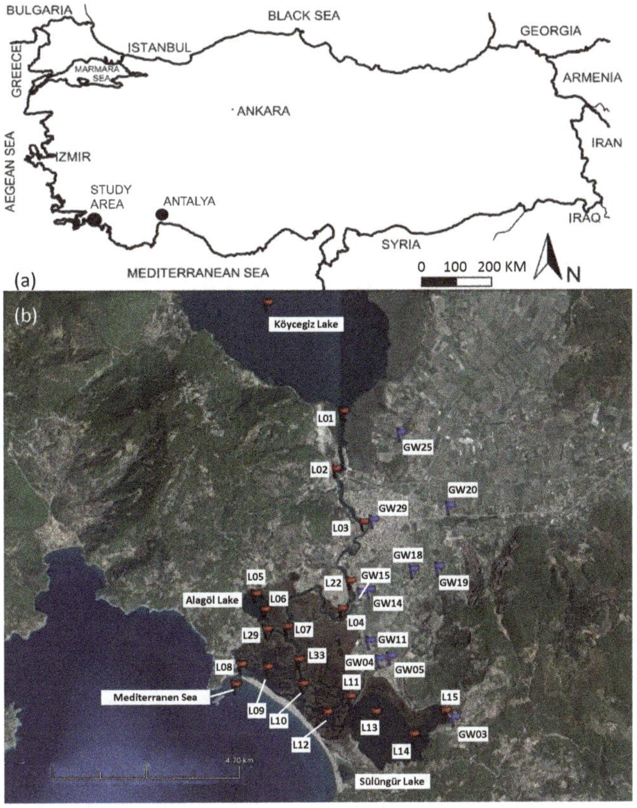

Figure 1. Geographic location of the Köycegiz–Dalyan coastal lagoon **(a)** and sampling locations **(b)**; source of modified satellite picture was Google Earth (2014).

2 Material and methods

2.1 Study area

Köycegiz–Dalyan coastal lagoon is located at the southwest of Turkey on the Mediterranean Sea coast within the province of Mugla (Fig. 1a). The geology in this region is mainly composed of allochthonous and authochthounous Flysch and karstic facies overlain by plio-quaternary sediments (Graciansky, 1968). Due to tectonic activities, several faults were formed in this area. Details about the geology and more maps can be found in Bayari et al. (1995).

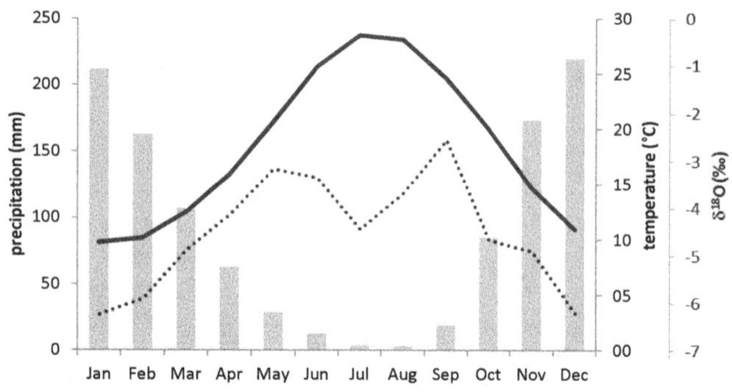

Figure 2. Long-term monthly data of average precipitation (grey bars) and air temperature (solid line) from Köycegiz meteorology station (1976–2010) and isotopic composition of precipitation in Antalya (dashed line). Data from Antalya are available at the IAEA database WISER (http://www-naweb.iaea.org/napc/ih/index.html, last access: 19 May 2014).

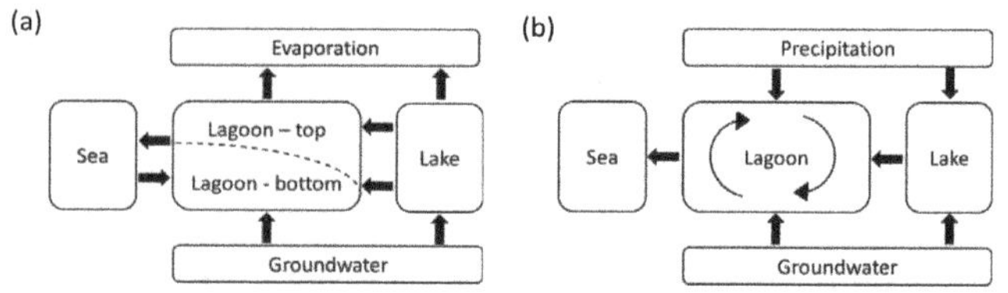

Figure 3. Conceptual model of flow connections between the lagoon and surrounding water bodies for **(a)** the dry and **(b)** wet season.

The total area of the watershed of Köycegiz Lake is approximately 830 km^2 and of the lagoon is 130 km^2. The upstream located Köycegiz Lake (2 m a.s.l.) is directly connected through surface water with the lagoon and further to the Mediterranean Sea by the lagoon and its various branches (Fig. 1b). The discharge from the Köycegiz Lake is 33 $m^3\,s^{-1}$ on average with up to 110 $m^3\,s^{-1}$ during winter times (Bayari et al., 2001). During winter, most of the branches in the wetland areas in the lagoon are connected. In summer, Köycegiz Lake water level decreases (−0.9 m) reducing the hydraulic gradient to the Mediterranean Sea considerably. The depth of the main Dalyan channel decreases from 5 m upstream near the lake to about 1 m downstream near the sea. In addition to the Dalyan Channel and its branches, the lagoon also includes the lakes Alagöl and Sülüngür. Maximum depths of these lakes are 4 and 13 m, respectively. Aquaculture activities are conducted in Sülüngür Lake. Both, Köycegiz Lake and the Köycegiz–Dalyan coastal lagoon are part of the area declared as a special protection area in 1988, as it is a unique and important ecosystem with a high diversity of species. It hosts one of the rare breeding and nesting sites for endangered sea turtles, *caretta caretta*, and possesses the ruins of Ancient City of Caunos and 4th century BC Lycian rock tombs that are found near the seaside by the river (Gurel et al., 2005). Groundwater is used as irrigation and drinking water in the area. We expect that the groundwater is mainly recharged locally from the surrounding forested mountains (up to 565 m a.s.l.; Fig. 1) of the karstic areas. The main sectors driving the economy in the watershed are agriculture, tourism and forestry. Aquaculture and capture fishing are among the important beneficial uses of the lagoon together with recreational activities.

The area is under the influence of typical Mediterranean climate characteristics, with a hot, dry summer season and a warm, rainy winter season with mean annual air temperatures of 18.3 °C and mean annual precipitation of 1083 mm. These data were taken in the study area from the State Meteorology Services of Turkish Republic for Köycegiz Meteorology Station covering the period 1976–2010, and monthly averages are presented in Fig. 2. Thus, precipitation usually occurs during the cold winter period and drought condition prevails during the hot summer period.

An environmental isotopic and hydrochemical study was conducted by Bayari et al. (1995) for determination of the dynamics of the upstream Köycegiz Lake. Köycegiz and Sultaniye are the two major basins that comprise Köycegiz Lake. According to their statements the important sources that feed the lake are mainly alluvial groundwater, streamwater (Namnam and Yuvarlakçay), and rain. The main components of outflow from the lake are discharge to Mediterranean Sea through the Dalyan Channel and evaporation from the lake surface. Their environmental isotopic data and chemical data indicate that rainfall and stream flow are low density waters and thermal groundwater is the high density water; complete annual mixing cannot be observed due to the density effects. The main geothermal inflow at the southern lake coast (Sultaniye Basin) is the Sultaniye spring. It is located at a depth of 8–10 m and about 4 km north-west of the lake exit into the Dalyan channel which is shallow (0–6 m) (Bayari et al., 2001); too shallow for receiving any geothermal influenced water from the Sultaniye Basin.

2.2 Conceptual model

Identifying different water sources in the lagoon we set up a conceptual model distinguishing between dry (Fig. 3a) and wet seasons (Fig. 3b). For the dry season our hypothesis was that evaporation results in low water tables in the lagoon favouring both fluxes from Köycegiz Lake and the sea into the lagoon. However, higher water levels maintain in the main Dalyan channel with freshwater flow from Köycegiz Lake to the sea. Thus, we expected density driven layering in the lagoon with (i) freshwater input from the lake in the top layer, which is influenced by evaporation, and (ii) saltwater input in the bottom layer mixed with groundwater (Fig. 3a). We further expected that the seawater influence decreases with distance to the coastline. For the wet season our hypothesis was that freshwater input, mainly from groundwater and lake during baseflow conditions and additionally from precipitation during events, results in high water tables in the lagoon favouring freshwater flow from the lake through the lagoon into the sea. We expected the lagoon water to be well mixed without distinct density driven layering (Fig. 3b). For both seasons, we excluded any direct influence of the geothermal Sultaniye spring to the lagoon, because the spring's influence was found only for the bottom layers of the Köycegiz Lake (Bayari et al., 1995) not outflowing into the shallow Dalyan channel and the lagoon but discharging northwards. Still, other unknown geothermal springs in the lagoon cannot be excluded.

2.3 Sampling campaigns

To quantify the different contributions of potential water sources like surface water, groundwater and seawater in the lagoon and how these water sources change over time and space, two sampling campaigns were conducted one representing the dry season (July 2011) and the other one representing the wet season (March 2012). Sampling in both seasons was without major antecedent rain events. Consequently, precipitation as major source in the lagoon can be neglected. Particularly in the wet season, water residence times in the lagoon are short due to high outflow rates from the lake (up to $110\,m^3\,s^{-1}$; Bayari et al., 2001) and which is also supported by modeling results of Ekdal (2008) indicating residence times < 2 days for the wet season in the main lagoon channel.

Water samples were taken in the lagoon along the main channel (L1, L2, L3, L22, L4, L7, L33, L10, L29, L9, L8), surrounding lakes (L5, L13, L14) and their inflow/outflow connections to the lagoon system (L6, L11, L12, L15) as well as in the Köycegiz Lake and Mediterranean Sea in two depths at the top (T), just below the surface, and at the very bottom (B). The samples were taken by boat used for transportation from Dalyan town to Iztuzu Beach, except for Sülüngür Lake. Since aquaculture activities are conducted in this lake boat of the fishing cooperative was used for sampling. Further samples were taken from surrounding groundwater wells. Groundwater samples were taken with the pump of the well, which is used for abstracting water. In total, samples were taken at 18 lagoon, 11 groundwater locations, as well as at one sea location and one lake location (Fig. 1b), which were further analysed for chemical analysis.

2.4 Water isotopes and chemical analysis

Water samples were analysed for $\delta^{18}O$ (± 0.15 ‰) and δ^2H (± 1 ‰) contents without any pre-treatment of the samples using a water isotope analyser (L2120-i, Picarro Inc., Santa Clara, CA, USA). The contents are given in the delta notation as δ-value (‰), which is the relative deviation of the sample from the V-SMOW (Vienna-Standard Mean Ocean Water). The results of the stable water isotope analysis from the observation area were compared to public available isotope contents in precipitation accessible through the IAEA (International Atomic Energy Agency) web database WISER (http://www-naweb.iaea.org/napc/ih/IHS_resources_isohis.html; 2014). Here, Antalya is the closest location of the Global Network of Isotopes in Precipitation (GNIP) having long-term isotope records in precipitation, which is 200 km east of the studied lagoon and 49 m asl. Based on these data, the local meteoric water line (LMWL; $\delta^2H = 8\ \delta^{18}O + 14.3$) and the annual weighed average isotope contents in precipitation ($\delta^{18}O = -4.9$ ‰; $\delta^2H = -24.9$ ‰) were calculated; monthly long-term weighed averages are shown in Fig. 2.

Chloride concentrations ($\pm 0.22\,mg\,L^{-1}$) were measured by using Merck test kits (catalog number 1.14897.0001). NaCl stock solution, which has 1 mg Cl^- in 1 mL, was used in order to prepare standard solutions for controlling the reliability of chloride measurements carried out with Merck test kits. Salinity measurements ($\pm 0.1\,mg\,L^{-1}$) were conducted in situ with a YSI 6600V2 Multiparameter Water Quality Sonde.

2.5 Endmember mixing analysis

Calculating different water fractions in the lagoon system (top and bottom), three endmembers were defined that differed in isotopic composition and chloride concentrations/salinity: (i) Köycegiz Lake water, (ii) groundwater, and (iii) Mediterranean seawater. The concentrations (C) of the endmembers were defined for both seasons separately. For lake (C_{LW}) and seawater (C_{SW}), the surface near water samples were taken and for groundwater an average concentration (C_{GW}) was calculated from all groundwater wells without considering GW011 due to increased chloride concentrations compared to other groundwater locations. Thus, the isotope contents (^{18}O) and chloride concentrations (Cl^-) or salinity (S) in the lagoon (C_{Lag}) were calculated from the three component mixing analysis:

$$C_{Lag_{18O}} = f_{GW} \cdot C_{GW_{18O}} + f_{LW} \cdot C_{LW_{18O}} + f_{SW} \cdot C_{SW_{18O}} \quad (1)$$

$$C_{Lag_{Cl}} = f_{GW} \cdot C_{GW_{Cl}} + f_{LW} \cdot C_{LW_{Cl}} + f_{SW} \cdot C_{SW_{Cl}} \quad (2)$$

$$C_{Lag_S} = f_{GW} \cdot C_{GW_S} + f_{LW} \cdot C_{LW_S} + f_{SW} \cdot C_{SW_S} \quad (3)$$

$$1 = f_{GW} + f_{LW} + f_{SW}, \quad (4)$$

where f refers to the fraction of groundwater (GW), lake water (LW), and seawater (SW), respectively. Getting information about the uncertainty of the method, we calculated the mixing ratios based on two different approaches considering simultaneously $\delta^{18}O$ (Eq. 1) and Cl^- (Eq. 2) or $\delta^{18}O$ (Eq. 1) and salinity (Eq. 3), both in combination with Eq. (4).

3 Results

3.1 Stable isotopes of water

Results of stable isotope analyses are presented in Table 1. All analysed water samples plotted close or below

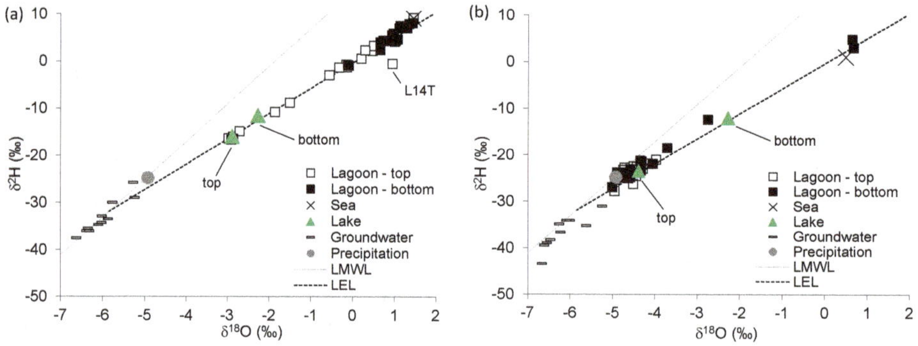

Figure 4. Dual isotope plot for **(a)** dry season and **(b)** wet season sampling campaign; LMWL and average precipitation taken from closest station of the GNIP data base i.e. Antalya.

the LMWL for both the dry (Fig. 4a) and wet seasons (Fig. 4b). Groundwater samples were the most depleted samples ranging from −6.7 to −5.2 ‰ for $\delta^{18}O$, and were even lower compared to average precipitation contents (−4.9 ‰ for $\delta^{18}O$). Assuming only negligible differences in isotopic composition of precipitation between Antalya and our observation area due to close proximity and similar location on the Mediterranean Sea, these differences support our assumption of higher altitude precipitation from surrounding mountains as major recharge source of groundwater. Average differences in elevation (400 m) and isotope contents (1.17 ‰ for $\delta^{18}O$; 9.9 ‰ for $\delta^{2}H$) give an altitude gradient of 0.29 ‰/100 m for $\delta^{18}O$ (2.5 ‰/100 m for $\delta^{2}H$). These gradients are in accordance with values reported for the southern Adriatic region (0.24 ‰/100 m; Vreca et al., 2006), the global and Italian gradients (0.2 ‰/100 m; Bowen and Wilkison, 2002; Longinelli and Selmo, 2003) and simulated values for the Mediterranean Sea region (Lykoudis and Argiriou, 2007).

In groundwater, more depleted contents were generally observed in the wet season compared to the dry season; however, absolute differences between seasons were small (0.21 ‰ for $\delta^{18}O$; 2.8 ‰ for $\delta^{2}H$). These differences can either result from a fraction of local seepage water with short residence times, from influence of seawater or from uncertainties of groundwater sampling. Well screening depths were unknown and therefore we expected some minor uncertainties when taking groundwater samples, i.e. water from same depths and taken with same flow rates during sampling.

Isotope contents of seawater were positive with more enriched contents in dry (1.5 ‰ for $\delta^{18}O$) compared to wet seasons (0.5 ‰ for $\delta^{18}O$). All Köycegiz Lake water samples plotted below the LMWL (Fig. 4) indicating enrichment due to evaporation and potential geothermal water origin as found in previous studies (Bayari et al., 1995, 2001). When considering isotope contents of reported geothermal origin in the area (−0.81, −4.87, −4–76 ‰, and −2.9, −30.0, −27.2 ‰ for $\delta^{18}O$ and $\delta^{2}H$, respectively; Bayari et al., 1995), it is evident that the geothermal origin is hidden in the evaporation signal and therefore these two sources cannot be distinguished considering isotope contents only. Additionally, a local evaporation line (LEL) was determined considering the top lake samples for both seasons only. The resulting LEL ($\delta^{2}H = 5.40\ \delta^{18}O - 0.3$) is similar to another Turkish lagoon ($\delta^{2}H = 5.29\ \delta^{18}O - 0.55$; Lecuyer et al. 2012). It intersects the LWML in $\delta^{18}O = -5.85$ ‰ ($\delta^{2}H = -31.9$ ‰), which is also close to the average groundwater contents ($\delta^{18}O = -6.08$ ‰; $\delta^{2}H = -34.84$ ‰) supporting assumption of higher elevation recharge area for the catchment.

Water samples from the lagoon mainly plotted on/below the LMWL and between groundwater and seawater samples. Distinct differences in isotopic contents were found (i) for the dry (Fig. 4a) and wet seasons (Fig. 4b) indicating a seasonally dynamic water body and (ii) for samples close to the surface (open squares, Fig. 4) and the bottom of the lagoon (closed square, Fig. 4) indicating a layered vs well mixed system in the dry and wet seasons, respectively. Particularly in the dry season, differences between top and bottom lagoon samples were obvious. Here, most interestingly, water samples at the bottom of the lagoon were more enriched compared to top water samples. This clearly indicates that the enrichment was not caused by evaporation but rather by mixing with enriched seawater which is more pronounced at the bottom due to saltwater density effects. In the wet season, similar isotope contents were found for top and bottom samples except for samples from Alagöl (L5; −2.7 ‰, $\delta^{18}O$) and Sülüngür Lake (L13, L14; +0.64–0.68 ‰, $\delta^{18}O$) which had more enriched isotope contents at the bottom only. Here, top water samples showed similar ranges in isotope contents (−4.5 to −4.0 ‰, $\delta^{18}O$) compared to other lagoon samples (−5.0 to −4.0 ‰, $\delta^{18}O$).

3.2 Chloride vs. stable isotopes of water

Results of geochemical analysis are given in Table 1. Chloride and salinity showed similar spatiotemporal results and therefore, chloride results are discussed in more detail only. Chloride concentrations were in line with the results of stable isotope of water. Chloride was lowest in groundwater

Table 1. Chemical analyses of water samples for the dry and wet seasons; asterisks indicate values used for endmember mixing analysis using either a three (3EMMA) or two (2EMMA) mixing approach.

	Dry season					Wet season				
Location	Depth	Chloride	Salinity	$\delta^{18}O$	$\delta^{2}H$	Depth	Chloride	Salinity	$\delta^{18}O$	$\delta^{2}H$
	(m)	($mg\,L^{-1}$)	($g\,L^{-1}$)	(‰)	(‰)	(m)	($mg\,L^{-1}$)	($g\,L^{-1}$)	(‰)	(‰)
L01T	0.1	2400	3.8	−2.90	−16.4	0.1	930	3.1	−4.70	−24.9
L02T	0.1	2600	3.8	−2.87	−16.0	0.1	930	3.1	−4.51	−26.3
L03T	0.1	2800	4.0	−2.97	−16.4	0.1	930	3.2	−4.78	−24.8
L04T	0.1	3700	7.0	−2.71	−14.9	0.1	940	3.0	−4.93	−25.4
L05T	0.1	11 400	23.6	−0.15	−1.3	0.1	2350	4.6	−4.50	−22.5
L06T	0.1	14 900	22.3	−0.16	−0.7	0.1	1500	4.2	−4.72	−22.8
L07T	0.1	7800	16.1	−1.86	−10.8	0.1	1050	3.2	−4.68	−24.8
L08T	0.1	18 600	37.9	1.45	9.2	0.1	1300	4.2	−4.74	−23.3
L09T	0.1	14 700	29.3	0.59	3.3	0.1	1300	4.5	−4.44	−24.5
L10T	0.1	14 700	27.6	0.47	2.2	0.1	840	3.4	−4.76	−24.7
L11T	0.1	15 800	27.6	0.30	2.4	0.1	2150	5.9	−4.95	−27.8
L12T	0.1	13 200	25.7	0.20	0.6	0.1	2500	6.9	−4.28	−22.9
L13T	0.1	18 200	30.6	1.00	4.2	0.1	1400	7.6	−4.17	−21.9
L14T	0.1	17 400	30.6	0.95	−0.6	0.1	1350	7.6	−3.97	−21.0
L15T	0.1	13 900	–	−0.33	−1.3	0.1	1200	7.3	−4.43	−22.5
L22T	0.1	8700	16.7	−1.51	−8.8	0.1	950	3.2	−4.73	−24.4
L29T	0.1	13 700	29.3	0.50	3.2	0.1	750	3.3	−4.62	−25.0
L33T	0.1	12 000	25.0	−0.56	−2.9	0.1	950	3.4	−4.76	−23.8
L01B	3.8	3300	26.3	−2.86	−16.4	3.8	940	3.1	−4.73	−24.7
L02B	4.4	3600	27.8	−0.12	−0.8	4.4	940	3.2	−4.7	−24.4
L03B	2.5	3700	31.8	−2.90	−16.6	2.4	950	3.2	−4.7	−24.8
L04B	2.1	20 000	32.7	0.73	4.3	2.0	970	3.1	−5.01	−27.0
L05B	3.3	22 300	38.2	1.43	8.1	3.2	7100	19.4	−2.7	−12.3
L06B	1.4	12 800	32.1	−0.09	−0.9	1.7	1600	4.7	−4.58	−23.1
L07B	2.0	21 400	35.8	1.13	7.5	1.9	1100	3.1	−4.90	−23.8
L08B*2EMMA	1.1	23 800	39.7	1.16	7.4	1.1	1300	4.3	−4.44	−23.9
L09B	1.3	24 200	39.0	1.35	7.9	1.2	1700	5.4	−4.33	−23.4
L10B	1.1	21 800	33.8	1.30	7.1	1.3	930	3.4	−4.78	−24.0
L11B	1.5	17 100	31.2	1.02	4.4	1.5	3500	7.5	−4.34	−21.3
L12B	1.5	14 300	34.6	0.66	2.4	1.5	3600	7.3	−4.31	−21.4
L13B	3.4	18 300	36.5	1.07	4.6	3.6	21 600	41.2	0.64	4.9
L14B	5.4	18 100	36.9	0.76	4.3	5.4	21 000	41.2	0.68	3.0
L15B	1.6	16 400	–	0.65	4.0	1.6	1320	8.0	−4.05	−21.9
L22B	3.0	22 100	35.9	0.97	5.9	3.0	980	3.3	−4.66	−24.8
L29B	1.8	17 500	35.5	0.93	5.5	1.8	850	3.3	−4.58	−24.8
L33B	3.8	19 800	38.8	1.11	7.0	3.8	3400	11.3	−3.71	−18.5
GW03	–	132	0.4	−5.27	−25.8	–	–	–	–	–
GW04	–	117	0.4	−6.10	−34.7	–	111	0.4	−6.08	−34.1
GW05	–	146	0.5	−6.03	−34.3	–	88	0.4	−6.25	−34.8
GW11	–	2300	1.3	−6.39	−36.1	–	460	1.1	−6.66	−43.4
GW14	–	69	0.3	−6.35	−35.5	–	41	0.3	−6.46	−38.3
GW15	–	41	0.3	−6.32	−36.0	–	40	0.3	−6.22	−36.6
GW18	–	42	0.4	−6.02	−32.9	–	16	0.5	−5.62	−35.2
GW19	–	25	0.3	−6.63	−37.6	–	–	0.3	−6.55	−38.9
GW20	–	56	0.4	−5.77	−30.0	–	18	0.2	−6.60	−39.5
GW25	–	57	0.6	−5.24	−29.0	–	50	0.5	−5.25	−31.0
GW29	–	46	0.4	−5.87	−33.5	–	26	0.4	−6.00	−34.1
GW*	–	73	0.4	−6.00	−32.9	–	49	0.4	−6.17	−36.6
Sea*3EMMA	0.1	20 800	40.0	1.45	9.1	0.1	21 700	39.2	0.49	1.1
Lake*	0.1	2200	3.7	−2.88	−15.9	0.1	920	3.2	−4.38	−23.4
Lake	12.8	4500	11.2	−2.26	−11.5	12.7	4800	13.6	−2.27	−12.0

samples for both sampling times suggesting no or negligible seawater influence for most of these groundwater locations. Only one sampling site (GW11) showed increased chloride concentrations (460 mg L^{-1} in wet season and 2300 mg L^{-1} in dry season), which was also accompanied by higher water isotope contents in the dry compared to the wet season (Table 1). If this was caused by mixing with seawater, it would result in an increased seawater contribution of 7 ± 5 % for the dry season in GW11. Another reason could be short residence times of recharge from the unsaturated zone. Consequently, chloride originating from agricultural activities (irrigation, pomegranates) would be leached and diluted by winter precipitation with low isotope contents in the wet season.

Chloride concentrations were similar during both sampling campaigns in the dry and wet seasons at the bottom of the Köycegiz Lake (4500 and 4800 mg L^{-1}), but differences were measured at the top (2200 and 920 mg L^{-1}). High chloride concentrations were measured in seawater with 21 700 and 20 800 mg L^{-1} during the wet and dry season, respectively.

In the lagoon, chloride concentrations were generally higher in the dry season compared to the wet season (Fig. 5, Table 1). In the dry season, a clear layering was also supported by the chloride concentrations which were higher at the bottom of the lagoon compared to its top. When looking at the chloride isotope relationship, lagoon samples were mainly plotting in the triangle of groundwater, Köycegiz Lake water and seawater samples suggesting three main endmembers in the system (Fig. 5a). In the wet season, high chloride concentrations were only measured in the lagoon lake systems that also had enriched isotope contents (Fig. 5b). All other lagoon samples had chloride concentrations lower than 5000 mg L^{-1} plotting in the triangle of groundwater, Köycegiz Lake water and seawater samples suggesting three main endmembers in the system (Fig. 5b).

3.3 Endmember mixing analysis

The three component endmember mixing analysis was calculated for (i) the wet and dry season and (ii) for the top and bottom layer. The selected endmembers are given in Table 1 (asterisks) and the resulting source fractions for each location and season are given in Table 2.

For the wet season, average fractions of water sources were similar in the top and bottom of the lagoon (Fig. 6b). The arithmetic average (median) of groundwater, lake and seawater contribution was 0.24 (0.25), 0.72 (0.73) and 0.04 (0.04) for the top and 0.21 (0.22), 0.62 (0.74), and 0.17 (0.02) for the bottom layer, respectively. Thus, the entire lagoon contained little seawater, and the main source was freshwater, either from the Köycegiz Lake or the groundwater. Certainly, we cannot exclude direct influence from precipitation having similar chemical composition compared to groundwater which will be further discussed below. High fractions of seawater were mainly found in the bottom of the lagoon lake systems (Fig. 7d, Table 2). The more shallow Alagöl lake (L5; 3.3 m) contained about 34 % seawater and 98 % seawater were calculated for the deeper Sülüngür Lake (L13, L14; 3.6–5.4 m) (Table 2). The branches of the lagoon showed slightly increased saltwater contributions (9 % top layer, 10 % bottom layer) compared the Dalyan channel locations (2 % top layer, 3 % bottom layer) (Fig. 7c, d). Besides, no variability in seawater and freshwater contribution

Table 2. Average results of the three component endmember mixing analysis giving the contributions of groundwater (f_{GW}), lake water (f_{LW}) and seawater (f_{SW}) in the lagoon top and bottom for dry and wet seasons.

	Dry season			Wet season		
	f_{GW}	f_{LW}	f_{SW}	f_{GW}	f_{LW}	f_{SW}
Location – TOP						
L01	0.020	0.975	0.005	0.210	0.780	0.010
L02	0.015	0.970	0.015	0.080	0.915	0.005
L03	0.070	0.905	0.025	0.265	0.720	0.015
L04	0.075	0.830	0.095	0.360	0.620	0.020
L05	0.000	0.559	0.441	0.255	0.675	0.070
L06	0.045	0.400	0.551	0.320	0.630	0.050
L07	0.140	0.530	0.335	0.210	0.775	0.015
L08	0.000	0.189	0.811	0.320	0.640	0.040
L09	0.000	0.431	0.569	0.130	0.835	0.035
L10	0.000	0.447	0.549	0.260	0.730	0.010
L11	0.000	0.335	0.665	0.605	0.290	0.105
L12	0.000	0.513	0.488	0.230	0.665	0.105
L13	0.000	0.307	0.693	0.135	0.790	0.070
L14	0.000	0.332	0.668	0.065	0.825	0.110
L15	0.030	0.650	0.320	0.250	0.665	0.085
L22	0.055	0.580	0.360	0.240	0.745	0.015
L29	0.000	0.468	0.532	0.150	0.840	0.010
L33	0.040	0.400	0.560	0.265	0.720	0.015
Location – BOTTOM						
L01	0.335	0.420	0.245	0.225	0.765	0.010
L02	0.020	0.645	0.335	0.220	0.775	0.005
L03	0.360	0.380	0.258	0.235	0.760	0.005
L04	0.088	0.140	0.772	0.425	0.555	0.020
L05	0.056	0.050	0.894	0.200	0.460	0.340
L06	0.100	0.250	0.650	0.250	0.695	0.050
L07	0.073	0.075	0.852	0.355	0.630	0.015
L08	0.145	0.000	0.855	0.115	0.865	0.020
L09	0.108	0.015	0.874	0.130	0.815	0.060
L10	0.061	0.168	0.770	0.280	0.705	0.015
L11	0.000	0.349	0.651	0.375	0.480	0.145
L12	0.030	0.305	0.674	0.350	0.505	0.145
L13	0.000	0.150	0.850	0.025	0.000	0.975
L14	0.060	0.060	0.880	0.025	0.000	0.975
L15	0.000	0.300	0.700	0.110	0.815	0.075
L22	0.107	0.055	0.838	0.205	0.785	0.010
L29	0.000	0.175	0.825	0.135	0.855	0.010
L33	0.045	0.005	0.950	0.150	0.675	0.175

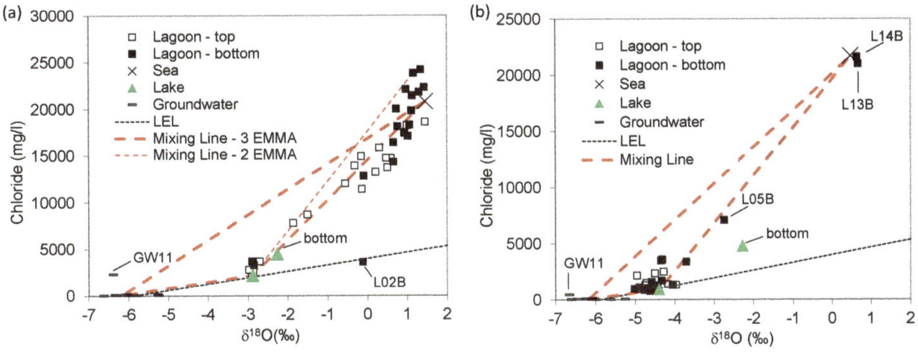

Figure 5. Chloride concentrations and $\delta^{18}O$ ratios for **(a)** dry season and **(b)** wet season sampling campaign; the dashed lines connect the three (bold) or two (light) endmembers used for the mixing analysis.

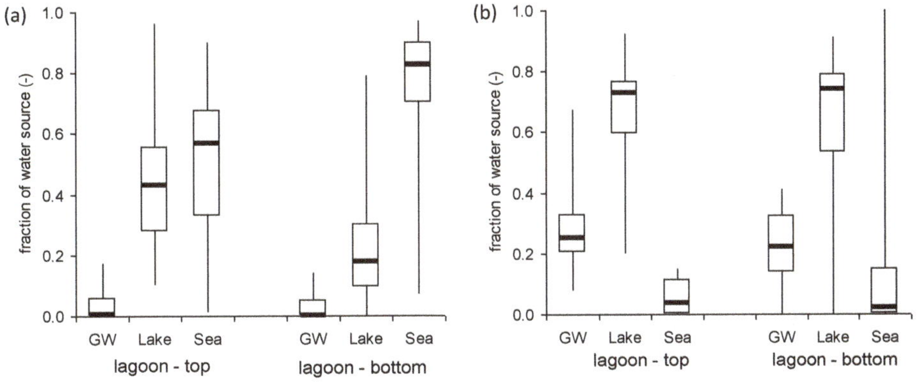

Figure 6. Fractions of different sources of the lagoon water for **(a)** dry and **(b)** wet season sampling campaign.

was found with distance from the shore line (Fig. 7c, d); the error bars in Fig. 7 indicate the variability of the results when using $\delta^{18}O$ and Cl^- or $\delta^{18}O$ and salinity as signatures for the endmember mixing analysis.

For the dry season, average fractions of water sources were different compared to the wet season, and more variability was found within the lagoon and when comparing top and bottom of the lagoon (Fig. 6a). The arithmetic average (median) of groundwater, lake and seawater contribution was 0.03 (0.01), 0.54 (0.43) and 0.43 (0.57) for the top and 0.09 (0.00), 0.20 (0.18), and 0.71 (0.83) for the bottom layer, respectively. Particularly the contribution of groundwater was little during the dry season (Table 2, Fig. 6a). The lagoon contained more seawater in the dry season compared to the wet season and at the bottom compared to the top layers. Further, there was a gradient of saltwater contribution in the lagoon with nearest distance to the shoreline (Fig. 7a, b). The closer to the sea, the higher is the fraction of seawater. Still, the lagoon lake systems contained on average higher fractions of saltwater (60 %, top; 88 %, bottom) compared to the Dalyan channel locations (35 %, top; 69 %, bottom) (Fig. 7a). The branches of the lagoon seem to be more mixed compared to lake and channel locations (Fig. 7a, b) containing on average 51 and 67 % of seawater on top and bottom, respectively.

4 Discussion

The results clearly indicated differences in contribution of various water sources in the dry and wet seasons. We proved that it is an extremely dynamic system dominated by seawater in the dry season (> 55 %) and freshwater in the wet season (> 95 %). Lecuyer et al. (2012) also found higher contribution of freshwater (62 %) compared to seawater (38 %) in the winter (the wet season) assuming seawater and stream water as sole endmembers. Still, their open water lagoon on the Turkish coast was generally more dominated by seawater throughout the rest of the year; particularly in summer and autumn freshwater contribution seemed to be mostly absent and hypersaline conditions formed due to evaporation of seawater. In contrast, our study site had freshwater influence even in the dry season. Here, the freshwater mainly came from the upstream lake and groundwater contribution was minor. Thus, the lagoon is groundwater influenced only in the wet season. In addition, we cannot exclude direct precipitation as additional water source for the wet season; due to little precipitation in summer (3 mm in average) its influence during the dry season was assumed to be negligible. We expect that winter precipitation has similar isotopic composition compared to the local groundwater and therefore, any

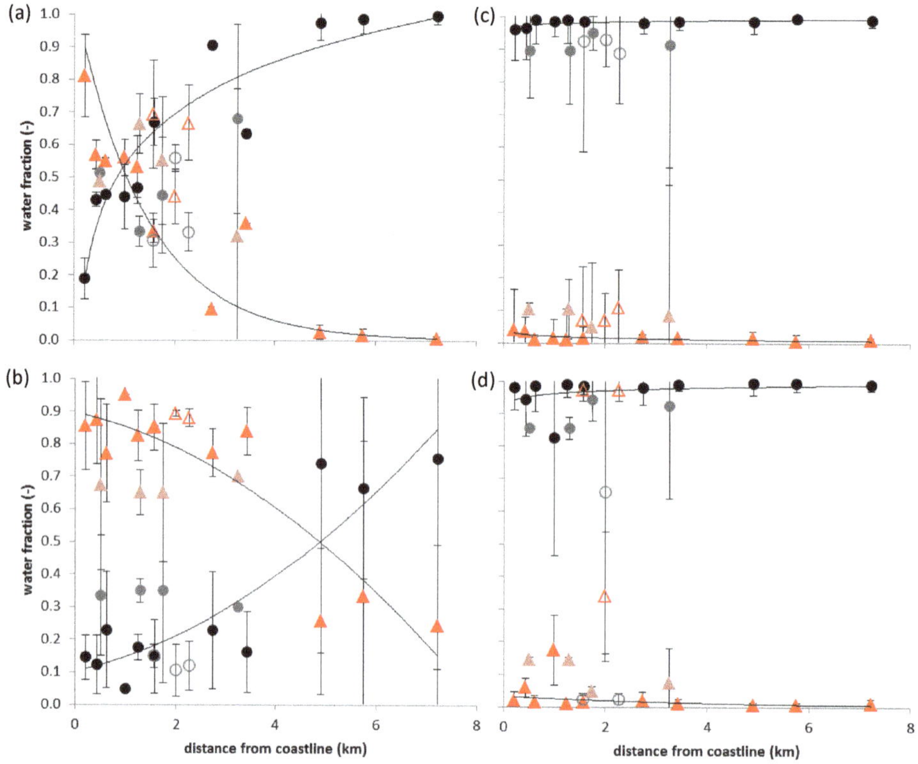

Figure 7. Changing fractions of freshwater (circles) and marine water (triangles) with distance from the coastline for **(a)** the top layer in the dry season, **(b)** bottom layer in the dry season, **(c)** top layer in the wet season, **(d)** bottom layer in the wet season; closed dark symbols indicate locations at the main lagoon channel, open symbols indicate surrounding lake locations and closed light symbols indicate their inflow/outflow connections to the lagoon system; error bars were determined from variability of endmember mixing analysis using salinity and chloride data individually in combination with $\delta^{18}O$.

contribution of direct precipitation was hidden in the groundwater term. However, this hidden precipitation is suggested to be little when looking at the upstream Köycegiz Lake water balance and the size of the entire catchment (960 km^2) compared to the lagoon size (130 km^2). In the Köycegiz Lake precipitation is for example more than 15 times smaller compared to its outflow into the lagoon (Bayari et al., 1995). Additionally, we sampled during a period without precipitation and therefore, our results are representative for base flow conditions in the lagoon system.

We found different dynamics for the bottom and top layers and also for the different locations in the lagoon. Particular seasonal changes were dramatic in the main Dalyan channel closer to the coast and at its bottom (Fig. 7b, d). We assume that the terrestrial water levels (groundwater, lake, lagoon) declined in the dry season influencing the hydraulic gradients and also density driven flow of the seawater further inland. Here, the intrusion reached up to 4 km inland at the bottom of the lagoon. A 50 : 50 mixing of salt and freshwater is expected for bottom layers at 4.9 km distance from the coast (Fig. 7b) and for top layer at 1 km. The freshwater (seawater) mixing relationship with distance from the shoreline was best approximated by logarithmic (exponential) function (Fig. 7). Still, the saltwater intrusion was mainly restricted to the lagoon system itself as the groundwater wells were unaffected by seawater influence in the dry season. Our findings are in agreement with previous studies on hydrodynamic modelling in this area (Ekdal et al., 2005; Erturk et al., 2003; Gönenc et al., 2004). In these studies, similar spatial and temporal dynamics were obtained concluding that intrusion causes strong stratification throughout almost the entire lagoon especially in the dry season. The flow direction in the upper layer was from Köycegiz Lake towards the Mediterranean Sea, while flow in the bottom layer was from the Mediterranean Sea towards the Köycegiz Lake. Barotrophy was found to be the driving force of the surface flow, whereas the bottom flow was baroclinic (Gönenc et al., 2004).

In the present study, the endmember mixing analysis yielded lower uncertainties in the wet compared to the dry season (Fig. 7), which is also obvious when looking at the endmember mixing triangles in Fig. 5. For the wet season, the composition of the seawater endmember was adequate (Fig. 5b). For the dry season though, higher chloride concentration as well as more enriched $\delta^{18}O$ were expected (Fig. 5a) and thus, samples plot outside of the mixing triangle. This indicates that either the endmember was chosen wrongly or/and evaporation is crucial. Evaporation of surface water explains an increase in salt concentrations and isotopic enrichment like observed in a close-by lagoon (Lecuyer et al., 2012). Even though evaporation was actually considered indirectly by the lake endmember, evaporation of lagoon water could be higher due to the smaller water volume compared to the lake. Therefore, a stronger enrichment of stable isotopes explains the deviations of top surface water samples located outside of the mixing triangle in the dry season (Fig. 5a). However, also enrichment of bottom samples was found in the dry season which is unusual and cannot be explained by evaporation only. Even hypersaline conditions in some of the bottom samples were found (compared to the seawater sample). Interestingly, the slope of the Cl^-–$\delta^{18}O$ relationship was steeper for bottom compared to top lagoon samples. It remained unknown whether an additional water source in the system has to be considered which was of geothermal origin as found for Köycegiz Lake (Bayari et al., 1995) and as common in this area due to geology and tectonic activity (Mutlu and Gülec, 1998).

Further assessing the two discussed uncertainties (i.e. choice of endmember and evaporation) and neglecting the small contributions of groundwater to the lagoon, a two component endmember mixing analysis was additionally conducted after correction of the data due to evaporation (Fig. 5a; 2 EMMA mixing line). First, the seawater surface sample was replaced by the deep lagoon sample at the very end of the Dalyan Channel exiting into the sea (L08B). Here, chloride concentrations and also isotopes were even higher compared to the seawater sample. It was measured in the depth and we expect it to be representative of the actual seawater not influenced by any freshwater compared to the actual seawater sample from the surface. Therefore, L08B could be used as endmember for the dry season being representative for seawater too. Second, all lagoon samples were forced onto the mixing line accounting for enrichment due to evaporation. Therefore, an Evaporation Line was calculated considering the top lake sample for both seasons only ($Cl^- = 670\ \delta^{18}O + 4000$). Here, 10 % increase in chloride was accompanied by 3.4 % increase in $\delta^{18}O$. This regression was used to correct the lagoon data back to the mixing line. Similar procedures were done with salinity and isotope data (data not shown). The difference between measured and corrected chloride concentrations (salinity) was further used to do a mass balance calculation. Thus, relative, average evaporation were estimated at all sites (Table 3); they have to be seen as relative because the actual surface Köycegiz Lake water already comprised evaporation which was estimated to 6.8 % (Bayari et al., 1995). The calculated evaporation in the lagoon ranged from 0 to 7 %. There was only one outliner L2B (Fig. 5a) with high chloride concentrations resulting in 54 % evaporation based on chloride data; but with 0.2 % evaporation based on salinity data. We attributed it to erroneous chloride analysis rather than to water influenced by geothermal origin because of differences in chemical and isotope characteristics compared to geothermal springs in this area (Bayari et al., 1995). The results of the two component endmember mixing approach yielded similar fractions of freshwater and marine water as the three component approach (Fig. 8, Table 3). Considering uncertainties of the methods (Fig. 7), no distinct differences in freshwater or saltwater sources were found hence. This suggested that for the dry season both groundwater and evaporation could be neglected in the system. Still, the correction of the data due to

Table 3. Average results of two-component endmember mixing analysis giving the contributions of lake water (f_{LW}) and seawater (f_{SW}) in the lagoon top and bottom for the dry season; average relative percentages of evaporation calculated for dry season based on data correction (details given in text).

Location – TOP	f_{LW}	f_{SW}	evaporation (%)
L01T	0.993	0.007	–
L02T	0.989	0.011	0.1
L03T	0.975	0.025	–
L04T	0.913	0.087	–
L05T	0.540	0.460	5.3
L06T	0.469	0.531	3.3
L07T	0.689	0.311	–
L08T	0.179	0.821	3.7
L09T	0.389	0.611	4.7
L10T	0.412	0.588	4.7
L11T	0.376	0.624	3.1
L12T	0.472	0.528	5.0
L13T	0.292	0.708	4.3
L14T	0.312	0.688	4.5
L15T	0.472	0.528	2.1
L22T	0.671	0.329	0.7
L29T	0.413	0.587	5.0
L33T	0.483	0.517	1.6
Location – BOTTOM			
L01B	0.598	0.402	–
L02B	0.667	0.333	0.2
L03B	0.494	0.506	–
L04B	0.198	0.802	1.4
L05B	0.075	0.925	1.9
L06B	0.365	0.635	2.4
L07B	0.126	0.874	1.7
L08B	0.000	1.000	–
L09B	0.016	0.984	0.8
L10B	0.156	0.844	2.9
L11B	0.312	0.688	4.8
L12B	0.315	0.685	3.6
L13B	0.194	0.806	2.6
L14B	0.181	0.819	1.4
L15B	0.374	0.626	4.2
L22B	0.100	0.900	0.8
L29B	0.226	0.774	2.7
L33B	0.118	0.882	1.5

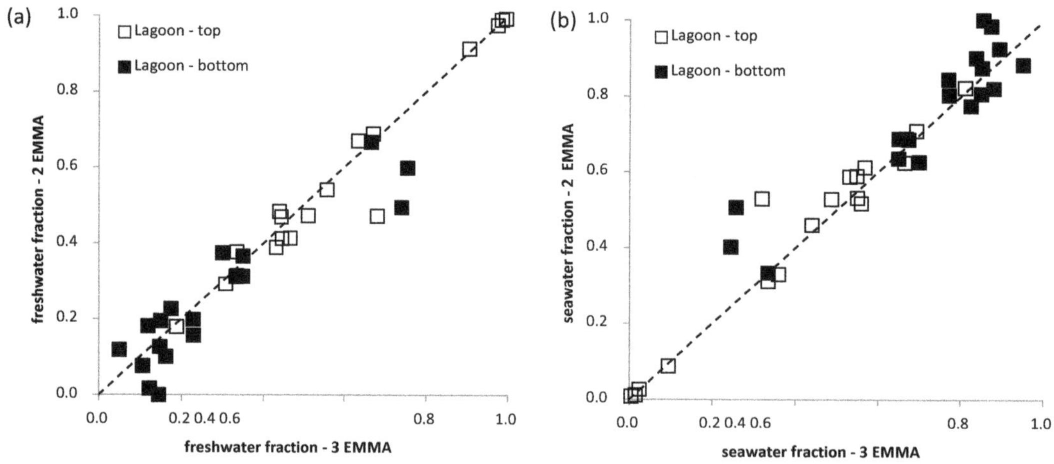

Figure 8. Fractions of freshwater **(a)** and seawater **(b)** contributions in the top and bottom lagoon samples calculated from two and three endmember mixing approaches; dashed line gives 1 : 1 line.

evaporation is kind of arbitrary, forcing all values onto the mixing line. Only knowing the actual evaporation at individual locations and in the lake would help to adequately correct the data which even might push some results into the three-endmember mixing triangle. Hence, fractions of groundwater even in the dry season would be underestimated by the current procedure of data correction.

Independent on the mixing approach, there were not only spatial differences in top and bottom layers for the main Dalyan channel, but also differences between different locations within the lagoon. The main channel responded quickly to changes and showed seasonal dynamics. The lake structures in the lagoon system were, however, responding differently. Here, the saltwater was found in the bottom layer even in the wet season indicating maintenance of stratification; particularly in the larger and deeper Sülüngür Lake. A partial mixing was found for the smaller and shallower Alagöl Lake where saltwater contribution was 34 % (±20 %). Also the side branches of the lagoon had less extreme changes as the main channel indicating higher water transit times in these areas and thus slower renewal. Particularly in the dry season, the contribution of both fresh and saltwater was about equal for the top layer and two-thirds to one-third for the bottom layer and independent on the distance to the coastline. These findings are in agreement with residence time calculations of a previous study (Ekdal, 2008) using the Water Quality Analysis Simulation Model. Average residence times of Sülüngür Lake (especially deeper parts of the lake) were considerably higher (16–700 d) when compared to other parts of the system (> 16 d). The residence time in Alagöl (5–16 d) was also high when compared to the main channel. The main channel had a low residence time (< 5 d), which showed the dynamic characteristics of the lagoon, and which is in agreement with the results of this study.

5 Conclusions

We showed that environmental tracers can be used not only to identify but also to quantify different water sources in a lagoon ecosystem. Freshwater and marine water sources were strongly dynamic and heterogeneous in time and space. We found different water sources and mixing ratios for dry and wet seasons and for top and bottom layers in the lagoon. In the wet season, freshwater was found in all locations and all depths except at the bottom of a larger lagoon lake. Generally, the freshwater was a mixture of upstream lake water and groundwater. The groundwater influence was, however, mainly restricted to the wet season and almost absent in the dry season. It was assumed that water levels decline and the input of seawater in the lagoon gets more pronounced; particularly in the main flow channel of the lagoon. Here, a clear stratification was observed in the dry season only, with higher saltwater contributions at the lagoon bottom compared to its top. At some of these locations, the lagoon changed from a complete freshwater system to a complete saltwater system which certainly has implications for the ecosystem which has to be highly adapted to such dynamic conditions. At side branches and lake structures in this wetland type lagoon, changes in water sources were less extreme and variable. From these findings, we conclude that the lagoon and the groundwater could be vulnerable to certain global change scenarios like sea level rise and decrease in precipitation. Consequently, water levels in the groundwater and lake would drop and the seawater influence would increase in the lagoon system affecting its ecosystem functions and probably also affecting the groundwater quality. In future, it needs to be analysed how the ecosystem itself reacts to changes of water sources to investigate the vulnerability of the ecosystem functions.

Acknowledgements. Funding by the EU GENESIS Project no. 226536 (FP7-ENV-2008-1) is kindly acknowledged. We would also like to acknowledge Melike Gürel, Nusret Karakaya, Ali Ertürk, Gökhan Cüceloglu and Çigdem Güzel for their contribution in taking the samples during the field studies. We thank one anonymous reviewer and K. Rozanski for their helpful comments that improved the quality of the manuscript.

Edited by: P. Wachniew

References

Alongi, D. M.: Coastal Ecosystem Processes, CRC Press, Boca Raton, 1998.

Anthony, A., Atwood, J., August, P., Byron, C., Cobb, S., Foster, C., Fry, C., Gold, A., Hagos, K., Heffner, L., Kellogg, D. Q., Lellis-Dibble, K., Opaluch, J. J., Oviatt, C., Pfeiffer-Herbert, A., Rohr, N., Smith, L., Smythe, T., Swift, J., and Vinhateiro, N.: Coastal Lagoons and Climate Change: Ecological and Social Ramifications in US Atlantic and Gulf Coast Ecosystems, Ecol. Soc., 14, 8, 2009.

Barnes, R. S. K.: Coastal lagoons, Cambridge University press, Cambridge, UK, 106 pp., 1980.

Bayari, S. C., Kazanci, N., Koyuncu, H., Çaglar, S. S., and Gökçe, D.: Determination of the origin of the waters of Köycegiz Lake, Turkey, J. Hydrol., 166, 171–191, doi:10.1016/0022-1694(94)02554-O, 1995.

Bayari, C. D., Kurittas, T., and Tezcan, L.: Dynamics of Lake Köycegiz, SW Turkey: An Environmental Isotopic and Hydrogeochemical Study, in: Use of isotope techniques in lake dynamics investigations, IAEA-TECDOC-1206, Vienna, Austria, 2001.

Ekdal, A.: Water Quality Modeling of Köycegiz – Dalyan Lagoon, Ph.D. thesis, Istanbul Technical University, Istanbul, Turkey, 236 pp., 2008.

Bowen, G. J. and Wilkinson, B.: Spatial distribution of $\delta^{18}O$ in meteoric precipitation, Geology, 30, 315–318, 2002.

Ekdal, A., Gurel, M., Erturk, A., and Tanik, A.: Hydrodynamic and Water Quality Modeling Approach for a Dynamic Lagoon System, Environmental Hydraulics and Sustainable Water Management, Proceedings, 15–18 December, Hong Kong, China, 621–627, 2005.

Erturk, A., Gurel, M., Koca, D., Ekdal, A., Tanik, A., Seker, D. Z., Kabdasli, S., and Gönenc, I. E.: Determination of Model Dimensions for a Complex Lagoon System – A Case Study From Turkey, Proceedings of XXX IAHR Congress Water Engineering and Research in a learning Society: Modern Developments and Traditional Concepts, 24–29 August 2003, Thessaloniki, Greece, 53–60, 2003.

Graciansky, P. C.: Stratigraphy of the overlapped units of the Lycien Nappes in the Teke Peninsula and their position within the Dinaro-Taurids, Bull. Miner. Res. Explor. Inst., Ankara, 71, 73–92, 1968 (in Turkish).

Gattacceca, J. C., Vallet-Coulomb, C., Mayer, A., Claude, C., Radakovitch, O., Conchetto, E., and Hamelin, B.: Isotopic and geochemical characterization of salinization in the shallow aquifers of a reclaimed subsiding zone: The southern Venice Lagoon coastland, J. Hydrol., 378, 46–61, doi:10.1016/j.jhydrol.2009.09.005, 2009.

Gönenc, I. E., Tanik, A., Seker, D. Z., Gurel, M., Erturk, A., Ekdal, A., Yuceil, K., Kose, C., Beyazgul, M., and Bilir, L. Z.: Ecosystem modeling for the sustainable management of lagoons, Final Report, The Scientific and Technical Research Council of Turkey, TUBITAK YDABCAG Project No: 100Y047, Ankara, 2004.

Google Earth, available at: http://www.google.com/earth/, last access: 25 May 2014.

Gurel, M., Tanik, A., Erturk, A., Dogan, E., Okus, E., Seker, D. Z., Ekdal, A., K., Y., Bederli Tumay, A., Karakaya, N., Beler Baykal, B., and Gönenc, I. E.: Köycegiz – Dalyan Lagoon: A case study for sustainable use and development, in: Coastal Lagoons: Ecosystem Processes and Modeling for Sustainable Use and Development, edited by: Gönenc, I. E. and Wolflin, J., CRC Press, Boca Raton, Florida, 440–474, 2005.

Kløve, B., Ala-Aho, P., Bertrand, G., Gurdak, J. J., Kupfersberger, H., Kværner, J., Muotka, T., Mykrä, H., Preda, E., Rossi, P., Uvo, C. B., Velasco, E., and Pulido-Velazquez, M.: Climate change impacts on groundwater and dependent ecosystems, J. Hydrol., 518, Part B, 250–266, 2014.

Lecuyer, C., Bodergat, A. M., Martineau, F., Fourel, F., Gurbuz, K., and Nazik, A.: Water sources, mixing and evaporation in the Akyatan lagoon, Turkey, Estua. Coastal Shelf Sci., 115, 200–209, doi:10.1016/j.ecss.2012.09.002, 2012.

Longinelli, A. and Selmo, E.: Isotopic composition of precipitation in Italy: a first overall map, J. Hydrol., 270, 75–88, 2003.

Lykoudis, S. P. and Argiriou, A. A.: Gridded data set of the stable isotopic composition of precipitation over the eastern and central Mediterranean, J. Geophys. Res.-Atmos., 112, D18107, doi:10.1029/2007JD008472, 2007.

Mutlu, H. and Gülec, N.: Hydrogeochemical outline of thermal waters and geothermometry applications in Anatolia (Turkey), J. Volcanol. Geotherm. Res., 85, 495–515, doi:10.1016/s0377-0273(98)00068-7, 1998.

Niencheski, L. F. H., Windom, H. L., Moore, W. S., and Jahnke, R. A.: Submarine groundwater discharge of nutrients to the ocean along a coastal lagoon barrier, Southern Brazil, Mar. Chem., 106, 546–561, doi:10.1016/j.marchem.2007.06.004, 2007.

Pérez-Ruzafa, A., Marcos, C., and Pérez-Ruzafa, I. M.: Mediterranean coastal lagoons in an ecosystem and aquatic resources management context, Phys. Chem. Earth, 36, 160–166, doi:10.1016/j.pce.2010.04.013, 2011a.

Pérez-Ruzafa, A., Marcos, C., Pérez-Ruzafa, I., and Pérez-Marcos, M.: Coastal lagoons: "transitional ecosystems" between transitional and coastal waters, J. Coast. Conserv., 15, 369–392, doi:10.1007/s11852-010-0095-2, 2011b.

Remane, A. and Schlieper, C.: Biology of Brackish Water, Wiley Interscience, New York, 372 pp., 1971.

Santos, I. R., Machado, M. I., Niencheski, L. F., Burnett, W., Milani, I. B., Andrade, C. F. F., Peterson, R. N., Chanton, J., and Baisch, P.: Major ion chemistry in a freshwater coastal lagoon from southern Brazil (Mangueira Lagoon): Influence of groundwater inputs, Aquat. Geochem., 14, 133–146, doi:10.1007/s10498-008-9029-0, 2008a.

Santos, I. R., Niencheski, F., Burnett, W., Peterson, R., Chanton, J., Andrade, C. F. F., Milani, I. B., Schmidt, A., and Knoeller, K.: Tracing anthropogenically driven groundwater discharge into a coastal lagoon from southern Brazil, J. Hydrol., 353, 275–293, doi:10.1016/j.jhydrol.2008.02.010, 2008b.

Schmidt, A., Santos, I. R., Burnett, W. C., Niencheski, F., and Knöller, K.: Groundwater sources in a permeable coastal barrier: Evidence from stable isotopes, J. Hydrol., 406, 66–72, doi:10.1016/j.jhydrol.2011.06.001, 2011.

Vreca, P., Bronic, I. K., Horvatincic, N., and Baresic, J.: Isotopic characteristics of precipitation in Slovenia and Croatia: Comparison of continental and maritime stations, J. Hydrol., 330, 457–469, 2006.

WISER, available at: http://www-naweb.iaea.org/napc/ih/IHS_resources_isohis.html, last access: 19 May 2014.

12

A review of the (Revised) Universal Soil Loss Equation ((R)USLE): with a view to increasing its global applicability and improving soil loss estimates

Rubianca Benavidez, Bethanna Jackson, Deborah Maxwell, and Kevin Norton

School of Geography, Environment, and Earth Sciences, Victoria University of Wellington, Wellington, 6012, New Zealand

Correspondence: Rubianca Benavidez (rubianca.benavidez@vuw.ac.nz)

Abstract. Soil erosion is a major problem around the world because of its effects on soil productivity, nutrient loss, siltation in water bodies, and degradation of water quality. By understanding the driving forces behind soil erosion, we can more easily identify erosion-prone areas within a landscape to address the problem strategically. Soil erosion models have been used to assist in this task. One of the most commonly used soil erosion models is the Universal Soil Loss Equation (USLE) and its family of models: the Revised Universal Soil Loss Equation (RUSLE), the Revised Universal Soil Loss Equation version 2 (RUSLE2), and the Modified Universal Soil Loss Equation (MUSLE). This paper reviews the different sub-factors of USLE and RUSLE, and analyses how different studies around the world have adapted the equations to local conditions. We compiled these studies and equations to serve as a reference for other researchers working with (R)USLE and related approaches. Within each sub-factor section, the strengths and limitations of the different equations are discussed, and guidance is given as to which equations may be most appropriate for particular climate types, spatial resolution, and temporal scale. We investigate some of the limitations of existing (R)USLE formulations, such as uncertainty issues given the simple empirical nature of the model and many of its sub-components; uncertainty issues around data availability; and its inability to account for soil loss from gully erosion, mass wasting events, or predicting potential sediment yields to streams. Recommendations on how to overcome some of the uncertainties associated with the model are given. Several key future directions to refine it are outlined: e.g. incorporating soil loss from other types of soil erosion, estimating soil loss at sub-annual temporal scales, and compiling consistent units for the future literature to reduce confusion and errors caused by mismatching units. The potential of combining (R)USLE with the Compound Topographic Index (CTI) and sediment delivery ratio (SDR) to account for gully erosion and sediment yield to streams respectively is discussed. Overall, the aim of this paper is to review the (R)USLE and its sub-factors, and to elucidate the caveats, limitations, and recommendations for future applications of these soil erosion models. We hope these recommendations will help researchers more robustly apply (R)USLE in a range of geoclimatic regions with varying data availability, and modelling different land cover scenarios at finer spatial and temporal scales (e.g. at the field scale with different cropping options).

1 Introduction

Soil erosion involves many processes, but the overall effect is of particles being transported and deposited from one location to another. Although it occurs naturally, soil erosion is often exacerbated by anthropogenic activities (Adornado et al., 2009). Soil erosion is affected by wind, rainfall and associated runoff processes, vulnerability of soil to erosion, and the characteristics of land cover and management (David, 1988; Aksoy and Kavvas, 2005; Panagos et al., 2015e). Understanding and mitigating erosion and associated degradation is critical because of its possible effects: nutrient loss, river and reservoir siltation, water quality degradation, and decreases in soil productivity (Bagherzadeh, 2014). In a review of the costs of soil erosion, Pimentel et al. (1995) reported soil erosion rates for regions around the world:

Asia, South America, and Africa with an average of 30 to 40 $t\,ha^{-1}\,year^{-1}$ and an average of 17 $t\,ha^{-1}\,year^{-1}$ for the United States of America and Europe. For comparison, the soil erosion rate for undisturbed forests was reported to range from 0.004 to 0.05 $t\,ha^{-1}\,year^{-1}$ globally (Pimentel et al., 1995). Within a landscape, erosion due to water can be caused by unconcentrated flow (sheet), occur within channels of concentrated flow (rills and gullies), and occur through raindrop impact and overland flow (inter-rill) (Aksoy and Kavvas, 2005; Morgan, 2005). Land management can be improved through understanding how these erosion processes occur and what areas are vulnerable to soil loss. Advances in technology such as the development of soil erosion models and increases in computing power for spatial analysis have assisted in making soil erosion modelling faster and more accurate.

Soil erosion models aid land management by helping elucidate the areas vulnerable to soil erosion in the baseline scenario, potential erosion rates, and possible causes of soil erosion. They range from relatively simple empirical models, and conceptual models, to more complicated physics-based models (Merritt et al., 2003). Like any other model, there are uncertainties associated with soil erosion models that cannot account for all the complex interactions of sediment delivery. Hence, unless extensive parameterisation and validation against observed data are accomplished, soil loss rates from models should be taken as best available estimates instead of absolute values (Wischmeier and Smith, 1978). Extensive reviews of soil erosion models of varying complexity have been done before but tend to focus on input requirements and applications (Aksoy and Kavvas, 2005; Merritt et al., 2003). A review by de Vente and Posen (2005) differs by focusing on semi-quantitative models that include different types of soil erosion in order to estimate basin sediment yield. Other reviews have focused on the use of different types of soil erosion models in particular places, such as Brazilian watersheds for de Mello et al. (2016).

One family of empirical soil loss models is the Universal Soil Loss Equation (USLE) suite of models, including the original USLE, the Revised Universal Soil Loss Equation (RUSLE), the Revised Universal Soil Loss Equation version 2 (RUSLE2), and the Modified Universal Soil Loss Equation (MUSLE). The USLE is an empirical model used to estimate the annual average rate of soil erosion (tons per unit area) for a given combination of crop system, management practice, soil type, rainfall pattern, and topography. It was originally developed at the plot scale for agricultural plots in the United States of America (Wischmeier and Smith, 1978). An updated form of USLE (RUSLE) was published to include new rainfall erosivity maps for the United States of America and improvements to the method of calculating the different USLE factors (Renard et al., 1997). RUSLE added changes in soil erodibility due to freeze–thaw and soil moisture, a method for calculating cover and management factors, changes to how the influence of topography is incorporated into the model, and updated values to represent soil conservation practices (Renard and Freimund, 1994). The RUSLE2 framework is a computer interface programmed to handle more complex field situations, including an updated database of factors (Foster et al., 2003). These three variations of (R)USLE measure soil loss per unit area at an annual timescale. The MUSLE is an extension to work at finer temporal resolution, using runoff and peak flow rate to estimate event-based soil loss (Sadeghi et al., 2014). These models have been used around the world due to their relative simplicity and seemingly low data requirements (Table A1).

This simplicity of the (R)USLE has been integrated into more complex soil erosion models to help with management and decision-making, including the Agricultural Non-Point Source model (AGNPS), the Chemical Runoff and Erosion from Agricultural Management Systems model (CREAMS), and the Sediment River Network model (SedNet) (Aksoy and Kavvas, 2005; de Vente and Poesen, 2005; Merritt et al., 2003). The AGNPS estimates upland erosion using the USLE and then uses sediment transport algorithms to simulate runoff and sediment and nutrient transport within watersheds (Aksoy and Kavvas, 2005). The usage of (R)USLE in large models is mainly for the purpose of assisting with decision-making, such as prioritising land use objectives in the Philippines (Bantayan and Bishop, 1998), scenario analysis for water quality in catchments in New Zealand (Rodda et al., 2001), or delineating unique soil landscapes in Australia (Yang et al., 2007).

Extensive reviews of soil erosion modelling and types of soil erosion models have been published that briefly discuss the (R)USLE as an empirical model, elements of which are commonly incorporated into more complex conceptual or physics-based soil erosion models (Aksoy and Kavvas, 2005; de Vente and Poesen, 2005; Merritt et al., 2003). This review is more specific to the (R)USLE and addresses the complexity of its different sub-factors, as well as the issues for researchers to consider before applying (R)USLE to their study area. These issues include equation choices, digital elevation model (DEM) resolution, granularity in land cover characteristics, scale, etc. The MUSLE is not included in this review because Sadeghi et al. (2014) have already done an extensive review of the model and event-scale estimates are beyond the scope of this paper. Annual estimates of soil loss are useful for understanding the baseline erosion in a catchment, but intra-annual and event-based soil loss estimates are useful for elucidating temporal variations in erosion. Performing event-based soil loss modelling is important for areas that frequently experience extreme events as these can cause large-scale sediment transport and mass wasting.

This paper discusses the advantages, disadvantages, and limitations of the USLE model family. Although alternative sub-factor equations are presented, we also discuss questions of suitability that future users should consider before applying these models. The main aim of this paper is to review the

(Revised) Universal Soil Loss Equation and its sub-factors through the following objectives:

- review the USLE and RUSLE literature to compile equations for the different sub-factors within the (R)USLE;
- provide guidance as to which datasets and equations are appropriate over a range of geoclimatic regions with varying levels of data availability;
- outline the limitations and caveats of the (R)USLE that future users must consider;
- outline potential future directions to overcome these limitations and to improve (R)USLE applications.

2 Universal Soil Loss Equation (USLE)

The principal equation for the USLE model family is below:

$$A = R \times K \times L \times S \times C \times P, \quad (1)$$

where A is mean annual soil loss (metric tons per hectare per year), R is the rainfall and runoff factor or rainfall erosivity factor (megajoule millimetres per hectare per hour per year), K[1] is the soil erodibility factor (metric ton hours per megajoules per millimetre), L is the slope length factor (unitless), S is the slope steepness factor (unitless), C is the cover and management factor (unitless), and P is the support practice factor (unitless).

The USLE was originally developed at the farm plot scale for agricultural land in the United States of America but has seen use in many other countries, at many other scales, and in many other geoclimatic regions. Although the name implies that the model can be applied to all soils, the original USLE is more accurate for soils with medium texture and slopes of less than 400 ft in length with a gradient ranging between 3 % and 18 %, and it is managed with consistent cropping practices that are well represented in plot-scale erosion studies (Wischmeier and Smith, 1978). Hence, applying the USLE family of models to soils and sites exceeding these limits requires careful parameterisation of the model and being mindful of the increased uncertainty in model predictions.

In the original development of the model, this farm plot is called the "unit plot" and is defined as a plot that is 22.1 m long, is 1.83 m wide, and has a slope of 9 % (Wischmeier and Smith, 1978). Although the model accounts for rill and interrill erosion, it does not account for soil loss from gullies or mass wasting events such as landslides (Thorne et al., 1985). The Appendix of this paper compiles a non-exhaustive list of studies that have applied the USLE and RUSLE models to watersheds around the world. The uncertainties in soil erosion modelling stem from the availability of long-term reliable data, including issues of temporal resolution (e.g. < 30 min resolution required for (R)USLE) and the availability of spatial data over a catchment. This issue is not unique to (R)USLE applications and is generally worse when applying more complex models with larger numbers of variables and more detailed data requirements (de Vente and Poesen, 2005; Hernandez et al., 2012). Hence, the ubiquitous usage of the (R)USLE can be attributed to its relatively low data requirements compared to more complex soil loss models, making it potentially easier to apply in areas with scarce data. Another limitation of the (R)USLE and arguably many erosion model applications is the lack of validation data with which to verify model outputs, which is discussed further in Sect. 4.

Although the application of the (R)USLE seems to be a simple linear equation at first glance, this review addresses the complex equations that go into calculating its sub-factors, such as rainfall erosivity, which requires detailed pluviographic data (< 30 min resolution).

2.1 Rainfall erosivity factor (*R*)

The R factor represents the effect that rainfall has on soil erosion and was included after observing sediment deposits after an intense storm (Wischmeier and Smith, 1978). The annual R factor is a function of the mean annual EI_{30} that is calculated from detailed and long-term records of storm kinetic energy (E) and maximum 30 min intensity (I_{30}) (Morgan, 2005; Renard et al., 1997). Due to the detailed data requirements for the standard (R)USLE calculation of rainfall erosivity, studies in areas with less detailed data have used alternative equations depending on the temporal resolution and availability of the rainfall data. These compiled studies have used long-term datasets with at least daily temporal resolution to construct their R-factor equation. Extensive work by Naipal et al. (2015) attempted to apply the (R)USLE at a coarse global scale (30 arcsec) by using USA and European databases to derive rainfall erosivity equations. These equations use a combination of annual precipitation (millimetres), mean elevation (metres), and simple precipitation intensity index (millimetres per day) to calculate the R factor for different Köppen–Geiger climate classifications (Naipal et al., 2015). Loureiro and Coutinho (2001) used 27 years of daily rainfall data from Portugal and the (R)USLE method of calculating EI_{30} to construct an equation that uses the number of days that received over 10 mm of rainfall and the amount of rainfall per month when the day's rainfall exceeded 10 mm. The Loureiro and Coutinho (2001) equation was modified by Shamshad et al. (2008) for use in tropical Malaysia by using long-term rainfall data to construct a regression equation relating monthly rainfall and annual rainfall with the R factor. Similarly, Sholagberu et al. (2016) used 23 years of daily rainfall data to create a regression equa-

[1] The RUSLE handbook by Renard et al. (1997) indicates that the K-factor metric units are metric tons per hectare per hour per megajoules per hectare per millimetre, but for mathematical correctness the hectare units cancel out.

tion relating annual rainfall and the *R* factor for the highlands of Malaysia. These simplified equations may be transferable to areas of similar climate that do not have the long-term detailed rainfall data required by the original (R)USLE. The imperial units of erosivity are in hundreds of foot ton-force (tonf) inch per acre per hour per year, and multiplying by 17.02 will give the SI units of megajoule millimetre per hectare per hour per year (Renard et al., 1997).

With the body of work that has been done in rainfall erosivity, some studies have managed to construct rainfall erosivity maps over large countries and regions. Panagos et al. (2017) have used pluviographic data from 63 countries to calculate rainfall erosivity and spatially interpolated the results to construct a global rainfall erosivity map at 30 arcsec resolution. Despite its coarse resolution, this global dataset can be used as a resource for rainfall erosivity in data-sparse regions. For the United States, Renard et al. (1997) details the procedure for obtaining rainfall erosivity values from their large national database. Renard et al. (1997) would be the recommended reference for study areas in the United States because of the extensive database that already exists for that country. For the European Union, Panagos et al. (2015d) constructed a rainfall erosivity map at 1 km resolution and published descriptive statistics for *R* values in each of the member countries. The interpolated map showed good agreement through cross-validation and to previous studies, but areas that had fewer rainfall stations and more diverse terrain caused higher prediction uncertainty (Panagos et al., 2015d). Using a large rainfall dataset, da Silva (2004) constructed a spatially interpolated map of *R* factors in Brazil whose trends showed agreement with previous work on rainfall erosivity in the country.

In areas that only have annual precipitation available, several equations and their studies can be used as a reference. In their global application, Naipal et al. (2015) published different *R*-factor equations depending on a study area's climate classification. One caveat is that the data for these equations had a large percentage of USA and European records, so resulting accuracy of *R* factors might be better for those locations (Naipal et al., 2015). In tropical areas such as Southeast Asia, the *R* factor by El-Swaify et al. (1987) as cited in Merritt et al. (2004) was used extensively in Thailand, the Philippines, and Sri Lanka. However, the units for the *R* factor in this equation are given as metric tons per hectare per year, which do not correspond to the original units used by (R)USLE (Merritt et al., 2004). This lack of consistency regarding units is not uncommon in the reviewed literature, which sometimes fails to explicitly report the units used for the different factors. For example, Renard and Freimund (1994) report that the units of *R*-factor equations by Arnoldus (1977) were presumed to be in metric units. By being clear and consistent about units in the (R)USLE literature, future researchers can be more certain about the accuracy of their borrowed *R*-factor equations instead of presuming the units to be the same as the original (R)USLE. Work by Bonilla and Vidal (2011) produced an *R*-factor equation for Chile and published erosivity values similar to those produced by work in areas of similar geography and geology. For New Zealand, Klik et al. (2015) proposed equations for calculating the annual *R* factor and seasonal *R* factor with coefficients that change depending on the study area's location within the country.

The usage of monthly precipitation data to determine the *R* factor is due to monthly rainfall data being more readily available compared to detailed storm records (Renard and Freimund, 1994). Although annual rainfall estimates are sufficient, using monthly rainfall data to construct sub-annual *R* factors and then aggregating those *R* factors to an annual scale is useful in sites with large temporal variability in rainfall. Renard and Freimund (1994) used data from 155 stations with known *R* factors based on the original USLE approach and related their *R* factors to observed annual and monthly precipitation. These equations developed by Renard and Freimund (1994) on the west coast of USA were used in Ecuador (Ochoa-Cueva et al., 2015), and Honduras and El Salvador (Kim et al., 2005). Work by Arnoldus (1980) developed *R*-factor equations in West Africa that use monthly and annual precipitation. However, as described earlier, these equations present a problem in terms of consistent units. In Southeast Asia, Shamsad et al. (2008) developed an *R*-factor equation in Malaysia that was used in the Philippines by Delgado and Canters (2012). In New Zealand, the monthly precipitation can be aggregated to seasonal precipitation and used in the equation for seasonal *R* factor derived by Klik et al. (2015).

Monthly or better precipitation records are very useful in (R)USLE applications because of the option of estimating soil loss at a monthly or seasonal scale, which can be useful in countries with high temporal variation of rainfall throughout the year. Monthly and seasonal erosion has been estimated by varying the *R* factor depending on the monthly precipitation while leaving all the other factors constant (Ferreira and Panagopoulos, 2014; Kavian et al., 2011). Klik et al. (2015) emphasised the need to understand the drivers of soil erosion, including whether rainfall intensity had a stronger effect compared to mean annual rainfall. In an assessment of spatial and temporal variations in rainfall erosivity over New Zealand, December and January were associated with higher erosivities, while August was associated with lowest erosivity (Klik et al., 2015). Similar work by Diodato (2004) has cited the use of monthly erosivity data to be more useful with respect to managing crop growing cycles and tillage practices, especially during seasons where high rainfall erosivity is expected. In locations where there is a large temporal variation in rainfall throughout the year, the seasonal approach of estimating soil erosion is more important for sustainable land management (Ferreira and Panagopoulos, 2014)

To examine how different *R*-factor equations affected predicted soil erosion rates over the same study site, Be-

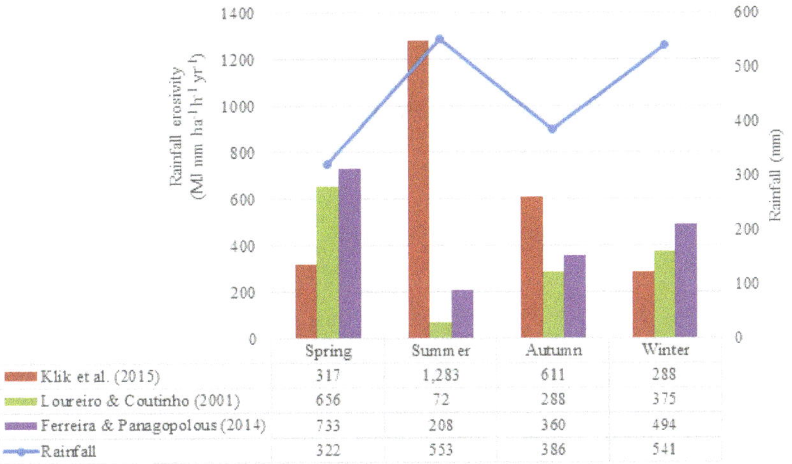

Figure 1. Graph of seasonal rainfall and estimates of erosivity in the Mangatarere watershed (Benavidez, 2018).

Table 1. Annual estimates (Benavidez, 2018) of erosivity in the Mangatarere watershed (megajoule millimetres per hectare per hour per year).

Equation source	Klik et al. (2015)	Loureiro and Coutinho (2001)	Ferreira and Panagopolous (2014)
Annual erosivity	2607	1391	1715

navidez (2018) tested three different equations over the ~ 157 km^2 Mangatarere watershed in New Zealand. The equations by Klik et al. (2015) developed in New Zealand, along with the equations by Loureiro and Coutinho (2001) and Ferreira and Panagopolous (2014) developed in Portugal, were used to estimate annual and seasonal erosivity (Fig. 1 and Table 1). All three equations consider and predict seasonal erosivity, are from similar latitudes, and were developed in temperate to semi-arid environments. For the same set of rainfall data, the three equations predicted different annual and seasonal values of erosivity. Regarding seasonal patterns of erosivity, Klik et al. (2015) predicted highest erosivity occurring during summer but lowest in winter and spring. This trend matches the national observations of the most erosive storms occurring during summer and the least erosive storms occurring during winter (Klik et al., 2015). By contrast, both Loureiro and Coutinho (2001) and Ferreira and Panagopolous (2014) predicted highest erosivity during spring and lowest during summer. This variation is thought to be due to the Portugal equations excluding days below 10.0 mm of rainfall, which introduces a bias towards the erosive effects of short, intense rainfall events while potentially excluding the erosive power of longer but less intense rainfall events.

It is unsurprising that the New Zealand approach performed best in a New Zealand climate, but it does demonstrate the risk of arbitrarily transferring equations between countries, even when geoclimatic conditions are not terribly dissimilar.

These differences highlight the importance of understanding the regional applicability of rainfall erosivity equations. In the reviewed (R)USLE studies for this section, a common occurrence was using equations derived in different countries and regions without much justification regarding why those equations were chosen with little consideration for their suitability. These studies also did not publish any testing of how different R factors produce different erosivity values from the same input dataset. The purpose of testing the different R factors is to illustrate this variation and encourages future users of (R)USLE to do the same sensitivity testing in their area.

In summary, there are many rainfall erosivity datasets and equations in the (R)USLE literature that can be used by new researchers applying the RUSLE to their study area. The erosivity dataset produced by Panagos et al. (2017) is recommended for areas with no rainfall data or in ungauged catchments since this is a raster dataset with global coverage (~ 30 arcsec resolution) and is freely available. For areas in the European Union, work by Panagos et al. (2015d) has produced a rainfall erosivity map with regional coverage at ~ 1 km resolution. These datasets can also be used to validate the erosivity factors calculated at the national or catchment scale. For study areas in which annual precipitation and the Köppen–Geiger classification are known, Naipal et al. (2015) has published rainfall erosivity equations and values for 17 different climate zones. Several studies have published erosivity equations for tropical areas: da Silva (2004) for Brazil, Shamshad et al. (2008) for Malaysia, and Jain and

Table 2. Summary of different studies that developed rainfall erosivity equations, original locations, and other studies that used their equations.

No.	Author	Original location	Resolution	Equation and requirements	Other studies
1	Wischmeier and Smith (1978) and Renard et al. (1997)	United States of America	Sub-daily	$R = \frac{\sum_{i=1}^{j} (EI_{30})_i}{N}$ $EI_{30} = E \times I_{30}$ $E = 916 + 331 \times \log_{10} I$ I: intensity (in h^{-1}) EI_{30i}: EI_{30} for storm i j: number of storms in an N-year period Units Imperial: 100 foot-ton in acre^{-1} h^{-1} year^{-1} Metric (multiply by 17.02): MJ mm ha^{-1} h^{-1} year^{-1}	Applied around the USA
2	Mihara (1951) and Hudson (1971) as cited in David (1988)	USA	Daily	$R = A \times \sum_{1}^{n} P_i^m$ $A = 0.002$ $M = 2$ P_i: precipitation total for day i when P exceeds 25 mm Units: not specified, likely to be original USLE imperial units	Watersheds around the Philippines (David, 1988)
3	Arnoldus (1980) as cited in Renard and Freimund (1994)	Morocco and other locations in West Africa	Monthly and annual	West Africa: $R = 4.79\text{MFI} - 142$ $R = 5.44\text{MFI} - 416$ Eastern USA: $R = 6.86\text{MFI} - 420$ Western USA: $R = 4.79\text{MFI} - 143$ Northwest USA: $R = 0.66\text{MFI} - 3$ $\text{MFI} = \sum_{i=1}^{12} \frac{P_i^2}{P}$ MFI: modified Fournier index P_i: monthly precipitation P: annual precipitation Units: t- cm ha^{-1} h^{-1} year^{-1} (Renard and Freimund, 1994)	Morocco, Turkey (Demirci and Karaburun, 2012); Morocco (Raissouni et al., 2016)
4	Renard and Freimund (1994)	West coast of USA	Monthly and annual	$R = 0.0483 \times P^{1.610}$ $R = 587.8 - 1.219P + 0.004105P^2$ Using MFI (Arnoldus, 1980): $R = 0.07397 \times \text{MFI}^{1.847}$ $R = 95.77 - 6.081\text{MFI} + 0.4770\text{MFI}^2$ P_i: monthly precipitation P: annual precipitation Units: MJ mm ha^{-1} h^{-1} year^{-1}	Central America (Kim et al., 2005); Iran (Zakerinejad and Maerker, 2015); Ecuador (Ochoa-Cueva et al., 2015)
5	Zhou et al. (1995) as cited in Li et al. (2014)	Southern China	Monthly	$R = \sum_{i=1}^{12} -1.15527 + 1.792P_i$ P_i: monthly precipitation Units: MJ mm ha^{-1} h^{-1} year^{-1}	China (Li et al., 2014)

Table 2. Continued.

No.	Author	Original location	Resolution	Equation and requirements	Other studies
6	Roose (1975) and Morgan (1974) as cited in Morgan (2005)	Peninsular Malaysia and Africa	Annual	Africa (Roose, 1975): $R = 0.5 \times P \times 17.3$ Peninsular Malaysia: $R = (9.28 \times P - 8838)\left(\frac{75}{1000}\right)$ P: mean annual precipitation (mm) Units: MJ mm ha^{-1} h^{-1} year^{-1}	Malaysia (Roslee et al., 2017); Vanuatu (Dumas and Fossey, 2009); Iran (Zakerinejad and Maerker, 2015)
7	El-Swaify et al. (1987) as cited in Merritt et al. (2004)	Possibly Thailand	Annual	$R = 38.5 + 0.35P$ P: mean annual precipitation Units: t ha^{-1} year^{-1} (all the other factors must have been developed to have no units so that the final soil loss is in t ha^{-1} year^{-1})	Thailand (Eiumnoh, 2000; Merritt et al., 2004); Philippines (Adornado and Yoshida, 2010; Adornado et al., 2009; Hernandez et al., 2012); Sri Lanka (Jayasinghe et al., 2010)
8	Land Development Department (2000) as cited in Nontananandh and Changnoi (2012)	Thailand	Annual	$R = 0.04669P - 12.1415$ P: mean annual rainfall Units: MJ mm ha^{-1} h^{-1} year^{-1}	Thailand (Nontananandh and Changnoi, 2012)
9	Loureiro and Coutinho (2001)	Portugal	Daily	$R = \frac{1}{N}\sum_{i=1}^{N}\sum_{m=1}^{12} EI_{30(\text{monthly})}$ $EI_{30\,(\text{monthly})} = 7.05\,\text{rain}_{10} - 88.92\,\text{days}_{10}$ Rain$_{10}$: monthly rainfall for days with ≥ 10.0 mm of rain Days$_{10}$: monthly number of days with rainfall ≥ 10.0 mm of rain N: number of years Units: MJ mm ha^{-1} h^{-1} year^{-1}	Spain (López-Vicente et al., 2008)
10	Fernandez et al. (2003), originally developed by the USDA-ARS (2002)	USA	Annual	$R = -823.8 + 5.213P$ P: annual precipitation Units: MJ mm ha^{-1} h^{-1} year^{-1}	USA (Fernandez et al., 2003); Greece (Jahun et al., 2015)
11	Ram et al. (2004) as cited in Jain and Das (2010)	India	Annual	$R = 81.5 + 0.38P$ P: annual precipitation for areas where annual precipitation ranges between 340 and 3500 mm Units: MJ mm ha^{-1} h^{-1} year^{-1}	India (Jain and Das, 2010)

Table 2. Continued.

No.	Author	Original location	Resolution	Equation and requirements	Other studies
12	Shamshad et al. (2008)	Malaysia	Monthly and annual	Based on Loureiro and Coutinho (2001) but for Malaysia: $R=\sum_{i=1}^{12} 6.97\,\text{rain}_{10}-11.23\,\text{days}_{10}$ $R=\sum_{i=1}^{12} 0.266\times\text{rain}_{10}^{2.071}\times\text{days}_{10}^{-1.367}$ $R=\sum_{i=1}^{12} 227\times\left(\frac{P_i^2}{P}\right)^{0.548}$ Rain_{10}: monthly rainfall for days with ≥ 10.0 mm of rain Days_{10}: monthly number of days with rainfall ≥ 10.0 mm of rain P_i: monthly precipitation P: annual precipitation Units: MJ mm $\text{ha}^{-1}\,\text{h}^{-1}\,\text{year}^{-1}$	Philippines (Delgado and Canters, 2012)
13	Irvem et al. (2007)	Turkey	Monthly and annual	$R=0.1215\times\text{MFI}^{2.2421}$ $\text{MFI}=\sum_{i=1}^{12}\frac{P_i^2}{P}$ P_i: monthly precipitation P: annual precipitation Units: MJ mm $\text{ha}^{-1}\,\text{h}^{-1}\,\text{year}^{-1}$	Turkey (Ozsoy et al., 2012)
14	Ferreira and Panagopolous (2014), similar to Loureiro and Coutinho (2001)	Portugal	Daily	$R=\sum_{i=1}^{12} 6.56\,\text{rain}_{10}-75.09\,\text{days}_{10}$ Rain_{10}: monthly rainfall for days with ≥ 10.0 mm of rain Days_{10}: monthly number of days with rainfall ≥ 10.0 mm of rain Units: MJ mm $\text{ha}^{-1}\,\text{h}^{-1}\,\text{year}^{-1}$	Portugal (Ferreira and Panagopoulos, 2014)
15	Nakil (2014) as cited in Nakil and Khire (2016)	India	Annual	$R=839.15\times e^{0.0008P}$ P: annual precipitation Units: MJ mm $\text{ha}^{-1}\,\text{h}^{-1}\,\text{year}^{-1}$	India (Nakil and Khire, 2016)
18	Naipal et al. (2015)	Global application, but original data from the USA and Europe	Annual	Various equations depending on Köppen–Geiger climate classification, including alternate equations if SDII is not available P: annual precipitation (mm) Z: mean elevation (m) SDII: simple precipitation intensity index (mm day^{-1})	
19	Klik et al. (2015)	New Zealand	Annual or seasonal	Annual or seasonal: $R=aP^b$ $R=aP+b$ P: annual precipitation (mm) or seasonal precipitation (mm) a and b: constants depending on region of New Zealand The equation used will depend on the region of New Zealand and the season Units: MJ mm $\text{ha}^{-1}\,\text{h}^{-1}$	
20	Sholagberu et al. (2016)	Malaysia	Annual	$R=0.0003P^{1.771}$ P: annual precipitation Units: MJ mm $\text{ha}^{-1}\,\text{h}^{-1}\,\text{year}^{-1}$	

Das (2010) for India. For arid areas, Arnoldus (1980) as cited in Renard and Freimund (1994) has derived erosivity equations for Morocco and other locations in West Africa. Many other equations are found in Table 2, and choosing several for sensitivity testing is recommended for future (R)USLE applications. It is also important to test against observed data or R factors derived by previous applications in the same study area or in study areas with similar climatic regimes.

2.2 Soil erodibility factor (K)

The K factor represents the influence of different soil properties on the slope's susceptibility to erosion (Renard et al., 1997). It is defined as the "mean annual soil loss per unit of rainfall erosivity for a standard condition of bare soil, recently tilled up-and-down slope with no conservation practice" (Morgan, 2005). The K factor essentially represents the soil loss that would occur on the (R)USLE unit plot, which is a plot that is 22.1 m long, is 1.83 m wide, and has a slope of 9 % (Lopez-Vicente et al., 2008).

Higher K-factor values indicate the soil's higher susceptibility to soil erosion (Adornado et al., 2009). In the (R)USLE, Wischmeier and Smith (1978) and Renard et al. (1997) use an equation that relates textural information, organic matter, information about the soil structure, and profile permeability with the K factor or soil erodibility factor. However, other soil classifications might not include soil structure and profile permeability information that matches the information required by (R)USLE nomograph. Hence, alternative equations have been developed that exclude the soil structure and profile permeability (Table 3). The question of which equation to use depends on the availability of soil data. Where only the textural class and organic matter content are known, Stewart et al. (1975) have approximated K-factor values based on these inputs. Similar to the R factor, the imperial units of soil erodibility are in ton acre hour per hundreds of acres per foot per tonf per inch. Multiplying by 0.1317 gives the erodibility in SI units of metric tons hectare hour per hectare per megajoule per millimetre (Renard et al., 1997).

Although seemingly relatively straightforward, the K factor equation proposed by Wischmeier and Smith (1978) comes with a few limitations regarding soil type. This equation was developed using data from medium-textured surface soils in the Midwestern USA, with an upper silt fraction limit of 70 % (Renard et al., 1997). An equation for volcanic soils in Hawaii was proposed by El-Swaify and Dangler (1976) as cited in Renard et al. (1997) but is only appropriate for soils similar to Hawaiian soils and not for all tropical soils. Despite these limitations, many studies outside the USA have used the original Wischmeier and Smith (1978) K-factor equation (Table 3). Being aware of the regional specificity of K-factor equations is important, and using different K-factor equations in one study area to find a range of soil erodibility could be a way of testing their applicability.

Similar to the sensitivity analysis of the R-factor equations, testing different K-factor equations to see the variation in erodibility values and then comparing these K factors with published values from similar soils would be a good way to test applicability. For the spatial coverage of the European Union, a soil erodibility raster dataset ($\sim$ 500 m resolution) is available for validation (Panagos et al., 2014). David (1988) and Dymond (2010) have published K-factor values for soils of different textural classes (e.g. clay, loam) that can be used if only soil texture is known (Tables 4 and 5). However, the values published by Dymond (2010) are broad and do not account for soils with mixed texture, while the values of David (1988) are based on soils in the Philippines. Like the R factor, it is important to check the derived K-factor values for the site-specific soil against previously published K-factor values for comparable sites and soil types.

2.3 Slope length (L) and steepness (S) factor

The LS factor represents the effect of the slope's length and steepness on sheet, rill, and inter-rill erosion by water, and it is the ratio of expected soil loss from a field slope relative to the original USLE unit plot (Wischmeier and Smith, 1978). The USLE method of calculating the slope length and steepness factor was originally applied at the unit plot and field scale, and the RUSLE extended this to the one-dimensional hill slope scale, with different equations depending on whether the slope had a gradient of more than 9 % (Renard et al., 1997; Wischmeier and Smith, 1978). Further research extends the LS factor to topographically complex units using a method that incorporates contributing area and flow accumulation (Desmet and Govers, 1996). The USLE and RUSLE method of calculating the LS factor uses slope length, angle, and a parameter that depends on the steepness of the slope in percent (Wischmeier and Smith, 1978).

One of the criticisms of the original USLE method of calculating LS factor is its limited applicability to complex topography. With advances in GIS technology, the method of determining the LS factor as a function of upslope contributing area or flow accumulation and slope has risen in popularity (Table 6). The use of DEMs to calculate the upslope contributing area and the resulting LS factor allows researchers to account for more topographically complex landscapes (Moore and Burch, 1986; Desmet and Govers, 1996). Desmet and Govers (1996) have also built on this method through showing its application in a GIS environment over topographically complex terrain when compared to the original method proposed by Wischmeier and Smith (1978). This method of using flow accumulation for slope length and steepness explicitly accounts for convergence and divergence of flow, which is important when considering soil erosion over a complex landscape (Wilson and Gallant, 2000). It is possible to use this method to calculate the LS factor over a large extent, but a high-resolution DEM is needed for accurate representation of the topography. The resolution required

Table 3. Summary of different studies with soil erodibility equations, original locations, and other studies that used their equations. All of the equations in Table 3 use imperial units of soil erodibility: ton acre hour per hundreds of acres per foot per tonf per inch. Multiply by 0.1317 for conversion into SI units of metric ton hours per megajoules per millimetre.

No.	Author	Original location	Data requirements	Equation	Other studies
1	Wischmeier and Smith (1978) and Renard et al. (1997)	USA	Very fine sand (%), clay (%), silt (%), organic matter (%), soil structure, profile permeability	$M = \text{Silt} \times (100 - \text{Clay})$ $K = \{[2.1 \times M^{1.14} \times (10^{-4}) \times (12 - a)] + [3.25 \times (b - 2)] + [2.5 \times (c - 3)]\} \div 100$ M: Particle-size parameter Silt: silt (%) as well as the percentage of very fine said (0.1 to 0.05 mm) Clay: clay (%) a: organic matter (%) b: soil structure code used in soil classification: 1: Very fine granular 2: Fine granular 3: Medium or coarse granular 4: Blocky, platy, or massive c: profile permeability class: 1: Rapid 2: Moderate to rapid 3: Moderate 4: Slow to moderate 5: Slow 6: Very slow	Thailand (Eiumnoh, 2000); Vanuatu (Dumas and Fossey, 2009); Philippines (Schmitt, 2009); India (Jain and Das, 2010); Turkey (Ozsoy et al., 2012); Iran (Bagherzadeh, 2014); Portugal (Ferreira and Panagopoulos, 2014); China (Li et al., 2014); European Union (Panagos et al., 2014)
2	Williams and Renard (1983) as cited in Chen et al. (2011)	USA	Sand (%), silt (%), clay (%), organic carbon (%)	$K = 0.2 + 0.3\exp\left(0.0256 \times \text{Sa} \times \left(1 - \frac{\text{Si}}{100}\right)\right) \times \left(\frac{\text{Si}}{\text{Cl+Si}}\right)^{0.3} \times \left(1.0 - \frac{0.25 \times \text{C}}{\text{C+exp}(3.72 - 2.95\text{C})}\right) \times \left(1.0 - \frac{0.7 \times \text{SN}}{\text{SN+exp}(-5.51 + 22.9\text{SN})}\right)$ Sa: sand (%) Si: silt (%) Cl: clay (%) $\text{SN} = 1 - (\text{Sa}/100)$ C: organic carbon	China (Chen et al., 2011)
3	David (1988), a simplified version of Wischmeier and Mannering (1969)	USA	Sand (%), clay (%), silt (%), organic matter (%), pH	$K = [(0.043 \times \text{pH}) + (0.62 \div \text{OM}) + (0.0082 \times \text{S}) - (0.0062 \times \text{C})] \times \text{Si}$ pH: pH of the soil OM: organic matter (%) S: sand content (%) C: clay ratio = % clay / (% sand + % silt) Si: silt content = % silt / 100	Philippines (David, 1988; Hernandez et al., 2012)
4	El-Swaify and Dangler (1976) as cited in Renard et al. (1997)	Hawaii, USA	Textural information, base saturation	$K = -0.03970 + 0.00311x_1 + 0.00043x_2 + 0.00185x_3 + 0.00258x_4 - 0.00823x_5$ x_1: unstable aggregate size fraction (<0.250 mm) (%) x_2 = modified silt (0.002–0.1 mm) (%) · modified sand (0.1–2 mm) (%) x_3: % base saturation x_4: silt fraction (0.002–0.050 mm) (%) x_5: modified sand fraction (0.1–2 mm) (%)	

Table 4. K-factor values from Dymond (2010) for soil textures in New Zealand.

Soil texture	K factor (Dymond, 2010)
Clay	0.20
Loam	0.25
Sand	0.05
Silt	0.35

Table 5. K-factor values from David (1988) for soil textures in the Philippines.

Soil texture	K factor (David, 1988)
Loamy fine sand	0.07
Clay	0.13–0.26
Clay loam	0.22–0.30
Loam	0.19–0.63
Sandy clay	0.09–0.20
Sandy loam	0.23–0.30
Silt loam	0.30–0.60
Silty clay	0.19–0.27
Silty clay loam	0.28–0.35

depends on the study area's scale. The relatively coarse globally available DEMs ($\sim$ 30 m at best) are less suited to field and sub-catchment scale studies where it may be important to capture effects of micro-topography.

The original equations for the LS factor assume that slopes have uniform gradients and any irregular slopes would have to be divided into smaller segments of uniform gradients for the equations to be more accurate (Wischmeier and Smith, 1978). At the plot or small field scale, this manual measurement of slopes and dividing into segments may be manageable, but it is less useful at larger scales. In terms of practicality, Desmet and Govers (1996) have reported studies of this method applied at a watershed scale with the disadvantages of it being time-consuming. Studies in Iran and the Philippines have implemented the (R)USLE methods within a GIS environment by calculating the LS factor for each raster cell in a DEM, essentially treating each pixel as its own segment of uniform slope (Bagherzadeh, 2014; Schmitt, 2009).

As explained above, the method of using flow accumulation, upslope contributing area, and slope in a GIS environment has gained popularity due to its ability to explicitly account for convergence and divergence of flow, thus capturing more complex topography (Wilson and Gallant, 2000). The flow accumulation method was applied at the scales of watersheds and regions (as shown in Table 6) and has even been applied by Panagos et al. (2015a) at the scale of the European Union using a 25 m DEM. The only thing limiting users is the availability of high-resolution DEMs and the trade-off between processing time and accuracy. The original (R)USLE methods require only slope angle and length, operate over a single cell in a DEM by treating it as a uniform slope, and take less processing time compared to the method using flow accumulation. However, the user must remember that this cannot capture the convergence and divergence of flow and thus sacrifices accuracy for time.

Additionally, the issue of limited vertical accuracy in global and many national DEMs confounds the uncertainties associated with coarse cell sizes. Further work on understanding the appropriate horizontal resolution and vertical accuracy of DEMs used for soil erosion predictions at the sub-catchment or field scales is suggested.

Benavidez (2018) investigated use of high-resolution DEMs (15 m and finer), finding the methods that only used slope length and steepness were adequate at delineating large vulnerable areas at the watershed scale. However, the methods using flow accumulation performed significantly better at the sub-watershed or field scale (Benavidez, 2018).

In summary, the choice of which LS-factor method to use is dependent on the spatial resolution of the DEM, availability of computing resources, and scale of the study site. DEms with spatial resolution coarser than $\sim$ 100 m do not accurately capture the flow network of a catchment (Panagos et al., 2015a). The LS-factor methods that account for only slope length and steepness are recommended for sites with such coarse DEMs. At the national, regional, or watershed scale, delineating large areas vulnerable to soil loss is more useful due to the ease of managing these areas at such large scales, and the methods that use only slope length and steepness are recommended. For sub-watershed or field studies and with sufficiently fine DEMs ($\sim$ 15 m or finer), using LS-factor methods that account for flow accumulation are more useful for identifying the most critical areas of vulnerability for targeted management approaches.

2.4 Cover and management factor (C)

The cover and management factor (C) is defined as the ratio of soil loss from a field with a particular cover and management to that of a field under "clean-tilled continuous fallow" (Wischmeier and Smith, 1978). The (R)USLE uses a combination of sub-factors such as impacts of previous management, canopy cover, surface cover and roughness, and soil moisture on potential erosion to produce a value for the soil loss ratio, which is used with the R factor to produce a value for the C factor (Renard et al., 1997). This method requires extensive knowledge of the study area's cover characteristics including agricultural management and may be suitable at the field or farm scale, but monitoring all these characteristics at the watershed scale may not be feasible.

A simpler method of determining the C factor is referencing studies that have reported values for similar land cover, or from studies done in the same area or region. Tables 8 and 9 give a broad overview of C factors for different cover types and common crops. Wischmeier and Smith (1987) also in-

Table 6. Summary of methods of calculating LS factor, original locations, and other studies that used these methods.

No.	Author	Original location	Data requirements	Equation	Other studies that utilised similar methods
1	Wischmeier and Smith (1978)	USA	Slope length and angle	$LS = (\frac{\lambda}{72.6})^m \times [(65.41 \times \sin^2\theta) + (4.56 \times \sin\theta) + 0.065]$ λ: slope length (ft) Θ: angle of slope m: dependent on the slope – 0.5 if slope > 5 % – 0.4 if slope is between 3.5 % and 4.5 % – 0.3 if slope is between 1 % and 3 % – 0.2 if slope is less than 1 %	Thailand (Eiumnoh, 2000; Merritt et al., 2004); Vanuatu (Dumas and Fossey, 2009); Iran (Bagherzadeh, 2014)
2	Renard et al. (1997)	USA	Slope length and angle	$L = \left(\frac{\lambda}{72.6}\right)^m$ $m = \frac{\beta}{1+\beta}$ $\beta = \frac{(\frac{\sin\theta}{0.0896})}{[3.0\times(\sin\theta)^{0.8}+0.56]}$ If slope is less than 9 %: $S = 10.8 \times \sin\theta + 0.03$ If slope is greater or equal to 9 %: $S = 16.8 \times \sin\theta - 0.50$ But if the slope is shorter than 15 ft: $S = 3.0 \times (\sin\theta)^{0.8} + 0.56$ λ: slope length (ft) Θ: angle of slope m: dependent on the slope – 0.5 if slope > 5 % – 0.4 if slope is between 3.5 % and 4.5 % – 0.3 if slope is between 1 % and 3 % – 0.2 if slope is less than 1 %	Philippines (Schmitt, 2009); China (Li et al., 2014); Thailand (Nontananandh and Changnoi, 2012); Turkey (Ozsoy et al., 2012)
3	David (1988), based on work by Madarcos (1985) and Smith and Whitt (1947)	Philippines, but based on work from the USA	Slope rise in percent	$LS = a + b \times S_L^{4/3}$ $a = 0.1$ $b = 0.21$ S_L: slope (%)	Philippines (David, 1988)
4	Morgan (2005) but previously published in earlier editions	Britain	Slope length and gradient in percent	$LS = \left(\frac{l}{22}\right)^{0.5}(0.065 + 0.045s + 0.0065s^2)$ l: slope length (m) s: slope steepness (%)	India (Nakil and Khire, 2016; Sinha and Joshi, 2012); Greece (Rozos et al., 2013)
5	Moore and Burch (1986) as cited in Mitasova et al. (1996) Desmet and Govers (1996); Mitasova et al. (2013);	USA	Upslope contributing area per unit width, which can be approximated through flow accumulation, cell size, slope	$LS = (m+1)\left(\frac{U}{L_0}\right)^m\left(\frac{\sin\beta}{S_0}\right)^n$ U ($m^2\ m^{-1}$): upslope contributing area per unit width as a proxy for discharge U = flow accumulation × cell size L_0: length of the unit plot (22.1) S_0: slope of unit plot (0.09) β: slope m (sheet) and n (rill) depend on the prevailing type of erosion (m = 0.4 to 0.6) and n (1.0 to 1.3)	Philippines (Adornado and Yoshida, 2010; Adornado et al., 2009); Sri Lanka (Jayasinghe et al., 2010); China (Chen et al., 2011); Iran (Zakerinejad and Maerker, 2015); Jordan (Farhan and Nawaiseh, 2015); Morocco (Raissouni et al., 2016); New Zealand (Fernandez and Daigneault, 2016). Similar methods from Moore and Burch (1986): India (Jain and Das, 2010); Portugal (Ferreira and Panagopoulos, 2014); Greece (Jahun et al., 2015); India (Nakil and Khire, 2016). Similar methods from Desmet and Govers (1996): USA (Boyle et al., 2011); Turkey (Demirci and Karaburun, 2012); Philippines (Delgado and Canters, 2012).

clude the effect of percent ground cover, reporting C-factor values for the same cover type over a range of cover percentage and condition. Morgan (2005) and David (1988) have reported values for the different growth stages of the same types of trees. A simple method of creating a C-factor layer is by using lookup tables to assign C-factor values to the land cover classes present in the study area. When using C factors from the literature, it is important to note that the definition of land cover type between two countries may vary. For example, land classified as forest in one country may be different in terms of vegetation cover or type compared to forest in another country (e.g. differences in pine forests and tropical forests). Therefore, it is crucial to understand the differences between land cover classifications before applying C-factor values from the literature. Van der Knijff et al. (2000) cites the large spatial and temporal variations in cover and crop over a large region such as the European Union as another

Table 7. *C*-factor equations that use NDVI.

No.	Author	Original location	Equation
1	Van der Knijff et al. (2000)	Europe	$C = \exp\left[\propto\left(\frac{\text{NDVI}}{\beta - \text{NDVI}}\right)\right]$ $\alpha = 2$ $\beta = 1$
2	Ma et al. (2010) as cited in Li et al. (2014)	China	$fg = \frac{\text{NDVI} - \text{NDVI}_{\text{min}}}{\text{NDVI}_{\text{max}} - \text{NDVI}_{\text{min}}}$ $C = \begin{cases} 1 & f_g = 0 \\ 0.6508 - 0.343 \times \log?(f_g) & 0 < f_g < 78.3\% \\ 0 & f_g \geq 78.3\% \end{cases}$

Table 8. *C* factors for general types of land cover compiled from various sources.

Cover	Dymond (2010) (New Zealand)	David (1988) (Philippines)	Morgan (2005) (various)	Fernandez et al. (2003) (USA)	Dumas and Fossey (2009) (Vanuatu)	Land Development Department (2002) as cited in Nontananandh and Changnoi (2012)
Bare ground	1	1	1			
Urban		0.2		0.03	0	0
Crop				0.128	0.01	0.255–0.525
Forest	0.005	0.001–0.006	0.001	0.001	0.001	0.003–0.048
Pasture	0.01		0.1			
Scrub	0.005	0.007–0.9	0.01	0.003	0.16	0.01–0.1

reason why using the lookup-table-based approach is inadequate and tedious.

To address this, another method of determining the *C* factor is through the normalized difference vegetation index (NDVI) estimated from satellite imagery. Although there are NDVI layers available, these are limited by geographical coverage, date of acquisition, and resolution. The MODIS NDVI dataset made by Caroll et al. (2004) at 250 m resolution covers the USA and South America[2]. NASA produced a global dataset of NDVI values at 1° resolution for the time span of July 1983 to June 1984, making it suitable for studying historical soil erosion but not necessarily for the current state of land cover[3].

In areas where ready-made NDVI products are unavailable, authors have used satellite imagery to obtain NDVI such as AVHRR or Landsat ETM (Van der Knijff et al., 2000; De Asis and Omosa, 2007; Ma et al., 2010, as cited in Li et al., 2014). De Asis and Omasa (2007) related the *C* factor and NDVI through fieldwork and image classification – determining the *C* factor at several points within the study area using the (R)USLE approach and relating it to the NDVI through regression correlation analysis. This may not be feasible in larger study areas such as the European Union, where Van der Knijff et al. (2000) determined NDVI from satellite imagery and created an equation based on its positive correlation with green vegetation (Table 7). This approach enabled them to create a *C*-factor map over the European Union. However, *C* factors were unrealistically high in some areas such as woodland and grassland, so values for those areas were taken from the literature.

An advantage of using NDVI is that researchers can determine sub-annual *C* factors if there is satellite imagery available, which can lead to understanding the contribution of cover to seasonal soil erosion and identifying critical periods within the year where soil erosion is a risk (Ferreira and Panagopoulos, 2014). Similar methods have been applied in Brazil by Durigon et al. (2014), Greece by Alexandridis et al. (2015), and Kyrgyzstan by Kulikov et al. (2016). Determining *C* factors at the seasonal scale is important because vegetation cover can change throughout the year due to agricultural and forestry practices. In study areas with a high temporal variation of rainfall throughout the year, seasonal vegetation can play a big part in exacerbating or mitigating soil erosion.

To summarise, the choice of which method to use depends on the scale of the study area, reported *C* factors for similar cover, and availability of high-resolution imagery. For small-scale studies, it is more feasible to determine the *C* factors through fieldwork. If previous (R)USLE studies have reported *C* factors for cover similar to the study area, those values can be used for the table-based approach. Lastly, high-resolution imagery can be used to determine the study area's

[2]http://glcf.umd.edu/data/ndvi/ (last access: 12 November 2018)

[3]https://data.giss.nasa.gov/landuse/ndvi.html (last access: 12 November 2018)

Table 9. *C* factors for specific types of land cover compiled from various sources.

Cover	Panagos et al. (2015b) (Europe)	David (1988) (Philippines)	Morgan (2005) (various)
Bananas		0.1–0.3	
Barley	0.21		
Chili			0.33
Cocoa			0.1–0.3
Coffee			0.1–0.3
Common wheat and spelt	0.2		0.1–0.4
Cotton seed	0.5	0.4–0.6	0.4–0.7
Dried pulses (legumes) and protein crop	0.32	0.3–0.5	0.04–0.7
Durum wheat	0.2		
Fallow land	0.5		
Grain – maize/corn	0.38	0.3–0.6	0.02–0.9
Groundnuts			0.3–0.8
Linseed	0.25		0.1–0.2
Oilseeds	0.28		
Palm with cover crops		0.05–0.3	0.1–0.3
Pineapple		0.2–0.5	0.01–0.4
Potatoes	0.34		0.1–0.4
Rape and turnip rape	0.3		
Rice	0.15	0.1–0.2	0.1–0.2
Rye	0.2		
Soya	0.28		0.2–0.5
Sugar beet	0.34		
Sugarcane			0.13–0.4
Sunflower seed	0.32		
Tobacco	0.49	0.4–0.6	
Yams			0.4–0.5

Table 10. Examples of where *C* factor accounts for crop management from Morgan (2005) and David (1988).

Crop	Management	*C* factor
Maize, sorghum, or millet	High productivity; conventional tillage	0.20–0.55
	Low productivity; conventional tillage	0.50–0.90
	High productivity; chisel ploughing into residue	0.12–0.20
	Low productivity; chisel ploughing into residue	0.30–0.45
	High productivity; no or minimum tillage	0.02–0.10
Coconuts	Tree intercrops	0.05–0.1
	Annual crops as intercrop	0.1–0.30

NDVI. At small scales and with a good understanding of differences in land cover classifications, pulling values from the literature may be the most efficient choice, but at larger regional scales this may become tedious.

At larger scales, high-resolution satellite imagery may be available to determine NDVI, but authors must be mindful of its acquisition date in relation to their study period, as well as data quality and im-age processing issues such as dealing with cloud cover and aggregating images from multiple satellite passes (Van der Knijff et al., 2000; Kulikov et al., 2016).

2.5 Support practice factor (*P*)

The support practice factor (*P*) is defined as the ratio of soil loss under a specific soil conservation practice (e.g. contouring, terracing) to that of a field with upslope and downslope tillage (Renard et al., 1997). The *P* factor accounts for management practices that affect soil erosion through modifying the flow pattern, such as contouring, strip cropping, or terracing (Renard et al., 1997). The more effective the conservation practice is at mitigating soil erosion, the lower the *P* factor (Bagherzadeh, 2014). Like the *C* factor, values for *P* factors can be taken from the literature; if there are no support practices observed, the P factor is 1.0 (Adornado et al., 2009). The P factor can also be estimated using sub-factors, but the difficulty of accurately mapping support practice factors or not observing support practices leads to many studies ignoring it by giving their P factor a value of 1.0 as seen in Appendix A1 (Adornado et al., 2009; Renard et al., 1997; Schmitt, 2009).

Another possible reason why studies may ignore the P factor is due to the nature of their chosen C factors. Some C factors already account for the presence of a support factor such as intercropping or contouring. For example, Morgan (2005) and David (1988) give C factors for one type of crop, but with different types of management (Table 10). Despite the P factor being commonly ignored, a number of studies have reported possible P factors for different kinds of tillage, terracing, contouring, and strip cropping (Table 11). The P factor has a significant impact on the estimation of soil loss. For example, a P factor of 0.25 for zoned tillage reflects the potential for this management factor to reduce soil by 75 % loss compared to conventional tillage (P factor: 1.00). At suitably detailed scales and with enough knowledge of farming practices, using these P factors may lead to a more accurate estimation of soil loss. Additionally, these P factors can be used in scenario analysis to understand how changing farming practices may mitigate or exacerbate soil loss. An application of (R)USLE in the Cagayan de Oro catchment in the Philippines showed, through scenario analysis, that soil conservation practices such as agroforestry and alley cropping could potentially lead to large decreases in soil loss compared to the baseline scenario (Benavidez, 2018).

In summary, including the P factor in (R)USLE applications is important because of the significant effects that some management practices can have on reducing soil loss compared to conventional tillage. The P factor is useful for studies where different management practices are being considered for the same site as it can elucidate which practices are more beneficial for soil conservation.

3 Limitations of (R)USLE

This section presents a few of the key limitations of the (R)USLE: regional applicability, uncertainties associated with the model, input data and validation, and representing other types of erosion.

The most commonly cited limitation of the (R)USLE models is their reduced applicability to regions outside of the United States of America (Aksoy and Kavvas, 2005; Naipal et al., 2015; Sadeghi et al., 2014). The original USLE was formulated based on soil erosion studies on agricultural land in the USA. When applied to different climate regimes and land cover conditions, this may lead to greater uncertainties associated with estimates of average annual soil loss (Kinnell, 2010). Since the (R)USLE parameters were developed based on small-scale studies of agricultural plots, there are also uncertainties associated with upscaling the original USLE to the catchment or regional scale (Nagle et al., 1999; Naipal et al., 2015). Wischmeier and Smith (1987) have also warned that using the (R)USLE in conditions extremely different from the agricultural conditions the model was formulated under may lead to extrapolation error. Of the studies reviewed for this paper (Table A1), most applications were done on catchments with predominantly agricultural land use, but under a range of different climatic conditions.

Sensitivity analysis and testing which (R)USLE subfactors suit particular study sites is one method of addressing the (R)USLE's regional applicability. Like the Mangatarere application method in Sect. 2.1, other studies have tested multiple R-factor equations on the same dataset to determine which equation was most appropriate for their study site (Eiumnoh, 2000; Benavidez, 2018). Their derived R-factor values were compared to the values for catchments with similar climate and rainfall, or to maps of the R factor at larger spatial scales (Panagos et al., 2017). To reduce uncertainty in accounting for land use, work by Post and Hartcher (2005) recommended using C-factor values for specific land cover classifications (e.g. specific crops, forest growth stages) instead of values for broad land cover categories (e.g. agriculture, forest). Although C-factor values can be taken from the literature or determined in situ, an extensive literature review compiling potential soil loss rates of different crop and forest covers compared to likely soil loss rates of bare soil can be used to determine likely C-factor values of a particular site. Improvements and modifications to the (R)USLE subfactors have made it applicable to larger spatial scales, including a coarse-resolution representation at the global scale (Naipal et al., 2015). The pan-European application by Panagos et al. (2015a) showed setting a maximum value for slope steepness of 50 % (26.6°) would prevent significantly large LS-factor values and account for the absence of soil on such steep slopes. Assembling published estimates of (R)USLE sub-factors from different climatic regions and soil types would help in sensitivity testing (R)USLE equations, deciding the most appropriate equation to use, and verifying the derived (R)USLE sub-factor values.

The uncertainties associated with the (R)USLE, and arguably soil erosion modelling in general, stem from several factors: the inability of models to capture the complex interactions involved in soil loss, the low availability of long-term reliable data for modelling, and the lack of soil erosion observational data for model validation, especially in data-scarce environments. The simplicity of the (R)USLE allows usage in locations where there are insufficient data for more complex models that require large input datasets (de Vente and Poesen, 2005; Hernandez et al., 2012). Of the studies reviewed, very few critically discuss the uncertainties associated with the (R)USLE, but those that do offer several ways to overcome these uncertainties.

Since the (R)USLE does not account for all the complex interactions associated with soil erosion, its predicted soil

Table 11. *P* factors for different types of agricultural management practices.

	David (1988)			
Tillage and residue management	*P* factor			
Conventional tillage	1.00			
Zoned tillage	0.25			
Mulch tillage	0.26			
Minimum tillage	0.52			
Slope (%)	Terracing		Contouring	Contour strip cropping
	Bench	Broad-based		
1–2	0.10	0.12	0.60	0.30
3–8	0.10	0.10	0.50	0.15
9–12	0.10	0.12	0.60	0.30
13–16	0.10	0.14	0.70	0.35
17–20	0.12	0.16	0.80	0.40
21–25	0.12	0.18	0.90	0.45
> 25	0.14	0.20	0.95	0.50
	Panagos et al. (2015c)			
Slope (%)	Contouring *P* factor			
9–12	0.6			
13–16	0.7			
17–20	0.8			
21–25	0.9			
> 25	0.95			

erosion rates should be taken as best estimates rather than absolute values (Wischmeier and Smith, 1987). Some applications have chosen to display their soil loss results as categorical to produce maps that show low, medium, or high areas of vulnerability instead of showing annual average amounts (Adornado et al., 2009; Schmitt, 2009). The (R)USLE is a good first attempt at identifying vulnerable areas and estimating soil loss for a landscape at the baseline scenario due to the model's relative simplicity and few data requirements (Aksoy and Kavvas, 2005). The (R)USLE is also useful for doing scenario analysis to check whether changing land use or management practices would either exacerbate or mitigate soil loss, making it useful for comparison purposes (Merritt et al., 2004; Nigel and Rughooputh, 2012).

Validating the soil erosion rates produced by the (R)USLE is difficult because of the lack of easily obtainable observational soil erosion records, especially in data-scarce environments. Out of the (R)USLE applications reviewed for this paper, $\sim 30\,\%$ presented explicit comparisons between their modelled soil loss from (R)USLE and observed soil loss, modelled soil loss from (R)USLE and other models (one study), and soil loss from multiple models and observed soil loss (one study).

One study compared the soil loss rates predicted by the RUSLE to estimates of the physically based WEPP (Water Erosion Prediction Project) model. Amore et al. (2004) compared RUSLE and WEPP and found that the ratio of modelled to observed soil loss of WEPP (0.7) was better than RUSLE (0.2) for the Trinità basin. However, both RUSLE and WEPP over-predicted sediment yield by up to 5 times the observed value for the nearby Ragoleto basin (Amore et al., 2004). Although WEPP also estimates rill and interrill erosion, WEPP is a continuous daily model that accounts for deposition and sediment delivery, which RUSLE does not predict (Aksoy and Kavvas, 2005).

Another study compared the soil loss estimates of the RUSLE and Unit Stream Power Erosion Deposition Model (USPED) to each other, and to observed data. In a comparison between the RUSLE and USPED, the ratio of modelled to observed of soil loss was almost unity for the USPED but 0.86 for the RUSLE (Aiello et al., 2014). The USPED model builds and improves on the RUSLE sub-factors through its ability to incorporate overland flow and sediment transport through the landscape (Aiello et al., 2014; Zakerinejad and Maerker, 2015).

Based on the remaining studies that reported comparisons of modelled RUSLE soil loss to observed soil loss, the modelled-to-observed ratio ranged from extreme under-prediction at 0.04 to over-prediction at over 3 times the observed values. The applications where RUSLE severely under-predicted soil loss cited the model's inability to account for gully erosion and mass wasting as one of the reasons for estimation errors, thus underscoring the importance of including these types of erosion in future improvements to RUSLE (Dabney et al., 2012; Gaubi et al., 2017). Another issue is differences in temporal and/or spatial resolution and sometimes differing timescales between modelled and observed estimates. Average observations based on occasional grab samples of sediment in streams may not well represent the monthly to annual sediment loads the (R)USLE is attempting to estimate. In another example, López-Vicente et al. (2008) compared observed to modelled values and had a ratio of modelled to observed soil loss of 0.62. However, the "observed" soil loss was based on ^{137}Cs measurements that were indicative of average soil loss values for the past 40 years, while the model values were based on 1997–2006 driving data. During this period, the study area experienced lower precipitation and thus had lower modelled soil loss measurements compared to the soil loss derived from the ^{137}Cs records (López-Vicente et al., 2008).

As stated earlier, the regional applicability of the RUSLE is a limitation that requires the sub-factors to be adjusted and modified based on the specific characteristics of the researcher's study site. Nakil and Khire (2016) and Abu Hammad et al. (2005) show this important practice in RUSLE applications in their studies. Through testing and refining their method of accounting for topography through the LS factor, the ratio of modelled to observed soil loss ranged from 0.8 to almost unity (Nakil and Khire, 2016). The initial application of RUSLE of Abu Hammad et al. (2005) over-estimated soil loss by a factor of 3, but with adjustments to the sub-factors based on local data on soil moisture, land cover, and support practices, the model error was reduced to 14 %. The importance of adjusting RUSLE with the availability of more detailed data was further shown in the pan-European study of Panagos et al. (2015e), where detailed soil, topography, land cover, and management practices allowed the researchers to refine their application where most of the ratios of modelled to observed soil loss were very good (0.9 to 1.3). In the validation areas where the soil loss comparisons were not good, further local testing and refining of the RUSLE sub-factors is seen as an area in which to improve the model results (Beskow et al., 2009; Ozsoy et al., 2012; Panagos et al., 2015e).

A global soil erosion study using RUSLE was accomplished by Borrelli et al. (2017) using the rainfall erosivity map generated by Panagos et al. (2017) that showed comparable results to regional and local soil erosion estimates, and good agreement with global soil erosion datasets such as the Global Assessment of Human-induced Soil Degradation (GLASOD) dataset[4].

[4]https://www.isric.online/projects/global-assessment-human-induced-soil-degradation-glasod (last access: 12 November 2018)

Future work in the soil erosion literature could include assembling a comprehensive database of global, regional, and national soil erosion rates to allow comparison between soil erosion modelling methods, not just (R)USLE results. A proxy for understanding soil erosion is water quality data such as total suspended solids (TSS) that includes sediment delivery and organic sources (Schmitt, 2009; Russo, 2015). However, TSS usually excludes the larger and heavier bed-load sediments that could be resulting from mass wasting events or erosion (Nagle et al., 1999). Nevertheless, water quality data are useful for inferring likely temporal patterns of soil erosion or the sediment yield during seasons of heavy rainfall or after extreme events. Ground truthing or analysis of satellite imagery is another useful method of validating the (R)USLE results, as the areas of extreme erosion risk can be checked for physical evidence of soil loss occurrence (De Asis and Omasa, 2007; Adornado and Yoshida, 2010; Nontananandh and Changnoi, 2012). The soil loss estimates can be validated against observations from similar catchments, recorded events of mass wasting, or larger-scale soil loss studies at the national or regional scale (Životić et al., 2012; Panagos et al., 2015e; Nakil and Khire, 2016).

Lastly, a frequently cited limitation is that the (R)USLE estimates soil loss through sheet and rill erosion, but not from other types of erosion such as gully erosion, channel erosion, bank erosion, or mass wasting events such as landslides (Nagle et al., 1999; Wischmeier and Smith, 1978). By excluding these types of erosion, the (R)USLE may underestimate the actual soil loss (Thorne et al., 1985). The model also does not account for deposition, leading to overestimation, or sediment routing (Desmet and Govers, 1996; Wischmeier and Smith, 1978). Since it does not predict the sediment pathways from hill slopes to water bodies, it is difficult to analyse possible effects on downstream areas, such as pollution or sedimentation (Jahun et al., 2015). One of the possible methods for linking the (R)USLE results to sediment delivery to streams is using the sediment delivery ratio (SDR), defined as "the ratio of the sediment delivered at a location in the stream system to the gross erosion from the drainage area above that point" (Yoon et al., 2009). This parameter varies depending on the gradient, slope shape, and length and can also be influenced by land cover, roughness, etc. (Wu et al., 2005). Given that it is influenced by characteristics similar to those of the (R)USLE, future work can include not only combining the (R)USLE with the SDR to estimate sediment delivery to streams but also avoiding possible double counting. These two limitations of deposition and routing are linked to the model's representation of more topographically complex terrain, and previous studies have attempted to address them by improving on the LS factor by incorporating upstream contributing area (Desmet and Govers, 1996; Moore et al., 1991). A more detailed discussion of addressing these limitations is in Sect. 4.1.

Despite these drawbacks, the USLE family of models is still widely used because of is relative simplicity and low data requirements compared to more complex physically based models. Studies around the world continue to improve (R)USLE parameterisation and application in different climate regimes and locations.

4 Future directions

Since the (R)USLE and its family of models are used over different geographic locations and climate types, it is important for future research to build on them and improve their representation of real-world soil loss. Some of the future directions include incorporating soil loss from other types of erosion, estimating soil loss at seasonal or sub-annual temporal scales, and improving the consistency of formulae and units in the scientific literature.

4.1 Representing other types of erosion

As previously discussed in Sect. 3, the (R)USLE does not account for all erosion types. This section mostly discusses possible extensions to include gully erosion, but further work to incorporate channel/bank erosion and mass wasting events must also be done.

The inability of (R)USLE to account for soil losses due to ephemeral gullies can lead to under-prediction of soil loss estimates (Thorne et al., 1985). These ephemeral gullies are small channels that form due to the erosive action of overland flow during a rainfall event (Momm et al., 2012). Gully erosion can contribute a significant amount of sediment loss; for example gully erosion is estimated to contribute between 30 % and 50 % of soil loss from a range of catchments in New Zealand (Basher et al., 2012). Desmet and Govers (1996) recommended that delineation of ephemeral gullies, such as through the Compound Topographic Index (CTI) developed by Thorne et al. (1985), combined with (R)USLE could improve the identification of vulnerable areas within a watershed. The CTI of Thorne et al. (1985) uses topographic analysis to predict locations and soil loss rates of ephemeral gullies based on upstream drainage area, slope, and the planform curvature. Hence, the combination of CTI and the (R)USLE is a promising direction for including gully erosion, but care must be taken in coupling these models because both already account for upstream drainage area and slope. Simply adding their soil loss rates could lead to "double counting" and requires further research to determine the threshold values of CTI and LS factor over which ephemeral gullying is likely (Benavidez, 2018).

Work along these lines, combining the effect of rill and sheet erosion with gully erosion was done by Momm et al. (2012) in Kansas and by Zakerinejad and Maeker (2015) in the Mazayjan watershed in Iran. Momm et al. (2012) combined several types of erosion – sheet and rill, gully, and bed and bank erosion – with the sheet and rill erosion estimated using the (R)USLE model. They used varying critical CTI thresholds to iteratively generate potential locations of ephemeral gullies and identify sub-watersheds prone to gully erosion, and then used scenario analysis to estimate reductions in sediment yields under conservation practices (Momm et al., 2012). One of the limitations of the Momm et al. (2012) study was that they only had a coarse-resolution DEM. Since ephemeral gullies are small features (typically a few metres wide and ~ 25 cm deep), higher-resolution DEMs such as those derived from lidar data would be better for analysis of these topographic features. The USPED, which is similar to the (R)USLE model, has also been used to estimate rill and sheet erosion rates with a stream power index (SPI) approach to estimate gully erosion rates (Zakerinejad and Maerker, 2015). Zakerinejad and Maerker (2015) estimated gully erosion in metric tons per hectare per year and combined it with the estimates from the USPED model to produce a map showing potential erosion and deposition within their study area. Hence, there are precedents as well as a need to combine erosion estimates from (R)USLE with a procedure that accounts for gully erosion for more effective land management.

4.2 Seasonal erosion vulnerability

(R)USLE applications usually estimate soil loss at the annual timescale, and the MUSLE estimates soil loss from a single storm event (Renard et al., 1997; Sadeghi et al., 2014). As seen in the review of methods of calculating rainfall erosivity, many different studies have attempted to estimate the R factor, underscoring its importance to soil erosion research. However, estimating the R factor at the annual timescale does not account for seasonal variations in rainfall. It is useful for land management to understand seasonal variations in soil erosion vulnerability because of the dual effect of rainfall and land cover on soil loss, and the effect of rainfall on land cover (Kulikov et al., 2016). For example, when a season of heavy rainfall coincides with low vegetation cover, the risk of soil erosion increases considerably (Ferreira and Panagopoulos, 2014). Thus, most of the studies around seasonal estimations of soil loss revolve around changes in land cover and rainfall. The soil erodibility (K factor) can vary too due to changes in permeability and the effects of freezing and thawing, but it is less frequently studied compared to variations in land cover and rainfall (López-Vicente et al., 2008).

Studies that incorporate seasonality in the (R)USLE commonly compute R factors and C factors at monthly or seasonal timescales. Lu and Yu (2002) computed monthly R factors in Australia, which were then used in a later study that computed C factors based on satellite imagery and the NDVI, to produce monthly maps of soil erosion vulnerability over the entire Australian continent (Lu et al., 2003; Lu and Yu, 2002). The method of estimating C factors using NDVI is popular due to the availability of remotely sensed imagery and the capability of processing datasets with relative expedience compared to time-consuming fieldwork. Other studies have used the NDVI and similar characteristics to estimate monthly and seasonal *C* factors in Brazil, Greece, and Kyrgyzstan (Alexandridis et al., 2015; Durigon et al., 2014; Ferreira and Panagopoulos, 2014; Kulikov et al., 2016; Panagos et al., 2012). The *C* factors can also be estimated monthly through the method recommended by (R)USLE, but this requires knowledge of prior land use, canopy cover, surface roughness, and soil moisture (López-Vicente et al., 2008).

Monthly or seasonal estimations of rainfall factors are more useful to land management planning around crop growth cycles and tillage practices (Diodato, 2004). Studies have used different methods to calculate *R* factors, with data requirements ranging from per-storm basis to annual averages. To estimate monthly and seasonal estimations, the required rainfall data can be as fine as individual storm intensity to use the (R)USLE method or be as coarse as average monthly rainfall. Diodato (2004) in Italy and Kavian et al. (2011) used the (R)USLE method to calculate storm energy and summed these up per month and season to obtain *R* factors. Other studies have used daily and monthly rainfall to calculate monthly *R* factors and combine them for seasonal *R* factors (Alexandridis et al., 2015; Kavian et al., 2011; López-Vicente et al., 2008; Lu et al., 2003; Panagos et al., 2015d; Shamshad et al., 2008). The results of these studies focused on identifying high and low periods of the landscape's vulnerability to soil erosion, depending on combinations of rainfall intensity and land cover.

To summarise, modelling the sub-annual variations of soil erosion and sediment yield is important because of the many temporal and spatial variations in the factors that influence annual soil loss. These variations include seasonal rainfall variability, changes in the spatial distribution of erosion-prone areas over crop growth and tillage cycles, and potential seasonal changes in water quality due to changes in seasonable distributions of extreme rainfall events. Seasons with higher heavy rainfall will have a higher possibility of soil loss and mass wasting events, which in turn have a degrading effect on water quality and can cause destruction of infrastructure, putting communities and lives in danger. Over the crop growth and tillage cycles, the potential sediment yields to streams will change, and this has implications for land management and farmers who must abide by water quality standards. Modelling at the annual timescale is insufficient to capture these seasonal or monthly changes in potential soil loss, which are more important to land management planning, and thus underscores the utility of doing modelling at the sub-annual scale. These sub-annual model results can then be aggregated into an annual estimate of soil loss that would be more accurate compared to modelling using only annual averages of rainfall or land cover conditions.

4.3 Consistency in units

The USLE was originally developed using imperial units; although the handbook provides conversion factors to convert to metric, there are still issues within the scientific literature regarding units. In the process of this review, it was noted that, although most studies used the metric units for *R* factor and *K* factor, there were other studies that did not report their units or had units that were not the imperial or metric units of (R)USLE. Since the original (R)USLE was formulated with US customary units, researchers must be careful to use the correct units and conversions to metric (Renard and Freimund, 1994). To convert from imperial to metric units, Renard et al. (1997) recommends a conversion factor of 17.02 for the *R* factor and 0.1317 for the *K* factor. As mentioned in Sect. 3, there are uncertainties associated with the (R)USLE, and publishing sub-factor values and soil loss estimates for future reference by other researchers is a potential way of reducing some of those uncertainties. The problem of unclear or inconsistent units causes problems for future researchers in terms of adapting the rainfall erosivity or soil erodibility equations for their own study sites, underscoring the need for clear and explicit reporting of units in the (R)USLE literature.

5 Summary and conclusion

At first glance, the USLE, along with its family of models, seem like a relatively straightforward linear model. However, this review shows the difficulty in finding the most appropriate method of calculating its sub-factors depending on location, availability of data, and previous studies done in nearby or similar regions. This paper reviewed the different components of the Universal Soil Loss Equation (USLE) and its updated form, the Revised Universal Soil Loss Equation (RUSLE). Different studies around the world were collected and analysed to compile how they adapted (R)USLE to their unique conditions, how they had estimated the (R)USLE sub-factors with limited data availability, and how these methods have been used by subsequent soil erosion studies. At the end of each sub-factor section, a brief summary is given outlining which datasets and equations would be useful for new users depending on their location and data availability. Each sub-factor section clarifies some of the assumptions and limitations associated with the original (R)USLE models and how users can overcome some of the uncertainties associated with these sub-factors. One common theme in the sub-factor reviews is that sensitivity testing of the sub-factors should be done by future (R)USLE applications by trialling several equations for one sub-factor before using it in the final soil erosion estimates.

This paper also presented the limitations of the (R)USLE, mainly the uncertainties associated with the simple empirical model, uncertainties with data availability and validation, and the model's inability to account for types of soil erosion other than rill or inter-rill erosion. Lastly, the paper outlined some key future directions for (R)USLE research: incorporating soil loss from other types of soil erosion, importance of estimating soil loss at sub-annual scales and recommended equations, validation of soil loss estimates, and consistency in reporting units in the future literature. To represent gully erosion, the Compound Topographic Index (CTI) was briefly discussed, while the sediment delivery ratio (SDR) was also presented to account for linking soil loss to sediment delivery to streams. The ability to predict sub-annual soil loss or seasonal erosion modelling is important in study areas having high temporal variation of rainfall throughout the year, and/or having varying crop growth and tillage cycles, both being factors that can impact potential soil loss. Land management policy and decisions might be more robust if they consider modelling scenarios that test the effect of different types of crop and support practices on soil erosion mitigation. These scenarios can include a myriad of options: expanded urban areas or development, changing crop rotation cycles, or applying support practices in steep or upland areas. Further, seasonal soil erosion has implications on water quality, and understanding the extent of the problem can help local government address potential sources of sediment delivery and be more proactive in land management. Validation of soil loss estimates is important in understanding the accuracy of the (R)USLE application, and future work could involve compiling an extensive global database of soil loss estimates derived from observations and models, including those models more complex than (R)USLE. This database would be useful for future researchers in comparing their results and assess the accuracy of model applications. Greater transparency in reporting the sub-factor units, sub-factor values, and soil loss estimates is important to decrease uncertainty when future (R)USLE applications borrow sub-factor equations and values from previous studies. The limitations section addresses the fourth objective of this review.

In conclusion, the choices made regarding applications of the (R)USLE depend on the kind of data that are available for a study area and how they can adapt or change information from other studies to suit their area's particular climate, soil type, topography, typical land cover, and support practices.

Data availability. The datasets mentioned in this review are available for download after registration or through contacting the authors. The following datasets were retrieved from the Joint Research Centre European Soil Data Centre (ESDAC): global rainfall erosivity (accessible at https://esdac.jrc.ec.europa.eu/content/global-rainfall-erosivity; Panagos et al., 2017) and rainfall erosivity in the EU and Switzerland (accessible at https://esdac.jrc.ec.europa.eu/content/rainfall-erosivity-european-union-and-switzerland; Panagos et al., 2015d). The rainfall erosivity map for Brazil was constructed by da Silva (2004). The MODIS normalized difference vegetation index (NDVI) is accessible at http://glcf.umd.edu/data/ndvi/ (Caroll et al., 2004). The global-scale NDVI dataset generated by NASA is accessible at https://data.giss.nasa.gov/landuse/ndvi.html (NASA, 2018). The Global Assessment of Human-induced Soil Degradation (GLASOD) is accessible at http://data.isric.org/geonetwork/srv/eng/catalog.search\T1\textbackslash#/metadata/9e84c15e-cb46-45e2-9126-1ca38bd5cd22 (Bridges and Oldeman, 1999).

Author contributions. The manuscript was mainly written by RB. Proofreading, additions to the text, and feedback were accomplished by BJ, DM, and KN.

Competing interests. The authors declare that they have no conflict of interest.

Acknowledgements. Much of the content of this paper was supported by the Victoria University of Wellington Doctoral Scholarship as part of the PhD research of Rubianca Benavidez.

Edited by: Uwe Ehret

References

Abu Hammad, A., Lundekvam, H., and Børresen, T.: Adaptation of RUSLE in the eastern part of the Mediterranean region, Environ. Manage., 34, 829–841, https://doi.org/10.1007/s00267-003-0296-7, 2005.

Adornado, H. A. and Yoshida, M: Assessing the Adverse Impacts of Climate Change: A Case Study in the Philippines, J. Dev. Sus. Agr., 5, 141–146, 2010.

Adornado, H. A., Yoshida, M., and Apolinares, H.: Erosion Vulnerability Assessment in REINA, Quezon Province, Philippines with Raster-based Tool Built within GIS Environment, J. Agric. Res. , 18, 24–31, https://doi.org/10.3173/air.18.24, 2009.

Aksoy, H. and Kavvas, M. L.: A review of hillslope and watershed scale erosion and sediment transport models, Catena, 64, 247–271, https://doi.org/10.1016/j.catena.2005.08.008, 2005.

Aiello, A., Adamo, M., and Canora, F.: Modelling Spatially–Distributed Soil Erosion through Remotely–Sensed Data and GIS, in: Computational Science and Its Applications – ICCSA 2014, edited by: Murgante, B., Misra, S., Rocha, A. M. A. C., Torre, C. M., Rocha, J. G., Falcão, M. I., Taniar, D., Apduhan, B. O., and Gervasi, O., 372–385, Springer, Cham, https://doi.org/10.1007/978-3-319-09147-1_27, 2014.

Alexandridis, T. K., Sotiropoulou, A. M., Bilas, G., Karapetsas, N., and Silleos, N. G.: The Effects of Seasonality in Estimating the *C*-Factor of Soil Erosion Studies, Land Degrad. Develop., 26, 596–603, https://doi.org/10.1002/ldr.2223, 2015.

Amore, E., Modica, C., Nearing, M. A., and Santoro, V. C.: Scale effect in USLE and WEPP application for soil erosion computation from three Sicilian basins, J. Hydrol., 293, 100–114, https://doi.org/10.1016/j.jhydrol.2004.01.018, 2004.

Arnoldus, H. M. J.: Methodology used to determine the maximum potential average annual soil loss due to sheet and rill erosion in Morocco, FAO Soils Bulletin, 34, 39–51, 1977.

Arnoldus, H. M. J.: An approximation of the rainfall factor in the Universal Soil Loss Equation, in: De Boodt, M. and Gabriels, D., Assessment of Erosion, 127–131, Chichester, UK, Wiley, 1980.

Bagherzadeh, A.: Estimation of soil losses by USLE model using GIS at Mashhad plain, Northeast of Iran, Arab. J. Geosci., 7, 211–220, https://doi.org/10.1007/s12517-012-0730-3, 2014.

Bantayan, N. C. and Bishop, I. D.: Linking objective and subjective modelling for landuse decision-making, Landsc. Urban Plan., 43, 35–48, https://doi.org/10.1016/S0169-2046(98)00101-7, 1998.

Basher, L., Douglas, G., Elliott, S., Hughes, A., Jones, H., McIvor, I., Page, M., Rosser, B., and Tait, A.: Impacts of climate change on erosion and erosion control methods – A critical review (Vol. 4), available at: https://www.mpi.govt.nz/document-vault/4074 (last access: 12 November 2018), 2012.

Benavidez, R. A.: Understanding the effect of changing land use on floods and soil erosion in the Cagayan de Oro catchment, (Unpublished doctoral dissertation), Victoria University of Wellington, New Zealand, 2018.

Beskow, S., Mello, C. R., Norton, L. D., Curi, N., Viola, M. R., and Avanzi, J. C.: Soil erosion prediction in the Grande River Basin, Brazil using distributed modeling, Catena, 79, 49–59, https://doi.org/10.1016/j.catena.2009.05.010, 2009.

Bonilla, C. A. and Vidal, K. L.: Rainfall erosivity in Central Chile, J. Hydrol., 410, 126–133, https://doi.org/10.1016/j.jhydrol.2011.09.022, 2011.

Borrelli, P., Robinson, D. A., Fleischer, L. R., Lugato, E., Ballabio, C., Alewell, C., Meusburger, K., Modugno, S., Schütt, B., Ferro, V., Bagarello, V., Van Oost, K., Montanarella, L., and Panagos, P.: An assessment of the global impact of 21st century land use change on soil erosion, Nat. Commun., 8, 1–13, https://doi.org/10.1038/s41467-017-02142-7, 2017.

Boyle, J. F., Plater, A. J., Mayers, C., Turner, S. D., Stroud, R. W., and Weber, J. E.: Land use, soil erosion, and sediment yield at Pinto Lake, California: Comparison of a simplified USLE model with the lake sediment record, J. Paleolimnol., 45, 199–212, https://doi.org/10.1007/s10933-010-9491-8, 2011.

Bridges, E. and Oldeman, L.: Global Assessment of Human-Induced Soil Degradation, Arid Soil Res. Rehabil., 13, 319–325, https://doi.org/10.1080/089030699263212, 1999 (data available at: http://data.isric.org/geonetwork/srv/eng/catalog.search\T1\textbackslash#/metadata/9e84c15e-cb46-45e2-9126-1ca38bd5cd22, last access: 23 November 2018).

Carroll, M. L., DiMiceli, C. M., Sohlberg, R. A., and Townshend, J. R. G.: 250 m MODIS Normalized Difference Vegetation Index, Collection 4, University of Maryland, College Park, Maryland, 2004 (data available at: http://glcf.umd.edu/data/ndvi/, last access: 23 November 2018).

Chen, L., Qian, X., and Shi, Y.: Critical Area Identification of Potential Soil Loss in a Typical Watershed of the Three Gorges Reservoir Region, Water. Resour. Manage., 25, 3445–3463, https://doi.org/10.1007/s11269-011-9864-4, 2011.

Dabney, S. M., Yoder, D. C., and Vieira, D. A. N.: The application of the Revised Universal Soil Loss Equation, Version 2, to evaluate the impacts of alternative climate change scenarios on runoff and sediment yield, J. Soil Water Conserv., 67, 343–353, https://doi.org/10.2489/jswc.67.5.343, 2012.

da Silva, A. M.: Rainfall erosivity map for Brazil, Catena, 57, 251–259, https://doi.org/10.1016/j.catena.2003.11.006, 2004.

David, W. P.: Soil and Water Conservation Planning: Policy Issues and Recommendations, J. Philipp. Dev., 15, 47–84, 1988.

de Asis, A. M. and Omasa, K.: Estimation of vegetation parameter for modeling soil erosion using linear Spectral Mixture Analysis of Landsat ETM data, ISPRS J. Photogramm. Remote. Sens., 62, 309–324, https://doi.org/10.1016/j.isprsjprs.2007.05.013, 2007.

de Mello, C. R., Norton, L. D., Pinto, L. C., Beskow, S., and Curi, N.: Agricultural watershed modeling: a review for hydrology and soil erosion processes, Ciencia E Agrotecnologia, 40, 7–25, https://doi.org/10.1590/S1413-70542016000100001, 2016.

de Vente, J. and Poesen, J.: Predicting soil erosion and sediment yield at the basin scale: Scale issues and semi-quantitative models, Earth Sci. Rev., 71, 95–125, https://doi.org/10.1016/j.earscirev.2005.02.002, 2005.

Delgado, M. E. M. and Canters, F:. Modeling the impacts of agroforestry systems on the spatial patterns of soil erosion risk in three catchments of Claveria, the Philippines, Agroforestry Syst., 85, 411–423, https://doi.org/10.1007/s10457-011-9442-z, 2012.

Demirci, A. and Karaburun, A.: Estimation of soil erosion using RUSLE in a GIS framework: A case study in the Buyukcekmece Lake watershed, northwest Turkey, Environ. Earth. Sci., 66, 903–913, https://doi.org/10.1007/s12665-011-1300-9, 2012.

Desmet, P. J. J. and Govers, G.: A GIS procedure for automatically calculating the USLE LS factor on topographically complex landscape units, J. Soil Water Conserv., 51, 427–433, 1996.

Diodato, N.: Estimating RUSLE's rainfall factor in the part of Italy with a Mediterranean rainfall regime, Hydrol. Earth Syst. Sci., 8, 103–107, https://doi.org/10.5194/hess-8-103-2004, 2004.

Duarte, L., Teodoro, A., Gonçalves, J., Soares, D., and Cunha, M.: Assessing soil erosion risk using RUSLE through a GIS open source desktop and web application, Environ. Monit. Assess., 188, 1–16, https://doi.org/10.1007/s10661-016-5349-5, 2016.

Dumas, P. and Fossey, M.: Mapping potential soil erosion in the Pacific Islands: A case study of Efate Island (Vanuatu), 11th Pacific Science Inter-Congress: Pacific Countries and Their Ocean, Facing Local and Global Changes, 2009.

Durigon, V. L., Carvalho, D. F., Antunes, M. A. H., Oliveira, P. T. S., and Fernandes, M. M.: NDVI time series for monitoring RUSLE cover management factor in a tropical watershed, Int. J. Remote Sens., 35, 441–453, https://doi.org/10.1080/01431161.2013.871081, 2014.

Dymond, J. R.: Soil erosion in New Zealand is a net sink of CO_2, Earth Surf. Proc. Land., 35, 1763–1772, https://doi.org/10.1002/esp.2014, 2010.

Eiumnoh, A.: Integration of Geographic Information Systems (GIS) and Satellite Remote Sensing (SRS) for Watershed Management, Technical Bulletin 150, 2000.

El-Swaify, S. A., Gramier, C. L., and Lo, A.: Recent advances in soil conservation in steepland in humid tropics, in: Proceedings of the International Conference on Steepland Agriculture in the Humid Tropics, 87–100, Kuala Lumpur, MADI, 1987.

El-Swaify, S. A. and Dangler, E. W: Erodibilities of selected tropical soils in relation to structural and hydrologic parameters, Soil Conserv. Soc. Am., Ankeny, Iowa, 1976.

Eltaif, N., Gharaibeh, M., Al-Zaitawi, F., and Alhamad, M.: Approximation of rainfall erosivity factors in north Jordan, Pedosphere, 20, 711–717, https://doi.org/10.1016/S1002-0160(10)60061-6, 2010.

Farhan, Y. and Nawaiseh, S.: Spatial assessment of soil erosion risk using RUSLE and GIS techniques, Environ. Earth. Sci., 74, 4649–4669, https://doi.org/10.1007/s12665-015-4430-7, 2015.

Fernandez, C., Wu, J., McCool, D., and Stoeckle, C.: Estimating water erosion and sediment yield with GIS, RUSLE, and SEDD, J. Soil Water Conserv., 58, 128–136, 2003.

Fernandez, M. A. and Daigneault, A.: Erosion mitigation in the Waikato District, New Zealand: economic implications for agriculture, Int. J. Food Agric. Econ., 48, 1–21, https://doi.org/10.1111/agec.12338, 2016.

Ferreira, V. and Panagopoulos, T.: Seasonality of soil erosion under Mediterranean conditions at the Alqueva dam watershed, Environ. Manage., 54, 67–83, https://doi.org/10.1007/s00267-014-0281-3, 2014.

Ferro, V., Giordano, G., and Iovino, M.: Isoerosivity and erosion risk map for Sicily, Hydrol. Sci. J., 36, 549–564, https://doi.org/10.1080/02626669109492543, 1991.

Flabouris, K.: Study of rainfall factor R on the RUSLE law, (Doctoral dissertation), Aristotle University of Thessaloniki, Greece, 2008.

Foster, G. R., Toy, T. E., and Renard, K. G.: Comparison of the USLE, RUSLE1.06c, and RUSLE2 for Application to Highly Disturbed Lands, in: First Interagency Conference on Research in Watersheds, 27–30 October, 154–160, United States Department of Agriculture, 2003.

Fu, G., Chen, S., and McCool, D.: Modeling the Impacts of no-till practice on Soil Erosion and Sediment Yield with RUSLE, SEDD, and ArcView GIS, Soil Till. Res., 85, 38–49, https://doi.org/10.1016/j.still.2004.11.009, 2006.

Gaubi, I., Chaabani, A., Ben Mammou, A., and Hamza, M. H.: A GIS-based soil erosion prediction using the Revised Universal Soil Loss Equation (RUSLE) (Lebna watershed, Cap Bon, Tunisia), Nat. Hazards, 86, 219–239, https://doi.org/10.1007/s11069-016-2684-3, 2017.

Hernandez, E. C., Henderson, A., and Oliver, D. P.: Effects of changing land use in the Pagsanjan–Lumban catchment on suspended sediment loads to Laguna de Bay, Philippines, Agric. Water Manag., 106, 8–16, https://doi.org/10.1016/j.agwat.2011.08.012, 2012.

Hudson, N. W.: Soil Conservation, Ithaca, New York, Cornell University Press, 1971.

Institute of Water Research: RUSLE – An online soil erosion assessment tool, available at: http://www.iwr.msu.edu/rusle/ (last access: 19 November 2018), 2015.

İrvem, A., Topaloğlu, F., and Uygur, V.: Estimating spatial distribution of soil loss over Seyhan River Basin in Turkey, J. Hydrol., 336, 30–37, https://doi.org/10.1016/j.jhydrol.2006.12.009, 2007.

Jahun, B. G., Ibrahim, R., Dlamini, N. S., and Musa, S. M.: Review of Soil Erosion Assessment using RUSLE Model and GIS, J. Biol. Agric. Healthc., 5, 36–47, 2015.

Jain, M. K. and Das, D.: Estimation of sediment yield and areas of soil erosion and deposition for watershed prioritization using GIS and remote sensing, Water Resour. Manage., 24, 2091–2112, https://doi.org/10.1007/s11269-009-9540-0, 2010.

Jayasinghe, P. K. S. C., Adornado, H. A., Yoshida, M., and Leelamanie, D. A. L.: A Web-Based GIS and Remote Sensing Framework for Spatial Information System (SIS): A Case Study in Nuwaraeliya, Sri Lanka, Agricultural Information Research, 19, 106–116, https://doi.org/10.3173/air.19.106, 2010.

Kavian, A., Fathollah Nejad, Y., Habibnejad, M., and Soleimani, K.: Modeling seasonal rainfall erosivity on a regional scale: A case study from Northeastern Iran, Int. J. Environ. Res., 5, 939–950, 2011.

Kim, J. B., Saunders, P., and Finn, J. T.: Rapid assessment of soil erosion in the Rio Lempa Basin, Central America, using the universal soil loss equation and geographic information systems, Environ. Manage., 36, 872–885, https://doi.org/10.1007/s00267-002-0065-z, 2005.

Kinnell, P. I. A.: Event soil loss, runoff and the Universal Soil Loss Equation family of models: A review, J. Hydrol., 385, 384–397, https://doi.org/10.1016/j.jhydrol.2010.01.024, 2010.

Klik, A., Haas, K., Dvorackova, A., and Fuller, I. C.: Spatial and temporal distribution of rainfall erosivity in New Zealand, Soil Res., 53, 815–825, 2015.

Kulikov, M., Schickhoff, U., and Borchardt, P.: Spatial and seasonal dynamics of soil loss ratio in mountain rangelands of south-western Kyrgyzstan, J. Mt. Sci., 13, 316–329, https://doi.org/10.1007/S11629-014-3393-6, 2016.

Land Development Department: Assessment of Soil Loss using the Equation of Soil Loss, Bangkok, Thailand, 2002.

Le Roux, J., Morgenthal, T., Malherbe, J., Pretorius, D., and Sumner, P.: Water erosion prediction at a national scale for South Africa, Water SA, 34, available at: http://www.scielo.org.za/scielo.php?script=sci_arttext&pid=S1816-79502008000300003 (last access: 19 November 2018), 2005.

Li, L., Wang, Y., and Liu, C.: Effects of land use changes on soil erosion in a fast developing area, Int. J. Environ. Sci. Technol., 11, 1549–1562, https://doi.org/10.1007/s13762-013-0341-x, 2014.

López-Vicente, M., Navas, A., and Machín, J.: Identifying erosive periods by using RUSLE factors in mountain fields of the Central Spanish Pyrenees, Hydrol. Earth Syst. Sci., 12, 523–535, https://doi.org/10.5194/hess-12-523-2008, 2008.

Loureiro, N. D. S. and Coutinho, M. D. A.: A new procedure to estimate the RUSLE EI30 index, based on monthly rainfall data and applied to the Algarve region, Portugal, J. Hydrol., 250, 12–18, https://doi.org/10.1016/S0022-1694(01)00387-0, 2001.

Lu, H., Prosser, I. P., Moran, C. J., Gallant, J. C., Priestley, G., and Stevenson, J. G.: Predicting sheetwash and rill erosion over the Australian continent, Soil Res., 41, 1037–1062, https://doi.org/10.1071/SR02157, 2003.

Lu, H. and Yu, B.: Spatial and seasonal distribution of rainfall erosivity in Australia, Soil Res., 40, 887, https://doi.org/10.1071/SR01117, 2002.

Ma, H. L., Wang, Z. L., and Zhou, X.: Estimation of soil loss based on RUSLE in Zengcheng, Guangdong Province, Yangtze River, 41, 90–93, 2010.

Madarcos, B. S.: Soil Erosion in Four Cashew-based Crops Under Two Slope Categories, University of the Philippines at Los Baños Library, 1985.

Merritt, W. S., Letcher, R. A., and Jakeman, A. J.: A review of erosion and sediment transport models, Environ. Model Softw., 18, 761–799, https://doi.org/10.1016/S1364-8152(03)00078-1, 2003.

Merritt, W. S., Croke, B. F. W., Jakeman, A. J., Letcher, R. A., and Perez, P.: A Biophysical Toolbox for assessment and management of land and water resources in rural catchments in Northern Thailand, Ecol. Modell., 171, 279–300, https://doi.org/10.1016/j.ecolmodel.2003.08.010, 2004.

Merzouk, A.: Relative erodibility of nine selected Moroccan soils as related to their physical, chemical, and mineralogical properties, (Doctoral dissertation), University of Minnesota, St. Paul, United States of America, 1985.

Mihara, H.: Raindrop and Soil Erosion, National Institute of Agricultural Science Bulletin Series A, 1, 1–59, National Institute of Agricultural Science, Tokyo, 1951.

Mitasova, H., Hofierka, J., Zlocha, M., and Iverson, L.: Modelling topographic potential for erosion and deposition using GIS, Int. J. Geogr. Inf. Sci., 10, 629–641, https://doi.org/10.1080/02693799608902101, 1996.

Mitasova, H., Barton, M., Ullah, I., Hofierka, J., and Harmon, R.: GIS-Based Soil Erosion Modeling, in: Treatise on Geomorphology, 3, 228–258, Academic Press, San Diego, Califronia, https://doi.org/10.1016/B978-0-12-374739-6.00052-X, 2013.

Momm, H. G., Bingner, R. L., Wells, R. R., and Wilcox, D.: Agnps GIS-based tool for watershed-scale identification and mapping of cropland potential ephemeral gullies, Appl. Eng. Agric., 28, 17–29, 2012.

Moore, I. D. and Burch, G. J.: Physical basis of the length-slope factor in the Universal Soil Loss Equation, Soil Sci. Soc. Am. J., 50, 1294–1298, 1986.

Moore, I. D., Grayson, R. B., and Ladson, A. R.: Digital Terrain Modeling?: A Review of Hydrological Geomorphological and Biological Applications, Hydrol. Process., 5, 3–30, https://doi.org/10.1002/hyp.3360050103, 1991.

Morgan, R. P. C.: Estimating regional variations in soil erosion hazard in Peninsular Malaysia, Malayan Nature Journal, 28, 94–106, 1974.

Morgan, R. P. C.: Soil Erosion and Conservation, Longman Group Ltd., Essex, 1986.

Morgan, R. P. C.: Soil Erosion and Conservation, National Soil Resources Institute, Cranfield University, https://doi.org/10.1002/9781118351475.ch22, 2005.

Nagle, G. N., Fahey, T. J., and Lassoie, J. P.: Management of sedimentation in tropical watersheds, Environ. Manage., 23, 441–452, https://doi.org/10.1007/s002679900199, 1999.

Naipal, V., Reick, C., Pongratz, J., and Van Oost, K.: Improving the global applicability of the RUSLE model – adjustment of the topographical and rainfall erosivity factors, Geosci. Model Dev., 8, 2893–2913, https://doi.org/10.5194/gmd-8-2893-2015, 2015.

Nakil, M.: Analysis of parameters causing water induced soil erosion, Unpublished Fifth Annual Progress Seminar, Indian Institute of Technology, Bombay, 2014.

Nakil, M. and Khire, M.: Effect of slope steepness parameter computations on soil loss estimation: review of methods using GIS, Geocarto Int., 31, 1078–1093, https://doi.org/10.1080/10106049.2015.1120349, 2016.

NASA: Normalized Difference Vegetation Index, New York City, USA, NASA Goddard Institute for Space Studies, available at: https://data.giss.nasa.gov/landuse/ndvi.html, last access: 23 November 2018.

Nigel, R. and Rughooputh, S. D. D. V.: Application of a RUSLE-based soil erosion modelling on Mauritius Island, Soil Res., 50, 645–651, 2012.

Nontananandh, S. and Changnoi, B.: Internet GIS, based on USLE modeling, for assessment of soil erosion in Songkhram watershed, Northeastern of Thailand, Kasetsart Journal – Natural Science, 46, 272–282, 2012.

Ochoa-Cueva, P., Fries, A., Montesinos, P., Rodríguez-Díaz, J. A., and Boll, J.: Spatial Estimation of Soil Erosion Risk by Land-cover Change in the Andes of Southern Ecuador, Land Degrad. Dev., 26, 565–573, https://doi.org/10.1002/ldr.2219, 2015.

Ozsoy, G., Aksoy, E., Dirim, M. S., and Tumsavas, Z.: Determination of soil erosion risk in the mustafakemalpasa river basin, Turkey, using the revised universal soil loss equation, geographic information system, and remote sensing, Environ. Manage., 50, 679–694, https://doi.org/10.1007/s00267-012-9904-8, 2012.

Panagos, P., Karydas, C. G., Gitas, I. Z., and Montanarella, L.: Monthly soil erosion monitoring based on remotely sensed biophysical parameters: a case study in Strymonas river basin towards a functional pan-European service, Int. J. Digit. Earth, 5, 461–487, https://doi.org/10.1080/17538947.2011.587897, 2012.

Panagos, P., Meusburger, K., Ballabio, C., Borrelli, P., and Alewell, C.: Soil erodibility in Europe: A high-resolution dataset based on LUCAS, Sci. Total Environ., 479–480, 189–200, https://doi.org/10.1016/j.scitotenv.2014.02.010, 2014.

Panagos, P., Borrelli, P., and Meusburger, K.: A New European Slope Length and Steepness Factor (LS-Factor) for Modeling Soil Erosion by Water, Geosciences, 5, 117–126, https://doi.org/10.3390/geosciences5020117, 2015a.

Panagos, P., Borrelli, P., Meusburger, K., Alewell, C., Lugato, E., and Montanarella, L.: Estimating the soil erosion cover-management factor at the European scale, Land Use Policy, 48, 38–50, https://doi.org/10.1016/j.landusepol.2015.05.021, 2015b.

Panagos, P., Borrelli, P., Meusburger, K., Van Der Zanden, E. H., Poesen, J., and Alewell, C.: Modelling the effect of support practices (P-factor) on the reduction of soil erosion by water at European scale, Environ. Sci. Policy, 51, 23–34, https://doi.org/10.1016/j.envsci.2015.03.012, 2015c.

Panagos, P., Ballabio, C., Borrelli, P., Meusburger, K., Klik, A., Rousseva, S., and Alewell, C.: Rainfall erosivity in Europe, Sci. Total Environ., 511, 801–814, https://doi.org/10.1016/j.scitotenv.2015.01.008, 2015d (data available at: https://esdac.jrc.ec.europa.eu/content/rainfall-erosivity-european-union-and-switzerland, last access: 23 November 2018).

Panagos, P., Borrelli, P., Poesen, J., Ballabio, C., Lugato, E., Meusburger, K., and Alewell, C.: The new assessment of soil loss by water erosion in Europe, Environ. Sci. Policy, 54, 438–447, https://doi.org/10.1016/j.envsci.2015.08.012, 2015e.

Panagos, P., Borrelli, P., Meusburger, K., Yu, B., Klik, A., Jae Lim, K., and Ballabio, C.: Global rainfall erosivity assessment based on high-temporal resolution rainfall records, Sci. Rep., 7, 4175, https://doi.org/10.1038/s41598-017-04282-8, 2017 (data available at: https://esdac.jrc.ec.europa.eu/content/global-rainfall-erosivity, last access: 23 November 2018).

Pimentel, D., Harvey, C., Resosudarmo, P., Sinclair, K., Kurz, D., McNair, M., and Blair, R.: Environmental and Economic Costs of Soil Erosion and Conservation Benefits, Science, 267, 1117–1123, https://doi.org/10.1126/science.267.5201.1117, 1995.

Post, D. A. and Hartcher, M. G.: Evaluating uncertainty in modelled sediment delivery in data-sparse environments: application to the Mae Chaem Catchment, Thailand, in: Proceedings of symposium S7 held during the Seventh IAHS Scientific Assembly, 80–89, Foz do Iguaçu, 2005.

Raissouni, A., Issa, L., Lech-Hab, K., and El Arrim, A.: Water Erosion Risk Mapping and Materials Transfer in the Smir Dam Watershed (Northwestern Morocco), JGEESI, 5, 1–17, https://doi.org/10.9734/JGEESI/2016/20061, 2016.

Ram, B., Dhyani, B. L., and Kumar, N.: Assessment of erodibility status and refined Iso-Erodent Map of India, Indian Journal of Soil Conservation, 32, 171–177, 2004.

Rango, A. and Arnoldus, H.: Aménagement des bassins versants, Cahiers techniques de la FAO, Food and Agriculture Organization of the United Nations, 1987.

Renard, K. and Freimund, J. R.: Using monthy precipitation data to estimate R-factor in the revised USLE, J. Hydrol., 157, 287–306, 1994.

Renard, K., Foster, G., Weesies, G., McCool, D., and Yoder, D.: Predicting soil erosion by water: a guide to conservation planning with the Revised Universal Soil Loss Equation (RUSLE), Agricultural Handbook No. 703, 65–100, https://doi.org/10.1201/9780203739358-5, 1997.

Rodda, H. J. E., Stroud, M. J., Shankar, U., and Thorrold, B. S.: A GIS based approach to modelling the effects of land-use change on soil erosion in New Zealand, Soil Use Manag., 17, 30–40, https://doi.org/10.1111/j.1475-2743.2001.tb00005.x, 2001.

Roose, E. J.: Erosion et ruissellement en Afrique de l'ouest: vingt années de mesures en petites parcelles expérimentales, Adiopodoumé, Ivory Coast, 1975.

Rosewell, C.: Potential Sources of Sediments and Nutrients: Sheet and Rill Erosion and Phosphorous Sources, in: State of the Environment Technical Paper Series (Inland Waters), Department of the Environment, Sport and Territories, Canberra, Australia, 1997.

Rosewell, C. and Turner, J.: Rainfall Erosivity in New South Wales, Issue 20 of Technical Report, Department of Conservation and Land Management, Sydney, 1992.

Roslee, R., Bidin, K., Musta, B., Tahir, S., Tongkul, F., and Norhisham, M. N.: GIS Application for Comprehensive Spatial Soil Erosion Analysis with MUSLE Model in Sandakan Town Area, Sabah, Malaysia, Geological Behavior, 1, 1–5, 2017.

Rozos, D., Skilodimou, H. D., Loupasakis, C., and Bathrellos, G. D.: Application of the revised universal soil loss equation model on landslide prevention. An example from N. Euboea (Evia) Island, Greece, Environ. Earth. Sci., 70, 3255–3266, https://doi.org/10.1007/s12665-013-2390-3, 2013.

Russo, A:. Applying the Revised Universal Soil Loss Equation model to land use planning for erosion risk in Brunei Darussalam, Australian Planner, 3682, 1–17, https://doi.org/10.1080/07293682.2014.957332, 2015.

Sadeghi, S. H. R., Gholami, L., Khaledi Darvishan, A., and Saeidi, P.: A review of the application of the MUSLE model worldwide, Hydrolog. Sci. J., 59, 365–375, https://doi.org/10.1080/02626667.2013.866239, 2014.

Sadeghifard, D., Jabari, E., and Ghayasian, H.: Rainfall erosivity zonation in Iran, in: The First National Conference on Civil Engineering, Sharif University of Technology, Iran, available at: https://www.civilica.com/Paper-NCCE01-226_2417394703.html (last access: 19 November 2018), 2004.

Schmitt, L. K.: Developing and applying a soil erosion model in a data-poor context to an island in the rural Philippines, Environ. Dev. Sustain., 11, 19–42, https://doi.org/10.1007/s10668-007-9096-1, 2009.

Shamshad, A., Azhari, M. N., Isa, M. H., Hussin, W. M. A. W., and Parida, B. P.: Development of an appropriate procedure for estimation of RUSLE EI30 index and preparation of erosivity maps for Pulau Penang in Peninsular Malaysia, Catena, 72, 423–432, https://doi.org/10.1016/j.catena.2007.08.002, 2008.

Sholagberu, A. T., Ul Mustafa, M. R., Wan Yusof, K., and Ahmad, M. H.: Evaluation of rainfall-runoff erosivity factor for Cameron highlands, Pahang, Malaysia, J. Ecol. Eng., 17, 1–8, https://doi.org/10.12911/22998993/63338, 2016.

Sinha, D. and Joshi, V. U.: Application of universal soil loss equation (USLE) to recently reclaimed badlands along the Adula and Mahalungi Rivers, Pravara Basin, Maharashtra, J. Geol. Soc. India, 80, 341–350, https://doi.org/10.1007/s12594-012-0152-6, 2012.

Smith, D. D. and Whitt, D. M.: Estimating Soil Loss from Field Area of Clay Pan Soils, Soil Sci. Soc. Am. J., 12, 485–490, 1947.

Stewart, B., Woolhiser, D., Wischmeier, W., Caro, J., and Frere, M. H.: Control of water pollution from cropland, United States Department of Agriculture, United States, 1975.

Thorne, C. R., Zevenbergen, L. W., Grissinger, E. H., and Murphey, J. B.: Calculator programe and nomograph for on-site predictions of ephemeral gully erosion, United States Department of Agriculture Soil Conservation Service, United States, 1985.

Torri, D., Poesen, J., and Borselli, L.: Predictability and uncertainty of the soil erodibility factor using a global dataset, CATENA, 46, 1–22, https://doi.org/10.1016/S0341-8162(97)00036-2, 1997.

USDA-ARS (U.S. Department of Agriculture – Agricultural Research Service): Revised Universal Soil Loss Equation, available at: https://www.ars.usda.gov/southeast-area/oxford-ms/national-sedimentation-laboratory/watershed-physical-processes-research/docs/revised-universal-soil-loss-equation-rusle-welcome-to-rusle-1 (last access: 19 November 2018), 2002.

van der Knijff, J. M., Jones, R. J. A., and Montanarella, L.: Soil Erosion Risk Assessment in Europe, available at: https://www.preventionweb.net/files/1581_ereurnew2.pdf (last access: 21 November 2018), 2000.

Williams, R. J. and Renard, K. G.: EPIC – a new method for assessing erosions effect on soil productivity, J. Soil Water Conserv., 38, 381–383, 1983.

Wilson, J. P. and Gallant, J. C.: Digital Terrain Analysis, edited by: Wilson, J. P. and Gallant, J. C., in: Terrain Analysis: Principles and Applications, 1–28, 2000.

Wischmeier, W. H. and Mannering, J. V.: Relation of Soil Properties to its Erodibility, Soil and Water Management and Conservation, 15, 131–137, https://doi.org/10.2136/sssaj1969.03615995003300010035x, 1969.

Wischmeier, W. H. and Smith, D. D.: Predicting rainfall erosion losses, Agriculture Handbook No. 537, 537, 285–291, https://doi.org/10.1029/TR039i002p00285, 1978.

Wu, S., Li, J., and Huang, G.: An evaluation of grid size uncertainty in empirical soil loss modeling with digital elevation models, Environ. Model. Assess., 10, 33–42, https://doi.org/10.1007/s10666-004-6595-4, 2005.

Yoon, K. S., Kim, C., and Woo, H.: Application of RUSLE for Erosion Estimation of Construction Sites in Coastal Catchments, J. Coastal Res., 2, 1696–1700, 2009.

Yang, X., Chapman, G. A., Gray, J. M., and Young, M. A.: Delineating soil landscape facets from digital elevation models using compound topographic index in a geopgraphic information system, Soil Res., 45, 569–576, 2007.

Zakerinejad, R. and Maerker, M.: An integrated assessment of soil erosion dynamics with special emphasis on gully erosion in the Mazayjan basin, southwestern Iran, Nat. Hazards, 79, 25–50, https://doi.org/10.1007/s11069-015-1700-3, 2015.

Zhou, F. J., Chen, M. H., Lin, F. X., Huang, Y. H., and Lu, C. L.: The rainfall erosivity index in Fujian Province, J. Soil Water Conserv., 9, 13–18, 1995.

Životić, L., Perović, V., Jaramaz, D., Dordević, A., Petrović, R., and Todorović, M.: Application of USLE, GIS, and remote sensing in the assessment of soil erosion rates in Southeastern Serbia, Pol. J. Environ. Stud., 21, 1929–1935, 2012.

Appendix A

Table A1. Summary of previous studies that have applied the USLE and RUSLE.

Author	Location	*R* factor	*K* factor	*LS* factor	*C* factor	*P* factor
David (1988)	Various watersheds in the Philippines	Mihara (1951) and Hudson (1971) as cited in David (1988)	Wischmeier and Mannering (1969)	Madarcos (1985) and Smith and Whitt (1947)	Literature	Literature
Eiumnoh (2000)	Sakae Krang watershed (Thailand)	El-Swaify et al. (1987) as cited in Merritt et al. (2004)	USLE method	USLE method	Literature	None observed ($P = 1$)
Fernandez et al. (2003)	Lawyers Creek watershed (USA)	USDA-ARS (2002)	From the SSURGO database (USDA)	Upslope contributing area method	Database from RUSLE software	Database from RUSLE software
Merritt et al. (2004)	Mae Chem watershed (Thailand)	El-Swaify et al. (1987) as cited in Merritt et al. (2004)	Previous studies in area	USLE method	Previous studies in area	Previous studies in area
Post and Hartcher (2005)	Mae Chem watershed (Thailand)	El-Swaify et al. (1987) as cited in Merritt et al. (2004)	Previous studies in area	$L = 1$ $S =$: derived from DEM	Previous studies in area	None observed ($P = 1$)
Dumas and Fossey (2009)	Efate Island (Vanuatu)	Roose (1975) and Morgan (1974) as cited in Morgan (2005)	USLE method	RUSLE method at pixel level	Literature	None observed ($P = 1$)
Adornado et al. (2009)	REINA (Philippines)	El-Swaify et al. (1987) as cited in Merritt et al. (2004)	Table by Stewart et al. (1975)	Upslope contributing area method	Literature	None observed ($P = 1$)
Schmitt (2009)	Negros Island (Philippines)	RUSLE method	USLE method	RUSLE method at pixel level	Literature	Previous studies
Jayasinghe et al. (2010)	Nuwaraeliya (Sri Lanka)	El-Swaify et al. (1987) as cited in Merritt et al. (2004)	Table by Stewart et al. (1975)	Upslope contributing area method	Literature	None observed ($P = 1$)
Jain and Das (2010)	Jharkhand (India)	Ram et al. (2004) as cited in Jain and Das (2010)	USLE method and previous studies	Upslope contributing area method	Literature	None observed ($P = 1$)
Adornado and Yoshida (2010)	Bukidnon (Philippines) and also REINA (Philippines)	El-Swaify et al. (1987) as cited in Merritt et al. (2004)	Table by Stewart et al. (1975)	Upslope contributing area method	Literature	None observed ($P = 1$)

Table A1. Continued.

Author	Location	*R* factor	*K* factor	*LS* factor	*C* factor	*P* factor
Boyle et al. (2011)	California (USA)	From previous studies	From previous studies	Upslope contributing area method	Literature	n/a
Chen et al. (2011)	Xiangxi watershed (China)	Wischmeier and Smith (1978)	Williams and Renard (1983) nomograph	Upslope contributing area method	Using NDVI	n/a
Demirci and Karaburun (2012)	Büyükçekmece Lake watershed (Turkey)	Arnoldus (1980)	Torri et al. (1997) equation	Upslope contributing area method	Using NDVI	None observed ($P = 1$)
Nontananandh and Changnoi (2012)	Songkhram watershed (Thailand)	Land Development Department (2000)	Values from Land Development Department (2000)	Modified RUSLE method	Literature	None observed ($P = 1$)
Ozsoy et al. (2012)	Mustafakemalpaşa River basin (Turkey)	From previous studies	USLE method	RUSLE method, using a third-party programme	Literature	None observed ($P = 1$)
Delgado and Canters (2012)	Claveria (Philippines)	Shamshad et al. (2008)	USLE method	RUSLE2 programme, using the upslope contributing area method	Literature	David (1988)
Hernandez et al. (2012; using SedNet, which has an USLE component)	Pagsanjan (Philippines)	El-Swaify et al. (1987) as cited in Merritt et al. (2004)	Wischmeier and Mannering (1969)	Algorithm within SedNet	Literature	n/a
Sinha and Joshi (2012)	Maharashtra (India)	Roose (1975)	USLE method	Morgan (1986)	Literature	Literature
Nigel and Rughooputh (2012)	Mauritius	Arnoldus (1980) as cited in Le Roux et al. (2005)	From previous studies	Upslope contributing area method	Literature	Literature
Životić et al. (2012)	Nišava River basin (Serbia)	Wischmeier and Smith (1978)	USLE method	RUSLE method	Using NDVI	None observed ($P = 1$)
Rozos et al. (2013)	Euboea Island (Greece)	Flabouris (2008)	Based on geological characteristics	Morgan (1986)	Literature	None observed ($P = 1$)
Bagherzadeh (2014)	Mashhad plain (Iran)	Wischmeier and Smith (1978)	USLE method	USLE method		None observed ($P = 1$)

Table A1. Continued.

Author	Location	*R* factor	*K* factor	*LS* factor	*C* factor	*P* factor
Ferreira and Panagopoulos (2014)	Alqueva (Portugal)	Similar to Loureiro and Coutinho (2001)	USLE method	Upslope contributing area method	Using NDVI	None observed ($P = 1$)
Li et al. (2014)	Guangdong (China)	Zhou et al. (1995)	USLE method	Similar to RUSLE method	Using NDVI	1 for wasteland and built-up area, 0.5 for forest, 0.2 for orchard land, 0.35 for cropland
Zakerinejad and Maerker (2015; using USPED, which has USLE components)	Mazayjan (Iran)	Ferro et al. (1991); Renard and Freimund (1994); Sadeghifard et al. (2004)	RUSLE method	Algorithm within USPED	Literature	None observed ($P = 1$)
Jahun et al. (2015)	Crete (Greece)	Fu et al. (2006)	RUSLE method	Upslope contributing area method	Using NDVI	Previous studies
Farhan and Nawaiseh (2015)	Wadi Kerak catchment (Jordan)	Eltaif et al. (2010)	Similar to USLE nomograph	Upslope contributing area method	Literature	Literature
Panagos et al. (2015e) and related papers	Europe	Rainfall Intensity Summarization Tool (RIST)	USLE method	3rd-party programme	Literature	Literature
Russo (2015)	Brunei Darussalam	Rosewell and Turner (1992)	Rosewell (1997)	RUSLE method	Based on ground covered	None observed ($P = 1$)
Nakil and Khire (2016)	Gangapur (India)	Nakil (2014)	USLE method	RUSLE method	Literature	Literature
Raissouni et al. (2016)	Smir Dam (Morocco)	Similar to Arnoldus (1980) methods	Merzouk (1985)	Upslope contributing area method	Literature	None observed ($P = 1$)
Fernandez and Daigneault (2016)	Waikato (New Zealand)	Institute of Water Research (2015)	Dymond et al. (2010)	Upslope contributing area method	Range between 1 (wood vegetation) and 10 (herbaceous vegetation or bare ground)	
Duarte et al. (2016)	Montalegre (Portugal)	Loureiro and Coutinho (2001)	USLE method	USLE method	Literature	Literature
Gaubi et al. (2017)	Lebna watershed (Tunisia)	Rango and Arnoldus (1987)	USLE method	Upslope contributing area method	Literature	None observed ($P = 1$)

n/a: not applicable.

13

Water yield following forest–grass–forest transitions

Katherine J. Elliott[1], **Peter V. Caldwell**[1], **Steven T. Brantley**[2], **Chelcy F. Miniat**[1], **James M. Vose**[3], **and Wayne T. Swank**[1]

[1]USDA Forest Service, Southern Research Station, Coweeta Hydrologic Lab, Otto, NC 28763, USA
[2]Joseph W. Jones Ecological Research Center, Ichauway, Newton, GA 31770, USA
[3]UDSA Forest Service, Southern Research Station, Center for Integrated Forest Science, Raleigh, NC 27695, USA

Correspondence to: Katherine J. Elliott (kelliott@fs.fed.us)

Abstract. Many currently forested areas in the southern Appalachians were harvested in the early 1900s and cleared for agriculture or pasture, but have since been abandoned and reverted to forest (old-field succession). Land-use and land-cover changes such as these may have altered the timing and quantity of water yield (Q). We examined 80 years of streamflow and vegetation data in an experimental watershed that underwent forest–grass–forest conversion (i.e., old-field succession treatment). We hypothesized that changes in forest species composition and water use would largely explain long-term changes in Q. Aboveground biomass was comparable among watersheds before the treatment (208.3 $Mg\,ha^{-1}$), and again after 45 years of forest regeneration (217.9 $Mg\,ha^{-1}$). However, management practices in the treatment watershed altered resulting species composition compared to the reference watershed. Evapotranspiration (ET) and Q in the treatment watershed recovered to pretreatment levels after 9 years of abandonment, then Q became less (averaging 5.4 % less) and ET more (averaging 4.5 % more) than expected after the 10th year up to the present day. We demonstrate that the decline in Q and corresponding increase in ET could be explained by the shift in major forest species from predominantly *Quercus* and *Carya* before treatment to predominantly *Liriodendron* and *Acer* through old-field succession. The annual change in Q can be attributed to changes in seasonal Q. The greatest management effect on monthly Q occurred during the wettest (i.e., above median Q) growing-season months, when Q was significantly lower than expected. In the dormant season, monthly Q was higher than expected during the wettest months.

1 Introduction

Forests play a critical role in regulating hydrological processes in headwater catchments by moderating the timing and magnitude of streamflow (Burt and Swank, 2002; Chang, 2003; Ice and Stednick, 2004; Ford et al., 2011b; Vose et al., 2011). Hydrological processes in forests are particularly sensitive to disturbances that reduce tree vigor or leaf area and thus decrease evapotranspiration (ET) (Aranda et al., 2012; Edburg et al., 2012; Brantley et al., 2013). Most efforts at studying the effects of disturbance on watershed hydrology have focused on quantifying the effects of forest harvesting practices on water yield (Q) (Bosch and Hewlett, 1982; Stednick, 1996; Burton, 1997; Brown et al., 2005; Wei and Zhang, 2010; Ford et al., 2011a; Zhang and Wei, 2012; Liu et al., 2015). Reviews have shown that, in general, harvesting $< 20\,\%$ of the basal area shows no detectable increase in annual Q; but Q increases thereafter as the percentage of basal area harvested increases (Bosch and Hewlett, 1982; Andréassian, 2004; Brown et al., 2005). However, recent work that aims to merge ecology of the resulting forest and species composition with traditional hydrology approaches (i.e., ecohydrology) has advanced our understanding greatly. For example, Brantley et al. (2013, 2015) showed that lasting changes in annual Q (lower) and persistent changes in peak flow ($> 20\,\%$, after the most intense storms) were observed with only about a 5 % basal area loss of eastern hemlock (*Tsuga canadensis* (L.) Carrière). Brantley et al. (2013) also suggested that a change in forest composition with less evergreen hemlock relative to deciduous trees could result in an increase in Q in winter months.

Most of the eastern US forests have been harvested at least once since the late 1800s (Yarnell, 1998; Foster et al., 2003; Thompson et al., 2013; Martinuzzi et al., 2015), and many forested areas have undergone forest-to-agriculture land-use changes, and then been abandoned to revert back to forest (i.e., abandoned agriculture or old-field succession) (Otto, 1983; Trimble et al., 1987; Wear and Bolstad, 1998; Bellemare et al., 2002; Alvarez, 2007; Thiemann et al., 2009; Ramankutty et al., 2010; Kirk et al., 2012). Land abandonment has also been prevalent and ongoing since the early 20th century in other countries (Cramer et al., 2008; García-Ruiz and Lana-Renault, 2011). Land-use and land-cover (LULC) changes, such as forest–grass–forest transitions, may alter the timing and quantity of Q. Because land-use conversion from forests to agriculture often includes a combination of changes in vegetation composition and soil physical attributes, it is difficult to separate the effects of vegetation changes from soil changes (see reviews by Neary et al., 2009; Zimmermann et al., 2010; Houlbrooke and Laurenson, 2013, Morris and Jackson, 2016). Land-cover conversion that requires heavy machinery or includes livestock grazing decreases soil infiltration and saturated hydraulic conductivity (e.g., Hassler et al., 2011; Price et al., 2011; Morris and Jackson, 2016), and can thus increase peak flow during storms, and flood frequency and severity (Reinhart, 1964; Hornbeck, 1973; Burt and Swank, 2002; Alila et al., 2009; Green and Alila, 2012). Without soil compaction and alteration of water flow pathways, forest trees typically use more water and extract water from deeper soil than shallower-rooted grasses (Zhang et al., 2001; Kulmatiski and Beard, 2013), which could result in higher ET and lower Q at the catchment scale.

Several studies have compared Q and ET of forests and pastures. Analyzing 250 catchments worldwide, Zhang et al. (2001) found that forested catchments had higher ET than grass pastures, with few exceptions. Replacing trees with grass cover generally increases Q by decreasing ET (Hibbert, 1969; Bosch and Hewlett, 1982; Farley et al., 2005), although not always (Brauman et al., 2012; Amatya and Harrison, 2016). In some basins, when agricultural land use is reduced and forest cover increased, Q is unchanged, and can be explained in part by the species-specific traits in water use (e.g., deciduous versus evergreen, and/or late season versus early season perennial grass) (Cruise et al., 2010), and the geomorphological differences among biomes (Zhou et al., 2015).

Large differences among tree species in their leaf and canopy conductance, transpiration per unit leaf area, and whole-tree water use for any given diameter exist in eastern US deciduous forests (Wullschleger et al., 2001; Ford and Vose, 2007; Ford et al., 2011a). This is especially true when comparing hardwoods within diffuse-porous and ring-porous xylem functional groups (Taneda and Sperry, 2008; Ford et al., 2011a; Von Allmen et al., 2015). *Liriodendron tulipifera*, a diffuse-porous species common to the eastern temperate deciduous forest biome, has among the highest transpiration rates of forest trees; while rates of *Acer rubrum* L. and *Betula lenta* L., also common diffuse-porous species, are lower than *L. tulipifera*. However, they have relatively high transpiration rates compared to common ring-porous *Quercus* species (Ford et al., 2011a).

Few studies have examined long-term changes in catchment hydrology through a forest–grass–forest transition, with specific attention focused on species compositional changes and their effect on ET and Q. A treated watershed within the Coweeta Hydrologic Laboratory, western North Carolina, experienced this LULC transition, and reported similar Q between forest and grass when the grass cover was well fertilized (Hewlett and Hibbert, 1961; Hibbert, 1969; Bosch and Hewlett, 1982; Burt and Swank, 1992). However, they did not investigate why Q was lower than expected after grass-cover abandonment and through the early successional development of the deciduous forest. Road construction could be a contributing factor because installing temporary roads to facilitate timber harvesting can affect hydrology (Harr et al., 1975; Alila et al., 2009), but only 3.3 % of the watershed area was in temporary roads (inactive for the least 50 years). In addition, roads comprising less than 6 % of the watershed area appear not to change storm hydrographs significantly (Harr et al., 1975; Swank et al., 2001; Alila et al., 2009). In a more recent study, Ford et al. (2011b) suggested that the decline in Q over time could be due to a shift in the dominant tree species in the treated, old-field succession watershed.

Working in the same experimental watershed as authors above, we compared the long-term changes (1934–2015) in the following: (1) aboveground biomass, leaf area index (LAI) and species and functional (xylem anatomy) group composition; (2) estimated growing-season mean daily water use (DWU); (3) annual water-balance derived ET; and (4) daily, monthly, and annual Q between the treated, old-field succession, watershed (WS6) and nearby reference (WS14, WS18) watersheds with an emphasis on the period of reforestation. We hypothesized that (1) a shift in species composition and their attendant DWU will largely explain long-term changes in Q, (2) annual Q would be lower in the treated WS6 through forest succession concurrent with greater DWU with additional changes in timing of Q due to altered species composition, and (3) monthly Q would be greater in the treated WS6 for wet periods (high or peak flows), and this effect would be greatest in the dormant season.

2 Materials and methods

2.1 Study area

The study was conducted at the USDA Forest Service Coweeta Hydrologic Laboratory, a 2185 ha forested basin in the Nantahala National Forest in North Carolina, USA

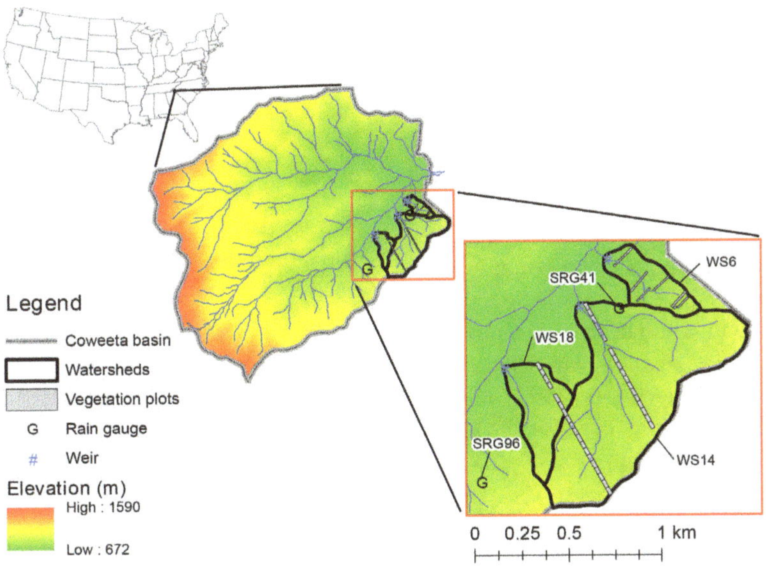

Figure 1. Map of the Coweeta Basin with treated WS6 and reference (WS14, WS18) watersheds with rain gauges, weirs, and vegetation plots.

(35°06′ N, 83°43′ W). Climate in the Coweeta Basin is classified as marine, humid temperate (Swift et al., 1988), with mild temperature (average 12.8 °C) and ample precipitation (average 1795 mm yr^{-1}). Three watersheds (WS6, WS14 and WS18) within 1 km of one another, similar in elevation, slope, and aspect were used in this study (Table 1). WS14 and WS18 are untreated reference watersheds (Fig. 1). WS6 experienced a disturbance regime similar to forest conversion to pasture and subsequent abandonment common across the region (see below). Soils in all watersheds are moderately permeable, well-drained, moderately deep to very deep, and with a saprolite layer up to 6 m deep (Thomas, 1996).

2.2 History of disturbance

Before 1842, the Coweeta Basin was burned semiannually (Douglass and Hoover, 1988). Between 1842 and 1900, light semiannual burning and grazing continued. From 1912 to 1923 heavy logging occurred (Douglass and Hoover, 1988). Loss of American chestnut (*Castanea dentata* (Marshall) Borkh.) in the 1930s (Woods and Shanks, 1959; Elliott and Swank, 2008) was followed by loss of *Tsuga canadensis* (L.) Carrière in the 2000s (Elliott and Vose, 2011).

The disturbance regime in WS6, the treated watershed, was extensive (Table S1 in the Supplement). In July 1941, 12 % of the catchment (1.06 ha area) along the stream was cut to determine how riparian vegetation affects Q (Dunford and Fletcher, 1947). In 1958, the entire watershed was clear-cut, merchantable timber was removed, and the residue was piled and burned. In 1959, surface soil was scarified and seeded to *Festuca octoflora* grass. In 1960, the watershed was treated with a one-time application of 1100 kg ha^{-1} lime, 110 kg ha^{-1} 30-10-10 NPK, and

Table 1. Characteristics of treated (WS6) and reference (WS14, WS18) watersheds at the Coweeta Hydrologic Laboratory in Otto, NC, USA. Mean annual precipitation (P) and water yield (Q) based on data collected over water years (WY, May–April) from 1934 to 2015.

		Watersheds		
Characteristic	Units	6	14	18
Area	ha	9.0	61.03	12.46
Mean elevation	m	824	878	823
Mean basin slope	%	50	50	55
Aspect		NW	NW	NW
Year of first complete flow record	WY	1939	1938	1938
Nearest rain gauge		SRG41	SRG41	SRG96
Mean precipitation (P)	mm yr^{-1}	1843	1843	2031
Mean water yield (Q)	mm yr^{-1}	866	997	1021
Mean evapotranspiration ($ET = P - Q$)	mm yr^{-1}	978	845	1010
Q/P		0.47	0.54	0.50

18.4 kg ha^{-1} granular 60 % potash. Between 1960 and 1965, *Kalmia latifolia* L., *Rhododendron maximum* L., and other hardwood sprouts were suppressed with spot applications of 2,4D [(2,4dichlorophenoxy) acetic acid] to maintain the watershed in grass cover (Hibbert, 1969). In 1965, the watershed was fertilized again, as above. In 1967, the grass was treated with the herbicides atrazine [2-chloro-(4-ethylamino)-6-9-isopropylamino)-Strizine], paraquat [1,1 dimethyll 4,4 bipyridinium ion (dichloride salt)], and 2,4D [(2,4dichlorophenoxy) acetic acid] (Douglass et al., 1969), and then left undisturbed (hereafter, old-field succession). Although the grass was not cut or grazed, the lime and fertilizer amendments with attendant high productivity and nutrient uptake by the grass make these applications somewhat similar to agricultural practices. The original objectives of the conversion from forest-to-grass were to compare water use of grass versus hardwoods (Hibbert, 1969; Swank and Crossley, 1988) and to determine how conversion to grass affects discharge characteristics (Burt and Swank, 1992).

Two adjacent forested watersheds (WS14, WS18) were selected as references to provide an adequate number of sample plots (described below) for analysis of changes in vegetation. These reference watersheds with similar physiography (Table 1) are characteristic of mature, second-growth hardwood forests, and have remained unmanaged since 1923 (Swank and Crossley, 1988). We considered forest age for WS14 and WS18 since the loss of *C. dentata* presently to be > 75 years old.

2.3 Measurements

2.3.1 Vegetation

The relative importance of woody species over time was characterized with repeated tree surveys. In treated WS6, surveys were conducted in 1934, 1958, 1982, 1995, and 2012. In 1934, only five 0.08 ha permanent plots were measured along the east side of the watershed; in all subsequent surveys, plots were placed across the entire watershed. In 1958, a pretreatment strip inventory sampled 25 % of the watershed area with 10 m wide strips approximately 40 m apart extending along transects from the ridge-top to the stream channel. This sampling method resulted in a total of 37 unequally sized plots (ranging from 0.02 to 0.14 ha), including the riparian corridor. In 1982, 34 0.02 ha plots were permanently marked continuously along five transects from ridge-top to near stream; these 34 plots were re-measured in 1995 and 2012.

In reference WS14, 31 0.08 ha permanent plots were surveyed in 1934, 1969, 1988–1992 (hereafter, 1992), and 2009–2010 (hereafter, 2009). In reference WS18, eight 0.08 ha permanent plots were surveyed in 1934, 1953, 1969, 1992, and 2009.

In all watersheds and for all survey periods, diameter of woody stems ≥ 2.54 cm at breast height (DBH, 1.37 m above ground) was measured by species and recorded into 2.54 cm DBH classes. In 1934, only percent cover was recorded for the two evergreen shrubs, *Rhododendron maximum* and *Kalmia latifolia*; for this reason, we do not estimate biomass and LAI (square meters of leaf area per square meter of ground area) for these species in 1934. In all other years, stem diameters of these evergreen shrubs were measured in the same manner as the tree diameters. Median DBH values were used to calculate basal area, aboveground biomass, leaf biomass, and LAI. We used species-specific allometric equations developed on-site to estimate the aboveground biomass, leaf biomass, and LAI contribution of each species in each watershed (McGinty, 1972; Santee and Monk, 1981; Martin et al., 1998; Ford and Vose, 2007; B. D. Kloeppel, unpublished data; C. F. Miniat, unpublished data). Species nomenclature follows Kirkman et al. (2007).

2.3.2 Water yield (Q) and evapotranspiration (ET)

We used both chronological-pairing (i.e., corresponding to the same meteorological input) and frequency-pairing (described below) analyses to detect potential hydrologic responses of Q and ET to LULC change. Both analyses used the paired watershed approach (Wilm, 1944, 1949). The chronological pairing approach allowed us to create a time series of estimated change in annual Q and ET over the period of record and to relate these changes to both the treatment and climate. In addition, this analysis allowed us to determine when a consistent change in Q began, enabling us to establish the time period of interest for the frequency pairing. The frequency-pairing approach allowed us to compare the post-treatment distribution of monthly and annual Q to that of the pretreatment period. We used WS18 and WS6 as the reference and treatment watersheds, respectively. We did not compare WS6 to WS14 because there were gaps in the WS14 flow record in the years immediately following the grass conversion and herbicide application. For both watersheds, a 5 min stream stage was used to estimate Q (Reinhard and Pierce, 1964; Swift et al., 1988).

We modeled WS6 annual Q and ET as a function of WS18, incorporating the effect of grass conversion and reforestation treatments over time. Annual Q was computed on a May–April water year to minimize the effects of year-to-year changes in storage, as soils are generally at their wettest by the beginning of May. The empirical chronological-pairing model was fit using PROC NLIN (SASv9.4, SAS Institute, Cary, NC) and had the following form:

$$\hat{Q}_{\mathrm{T}} = a + bQ_{\mathrm{R}} + eM_1t_1 + \left[M_2c\left(h - \frac{1}{1+\exp^{-t_2}}\right)\right], \quad (1)$$

where $\hat{Q}_{\mathrm{T}}$ = predicted Q from treated watershed WS6 (mm yr^{-1}), Q_{R} = measured Q from reference watershed WS18 (mm yr^{-1}), M_1 = management representing grass conversion, ($M_1 = 1$ for water years between and including 1960 and 1966; $M_1 = 0$ otherwise), t_1 = time since grass fertilization (t_1 = water year − fertilization year for water years between and including 1960 and 1966 where fertilization years include water years 1959, 1961, and 1966; $t_1 = 0$ otherwise), M_2 = management representing reforestation after grass conversion ($M_2 = 1$ for water years greater than or equal to 1967; $M_2 = 0$ otherwise), and t_2 = time since reforestation after grass conversion (t_2 = water year − 1967 for water years greater than or equal to 1967; $t_2 = 0$ otherwise); a, b, c, e, and h are fitted parameters.

This overall modeling approach has been used in prior studies to assess the impact of forest management on Q (Ford et al., 2011b; Kelly et al., 2016). The $a + bQ_{\mathrm{R}}$ term in EQ1 reflects the relationship between reference and treatment watersheds assuming no treatment. The increasing linear Q response (eM_1t_1 term in EQ1) accounts for the decline in annual grass production after fertilization and water use, as noted by Hibbert (1969). The $M_2c(h - \frac{1}{1+\exp^{-t_2}})$ term in EQ1 accounts for the exponential decline in Q as the forest regenerates, which has been observed in numerous paired watershed experiments (Swank et al., 1988).

As in Ford et al. (2011b), we define the Q treatment response, D_Q, as the difference between the observed Q in the treated watershed (Q_T) and that predicted by the model assuming no treatments had taken place ($\hat{Q}_T$):

$$D_Q = Q_T - \left(\hat{Q}_T; M_1, M_2 = 0\right). \tag{2}$$

The proportion of the variability explained by the model was quantified using the ratios of the error-to-total sum of squares and the total-to-error degrees of freedom as follows:

$$R^2_{\text{adjusted}} = 1 - \frac{\text{SS}_\text{E}}{\text{SS}_\text{T}} \times \frac{\text{df}_\text{T}}{\text{df}_\text{E}}. \tag{3}$$

Parameter estimates were interpreted as statistically significant at $\alpha = 0.05$. Observed annual ET was computed as precipitation $(P) - Q_T$ while expected ET with no treatment was computed as $P - \hat{Q}_T$, both assuming the largely impermeable bedrock underlying the basin that results in negligible deep groundwater losses (Douglass and Swank, 1972). Watershed P was estimated using a nearby 8 in. (20.3 cm) National Weather Service standard rain gauge, SRG 96 (Laseter et al., 2012). The ET treatment response, D_{ET}, is then as follows:

$$D_{\text{ET}} = [P - Q_T] - \left(\left[P - \hat{Q}_T\right]; M_1, M_2 = 0\right). \tag{4}$$

2.3.3 Frequency-pairing flow distributions

We used the frequency-pairing method (Alila et al., 2009; Brantley et al., 2015) to detect differences in frequency between observed and predicted annual and monthly Q after treatment. Briefly, frequency pairing is an analytical method that quantifies differences in observed and predicted Q parameters based on the probability of occurrence of a given Q (or flow at a given probability) rather than based on occurrence at a discreet time (i.e., chronological pairing). This accounts for rainfall amount and antecedent soil conditions. We used pre-treatment Q during water years 1939–1941, 1948–1951, and 1956–1958, to estimate the expected cumulative distribution functions (CDFs, F_Y) for observed and predicted Q in the treatment watershed using the following linear regression equation:

$$\hat{Y}_i = b_0 + b_1 X_i, \tag{5}$$

where X_i is the observed Q in the reference watershed for period i (day of year) and $\hat{Y}_i$ is the expected Q for the treatment watershed under undisturbed conditions for the same period. We used PROCMODEL (SAS v9.3, SAS Institute, Cary, NC) to predict monthly post-treatment Q in the treatment watershed from May 1979 to April 2015 and annual post-treatment Q for water years 1980–2015. To model monthly Q, we separated the data by calendar month and created twelve separate regression equations. Using separate regression equations for each month helped account for variations in paired watershed Q relationships among months and helped to distinguish differences in effects among seasons.

Observed and predicted Q values were then plotted as an estimate of the probability of occurrence for ranked event $Y_{(i)}$ during any time period i. The exceedance probability, $1 - p$, was estimated for each period using the following equation:

$$1 - F_Y\left[\hat{Y}_i\right] = \frac{m - 0.40}{n + 0.20}, \tag{6}$$

where m is the rank for a given flow and n is the total number of flow periods in the distribution. This function provided an empirical estimate of the quantile for a given flow value (Cunnane, 1978; Stedinger et al., 1993). Confidence limits for each predicted flow at each probability of occurrence were estimated as follows:

$$Y_m \pm z_{1-\frac{\alpha}{2}} \sqrt{(\text{Var}_1[Y_m] + \text{Var}_2[Y_m])}. \tag{7}$$

We used a pair of Monte Carlo simulations to estimate the variability associated with the predictive uncertainty in Eq. (7) (Var_1), and the uncertainty associated with the sampling variability at each rank (Var_2). For these analyses, we used 1000 iterations for each simulation. We used the raw, expected post-treatment values from Eq. (5) to correct for the loss of variability in the upper tails of the distribution (Alila et al., 2009). The CDFs were then used to construct flow duration curves to assess changes in untransformed Q at monthly and annual intervals by comparing the change in magnitude for a given probability or the change in probability for a given magnitude (Alila et al., 2009; Green and Alila, 2012; Brantley et al., 2015).

2.3.4 Growing-season daily water use (DWU)

Plant water loss was estimated by scaling up sap flux measurements of numerous species and stem diameter sizes at Coweeta Hydrologic Laboratory (Ford and Vose, 2007; Ford et al., 2011b; Brantley et al., 2013; Miniat, unpublished) using methods outlined in Ford et al. (2011a). We fitted the observed growing-season mean DWU (kg day^{-1}) to stem DBH (cm) using a power function of the following form:

$$\text{DWU} = b_0 \cdot \text{DBH}^{b_1}. \tag{8}$$

Species were grouped into xylem functional types (diffuse-porous, ring-porous, semi-ring porous, evergreen shrub, or tracheid) and growing-season DWU models were developed for each xylem functional type. For example, *Carya*

Table 2. Summary of growing-season daily water use (DWU, kg day^{-1}) models for each xylem functional group as a function of stem diameter at breast height (DBH, cm); DWU $= b_0 \cdot \mathrm{DBH}_1^b$.

Xylem group	N	Min DBH (cm)	Max DBH (cm)	b_0	b_1	Adjusted R^2	SE of estimate
Diffuse-porous	95	7.4	61.8	0.1428	1.7676	0.70	30.3
Evergreen Shrub	16	5.3	16.3	0.6445	0.7002	0.00	2.5
Ring-porous	38	23.9	86.7	0.2392	1.1488	0.55	9.1
Semi-ring-porous	18	20.2	55.7	0.0009	2.8557	0.88	8.8
Tracheid	116	9.5	67.5	0.0005	2.8411	0.73	8.6

spp. have semi-ring porous xylem; *Quercus* spp. and *Oxydendrum arboreum* have ring-porous xylem; and *Betula lenta*, *Liriodendron tulipifera*, and *Acer rubrum* have diffuse-porous xylem (Table S2). Because *R. pseudoacacia* behaves more like a diffuse-porous species, its measured values of DWU and DBH were combined with the diffuse-porous model. Even though *Robinia pseudoacacia* has ring-porous xylem, it is isohydric (i.e., maintaining stable leaf water potentials as soil water potentials drop, Klein, 2014) and has higher DWU than *Quercus* or *Carya* (Miniat and Hubbard, unpublished). For the two understory evergreen species, *Kalmia latifolia* and *Rhododendron maximum*, we applied the mean DWU value from 16 instrumented shrubs because DWU models based on DBH alone provided limited predictive power (Table 2). We estimated growing-season mean-plot DWU by modeling DWU by functional type and vegetation surveys by diameter for all watersheds. We did not estimate DWU for the 1934 survey, when *C. dentata* was most abundant, because most of the trees had been affected by chestnut blight, compromising their functional xylem.

3 Results

3.1 Vegetation dynamics

Prior to treatment, species composition and aboveground biomass among the watershed were similar (Fig. 2). In 1934, aboveground biomass was comparable among the treated WS6 and references WS14 and WS18, averaging 200 Mg ha^{-1} ($p = 0.706$) (Fig. 2a, Table S3). Biomass declined in WS6 (99.51 Mg ha^{-1}) from 1934 to 1958 prior to conversion to grass, and in WS18 (148.42 Mg ha^{-1}) from 1934 to 1953 (Table S3). The decline in biomass and LAI between 1934 and the 1950s was primarily due to the loss of *Castanea dentata* (Fig. 2a–c). In 1934, *C. dentata* occupied from 40 to 54 % of the biomass and 29–43 % of the LAI across the three watersheds (Fig. S1 in the Supplement).

The grass cover in the treated watershed was highly productive, but following the herbicide treatment (i.e., old-field succession), early successional vegetation rapidly established (Fig. S1a). During the 5 years when WS6 was maintained in grass, biomass ranged from 5.67 to 7.30 Mg ha^{-1}. In 1968, 1 year after cessation of treatment, the aboveground biomass was 3.92 Mg ha^{-1} in WS6. At that time, the 1-year-old field was dominated by *Erechtites hieracifolia* (L.) Raf., *Phytolacca americana* L., *Eupatorium* spp., and *Equisetum arvense* L. and had remnants of *Festuca octoflora*. In the years between 1968 and 1982, WS6 was rapidly colo-

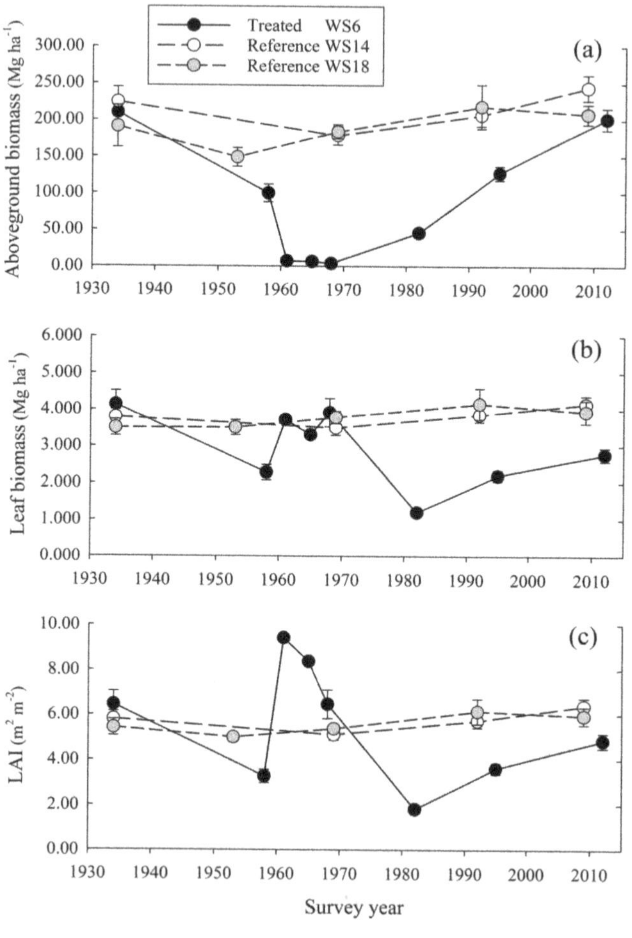

Figure 2. Mean (± SE bars) **(a)** aboveground biomass, **(b)** leaf biomass, and **(c)** leaf area index (LAI) for the treated WS6 and reference (WS14, WS18) watersheds over time.

nized by *Robinia pseudoacacia* and *Liriodendron tulipifera* (Fig. S1a), whereas the most abundant species in the reference watersheds in the years following the loss of *C. dentata* (1969 to 2010s) were *Quercus* spp. and *Acer rubrum* (Fig. S1b–c; Tables S4–S6).

Forest composition following grass cover was biased towards tree species with deep functional sapwood and diffuse-porous xylem. In 1934, all watersheds were dominated by species with semi-ring-porous (*C. dentata* and *Carya*) or ring-porous (*Quercus*) xylem, accounting for more than 80 % of the aboveground biomass (Fig. 3a–c) and 80 % of the LAI (Tables S4–S6). Although species with semi-ring-porous xylem declined in all watersheds over time, the increase in species with diffuse-porous xylem was greater in the treated watershed compared to reference watersheds (Fig. 3a–c). As the young forest developed following grass herbicide and abandonment, species with diffuse-porous xylem and *R. pseuodoacacia* dominated forest biomass, while species with ring-porous xylem were only 2.7 %. By 2012, 93 % of vegetation in the treatment watershed was comprised of species with diffuse-porous xylem (Fig. 3a), while the reference watersheds were about half of the species with ring-porous xylem (Fig. 3b and c).

3.2 Water yield (Q) and evapotranspiration (ET)

The forest–grass–forest treatment of WS6 resulted in significant effects on Q over time. Models of annual Q explained more than 98 % of the variability in Q over the period of record. Initial harvesting increased Q by 99 mm (10.5 % above the expected Q) in 1960 (Fig. 4), and Q remained higher than expected during the grass conversion period except in 1959, 1961, and 1966 when grass production was highest due to fertilizer application. The largest treatment effect occurred in 1967 when herbicide was applied to the watershed, resulting in a Q increase of 259 mm (31 % above the expected Q) (Fig. 4). Q remained higher than expected for approximately 9 years after the herbicide treatment as the vegetation re-established. Beginning in 1977 and continuing through 2015, Q was less than expected in 32 of 35 years (Fig. 4), suggesting that the new forest used more water (i.e., had higher ET) than expected had it not undergone treatment. Since 1980, on average, annual Q decreased by 6.1 %, ranging from a Q increase in 1981 of 30 mm (+5.5 %) to a decrease of 142 mm (16 %) in 2003. ET increased by 4.5 % on average, relative to what was expected in the absence of management (Fig. S2).

3.3 Changes in flow distribution

In addition to the forest–grass–forest treatment changing the amount of Q, it also fundamentally changed the distribution of Q, with the most pronounced changes at the height of the growing and dormant seasons. The annual and monthly Q relationships between the reference and treatment

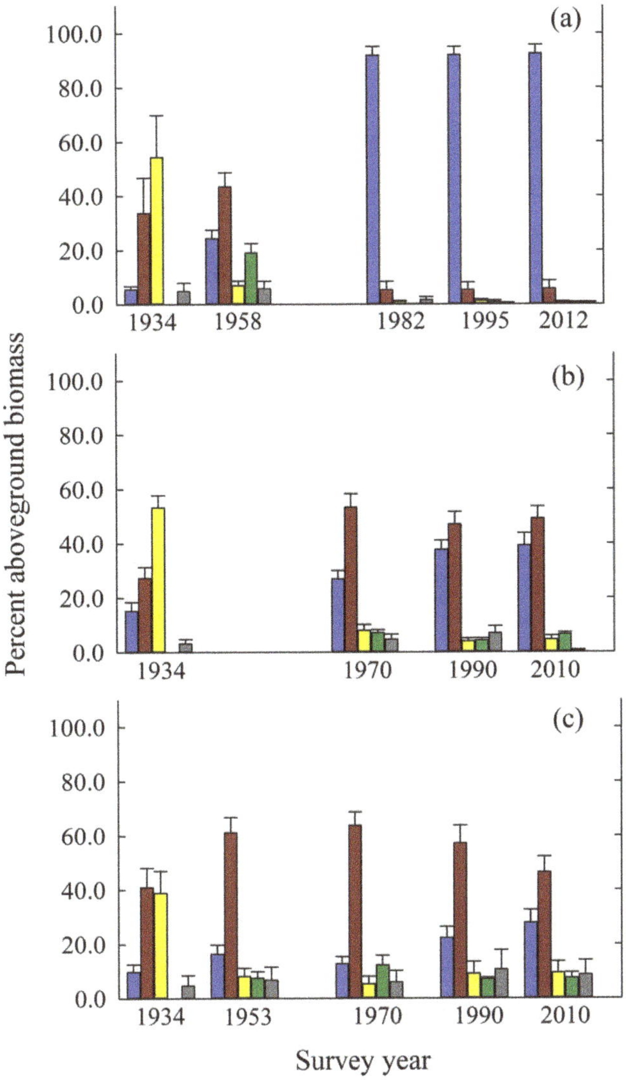

Figure 3. Percentage (± SE bars) of aboveground biomass for the xylem functional groups (diffuse-porous, ring-porous, semi-ring porous, tracheid, and evergreen shrub) in the **(a)** treated WS6, **(b)** reference WS14, and **(c)** reference WS18 over time.

watersheds for the pre-treatment period were highly significant (annual: $n = 10$, $r^2 = 0.97$, $p < 0.001$; monthly: $n = 10$, $r^2 = 0.94$, $p < 0.001$) using the frequency-pairing method. Annual Q was unchanged at low and high probabilities of non-exceedance (< 10 %), but was lower in some years between the 30 and 60 % probability of non-exceedance (Fig. 5a). Monthly Q was higher than expected at high probability of non-exceedance in February (Fig. 5b), whereas monthly Q was lower than expected at the high probability of non-exceedance in July (Fig. 5c). Median monthly Q was lower than expected for only January (−14.8 %) and May (−13.4 %) (Table 3). At wetter periods (above median Q), monthly Q was lower than expected for several months and immediately following the growing season

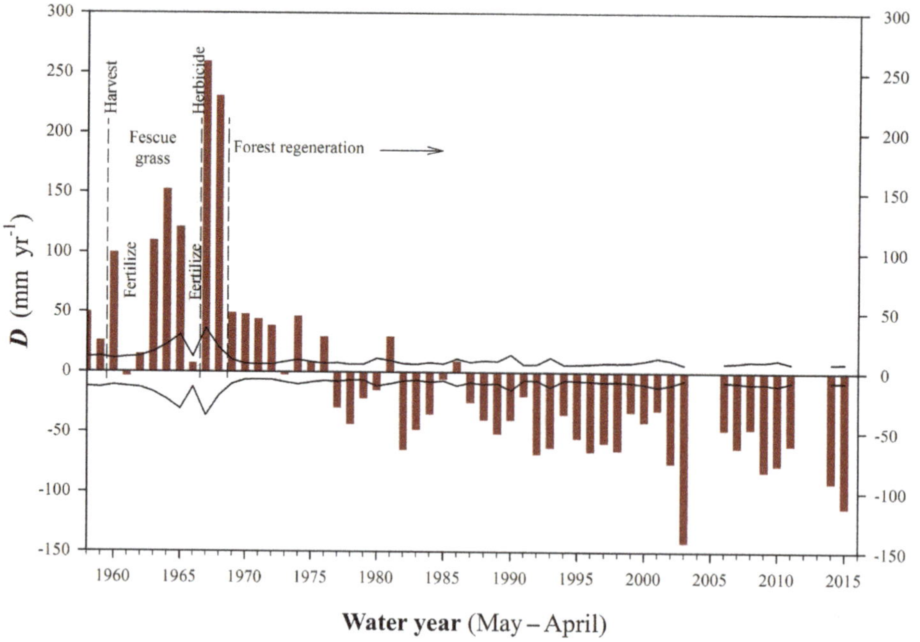

Figure 4. Changes in water yield (Q, $D = Q_T - (\hat{Q}_T; M_1, M_2 = 0)$) for the treated WS6 over time (bars). Solid lines are the standard errors of the mean prediction. We used the paired-watershed approach with WS18 as the reference. The year of harvest, conversion to Kentucky-31 fescue grass (*Festuca octoflora*) cover, fertilize, herbicide, and abandonment to allow forest regeneration are denoted by dashed lines.

(June–October, December; Table 3); whereas, for February–April, monthly Q was higher than expected. At drier periods (below median Q), February, March, and September had lower-than-expected monthly Q (Table 3). No significant changes in monthly Q distributions were observed in November.

3.4 Daily water use (DWU)

Growing-season DWU differed among species for any given DBH largely dependent on xylem anatomy (Table 2, Fig. 6). For example, DWU for a tree with 50 cm DBH could be 6.5 times higher with diffuse-porous xylem compared to ring-porous xylem (Fig. 5). *Liriodendron tulipifera*, *Betula lenta* and *Nyssa sylvatica* had the highest DWU; *Acer rubrum* and *Carya* were intermediate; and *Quercus alba*, *Q. montana*, and *Q. rubra* had the lowest estimated DWU compared to all other species (Ford et al., 2011b). *Robinia pseudoacacia* had higher DWU than *Quercus* or *Carya* (Miniat and Hubbard, unpublished). Models based on DBH and xylem anatomy explained 55–88 % of the variability in DWU among tree species (Table 2). For the evergreen understory species, however, DBH explained little variation in DWU, even though the standard errors were quite low.

Mean growing-season DWU for each catchment increased over time, but the treated watershed showed the greatest increase (Fig. 7a). In the 2010s, the 45-year-old forest in WS6 had 25–43 % higher DWU than the > 75-year-old reference forests (Fig. 7a–d), despite lower leaf area than the reference watersheds at that time (Fig. 2c). In reference WS14, tree species with diffuse-porous xylem contributed 48–63 % of the total water use between 1969 (age 35) and 2009 (age 75+), while evergreen shrubs contributed 20–23 %, and tree species with ring-porous xylem contributed 13 % or less to the total water use (Fig. 7c). Since the grass cover was

Table 3. Relative changes in monthly water yield (Q) for different parts of the cumulative distribution function for the period May 1979 to April 2015 for the treated WS6 using the frequency-pairing method. "Lower" and "Higher" denote direction and significance ($p < 0.05$) of change; NS = not significant.

Month	Change in median monthly Q (%)	Change in Q below the median Q	Change in Q above the median Q
Jan	Lower (−14.8)	NS	NS
Feb	NS	Lower	Higher
Mar	NS	Lower	Higher
Apr	NS	NS	Higher
May	Lower (−13.4)	NS	NS
Jun	NS	NS	Lower
Jul	NS	NS	Lower
Aug	NS	NS	Lower
Sep	NS	Lower	Lower
Oct	NS	NS	Lower
Nov	NS	NS	NS
Dec	NS	NS	Lower

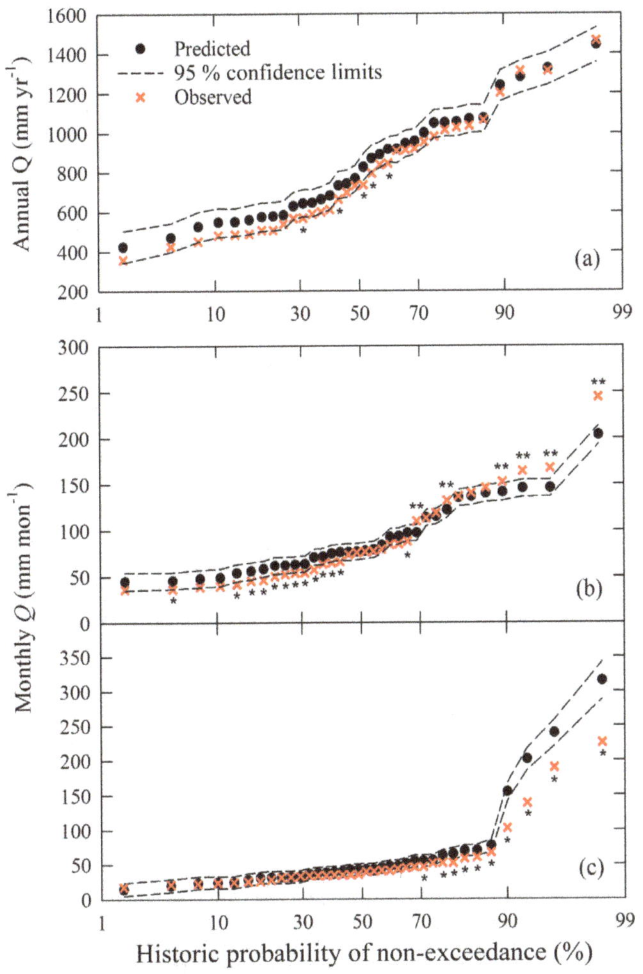

Figure 5. Changes in the cumulative distribution function (CDF) expressed as historic probability of non-exceedance for **(a)** annual water yield (Q), and monthly Q for **(b)** February and **(c)** July. * ($p < 0.05$) and ** ($p < 0.01$) denote years in the distribution functions when Q was significantly lower or higher than predicted.

abandoned in WS6, tree species with diffuse-porous xylem alone have contributed more than 90 % of the total daily water use in that watershed (Fig. 7b).

4 Discussion

We hypothesized that a shift in species composition and the resulting shift in DWU would largely explain long-term changes in Q in the treated watershed as the forest regenerated following grass abandonment. We found that forest species composition in the treated watershed shifted from dominance by species with ring-porous xylem prior to grass conversion to species with diffuse-porous xylem through old-field succession. With this major shift in species composition, DWU increased from 1982 to 2012 in the treated watershed, and it was much higher than that in the older refer-

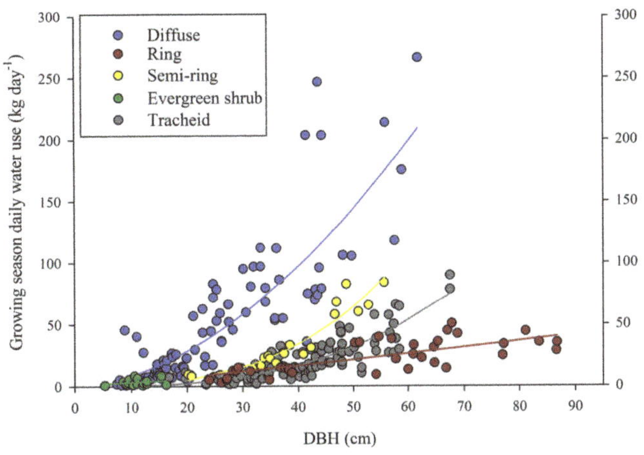

Figure 6. Growing-season daily water use of tree species by xylem functional group (diffuse-porous, ring-porous, semi-ring porous, evergreen shrub and tracheid) and DBH (diameter at 1.37 m above ground).

ence watersheds. These changes in species composition and DWU correspond with the long-term trend in lower-than-expected Q over that time period. Seasonal variation in Q helped to explain this long-term pattern.

4.1 Vegetation dynamics

Species composition has changed dramatically in the treated watershed through old-field succession following the forest–grass–forest transition. Prior to conversion to grass (1958), the forest was dominated by *Quercus montana* and *Q. coccinea*, similar to the reference watersheds at that time. After the grass was treated with herbicide, and the forest was allowed to reestablish, the forest shifted to dominance by *Liriodendron tulipifera* and *Robinia pseudoacacia*. Other studies have found that shade-intolerant *R. pseudoacacia* and *L. tulipifera* respond and grow rapidly following clear-cutting or other disturbances that create large canopy gaps (Elliott and Swank, 1994; Elliott et al., 1997, 1998; Shure et al., 2006; Boring et al., 2014). During grass dominance all woody species were eliminated with spot herbicide application. This treatment killed stump sprouts, and during forest succession recruitment favored small, wind-dispersed seeds, and discriminated against large-seeded and slow growing species such as *Quercus*, *Carya*, *Tilia*, and *Aesculus* (Elliott et al., 1997, 2002). Aboveground biomass approached pretreatment levels after 45 years of forest growth; however, LAI remained lower than that of the pretreatment or reference watershed conditions. The lower LAI could be attributed to the differences among species in the ratio of leaf area per total aboveground biomass and crown structure, where shade-intolerant *R. pseudoacacia* and *L. tulipifera* have lower ratios and concentrate foliage to the uppermost crown more than intermediate shade-tolerant *Quercus* (Kato et al., 2009).

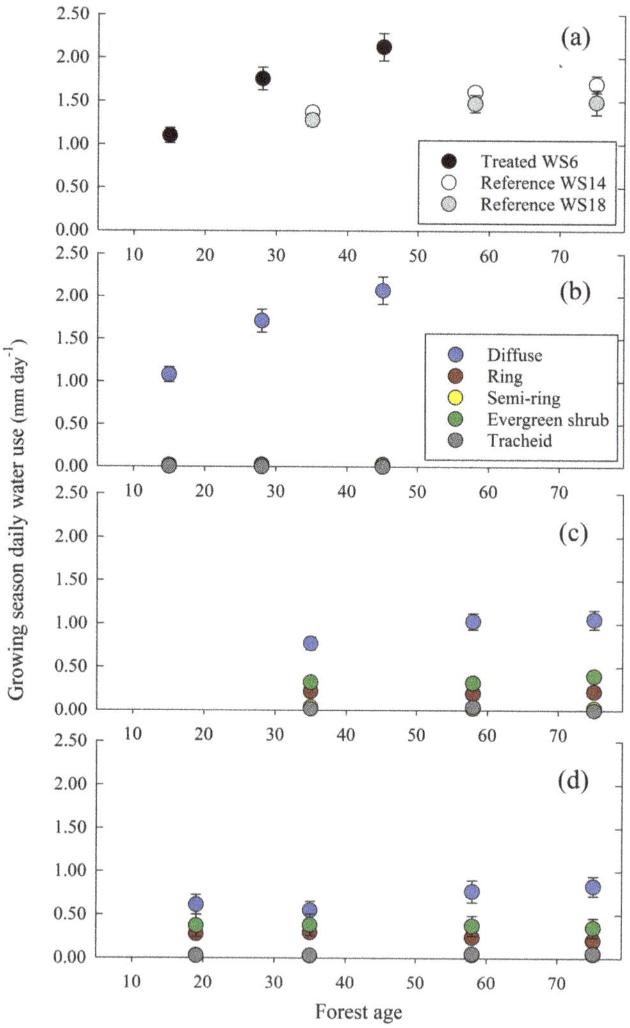

Figure 7. **(a)** Mean (±SE bars) growing-season daily water use (DWU) versus forest age in the treated WS6 and reference (WS14, WS18) watersheds; **(b)** DWU versus forest age in treated WS6 by xylem functional group (diffuse-porous, ring-porous, semi-ring porous, evergreen shrub, and tracheid); **(c)** DWU versus forest age in reference WS14 by xylem functional group; and **(d)** DWU versus forest age in reference WS18 by xylem functional group.

Many studies have investigated forest growth following harvesting (e.g., Palik et al., 2012; Boring et al., 2014; Loftis et al., 2014; Stanturf et al., 2014; Boggs et al., 2016), and the hardwood species composition that succeeds following harvest depends largely on the severity of disturbance, i.e., partial harvest, retention harvest, or clear-cutting, as well as the geographical region (Halpin and Lorimer, 2016). In northern Appalachian forests, *Prunus pensylvanica* and *Betula papyrifera* are common pioneer species that assume early dominance following clear-cutting (Hornbeck et al., 2014). In central Appalachian forests, *Prunus serotina*, *Acer rubrum*, *Betula lenta*, and *Fagus grandifolia* dominate following extensive harvests (Kochenderfer, 2006; D'Amato et al., 2015). *Robinia pseudoacacia* and *L. tulipifera*, two species that recruit and grow rapidly after clear-cutting, are much more abundant in the southern Appalachians (Elliott and Vose, 2011; Boring et al., 2014) than in the central Appalachians (Kochenderfer, 2006), and are absent in the northern Appalachians (Campbell et al., 2007; Hornbeck et al., 2014).

4.2 Species effects on water yield (Q) and evapotranspiration (ET)

We found that annual Q declined and ET increased through old-field succession relative to the time prior to the grass conversion. After 1980, 13 years following herbicide application, Q was consistently lower than expected for the next 35 years. Q was reduced by 6.5 % averaged over this time period; however, in 16 of those years, Q was greatly reduced (> 50 mm, 9.2 %). In 2003 and 2015, Q was reduced by 142 mm (16 %) and 113 mm (17 %), respectively. This supports our hypothesis that changes in ET and Q have occurred as a result of a shift in species composition. We also found that species effects were seasonal and influenced certain parts of the flow regime.

The range of changes in Q after treatment suggests that species composition affects storage and use of water under a wide range of precipitation conditions that play out over monthly and annual scales. For example, in 2003, when the decrease in Q was greatest (−142 mm; 16 %), P was 6 % greater than the long-term (1939–2015) average but this followed 4 years of below-average P. Average P for 1999–2002 was 23 % below the long-term average. In this case, the vegetation in the old-field succession watershed may have used more of the available water in 2003, following the dry period, than the vegetation in reference watersheds. As a result, less of the available water served to refill soil storage in the treated watershed compared to the reference watershed, resulting in a larger predicted decrease in Q in 2003. In 1981 when Q was higher than expected (+30 mm; 5.5 %), P was 29 % lower than the long-term average but this followed the second highest annual P in 1980 (+27 % greater than the 1939–2015 average). Much of the excess rainfall occurred at the end of the 1980 water year and the beginning of water year 1981. Precipitation during March and April of water year 1980, and May of water year 1981 was 123, 35, and 39 % greater than the long-term (1939–2015) average for those months, respectively. Given that the Q for the treated watershed was higher than expected in wetter months (those above median Q) of the dormant season, these wet months resulted in a higher-than-expected annual Q.

Our monthly analysis showed that changes in ET and Q varied seasonally. First, changes in monthly distribution of Q suggest that old-field succession and the consequent species changes have lowered streamflow during the growing season during wetter months. We observed that Q was lower than expected in September during both drier (below median Q) and wetter periods suggesting that changes in soil storage at the end of the growing season highly influences base flow. Others have found that forest clear-cutting had a longer-lasting influence on streamflow distribution, even when annual Q returned to baseline conditions within a few years (Burt et al., 2015; Kelly et al., 2016).

Second, changes in monthly distribution of Q suggest that there is a potential for increased frequency and severity of high flows in dormant season months under wet conditions. This could be particularly concerning during severe tropical storms. However, for the Appalachian region most tropical storms occur later in the year (September–December) (Holland and Webster, 2007; Smith et al., 2011). Interestingly, we found lower-than-expected Q during wetter periods for September–December months. If timing of large storms remains unchanged, then shifting species composition from those that have conservative water use (i.e., ring-porous xylem) to those that are less conservative (i.e., diffuse-porous xylem) could mitigate the effects of high flows during large storms.

The observed changes in monthly Q during the dormant season indicate a likelihood of soil saturation during the wettest periods. Higher-than-expected Q in the dormant season is likely a result of lower ET and higher soil moisture at that time of year (Berghuijs et al., 2014; Burt et al., 2015), rather than reduced infiltration capacity. For example, in an earlier study, Burt and Swank (1992) reported that the dead grass was not removed following herbicide application on the treated watershed and so the infiltration capacity remained high throughout 1967 and 1968. More likely the higher-than-expected Q in the dormant season is due to the lack of evergreen species in the treated watershed. Where evergreen species are a component of forested watersheds, they can affect ET and Q in the dormant season (Brantley et al., 2013, 2015); they transpire during dormant season months as long as environmental conditions are suitable (Ford and Vose, 2007; Ford et al., 2011a; Brantley, unpublished data) and they intercept precipitation during the dormant season because they retain their foliage. Even though evergreens (shrubs + tracheids) were a relatively small component (13.8 % of total aboveground biomass) of the old-field succession watershed before treatment, after treatment there were no evergreen shrubs due to the severity of the treatment. Yet, they remain a component (6.0 and 15.9 % for WS14 and WS18, respectively) of the reference watersheds. Thus, evergreen species reduce soil moisture storage and have the potential to mitigate spring flooding because of their contribution to ET and their location within riparian zones (Brantley et al., 2015).

Our results demonstrate that species changes largely explain the decreasing trend in Q following old-field abandonment based on modeled growing-season DWU over time, and enable us to assess the effects of forest structure and species composition on Q. For example, the estimates of DWU (Fig. 6) are consistent with the differences in temporal patterns of Q between the old-field succession WS6 and reference watersheds (Fig. 3). The mean DWU in WS6 was greater in 1995 than DWU in the reference watersheds in 1969 or 1992, suggesting that Q in WS6 became less than expected between these years due to altered DWU. Similarly, mean DWU in the 45-year-old old-field succession WS6 was greater still in 2012 than the > 75-year-old reference watersheds, WS14 or WS18, in 2010. Indeed, Q was consistently less than expected during this period, and was significantly less in 32 of the 35 years (including 1995, 2010, and 2012).

Few studies have examined the consequence of shifts in hardwood species composition on the hydrologic cycle (Swank et al., 2014; Caldwell et al., 2016). Changes in forest composition, structure and age as well as climate will interact to induce long-term changes in Q from forested mountain watersheds (Von Allmen et al., 2015; Caldwell et al., 2016). We found stronger and longer-lasting decreases in annual and monthly Q through old-field succession, than found by clear-cutting alone followed by forest succession (Reinhart, 1964; Hornbeck, 1973; Swank et al., 2001, 2014; Troendle et al., 2001; Adams and Kochenderfer, 2014; Hornbeck et al., 2014). For example, researchers at the Fernow Experimental Forest in West Virginia examined changes in annual Q following clear-cutting (Adams and Kochenderfer, 2014); there, the initial increase in Q returned to pretreatment levels within 3–4 years. In another treated watershed (WS7) in Coweeta that was allowed to regenerate naturally after the clear-cut, there was only 1 year when observed Q was significantly lower than predicted (Swank et al., 2014).

The observed changes in monthly and annual Q for the treated WS6 were largely a result of a rapid response of co-dominant species with less conservative transpiration rates (Wullschleger et al., 2001; Ford et al., 2012; Boggs et al., 2015; Brantley et al., 2015). Under similar environmental conditions, both *L. tulipifera* and *R. pseudoacacia* have much higher daily water use than species with ring-porous and semi-ring-porous xylem, such as *Quercus* and *Carya* (Ford et al., 2011b; Vose et al., 2016a, b). Overall, we estimated that growing-season daily water use increased significantly following old-field abandonment, and it was much higher in the 45-year-old treated watershed than the > 75-year-old reference watersheds.

5 Conclusions

Our long-term results are relevant to land areas that are currently in pasture and those that have reverted back to forests. In many parts of the world, pasture land and cropland area have increased since the 1990s as natural landscapes have been converted to agricultural ecosystems (e.g., Scanlon et al., 2007; Rodriguez et al., 2010); in other areas agricultural land has been abandoned (see review Rey Benayas et al., 2007). In general, grass pastures transpire less water and have lower interception loss than forests, resulting in greater Q for this LULC type (e.g., Wang et al., 2008; Holdo and Nippert, 2015). In the forest–grass–forest watershed, for 2 of the 5 years when the watershed was in grass cover, Q was equivalent to the pre-conversion hardwood forest, while for the other 3 years Q was greater under grass cover. Fertilizer application in 2 of the 5 years resulted in high grass productivity (Hewlett, 1961; Hewlett and Hibbert, 1966; Burt and Swank, 1992) such that LAI was maximized allowing for ET similar to that of the reference forested watershed. Q increased initially once herbicide was applied to the grass, quickly returned to expected levels, and then declined relative to expected levels as the abandoned old-field was allowed to regenerate to forest.

We found that within a deciduous forest, species identity matters in terms of how much precipitation leaves the watershed as ET versus Q. Through old-field succession, the treated watershed was dominated by water-demanding species with higher DWU than the pretreatment forest. We demonstrate that a shift in tree species composition from dominance by species with ring-porous xylem to dominance by species with diffuse-porous xylem can increase DWU, and in turn, produce a long-term reduction in Q.

Even within unmanaged watersheds, hydrologic parameters are not stationary (sensu Milly et al., 2008; Burt et al., 2015) and subtle changes in species composition can influence Q, particularly in dry years (Caldwell et al., 2016). Species-specific ecohydrological models (e.g., Novick et al., 2016) are increasingly vital in predicting long-term changes in ET and Q (Sun et al., 2016; Vose et al., 2016a, b). If drought frequency and severity increase as expected (Allen et al., 2010; Ayres et al., 2014; Peters et al., 2015; Swain and Hayhoe, 2015), then understanding the interaction of land use, species, and climate change on water resources will become even more important in the future (Grant et al., 2013; Clark et al., 2016; Kelly et al., 2016; Vose et al., 2016b). As previously outlined as a critical research need (Vose et al., 2016b), our results provide an example of scaling DWU from tree level, plots, and small watersheds in order to understand the species-specific influences on water balance and streamflow dynamics in diverse eastern US deciduous forests.

6 Data availability

All data in this manuscript are archived at USDA Forest Service, Southern Research Station, Coweeta Hydrologic Laboratory, Otto, NC, 28763.

Author contributions. Wayne T. Swank and Katherine J. Elliott conceived the study and designed the experiment. Katherine J. Elliott and Peter V. Caldwell wrote the paper. Katherine J. Elliott, Peter V. Caldwell, Chelcy F. Miniat, and Steven T. Brantley performed the analyses. Wayne T. Swank, James M. Vose, Chelcy F. Miniat, and Katherine J. Elliott performed the research. Katherine J. Elliott, Peter V. Caldwell, Steven T. Brantley, Chelcy F. Miniat, James M. Vose, and Wayne T. Swank contributed to discussions and editing.

Competing interests. The authors declare that they have no conflict of interest.

Acknowledgements. This study was supported by the USDA Forest Service, Southern Research Station, and by NSF grants DEB0218001 and DEB0823293 to the Coweeta LTER program at the University of Georgia. Any opinions, findings, conclusions, or recommendations expressed in the material are those of the authors and do not necessarily reflect the views of the National Science Foundation or the University of Georgia. We acknowledge the support of the long-term climate and hydrologic data network at Coweeta Hydrologic Laboratory as well as many individuals, past and present, especially Patsy Clinton, Charles Marshall, and Stephanie Laseter for field and climate data collection and processing.

Edited by: K. Bishop

References

Adams, M. B. and Kochenderfer, J. N.: Recovery of Central Appalachian forested watersheds: Comparison of Fernow and Coweeta, in: Long-Term Response of a Forest Watershed Ecosystem: Clearcutting in the Southern Appalachians (The Long-term Ecological Research Network Series), edited by: Swank, W. T. and Webster, J. R., Oxford University Press, Oxford, UK, 194–212, 2014.

Alila, Y., Kuraś, P. K., Schnorbus, M., and Hudson, R.: Forests and floods: a new paradigm sheds light on age-old controversies, Water Resour. Res., 45, W08416, doi:10.1029/2008WR007207, 2009.

Allen, C. D., Macalady, A. K., Chenchouni, H., Bachelet, D., McDowell, N., Vennetier, M., Kitzberger, T., Rigling, A., Breshears, D. D., Hogg, E. H., Gonzalez, P., Fensham, R., Zhang, Z., Castro, J., Demidova, N., Lim, J.-H., Allard, G., Running, S. W., Semerci, A., and Cobb, N.: A global overview of drought and heat-induced tree mortality reveals emerging climate change risks for forests, Forest Ecol. Manage., 259, 660–684, doi:10.1016/j.foreco.2009.09.001, 2010.

Alvarez, M.: The State of America's forests, DIANE Publishing, Darby, Pennsylvania, 2007.

Amatya, D. M. and Harrison, C. A.: Grass and forest potential evapotranspiration comparison using five methods in the Atlantic coastal plain, J. Hydrol. Eng., 21, 05016007, doi:10.1061/(ASCE)HE.1943-5584.0001341, 2016.

Andréassian, J. H.: Waters and forests: from historical controversy to scientific debate, J. Hydrol., 291, 1–27, doi:10.1016/j.jhydrol.2003.12.015, 2004.

Aranda, I., Forner, A., Cuesta, B., and Valladares, F.: Species-specific water use by forest tree species: From the tree to the stand, Agr. Water Manage., 114, 67–77, doi:10.1016/j.agwat.2012.06.024, 2012.

Ayres, M. P., Hicke, J. A., Kerns, B. K., McKenzie, D., Littell, J. S., Band, L. E., Luce, C. H., Weed, A. S., and Raymond, C. L.: Disturbance regimes and stressors, in: Climate Change and United States Forests, edited by: Peterson, D. L., Vose, J. M., and Patel-Weynand, T., Springer, New York, 55–92, 2014.

Bellemare, J., Motzkin, G., and Foster, D. R.: Legacies of the agricultural past in the forested present: an assessment of historical land-use effects on rich mesic forests, J. Biogeogr., 29, 1401–1420, doi:10.1046/j.1365-2699.2002.00762.x, 2002.

Berghuijs, W. R., Sivapalan, M., Woods, R. A., and Savenije, H. H. G.: Patterns of similarity of seasonal water balances: A window into streamflow variability over a range of time scales, Water Resour. Res., 50, 5638–5661, doi:10.1002/2014WR015692, 2014.

Boggs, J., Sun, G., Domec, J.-C., McNulty, S., and Treasure, E.: Clearcutting upland forest alters transpiration of residual trees in the riparian buffer zone, Hydrol. Process., 29, 4979–4992, doi:10.1002/hyp.10474, 2015.

Boggs, J., Sun, G., and McNulty, S.: Effects of timber harvest on water quantity and quality in small watersheds in the Piedmont of North Carolina, J. Forest., 114, 27–40, doi:10.5849/jof.14-102, 2016.

Boring, L. R., Elliott, K. J., and Swank, W. T.: Successional forest dynamics 30 years following clearcutting, in: Long-Term Response of a Forest Watershed Ecosystem: Clearcutting in the Southern Appalachians (The Long-term Ecological Research Network Series), edited by: Swank, W. T. and Webster, J. R., Oxford University Press, Oxford, UK, 11–35, 2014.

Bosch, J. M. and Hewlett, J. D.: A review of catchment experiments to determine the effect of vegetation changes on water yield and evapotranspiration, J. Hydrol., 33, 3–23, doi:10.1016/0022-1694(82)90117-2, 1982.

Brantley, S., Ford, C. R., and Vose, J. M.: Future species composition will affect forest water use after loss of hemlock from southern Appalachian forests, Ecol. Appl., 23, 777–790, doi:10.1890/12-0616.1, 2013.

Brantley, S. T., Miniat, C. F., Elliott, K. J., Laseter, S. H., and Vose, J. M.: Changes to southern Appalachian water yield and stormflow after loss of a foundation species, Ecohydrology, 8, 518–528, doi:10.1002/eco.1521, 2015.

Brauman, K. A., Freyberg, D. L., and Daily, G. C.: Potential evapotranspiration from forest and pasture in the tropics: A case study in Kona, Hawaii, J. Hydrol., 440–441, 52–61, doi:10.1016/j.jhydrol.2012.03.014, 2012.

Brown, A. E., Zhang, L., McMahon, T. A., Western, A. W., and Vertessy, R. A.: A review of paired catchment studies for determining changes in water yield resulting from alterations in vegetation, J. Hydrol., 310, 28–61, doi:10.1016/j.jhydrol.2004.12.010, 2005.

Burt, T. and Swank W. T.: Forests or floods?, Geogr. Rev., 15, 37–41, 2002.

Burt, T. P. and Swank, W. T.: Flow frequency responses to hardwood-to-grass conversion and subsequent succession, Hydrol. Process., 6, 179–188, doi:10.1002/hyp.3360060206, 1992.

Burt, T. P., Howden, N. J. K., McDonnell, J. J., Jones, J. A., and Hancock, G. R.: Seeing the climate through the trees: observing climate and forestry impacts on streamflow using a 60-year record, Hydrol. Process., 29, 473–480, doi:10.1002/hyp.10406, 2015.

Burton, T. A.: Effects of basin-scale timber harvest on water yield and peak streamflow, J. Am. Water Resour. Assoc., 33, 1187–1194, doi:10.1111/j.1752-1688.1997.tb03545.x, 1997.

Caldwell, P. V., Miniat, C. F., Elliott, K. J., Swank, W. T., Brantley, S. T., and Laseter, S. H.: Declining water yield from forested mountain watersheds in response to climate change and forest mesophication, Global Change Biol., 22, 2997–3012, doi:10.1111/gcb.13309, 2016.

Campbell, J. L., Driscoll, C. T., Eagar, C., Likens, G. E., Siccama, T. G., Johnson, C. E., Fahey, T. J., Hamburg, S. P., Holmes, R. T., Bailey, A. S., and Buso, D. C.: Long-term trends from ecosystem research at the Hubbard Brook Experimental Forest, GTR-NRS-17, USDA Forest Service, Northern Research Station, Newtown Square, Pennsylvania, 2007.

Chang, M.: Forest Hydrology: An Introduction to Forests and Water, CRC Press, Boca Raton, Florida, 2003.

Clark, J. S., Iverson, L., Woodall, C. W., Allen, C. D., Bell, D. M., Bragg, D. C., D'Amato, A. W., Davis, F. W., Hersh, M. H., Ibanez, I., Jackson, S. T., Matthews, S., Pederson, N., Peters, M., Schwartz, M. W., Waring, K. M., and Zimmermann, N. E.: The impacts of increasing drought on forest dynamics, structure, and biodiversity in the United States, Global Change Biol., 22, 2329–2352, doi:10.1111/gcb.13160, 2016.

Cramer, V. A., Hobbs, R. J., and Standish, R. J.: What's new about old fields: Land abandonment and ecosystem assembly, Trends Ecol. Evol., 23, 104–112, doi:10.1016/j.tree.2007.10.005, 2008.

Cruise, J. F., Laymon, C. A., and Al-Hamdan, O. A.: Impact of 20 years of land-cover change on the hydrology of streams in the southeastern United States, J. Am. Water Resour. Assoc., 46, 1159–1170, doi:10.1111/j.1752-1688.2010.00483.x, 2010.

Cunnane, C.: Unbiased plotting positions – a review, J. Hydrol., 37, 205–222, 1978.

D'Amato, A. W., Catanzaro, P. F., and Fletcher L. S.: Early regeneration and structural response to patch selection and structural retention in second-growth northern hardwoods, Forest Sci., 61, 183–189, doi:10.5849/forsci.13-180, 2015.

Douglass, J. E. and Hoover, M. D.: History of Coweeta, in: Forest Hydrology and Ecology at Coweeta, Ecological Studies, edited by: Swank, W. T. and Crossley Jr., D. A., Springer-Verlag, New York, 17–31, 1988.

Douglass, J. E. and Swank, W. T.: Streamflow modification through management of eastern forests, Res. Paper SE-94, USDA Forest Service, Southeastern Forest Experiment Station, Asheville, North Carolina, 1972.

Douglass, J. E., Cochrane, D. R., Bailey, G. W., Teasley, J. I., and Hill, D. W.: Low herbicide concentration found in streamflow after a grass cover is killed, Res. Note SE-108, USDA Forest Service, Southeastern Forest Experiment Station, Asheville, North Carolina, 1969.

Dunford, E. G. and Fletcher, P. W.: Effect of removal of streambank vegetation upon water yield, T. Am. Geophys. Un., 28, 105–110, 1947.

Edburg, S. L., Hicke, J. A., Brooks, P. D., Pendall, E. G., Ewers, B. E., Norton, U., Gochis, D., Gutmann, E. D., and Meddens, A. J. H.: Cascading impacts of bark beetle-caused tree mortality on coupled biogeophysical and biogeochemical processes, Front. Ecol. Environ., 10, 416–424, doi:10.1890/110173, 2012.

Elliott, K. J. and Swank, W. T.: Changes in tree species diversity after successive clearcuts in the southern Appalachians, Vegetatio, 115, 11–18, doi:10.1007/BF00119382, 1994.

Elliott, K. J. and Swank, W. T.: Long-term changes in forest composition and diversity following early logging (1919–1923) and the decline of American chestnut (*Castanea dentata* (Marshall) Borkh.), Plant Ecol., 197, 155–172, doi:10.1007/s11258-007-9352-3, 2008.

Elliott, K. J. and Vose, J. M.: The contribution of the Coweeta Hydrologic Laboratory to developing an understanding of long-term (1934–2008) changes in managed and unmanaged forests, Forest Ecol. Manage., 261, 900–910, doi:10.1016/j.foreco.2010.03.010, 2011.

Elliott, K. J., Boring, L. R., Swank, W. T., and Haines, B. R.: Successional changes in diversity and composition in a clearcut watershed in Coweeta Basin, North Carolina, Forest Ecol. Manage., 92, 67–85, doi:10.1016/S0378-1127(96)03947-3, 1997.

Elliott, K. J., Boring, L. R., and Swank, W. T.: Changes in vegetation structure and diversity following grass-to-forest succession, Am. Midl. Nat., 140, 219–232, doi:10.1674/0003-0031(1998)140[0219:CIVSAD]2.0.CO;2, 1998.

Elliott, K. J., Boring, L. R., and Swank, W. T.: Aboveground biomass and nutrient accumulation 20 years after clear-cutting a southern Appalachian watershed, Can. J. Forest Res., 32, 667–683, doi:10.1139/x02-009, 2002.

Farley, K. A., Jobbagy, E. G., and Jackson, R. B.: Effects of afforestation on water yield: A global synthesis with implications for policy, Global Change Biol., 11, 1565–1576, doi:10.1111/j.1365-2486.2005.01011.x, 2005.

Ford, C. R. and Vose, J. M.: *Tsuga canadensis* (L.) Carr. mortality will impact hydrologic processes in southern Appalachian forest ecosystems, Ecol. Appl., 17, 1156–1167, doi:10.1890/06-0027, 2007.

Ford, C. R., Hubbard, R. M., and Vose, J. M.: Quantifying structural and physiological controls on canopy transpiration of planted pine and hardwood stands in the southern Appalachians, Ecohydrology, 4, 183–195, doi:10.1002/eco.136, 2011a.

Ford, C. R., Laseter, S. H., Swank, W. T., and Vose, J. M.: Can forest management be used to sustain water-based ecosystem services in the face of climate change?, Ecol. Appl., 21, 2049–2067, doi:10.1890/10-2246.1, 2011b.

Ford, C. R., Elliott, K. J., Clinton, B. D., Kloeppel, B. D., and Vose, J. M.: Forest dynamics following hemlock mortality in the southern Appalachians, Oikos, 121, 523–536, doi:10.1111/j.1600-0706.2011.19622.x, 2012.

Foster, D., Swanson, F., Aber, J., Burke, I., Brokaw, N., Tilman, D., and Knapp, A.: The importance of land-use legacies to ecology and conservation, BioScience, 53, 77–88, doi:10.1641/0006-3568(2003)053[0077:TIOLUL]2.0.CO;2, 2003.

García-Ruiz, J. and Lana-Renault, N.: Hydrological and erosive consequences of farmland abandonment in Europe, with special reference to the Mediterranean region – A review, Agr. Ecosyst. Environ., 140, 317–338, doi:10.1016/j.agee.2011.01.003, 2011.

Grant, G. E., Tague, C. L., and Allen, C. D.: Watering the forest for the trees: an emerging priority for managing water in forest landscapes, Front. Ecol. Environ., 11, 314–321, doi:10.1890/120209, 2013.

Green, K. C. and Alila, Y.: A paradigm shift in understanding and quantifying the effects of forest harvesting on floods in snow environments, Water Resour. Res., 48, W10503, doi:10.1029/2012WR012449, 2012.

Halpin, C. R. and Lorimer, C. G.: Trajectories and resilience of stand structure in response to variable disturbance severities in northern hardwoods, Forest Ecol. Manage., 365, 69–82, doi:10.1016/j.foreco.2016.01.016, 2016.

Harr, R. D., Harper, W. C., Krygier, J. T., and Hsieh, F. S.: Changes in storm hydrographs after road building and clear-cutting in the Oregon Coast Range, Water Resour. Res., 11, 436–444, doi:10.1029/WR011i003p00436, 1975.

Hassler, S. K., Zimmermann, B., van Breugel, M., Hall, J. S., and Elsenbeer, H.: Recovery of saturated hydraulic conductivity under secondary succession on former pasture in the humid tropics, Forest Ecol. Manage, 261, 1634–1642, doi:10.1016/j.foreco.2010.06.031, 2011.

Hewlett, J. D.: Response of fescue to natural moisture gradient on an artificial slope, Res. Note SE-152, USDA Forest Service, Southeastern Forest Experiment Station, Asheville, North Carolina, 1961.

Hewlett, J. D. and Hibbert, A. R.: Increases in water yield after several types of forest cutting, Q. Bull. Int. Assoc. Sci. Hydrol., 6, 5–17, 1961.

Hewlett, J. D. and Hibbert, A. R.: Factors affecting the response of small watersheds to precipitation in humid areas, in: Forest Hydrology, edited by: Sopper, W. E. and Lull, H.W., Pergamon Press, Oxford, UK, 275–290, 1966.

Hibbert, A. R.: Water yield changes after converting a forested catchment to grass, Water Resour. Res., 5, 634–640, 1969.

Holdo, R. M. and Nippert, J. B.: Transpiration dynamics support resource partitioning in African savanna trees and grasses, Ecology, 96, 1466–1472, doi:10.1890/14-1986.1, 2015.

Holland, G. J. and Webster, P. J.: Heightened tropical cyclone activity in the North Atlantic: natural variability of climate trend?, Philos. T. Roy. Soc. A, 365, 2695–2716, doi:10.1098/rsta.2007.2083, 2007.

Hornbeck, J. W.: Storm flow from hardwood-forested and cleared watersheds in New Hampshire, Water Resour. Res., 9, 346–354, 1973.

Hornbeck, J. W., Bailey, A. S., Eagar, C., and Campbell, J. L.: Comparisons with results from the Hubbard Brook Experimental Forest in the northern Appalachians, in: Long-Term Response of a Forest Watershed Ecosystem: Clearcutting in the Southern Appalachians (The Long-term Ecological Research Network Series), edited by: Swank, W. T. and Webster, J. R., Oxford University Press, Oxford, UK, 213–228, 2014.

Houlbrooke, D. J. and Laurenson, S.: Effect of sheep and cattle treading damage on soil microporosity and soil water holding capacity, Agr. Water Manage., 121, 81–84, doi:10.1016/j.agwat.2013.01.010, 2013.

Ice, G. G. and Stednick, J. D.: A Century of Forest and Wildland Watershed Lessons, Society of American Foresters, Bethesda, Maryland, 2004.

Kato, A., Moskal, L. M., Schiess, P., Swanson, M. E., Calhoun, D., and Stuetzle, W.: Capturing tree crown formation through implicit surface reconstruction using airborne lidar data, Remote Sens. Environ., 113, 1148–1162, doi:10.1016/j.rse.2009.02.010, 2009.

Kelly, C. N., McGuire, K. J., Miniat, C. F., and Vose, J. M.: Streamflow response to increasing precipitation extremes altered by forest management, Geophys. Res. Lett., 43, 3727–3736, doi:10.1002/2016GL068058, 2016.

Kirk, R. W., Bolstad, P. V., and Manson, S. M.: Spatio-temporal trend analysis of long-term development patterns (1900–2030) in a Southern Appalachian county, Landsc. Urban Plan., 104, 47–58, doi:10.1016/j.landurbplan.2011.09.008, 2012.

Kirkman, L. K., Brown, C. L., and Leopold, D. J.: Native Trees of the Southeast, Timber Press, Portland, Oregon, 2007.

Klein, T.: The variability of stomatal sensitivity to leaf water potential across tree species indicates a continuum between isohydric and anisohydric behaviours, Funct. Ecol., 28, 1313–1320, doi:10.1111/1365-2435.12289, 2014.

Kochenderfer, J. N.: Fernow and the Appalachian hardwood region, in: The Fernow Watershed Acidification Study, edited by: Adams, M. B., DeWalle, D. R., and Hom, H. L., Springer, Dordrecht, the Netherlands, 17–39, 2006.

Kulmatiski, A. and Beard, K. H.: Woody plant encroachment facilitated by increased precipitation intensity, Nat. Clim. Change, 3, 833–837, doi:10.1038/nclimate1904, 2013.

Laseter, S. H., Ford, C. R., Vose, J. M., and Swift, L. W. Jr.: Long-term temperature and precipitation trends at the Coweeta Hydrologic Laboratory, Otto, North Carolina, USA, Hydrol. Res., 43, 890–901, doi:10.2166/nh.2012.067, 2012.

Liu, W., Wei, X., Liu, S., Liu, Y., Fan, H., Zhang, M., Yin, J., and Zhan, M.: How do climate and forest changes affect long-term streamflow dynamics? A case study in the upper reach of Poyang River basin, Ecohydrology, 8, 46–57, doi:10.1002/eco.1486, 2015.

Loftis, D. L., Schweitzer, C. J., and Keyser, T. L.: Structure and species composition of upland hardwood communities after regeneration treatments across environmental gradients, in: Sustaining Young Forest Communities: Ecology and management of early successional habitats in the central hardwood region, USA, edited by: Greenberg, C., Collins, B., and Thompson III, F., Springer, New York, 59–71, 2014.

Martin, J. G., Kloeppel, B. D., Schaefer, T. L., Kimbler, D. L., and McNulty, S. G.: Aboveground biomass and nitrogen allocation of ten deciduous southern Appalachian tree species, Can. J. Forest Res., 28, 1648–1659, doi:10.1139/x98-149, 1998.

Martinuzzi, S., Bavier-Pizarro, G. I., Lugo, A. E., and Radeloff, V. C.: Future land-use changes and the potential for novelty in ecosystems of the United States, Ecosystems, 18, 1332–1342, doi:10.1007/s10021-015-9901-x, 2015.

McGinty, D. T.: The ecological roles of *Kalmia latifolia* L. and *Rhododendron maximum* L. in the hardwood forest at Coweeta, MS thesis, University of Georgia, Athens, Georgia, 1972.

Milly, P. C. D., Betancourt, J., Falkenmark, M., Hirsch, R. M., Kundzewicz, Z. W., Lettenmaier, D. P., and Stouffer, R. J.: Climate change – Stationarity is dead: Whither water management?, Science, 319, 573–574, doi:10.1126/science.1151915, 2008.

Morris, L. A. and Jackson, C. R.: Interactions of soils and land uses with water quantity and quality, in: Soil Ecosystems Services, edited by: Stromberger, M., Comerford, N., and Lindbo, D., Soil Sci. Soc. Am., Madison, WI, doi:10.2136/2015.soilecosystemsservices.2014.0077, 2016.

Neary, D. G., Ice, G. G., and Jackson, C. R.: Linkages between forest soils and water quality and quantity, Forest Ecol. Manage., 258, 2269–2281, doi:10.1016/j.foreco.2009.05.027, 2009.

Novick, K. A., Miniat, C. F., and Vose, J. M.: Drought limitations to leaf-level gas exchange: results from a model linking stomatal optimization and cohesion-tension theory, Plant Cell Environ., 39, 583–596, doi:10.1111/pce.12657, 2016.

Otto, J. S.: The decline of forest farming in southern Appalachia, J. Forest., 27, 18–27, 1983.

Palik, B., Martin, M., Zenner, E., Blinn, C., and Kolka, R.: Overstory and regeneration dynamics in riparian management zones of northern Minnesota forested watersheds, Forest Ecol. Manage., 271, 1–9, doi:10.1016/j.foreco.2012.01.045, 2012.

Peters, M. P., Iverson, L. R., and Matthews, S. N.: Long-term droughtiness and drought tolerance of eastern US forests over five decades, Forest Ecol. Manage., 345, 56–64, doi:10.1016/j.foreco.2015.02.022, 2015.

Price, K., Jackson, C. R., Parker, A. J., Reitan, T., Dowd, J., and Cyterski, M.: Effects of watershed land use and geomorphology on stream low flows during severe drought conditions in the southern Blue Ridge Mountains, Georgia and North Carolina, United States, Water Resour. Res., 47, W02516, doi:10.1029/2010WR009340, 2011.

Ramankutty, N., Heller, E., and Rhemtulla, J.: Prevailing myths about agricultural abandonment and forest regrowth in the United States, Ann. Assoc. Am. Geogr., 100, 502–512, doi:10.1080/00045601003788876, 2010.

Reinhart, K. G.: Effect of a commercial clearcutting in West Virginia on overland flow and storm runoff, J. Forest., 62, 167–171, 1964.

Reinhart, K. G. and Pierce, R. S.: Stream-gaging stations for research on small watersheds, USDA Forest Service, Washington, D.C., 1964.

Rey Benayas, J. M., Martins, A., Nicolau, J. M., and Schulz, J. J.: Abandonment of agricultural land: an overview of drivers and consequences, Perspec. Agric. Vet. Sci. Nutr. Nat. Resour. No. 057, CABI Publishing, Oxfordshire, UK, doi:10.1079/PAVSNNR20072057, 2007.

Rodriguez, D. A., Tomasella, J., and Linhares, C.: Is the forest conversion to pasture affecting the hydrological response of Amazonian catchments: Signals in the Ji-Paraná Basin, Hydrol. Process., 24, 1254–1269, doi:10.1002/hyp.7586, 2010.

Santee, W. R. and Monk, C. D.: Stem diameter and dry weight relationships in *Tsuga canadensis* (L.) Carr., Bull. Torrey Bot. Club, 108, 320–323, 1981.

Scanlon, B. R., Jolly, I., Sophocleaous, M., and Zhang, L.: Global impacts of conversions from natural to agricultural ecosystems on water resources: Quantity versus quality, Water Resour. Res., 43, W03437, doi:10.1029/2006WR005486, 2007.

Shure, D. J., Phillips, D. L., and Bostick, P. E.: Gap size and succession in cutover southern Appalachian forests: an 18 year study of vegetation dynamics, Plant Ecol., 185, 299–318, doi:10.1007/s11258-006-9105-8, 2006.

Smith, J. A., Villarini, G., and Baeck, M. L.: Mixture distributions and the hydroclimatology of extreme rainfall and flooding in the Eastern United States, J. Hydrometeorol., 12, 294–309, doi:10.1175/2010JHM1242.1, 2011.

Stanturf, J. A., Palik, B. J., and Dumroese, R. K.: Contemporary forest restoration: A review emphasizing function, Forest

Ecol. Manage., 331, 292–323, doi:10.1016/j.foreco.2014.07.029, 2014.
Stedinger, J. R., Vogel, R. M., and Foufoula-Georgiou, E.: Frequency analysis of extreme events, in: Handbook of Hydrology, edited by: Maidment, D. R., McGraw-Hill, New York, 18.10–18.66, 1993.
Stednick, J. D.: Monitoring the effects of timber harvest on annual water yield, J. Hydrol., 176, 79–95, doi:10.1016/0022-1694(95)02780-7, 1996.
Sun, S., Sun, G., Cohen, E., McNulty, S. G., Caldwell, P. V., Duan, K., and Zhang, Y.: Projecting water yield and ecosystem productivity across the United States by linking an ecohydrological model to WRF dynamically downscaled climate data, Hydrol. Earth Syst. Sci., 20, 935–952, doi:10.5194/hess-20-935-2016, 2016.
Swain, S. and Hayhoe, K.: CMIP5 projected changes in spring and summer drought and wet conditions over North America, Clim. Dynam., 44, 2737–2750, doi:10.1007/s00382-014-2255-9, 2015.
Swank, W. T. and Crossley, D. A. Jr.: Introduction and site description, in: Forest Hydrology and Ecology at Coweeta, Ecological Studies, edited by: Swank, W. T. and Crossley Jr., D. A., Springer-Verlag, New York, 3–16, 1988.
Swank, W. T., Vose, J. M., and Elliott, K. J.: Long-term hydrologic and water quality responses following commercial clearcutting of mixed hardwoods on a southern Appalachian catchment, Forest Ecol. Manage., 143, 163–178, doi:10.1016/S0378-1127(00)00515-6, 2001.
Swank, W. T., Knoepp, J. D., Vose, J. M., Laseter, S. N., and Webster, J. R.: Response and recovery of water yield and timing, stream sediment, abiotic parameters, and stream chemistry following logging, in: Long-Term Response of a Forest Watershed Ecosystem: Clearcutting in the Southern Appalachians (The Long-term Ecological Research Network Series), edited by: Swank, W. T. and Webster, J. R., Oxford University Press, Oxford, UK, 36–56, 2014.
Swift Jr., L. W., Cunningham, G. B., and Douglass, J. E.: Climate and hydrology, in: Forest Hydrology and Ecology at Coweeta, Ecological Studies, edited by: Swank, W. T. and Crossley Jr., D. A., Springer-Verlag, New York, 35–55, 1988.
Taneda, H.and Sperry, J. S.: A case-study of water transport in co-occurring ring- versus diffuse-porous trees: contrasts in water status, conducting capacity, cavitation and vessel refilling, Tree Physiol., 28, 1641–1652, doi:10.1093/treephys/28.11.1641, 2008.
Thiemann, J. A., Webster, C. R., Jenkins, M. A., Hurley, P. M., Rock, J. H., and White, P. S.: Herbaceous-layer impoverishment in a post-agricultural Southern Appalachian landscape, Am. Midl. Nat., 162, 148–168, doi:10.1674/0003-0031-162.1.148, 2009.
Thomas, D. J.: Soil survey of Macon County, North Carolina, USDA Natural Resource Conservation Service, US Government Printing Office, Washington, D.C., 1996.
Thompson, J. R., Carpenter, D. N., Cogbill, C. V., and Foster, D. R.: Four centuries of change in northeastern United States forests, PloS One, 8, e72540, doi:10.1317/journal.pone.0072540, 2013.
Trimble, S. W., Weirich, F. H., and Hoag, B. L.: Reforestation and the reduction of water yield on the southern piedmont since circa 1940, Water Resour. Res., 23, 425–437, doi:10.1029/WR023i003p00425, 1987.
Troendle, C. A., Wilcox, M. S., Bevenger, G. S., and Porth, L. S.: The Coon Creek water yield augmentation project: Implementation of timber harvesting technology to increase streamflow, Forest Ecol. Manage., 143, 179–187, doi:10.1016/S0378-1127(00)00516-8, 2001.
Von Allmen, E. I., Sperry, J. S., and Bush, S. E.: Contrasting whole-tree water use, hydraulics, and growth in a co-dominant diffuse-porous vs. ring-porous species pair, Trees, 29, 717–728, doi:10.1007/s00468-014-1149-0, 2015.
Vose, J. M., Sun, G., Ford, C. R., Bredemeier, M., Ostsuki, K., Wei, A., Zhang, Z., and Zhang, L.: Forest ecohydrological research in the 21st century: what are the critical needs?, Ecohydrology, 4, 146–158, doi:10.1002/eco.193, 2011.
Vose, J. M., Miniat, C. F., Luce, C. H., Asbjornsen, H., Caldwell, P. V., Campbell, J. L., Grant, G. E., Isaak, D. J., Loheide, S. P., and Sun, G.: Ecohydrological implications of drought, in: Effects of Drought on Forests and Rangelands in the United States: A comprehensive science synthesis, edited by: Vose, J. M., Clark, J. S., Luce, C. H., and Patel-Weynand, T., GTR-WO-93b, USDA Forest Service, Washington Office, Washington, D.C., 231–251, 2016a.
Vose, J. M., Miniat, C. F., Luce, C. H., Asbjornsen, H., Caldwell, P. V., Campbell, J. L., Grant, G. E., Isaak, D. J., Loheide, S. P., and Sun, G.: Ecohydrological implications of drought for forests in the United States, Forest Ecol. Manage., 380, 335–345, doi:10.1016/j.foreco.2016.03.025, 2016b.
Wang, E., Cresswell, H., Paydar, Z., and Gallant, H.: Opportunities for manipulating catchment water balance by changing vegetation type on a topographic sequence: a simulation study, Hydrol. Process., 22, 736–749, doi:10.1002/hyp.6655, 2008.
Wear, D. N. and Bolstad, P.: Land-use changes in Southern Appalachian landscapes: spatial analysis and forecast evaluation, Ecosystems, 1, 575–594, doi:10.1007/s100219900052, 1998.
Wei, X. and Zhang, M.: Quantifying streamflow change caused by forest disturbance at a large spatial scale: A single watershed study, Water Resour. Res., 46, W12525, doi:10.1029/2010WR009250, 2010.
Wilm, H. G.: Statistical control of hydrologic data from experimental watersheds, T. Am. Geophys. Un., 2, 618–622, 1944.
Wilm, H. G.: How long should experimental watersheds be calibrated?, T. Am. Geophys. Un., 30, 272–278, 1949.
Woods, F. W. and Shanks, R. E.: Natural replacement of chestnut by other species in the Great Smoky Mountains National Park, Ecology, 40, 349–361, 1959.
Wullschleger, S. D., Hanson, P. J., and Todd, D. E.: Transpiration from a multi-species deciduous forest as estimated by xylem sap flow techniques, Forest Ecol. Manage., 143, 205–213, doi:10.1016/S0378-1127(00)00518-1, 2001.
Yarnell, S. L.: The southern Appalachians: a history of the landscape, GTR-SRS-18, USDA Forest Service, Southern Research Station, Asheville, North Carolina, 1998.
Zhang, L., Dawes, W. R., and Walker, G. R.: Response of mean annual evapotranpiration to vegetation changes at catchment scale, Water Resour. Res., 37, 701–708, doi:0043-1397/01/2000WR90032509.00, 2001.
Zhang, M. and Wei, X.: The effects of cumulative forest disturbance on streamflow in a large watershed in the central interior of British Columbia, Canada, Hydrol. Earth Syst. Sci., 16, 2021–2034, doi:10.5194/hess-16-2021-2012, 2012.

PERMISSIONS

All chapters in this book were first published in HESS, by Copernicus Publications; hereby published with permission under the Creative Commons Attribution License or equivalent. Every chapter published in this book has been scrutinized by our experts. Their significance has been extensively debated. The topics covered herein carry significant findings which will fuel the growth of the discipline. They may even be implemented as practical applications or may be referred to as a beginning point for another development.

The contributors of this book come from diverse backgrounds, making this book a truly international effort. This book will bring forth new frontiers with its revolutionizing research information and detailed analysis of the nascent developments around the world.

We would like to thank all the contributing authors for lending their expertise to make the book truly unique. They have played a crucial role in the development of this book. Without their invaluable contributions this book wouldn't have been possible. They have made vital efforts to compile up to date information on the varied aspects of this subject to make this book a valuable addition to the collection of many professionals and students.

This book was conceptualized with the vision of imparting up-to-date information and advanced data in this field. To ensure the same, a matchless editorial board was set up. Every individual on the board went through rigorous rounds of assessment to prove their worth. After which they invested a large part of their time researching and compiling the most relevant data for our readers.

The editorial board has been involved in producing this book since its inception. They have spent rigorous hours researching and exploring the diverse topics which have resulted in the successful publishing of this book. They have passed on their knowledge of decades through this book. To expedite this challenging task, the publisher supported the team at every step. A small team of assistant editors was also appointed to further simplify the editing procedure and attain best results for the readers.

Apart from the editorial board, the designing team has also invested a significant amount of their time in understanding the subject and creating the most relevant covers. They scrutinized every image to scout for the most suitable representation of the subject and create an appropriate cover for the book.

The publishing team has been an ardent support to the editorial, designing and production team. Their endless efforts to recruit the best for this project, has resulted in the accomplishment of this book. They are a veteran in the field of academics and their pool of knowledge is as vast as their experience in printing. Their expertise and guidance has proved useful at every step. Their uncompromising quality standards have made this book an exceptional effort. Their encouragement from time to time has been an inspiration for everyone.

The publisher and the editorial board hope that this book will prove to be a valuable piece of knowledge for researchers, students, practitioners and scholars across the globe.

LIST OF CONTRIBUTORS

M. Bechmann
Friedrich Schiller University, Jena, Germany, Institute of Geosciences, Burgweg 11, 07749 Jena, Germany

C. Schneider and S. Attinger
Helmholtz Centre for Environmental Research, Leipzig, Germany, Department Computational Hydrosystems, Permoser Straße 15, 04318 Leipzig, Germany

A. Carminati
Georg August University, Göttingen, Germany, Faculty of Agricultural Sciences, Department of Crop Sciences, Büsgenweg 2, 37077 Göttingen, Germany

D. Vetterlein
Helmholtz Centre for Environmental Research, Halle, Germany, Department of Soil Physics, Theodor-Lieser-Strasse 4, 06120 Halle/Saale, Germany

A. Hildebrandt
Friedrich Schiller University, Jena, Germany, Institute of Geosciences, Burgweg 11, 07749 Jena, Germany
Max Planck Institute for Biogeochemistry, Jena, Germany, Hans-Knöll-Str. 10, 07745 Jena, Germany

Irina Y. Petrova
Department of the Land in the Earth System, Max Planck Institute for Meteorology, Hamburg, Germany
Laboratory of Hydrology and Water Management, Ghent University, Ghent, Belgium

Cathy Hohenegger
Department of the Land in the Earth System, Max Planck Institute for Meteorology, Hamburg, Germany

Chiel C. van Heerwaarden
Meteorology and Air Quality Group, Wageningen University, Wageningen, the Netherlands

Françoise Guichard
CNRM (UMR 3589 CNRS and Météo-France), Toulouse, France

Haimanote K. Bayabil and Tigist Y. Tebebu
Department of Biological and Environmental Engineering, Cornell University, Ithaca, New York, USA

Cathelijne R. Stoof
Department of Biological and Environmental Engineering, Cornell University, Ithaca, New York, USA
Soil Geography and Landscape Group, Wageningen University, the Netherlands

Tammo S. Steenhuis
Department of Biological and Environmental Engineering, Cornell University, Ithaca, New York, USA
School of Civil and Water Resources Engineering, Bahir Dar University, Bahir Dar, Ethiopia

Abraham Endalamaw and W. Robert Bolton
International Arctic Research Center, University of Alaska Fairbanks, Fairbanks, AK, USA

Jessica M. Young-Robertson
School of Natural Resources and Extension, University of Alaska Fairbanks, Fairbanks, AK, USA

Don Morton
Boreal Scientific Computing LLC, Fairbanks, AK, USA

Larry Hinzman
University of Alaska Fairbanks, Fairbanks, AK, USA

Bart Nijssen
Civil and Environmental Engineering, University of Washington, Seattle, WA, USA

Guiomar Ruiz-Pérez
Department of Crop Production Ecology, Swedish University of Agricultural Sciences, Uppsala, Sweden
Research Group of Hydrological and Environmental Modelling (GIHMA), Research Institute of Water and Environmental Engineering, Universitat Politècnica de València, Valencia, Spain

Julian Koch
Department of Hydrology, Geological Survey of Denmark and Greenland, Copenhagen, Denmark
Department of Geosciences and Natural Resources Management, University of Copenhagen, Copenhagen Denmark

Salvatore Manfreda
Department of European and Mediterranean Cultures, University of Basilicata, Matera, Italy

Kelly Caylor
Bren School of Environmental Science and Management, Department of Geography, UC Santa Barbara, Santa Barbara, California, USA

Félix Francés
Research Group of Hydrological and Environmental Modelling (GIHMA), Research Institute of Water and Environmental Engineering, Universitat Politècnica de València, Valencia, Spain

C. Duku and L. Hein
Environmental Systems Analysis Group, Wageningen University, 6700AA, Wageningen, the Netherlands

H. Rathjens
Department of Earth, Atmospheric, and Planetary Sciences, Purdue University, 550 Stadium Mall Drive, West Lafayette, IN 47907-2093, USA

S. J. Zwart
Africa Rice Center, 2031 PB, Cotonou, Benin

Anoop Kumar Shukla, Shray Pathak, Lalit Pal, Chandra Shekhar Prasad Ojha and Rahul Dev Garg
Department of Civil Engineering, Indian Institute of Technology Roorkee, Uttarakhand, India

Ana Mijic
Department of Civil and Environmental Engineering, Imperial College London, London, UK

Tobias Karl David Weber, Sascha Christian Iden and Wolfgang Durner
Soil Science and Soil Physics Division, Institute of Geoecology, Technische Universität Braunschweig, Langer Kamp 19c, 38106 Braunschweig, Germany
Institute of Soil Science and Land Evaluation, University of Hohenheim, 70593 Stuttgart, Germany

Inne Vanderkelen
Department of Hydrology and Hydraulic Engineering, Vrije Universiteit Brussel, Brussels, Belgium

Nicole P. M. van Lipzig
Department of Earth and Environmental Sciences, KU Leuven, Leuven, Belgium

Wim Thiery
Department of Hydrology and Hydraulic Engineering, Vrije Universiteit Brussel, Brussels, Belgium
Institute for Atmospheric and Climate Science, ETH Zurich, Zurich, Switzerland

Yuanyuan Yin, Xingcai Liu and Xuejun Zhang
Key Laboratory of Water Cycle and Related Land Surface Processes, Institute of Geographic Sciences and Natural Resources Research, Chinese Academy of Sciences, Beijing, China

Qiuhong Tang
Key Laboratory of Water Cycle and Related Land Surface Processes, Institute of Geographic Sciences and Natural Resources Research, Chinese Academy of Sciences, Beijing, China
University of Chinese Academy of Sciences, Beijing, China

C. Stumpp and P. Maloszewski
Helmholtz Zentrum München, German Research Center for Environmental Health, Institute of Groundwater Ecology, Neuherberg, Germany

A. Ekdal
Istanbul Technical University, Environmental Engineering Department, Maslak, Istanbul, Turkey

I. E. Gönenc
IGEM Research & Consulting Co., Istanbul, Turkey

Rubianca Benavidez, Bethanna Jackson, Deborah Maxwell and Kevin Norton
School of Geography, Environment, and Earth Sciences, Victoria University of Wellington, Wellington, 6012, New Zealand

Katherine J. Elliott, Peter V. Caldwell, Chelcy F. Miniat and Wayne T. Swank
USDA Forest Service, Southern Research Station, Coweeta Hydrologic Lab, Otto, NC28763, USA

Steven T. Brantley
Joseph W. Jones Ecological Research Center, Ichauway, Newton, GA31770, USA

James M. Vose
UDSA Forest Service, Southern Research Station, Center for Integrated Forest Science, Raleigh, NC27695, USA

Index